Lecture Notes in Mathematics

Edited by A. Dold and B. Eckmann

747

Complex Analysis Joensuu 1978

Proceedings of the Colloquium on Complex Analysis, Joensuu, Finland, August 24–27, 1978

Edited by
Ilpo Laine, Olli Lehto, and Tuomas Sorvali

Springer-Verlag
Berlin Heidelberg New York 1979

Editors

Ilpo Laine
University of Joensuu
Department of Mathematics
and Physics
SF-80101 Joensuu 10 / Finland

Olli Lehto
University of Helsinki
Department of Mathematics
SF-00100 Helsinki 10 / Finland

Tuomas Sorvali
University of Joensuu
Department of Mathematics
and Physics
SF-80101 Joensuu 10 / Finland

AMS Subject Classifications (1970): 30 A 20, 30 A 30, 30 A 38, 30 A 40, 30 A 46, 30 A 50, 30 A 52, 30 A 58, 30 A 60, 30 A 70, 30 A 74, 30 A 82, 30 A 96, 31 A 05, 31 B 05, 31 D 05, 32 A 20, 32 G 15, 32 H 25, 34 A 20, 44 A 15, 46 C 05, 47 A 20

ISBN 3-540-09553-5 Springer-Verlag Berlin Heidelberg New York
ISBN 0-387-09553-5 Springer-Verlag New York Heidelberg Berlin

Library of Congress Cataloging in Publication Data
Colloquium on Complex Analysis, Joensuu, Finland, 1978.
Complex analysis, Joensuu 1978.
(Lecture notes in mathematics; 747)
Bibliography: p.
Includes index.
1. Functions of comply variables--Congresses. 2. Mathematical analysis--Congresses.
I. Laine, Ilpo. II. Lehto, Olli. III. Sorvali, Tuomas, 1944- IV. Title. V. Series.
QA3.L28 no. 747 [QA331] 510'.8s [515'.9] 79-21285
ISBN 0-387-09553-5

© by Springer-Verlag Berlin Heidelberg 1979
Printed in Germany

Printing and binding: Beltz Offsetdruck, Hemsbach/Bergstr.
2141/3140-543210

PREFACE

This volume consists of papers presented at the Colloquium on Complex Analysis held at the University of Joensuu, August 24 - 27, 1978. The IV Romanian-Finnish Seminar on Complex Analysis was organized as a part of this Colloquium. The major part of the contributions in this volume is related to the theory of quasiconformal and quasiregular mappings, Nevanlinna theory and complex differential equations, Riemann surfaces and potential theory.

We wish to thank the staff of the Department of Mathematics and Physics in the University of Joensuu for their cooperation in organizing these meetings and preparing this volume, Springer-Verlag for their willingness to publish this volume and, finally, Eija Faari and Riitta Laakkonen for their patient job of typing the manuscript.

Joensuu and Helsinki, April 1979,

Ilpo Laine   Olli Lehto   Tuomas Sorvali

CONTENTS

(*) A contribution for the IV Romanian-Finnish Seminar on Complex Analysis

OTHER LECTURES GIVEN AT THE COLLOQUIUM(§)

Ahlfors, L. V.: Beltrami differentials in several dimensions (Enseignement math., II. Sér. 24 (1978), 225 - 236)

Apostol, C.: Comments on a theorem on invariant subspaces by Scott Brown(*)

Blatter, Chr.: A two variables distortion theorem for univalent functions (Comm. math. Helv. 53 (1978), 651 - 659)

Blevins, D. K.: Conformal mappings and quasicircles

Boboc, N.: Standard H-cones(*)

Bojarski, B.: Analytic methods in the quasiconformal theory in R^n

Bshouty, D.: Löwner differential equation and quasiconformal extensions of conformal mappings

Bucur, Gh.: Standard H-cones(*)

Cegrell, U.: Construction of capacities on C^n

Colojoară, I.: On a functional calculus based on Cauchy-Pompeiu's formula(*)

Douady, A.: Projective structures on Riemann surfaces

Essén, M., Shea, D.: On the case of equality in some inequalities of A. Baernstein (to appear in Ann. Acad. Sci. Fenn.)

Fuchs, W.: On the nodes of best approximation by polynomials in the Chebychev sense

Gackstatter, F., Laine, I.: Zur Theorie der gewöhnlichen Differentialgleichungen im Komplexen (to appear in Ann. Polon. Math.)(*)

Gehring, F. W.: Remarks on the Schwarzian derivative

Gowrisankaran, K.: Construction of inner functions of polydiscs (to appear in Ann. Inst. Fourier)

(§) A reference indicates a related article.

(*) A contribution for the IV Romanian-Finnish Seminar on Complex Analysis

Hengartner, W., Gauthier, P.: Uniform approximation and simultaneous interpolation

Huber, A.: Isometric and conformal sewing (Comm. Math. Helv. 50 (1975), 179 - 186, ibid. 51 (1976), 319 - 331, and a forthcoming article)

Kiselman, Chr.: On the density of plurisubharmonic functions: a short proof of Siu's theorem (to appear in Bull. Soc. math. France)

Lelong, P.: An inverse function theorem for plurisubharmonic functions (to appear in Séminaire P. Lelong)

Matsuda, M.: Algebraic differential equations of the first order free from parametric singularities from the differential-algebraic standpoint (to appear in J. math. Soc. Japan)

Menke, K.: Näherung der Lösung des Dirichlet Problems durch ein Interpolationsverfahren

Meyer, G.: On the zeros of exponential polynomials (to appear in Arch. Math.)

Mues, E.: Über die Werteverteilung von Differentialpolynomen

Netanyahu, E., Schiffer, M. M.: On the monotonicity of some functionals in the family of univalent functions (to appear in Israel J. Math.)

Ohtsuka, M.: On type problem of Riemann surfaces

Osgood, B.: A univalent criterion for multiply-connected domains

Palka, B.: Quasiconformally homogeneous domains

Rickman, S.: Omitted values, counting function and equidistribution of quasiregular mappings (to appear in Acta math.)(*)

Rubel, L. A.: First-order conformal invariants

Sakai, M.: Analytic functions with finite Dirichlet integrals on Riemann surfaces (to appear in Acta math.)

Schwarz, B.: Disconjugacy of complex second-order matrix differential systems

Siciak, J.: On holomorphic extendability of functions on generic real analytic submanifolds (to appear in Bull. Acad. Polon. Sci.)

Siddiqi, J. A.: Nonquasianalytic classes of functions and uniform approximation on arcs by exponential sums

Sontag, A.: On the existence of substantial boundary points for extremal quasiconformal maps with angular dilatation

Stoica, L.: Axiomatic approach to potential theory associated with elliptic degenerated operators(*)

Vaaler, J.: An inequality for the volume of a centrally sliced cube in R^n (to appear in Pacific J. Math.)

Voiculescu, D., Bercovici, H.: Tensor operations on characteristic functions of C_o-contractions (to appear in Acta Sci. math.)(*)

Winkler, J.: Zur Existenz ganzer Funktionen bei vorgegebener Menge der Nullstellen und Einsstellen

COLLOQUIUM PARTICIPANTS

Ahlfors, L. V.	Harvard University, Cambridge, MA, U.S.A.
Alastalo, Hannu(*)	University of Joensuu, Finland
Anderson, Glen D.	Michigan State University, East Lansing, MI, U.S.A.
Andreian Cazacu, Cabiria(*)	Universitatea din Bucureşti, Romania
Apostol, Constantin(*)	INCREST, Bucureşti, Romania
Arsene, Grigore(*)	INCREST, Bucureşti, Romania
Aulaskari, Rauno(*)	University of Joensuu, Finland
Bakken, Ivar	Universitetet i Tromsø, Norway
Bayoumi, Aboubakr	Uppsala Universitet, Sweden
Becker, Jochen	Technische Universität Berlin, Federal Republic of Germany
Berg, Christian	Københavns Universitet, Denmark
Blatter, Christian	Eidgenössische Technische Hochschule, Zürich, Switzerland
Blevins, Donald K.	University of Florida, Gainesville, FL, U.S.A.
Boboc, Nicu(*)	Universitatea din Bucureşti, Romania
Bojarski, Bogdan	Uniwersytet Warszawski, Warszawa, Poland
Brennan, James	University of Kentucky, Lexington, KY, U.S.A.
Bshouty, Daoud	Eidgenössische Technische Hochschule, Zürich, Switzerland
Bucur, Gheorghe(*)	INCREST, Bucureşti, Romania
Campbell, Douglas	Brigham Young University, Provo, UT, U.S.A.
Caraman, Petru(*)	Universitatea "Al. I. Cuza", Iaşi, Romania
Cegrell, Urban	Uppsala Universitet, Sweden
Coeuré, G.	Université de Lille, France
Colojoară, Ion(*)	Universitatea din Bucureşti, Romania
Cornea, Aurel(*)	INCREST, Bucureşti, Romania
Douady, Adrien	Université Paris-Sud, Orsay, France
Drasin, David	Purdue University, West Lafayette, IN, U.S.A.
Earle, Clifford J.	Cornell University, Ithaca, NY, U.S.A.
Erkama, Timo(*)	University of Helsinki, Finland
Essén, Matts	Kungliga Tekniska Högskolan, Stockholm, Sweden
Farkas, Hershel	Hebrew University, Jerusalem, Israel
Flondor, Paul(*)	Institutul Politehnic Bucureşti, Romania

(*) A member of the IV Romanian-Finnish Seminar on Complex Analysis

Frank, Günter	Fernuniversität Hagen, Federal Republic of Germany
Fuchs, Wolfgang	Cornell University, Ithaca, NY, U.S.A.
Fuglede, Bent	Københavns Universitet, Denmark
Fuji'i'e, Tatsuo	Kyoto University, Japan
Gackstatter, Fritz	Rheinisch-Westfälische Technische Hochschule, Aachen, Federal Republic of Germany
Gauld, David	University of Auckland, New Zealand
Gauthier, Paul	Université de Montréal, Canada
Geatti, Laura	Università di Pisa, Italy
Gehring, Frederick W.	University of Michigan, Ann Arbor, MI, U.S.A.
Ghişa, Dorin(*)	Universitatea din Timişoara, Romania
Göktürk, Zerrin	Boğaziçi Üniversitesi, Bebek-Istanbul, Turkey
Gowrisankaran, Kohur	McGill University, Montreal, Canada
Granlund, Seppo(*)	Helsinki University of Technology, Finland
Grassman, E. G.	University of Calgary, Alberta, Canada
Grunsky, Helmut	Universität Würzburg, Federal Republic of Germany
Gussi, Gheorghe(*)	INCREST, Bucureşti, Romania
Haario, Heikki(*)	University of Helsinki, Finland
Hag, Kari	Norges Tekniske Høgskole, Trondheim, Norway
Hag, Per	Norges Laererhøgskole, Trondheim, Norway
Harmelin, Reuven	Technion-Israel Institute of Technology, Haifa, Israel
Hellerstein, Simon	University of Wisconsin, Madison, WI, U.S.A.
Hengartner, Walter	Université Laval, Québec, Canada
Hennekemper, Wilhelm	Universität Dortmund, Federal Republic of Germany
Hubbard, John	Cornell University, Ithaca, NY, U.S.A.
Huber, Alfred	Eidgenössische Technische Hochschule, Zürich, Switzerland
Hyvönen, Jaakko(*)	University of Joensuu, Finland
Ikegami, Teruo	Osaka City University, Japan
Ivaşcu, Dumitru(*)	Universitatea din Bucureşti, Romania
Jackson, Howard L.	McMaster University, Hamilton, Canada
Janßen, Klaus	Universität Düsseldorf, Federal Republic of Germany
Johnson, Raymond	Howard University, Washington, DC, U.S.A.
Kahramaner, Suzan	İstanbul Üniversitesi, Turkey
Kangasaho, Jukka(*)	University of Joensuu, Finland

Kiikka, Maire(*)	University of Helsinki, Finland
Kiltinen, John O.	Northern Michigan University, Marquette, MI, U.S.A.
Kiselman, Christer	Uppsala Universitet, Sweden
Knab, Otto	Universität Karlsruhe (TH), Federal Republic of Germany
Korevaar, Jacob	Universiteit van Amsterdam, Netherlands
Kortram, Ronald	Universiteit te Nijmegen, Netherlands
Kotman, Larry	University of Wisconsin, La Crosse, WI, U.S.A.
Kra, Irwin	State University of New York, Stony Brook, NY, U.S.A.
Kuusalo, Tapani(*)	University of Jyväskylä, Finland
Labrèche, Martine	Université de Montréal, Canada
Lahtinen, Aatos(*)	University of Helsinki, Finland
Laine, Ilpo(*)	University of Joensuu, Finland
Latvamaa, Esko(*)	Technical College of Joensuu, Finland
Launonen, Eero(*)	Technical College of Kuopio, Finland
Ławrynowicz, Julian	Instytut Matematyczny PAN, Łódź, Poland
Lelong, Pierre	Université Paris VI, France
Lindfors, Ilpo(*)	University of Joensuu, Finland
Louhivaara, I. S.(*)	Freie Universität Berlin, Federal Republic of Germany and University of Jyväskylä, Finland
Maeda, Fumi-Yuki	Hiroshima University, Japan
Martens, Henrik H.	Norges Tekniske Høgskole, Trondheim, Norway
Martio, Olli(*)	University of Helsinki, Finland
Matsuda, Michihiko	Osaka University, Japan
Mattila, Pertti(*)	University of Helsinki, Finland
Menke, Klaus	Universität Dortmund, Federal Republic of Germany
Meyer, Gottfried	Universität Würzburg, Federal Republic of Germany
Miles, Joseph	University of Illinois at Urbana-Champaign, Urbana, IL, U.S.A.
Miniowitz, Ruth	Technion-Israel Institute of Technology, Haifa, Israel
Mues, Erwin	Universität Karlsruhe (TH), Federal Republic of Germany
Nevanlinna, Rolf(*)	Academy of Finland, Helsinki, Finland
Netanyahu, Elisha	Technion-Israel Institute of Technology, Haifa, Israel
Niemi, Hannu(*)	University of Helsinki, Finland
Nguyen-Xuan-Loc	Université Paris-Sud, Orsay, France
Näätänen, Marjatta(*)	University of Helsinki, Finland

Ohtsuka, Makoto	Hiroshima University, Japan
Oja, Kirsti(*)	Helsinki University of Technology, Finland
Ortel, Marvin	University of Hawaii, Honolulu, HI, U.S.A.
Osgood, Brad	University of Michigan, Ann Arbor, MI, U.S.A.
Palka, Bruce	University of Texas, Austin, TX, U.S.A.
Pearcy, Carl	University of Michigan, Ann Arbor, MI, U.S.A.
Peschl, Ernst	Universität Bonn, Federal Republic of Germany
Pesonen, Martti(*)	University of Joensuu, Finland
Pfluger, Albert	Eidgenössische Technische Hochschule, Zürich, Switzerland
Piranian, George	University of Michigan, Ann Arbor, MI, U.S.A.
Pirinen, Aulis(*)	Helsinki University of Technology, Finland
Radu, Nicolae(*)	Universitatea din Bucureşti, Romania
Reich, Edgar	University of Minnesota, Minneapolis, MN, U.S.A.
Reimann, Martin	Universität Bern, Switzerland
von Renteln, Michael	Universität Giessen, Federal Republic of Germany
Rickman, Seppo(*)	University of Helsinki, Finland
Riihentaus, Juhani(*)	Technical College of Oulu, Finland
Rubel, Lee A.	University of Illinois at Urbana-Champaign, Urbana, IL, U.S.A.
Räsänen, Sisko(*)	University of Joensuu, Finland
Sakai, Makoto	Hiroshima University, Japan
Sarvas, Jukka(*)	University of Helsinki, Finland
Schober, Glenn	Indiana University, Bloomington, IN, U.S.A.
Schwarz, Binyamin	Technion-Israel Institute of Technology, Haifa, Israel
Seppälä, Mika(*)	Helsinki School of Economics, Finland
Shea, Daniel	University of Wisconsin, Madison, WI, U.S.A.
Siciak, Józef	Uniwersytet Jagielloński, Kraków, Poland
Siddiqi, Jamil A.	Université Laval, Québec, Canada
Sontag, Alexia	Wellesley College, MA, U.S.A.
Sorvali, Esko(*)	Technical College of Joensuu, Finland
Sorvali, Tuomas(*)	University of Joensuu, Finland
Steinmetz, Norbert	Universität Karlsruhe (TH), Federal Republic of Germany
Stoica, Lucreţiu(*)	INCREST, Bucureşti, Romania
Strebel, Kurt	Universität Zürich, Switzerland
Suciu, Ion(*)	INCREST, Bucureşti, Romania

Sung, Chen-Han	Purdue University, West Lafayette, IN, U.S.A.
Tammi, Olli(*)	University of Helsinki, Finland
Tanaka, Hiroshi	Hokkaido University, Sapporo, Japan
Tietz, Horst	Technische Universität Hannover, Federal Republic of Germany
Vaaler, Jeffrey	University of Texas, Austin, TX, U.S.A.
Valuşescu, Ilie(*)	INCREST, Bucureşti, Romania
Vasilescu, Florin-Horia(*)	INCREST, Bucureşti, Romania
Voiculescu, Dan(*)	INCREST, Bucureşti, Romania
Vuorinen, Matti(*)	University of Helsinki, Finland
Wallin, Hans	Umeå Universitet, Sweden
Weitsman, Allen	Purdue University, West Lafayette, IN, U.S.A.
Winkler, Jörg	Technische Universität Berlin, Federal Republic of Germany

AN EXTREMAL DISPLACEMENT MAPPING IN n-SPACE

G. D. Anderson and M. K. Vamanamurthy

1. Introduction.

1.1. Statement of problem. In this paper we solve the extremal problem of finding a self-homeomorphism F_n of the unit ball B^n in R^n satisfying the following conditions: For fixed r, $0 < r < 1$,

a) F_n keeps the boundary $\partial B^n = S^{n-1}$ pointwise fixed,

b) $F_n(0,0,\ldots,0) = (-r,0,\ldots,0)$,

c) F_n maps a 2-dimensional plane section $R^2 \cap B^n$ containing $B^1 = \{(x_1,0,\ldots,0)\} : |x_1| < 1\}$ onto another such,

d) F_n is quasiconformal with minimum linear dilatation

$$K(F_n) = \operatorname*{ess\,sup}_{x \in B^n} \frac{L_n(x)}{\ell_n(x)} ,$$

where

$$L_n(x) = \limsup_{y \to x} \frac{|F_n(y) - F_n(x)|}{|y - x|}, \quad \ell_n(x) = \liminf_{y \to x} \frac{|F_n(y) - F_n(x)|}{|y - x|}$$

denote the maximum and minimum stretchings at x, respectively. We shall call F_n an extremal displacement mapping in n-space.

1.2. Acknowledgement. The authors wish to thank Professor F. W. Gehring for conversations about this problem.

1.3. Description of the mapping. Since the extremal problem was solved by Teichmüller for $n = 2$ [4] and since we make use of his result, we begin with a brief description of the extremal displacement mapping F_2 of Teichmüller.

The plane mapping

$$x_1 + ix_2 = f_1(u_1 + iu_2) = r\,\mathrm{tn}^2\left(\frac{2K}{\pi} \sinh^{-1} \frac{u_1 + iu_2}{2}, r'\right)$$

maps the quarter ellipse $u_1^2/b^2 + u_2^2/a^2 < 1$, $u_1 > 0$, $u_2 > 0$ conformally onto the upper half disk $|x_1 + ix_2| < 1$, $x_2 > 0$. Here K and K' are the complete elliptic integrals of the first kind defined by

$$K = K(r) = \int_0^1 [(1 - t^2)(1 - r^2t^2)]^{-1/2} dt ,$$

$$K' = K(r'),\ r' = (1 - r^2)^{1/2} ,$$

tn denotes the Jacobian elliptic tangent function, and

$$a = R + R^{-1}, \quad b = R - R^{-1}, \quad R = \exp \frac{\pi K'}{4K} . \tag{1}$$

Likewise

$$y_1 + iy_2 = f_2(v_1 + iv_2) = r\,\mathrm{tn}^2\left(\frac{2K}{\pi} \cosh^{-1} \frac{v_1 + iv_2}{2}, r'\right)$$

maps the quarter ellipse $v_1^2/a^2 + v_2^2/b^2 < 1$, $v_1 > 0$, $v_2 > 0$ conformally onto the upper half disk $|y_1 + iy_2| < 1$, $y_2 > 0$. Then the extremal displacement mapping F_2 is given by

$$F_2 = f_2 \circ \varphi \circ f_1^{-1}$$

for $x_2 > 0$, where φ is the affine mapping

$$v_1 + iv_2 = \varphi(u_1 + iu_2) = \frac{a}{b}u_1 + i\,\frac{b}{a}u_2 .$$

Finally the mapping F_2 is extended by reflection in the x_1-axis to the unit x_1x_2-disk.

Now for each $n \geqq 3$ let F_n be the self-mapping of B^n obtained by rotating F_{n-1} about R^{n-2} in R^n (see §3 below). Then F_n has the above required properties, and we shall prove the following

<u>Theorem 1.</u> For $0 < r < 1$ and $n \geqq 2$ the mapping F_n described above is an extremal quasiconformal self-mapping of B^n with $K(F_n) = K(F_2) = \coth^2 \frac{\pi K'(r)}{4K(r)}$.

<u>Conjecture.</u> Condition c) in §1.1 above can be removed.

2. Proof of theorem for n = 3.

First take $n = 3$ and let P_1 be any point in B^3. By symmetry we may obviously assume that $P_1 = (x_1, x_2, 0)$, where $x_2 \geqq 0$. Let $P_2 = (y_1, y_2, 0)$ and $Q_1 = (u_1, u_2)$, $Q_2 = (v_1, v_2)$, where

$$y_1 + iy_2 = F_2(x_1 + ix_2)$$

$$u_1 + iu_2 = f_1^{-1}(x_1 + ix_2) ,$$

$$v_1 + iv_2 = f_2^{-1}(y_1 + iy_2) .$$

Then the maximum stretching $L_2(P_1) = L_2$ and the minimum stretching $\ell_2(P_1) = \ell_2$ of F_2 at P_1 are

$$L_2 = \frac{a|f_2'(Q_2)|}{b|f_1'(Q_1)|} , \quad \ell_2 = \frac{b|f_2'(Q_2)|}{a|f_1'(Q_1)|} .$$

Hence

$$K(F_2) = \frac{L_2}{\ell_2} = \frac{a^2}{b^2} = \coth^2 \frac{\pi K'}{4K} .$$

The three stretchings of F_3 at P_1 are

$$\begin{cases} L_2,\ \ell_2,\ y_2/x_2 & \text{for } x_2 > 0 , \\ L_2,\ \ell_2,\ \ell_2 & \text{for } x_2 = 0,\ 0 < x_1 < 1 , \\ L_2,\ L_2,\ \ell_2 & \text{for } x_2 = 0,\ -1 < x_1 < 0 . \end{cases}$$

Now $\ell_3 \leqq \ell_2 \leqq L_2 \leqq L_3 \Rightarrow K(F_3) = \text{ess sup } \frac{L_3}{\ell_3} \geqq \frac{L_2}{\ell_2} = K(F_2)$. If we can show that

$$\ell_2 < \frac{y_2}{x_2} < L_2 \tag{2}$$

for $x_2 > 0$ and $0 < r < 1$, then it will follow that $\ell_3 = \ell_2$ and $L_3 = L_2$, whence $K(F_3) = L_2/\ell_2 = K(F_2)$ and F_3 is extremal.

Now (2) is equivalent to

$$\frac{b}{a} < \frac{x_2}{y_2} \frac{|f_2'(Q_2)|}{|f_1'(Q_1)|} < \frac{a}{b} . \tag{3}$$

By elementary computations it can be shown that f_1^{-1} and f_2^{-1} can be extended by reflection so that, after extension, f_1^{-1} maps the half plane $x_2 > 0$ conformally onto the quarter ellipse

$$D_1 : \frac{u_1^2}{(ab)^2} + \frac{u_2^2}{(a^2 - 2)^2} < 1,\ u_1 > 0,\ u_2 > 0$$

and f_2^{-1} maps the half plane $y_2 > 0$ conformally onto the quarter ellipse

$$D_2 : \frac{v_1^2}{(a^2 - 2)^2} + \frac{v_2^2}{(ab)^2} < 1,\ v_1 > 0,\ v_2 > 0 .$$

Then the problem reduces to showing that

$$\frac{b}{a} < \frac{\rho(D_2,Q_2)}{\rho(D_1,Q_1)} < \frac{a}{b} , \tag{4}$$

where ρ denotes hyperbolic density. Clearly (4) is implied by the inequalities

$$\frac{b}{a} < \frac{\rho(D_2,Q_2)}{\rho(D_1,Q_1)} < \frac{a}{b} \quad \text{for } |f_1(Q_1)| < 1,\ 0 < u_2 \leq \frac{a}{b}u_1 \tag{5}$$

and

$$\frac{b}{a} < \frac{\rho(D_2,Q_2)}{\rho(D_1,Q_1)} < \frac{a}{b} \quad \text{for} \quad |f_1(Q_1)| < 1, \ \frac{a}{b}u_1 \leq u_2 , \tag{6}$$

where $Q_1 = (u_1,u_2)$. We shall prove only (5), since (6) may be proved similarly.

Let g_1, g_2 be the similarity mappings

$$u_1 + iu_2 = g_1(U_1 + iU_2) = bU_1 + ibU_2$$

$$v_1 + iv_2 = g_2(V_1 + iV_2) = aV_1 + iaV_2 .$$

Then $E_1 = g_1^{-1}(D_1)$ and $E_2 = g_2^{-1}(D_2)$ are first quadrants of elliptic domains in the U_1U_2- and V_1V_2-planes, respectively. If we let $M_j = g_j^{-1}(Q_j)$ for $j = 1,2$, it is easy to see that (5) is equivalent to

$$1 < \frac{\rho(E_2,M_2)}{\rho(E_1,M_1)} < (\frac{a}{b})^2 \quad \text{for} \quad |f_1g_1(M_1)| < 1, \ 0 < U_2 \leq \frac{a}{b}U_1 , \tag{7}$$

where $M_1 = (U_1,U_2)$. We shall show that the first inequality in (7) holds if $U_2 \leq 2^{-1/2}$ a/b.

Now identify the U_1U_2- and the V_1V_2-planes, making the corresponding axes coincide. Then $E_2 \subset E_1$, $E_2 \neq E_1$, and M_1 lies vertically above M_2. If we translate E_2 vertically upward by a distance $|M_1 - M_2|$, then M_2 coincides with M_1, and

$$\rho(E_2,M_2) = \rho(E_2',M_1) ,$$

where E_2' is the translate of E_2. By virtue of the Schwarz Lemma in terms of hyperbolic density, to prove the first half of (7) it is sufficient to prove that E_2' is properly contained in E_1. This will be true if $|M_1 - M_2| < d$, where d is the minimum vertical distance between the two elliptic arcs

$$\frac{U_1^2}{(\frac{a^2-2}{a})^2} + \frac{U_2^2}{b^2} = 1, \ U_1 > 0, \ U_2 > 0$$

and

$$\frac{U_1^2}{a^2} + \frac{U_2^2}{(\frac{a^2-2}{b})^2} = 1, \ U_1 > 0, \ U_2 > 0 .$$

By elementary differential calculus it is easy to show that

$$d = \frac{2(a^2-1)^{1/2}[(a^2-2)^4 - a^4b^4]^{1/2}}{a^2b(a^2-2)} ,$$

and that the minimum occurs when

$$U_1 = \frac{[(a^2-2)^6 - a^8b^4]^{1/2}}{[(a^2-2)^4a^2 - a^6b^4]^{1/2}} .$$

If $\operatorname{Im} M_1 \leqq 2^{-1/2}\, a/b$, then

$$|M_1 - M_2| = U_2(1 - b^2/a^2) < 2^{-1/2}(a/b)(1 - b^2/a^2) = \frac{2\sqrt{2}}{ab}$$

since (1) implies that $a^2 - b^2 = 4$. Thus it is enough to prove that

$$\frac{2\sqrt{2}}{ab} < d \quad \text{for} \quad a \geqq 2,\ a^2 - b^2 = 4 . \tag{8}$$

By using straightforward algebraic manipulations, we may show that (8) is equivalent to the inequality

$$3t^3 - 16t^2 + 20t - 8 > 0 \quad \text{for} \quad t \geqq 4 , \tag{9}$$

where $t = a^2$. When $t = 4$ the left side of (9) is 8, and the derivative is positive for $t \geqq 4$. This proves (8), and hence the left side of (5).

Next, using the differential equation $\nabla^2 \log \rho = 4\rho^2$ satisfied by hyperbolic density, together with the inequality just proved, we have

$$\nabla^2 \log \frac{\rho(E_2', M_1)}{\rho(E_1, M_1)} = 4[\rho^2(E_2', M_1) - \rho^2(E_1, M_1)] > 0$$

for such M_1, so that

$$\log \frac{\rho(E_2', M_1)}{\rho(E_1, M_1)}$$

is subharmonic as a function of M_1 in the set $\{(U_1,U_2) : |f_1 g_1(U_1,U_2)| < 1,\ 0 < U_2 \leqq \frac{a}{b}U_1\}$. Hence,

$$h(u_1,u_2) = \log \frac{\rho(D_2, \varphi(u_1,u_2))}{\rho(D_1, (u_1,u_2))}$$

is subharmonic in the set

$$S = \{(u_1,u_2) : |f_1(u_1,u_2)| < 1,\ 0 < u_2 \leqq \tfrac{a}{b}u_1\} .$$

We wish to determine the boundary values of h on S. By a clockwise rotation of D_2 through a right angle and then a reflection in the real axis, we may make the quarter ellipse $f_2^{-1}(B_2^+)$ coincide with $f_1^{-1}(B_2^+)$, where B_2^+ denotes the upper half of the unit disk. Thus it is easy to see that

$$\rho(D_2, \varphi(u_1,u_2)) = \rho(D_1, (\tfrac{b}{a}u_2, \tfrac{a}{b}u_1)) ,$$

and that

$$h(u_1,u_2) = \log \frac{\rho(D_1,(\frac{b}{a}u_2,\frac{a}{b}u_1))}{\rho(D_1,(u_1,u_2))} .$$

Then $h(u_1,u_2) = 0$ when $(u_1,u_2) = (\frac{b}{a}u_2,\frac{a}{b}u_1)$, that is, on the line $u_2 = \frac{a}{b}u_1$.

Next, for $u_2 \to 0$,

$$\frac{y_2}{x_2} = \frac{y_2}{v_2} \cdot \frac{v_2}{u_2} \cdot \frac{u_2}{x_2} \sim \frac{b}{a} \frac{|f_2'(Q_2)|}{|f_1'(Q_1)|}$$

so that h has constant boundary value $\log \frac{a}{b}$ along the line $u_2 = 0$.

Finally, taking into account the definition of hyperbolic density and using the fact that the extremal displacement mapping F_2 leaves boundary points fixed, we may easily show that h has the boundary values

$$h(u_1,u_2) = \log \left| \frac{\frac{a}{b}u_1 + i\frac{b}{a}u_2}{u_1 + iu_2} \right| \tag{10}$$

of the curved portion of ∂S. Parametrizing this arc by means of

$$u_1 + iu_2 = b \cos \theta + i\, a \sin \theta, \ 0 \leq \theta \leq \pi/4 ,$$

and employing the identity $a^2 - b^2 = 4$, we may simplify (10) to

$$h(u_1,u_2) = \frac{1}{2} \log \frac{a^2 - 4\sin^2\theta}{b^2 + 4\sin^2\theta}, \ 0 \leq \theta \leq \pi/4 .$$

As θ increases from 0 to $\pi/4$, the numerator decreases monotonically from a^2 to $a^2 - 2$, while the denominator increases monotonically from b^2 to $b^2 + 2$. Since $b^2 + 2 = a^2 - 2$, it follows that the boundary values of $h(u_1,u_2)$ decrease from $\log a/b$ to 0 along the curved portion of ∂S. Thus the boundary values of h lie between 0 and $\log a/b$, except possibly at $(u_1,u_2) = (0,0)$.

In order to invoke the Phragmén-Lindelöf principle we now establish an upper bound for h in the set S. First, by partitioning S into two sets by the line $u_2 = u_1$ and then considering the corresponding cases we may easily show that

$$\frac{b}{a} \leq \frac{\min(\frac{b}{a}u_2,\frac{a}{b}u_1)}{\min(u_1,u_2)} \leq 1$$

for all $(u_1,u_2) \in S$. In a small neighborhood of $(0,0)$ relative to S, $d(u_1,u_2) = \min(u_1,u_2)$, where $d(u_1,u_2)$ denotes distance from the point (u_1,u_2) to ∂D_1. Hence

$$\frac{b}{a} \leqq \frac{d(\frac{b}{a}u_2, \frac{a}{b}u_1)}{d(u_1,u_2)} \leqq 1 \tag{11}$$

for all (u_1,u_2) in this relative neighborhood. By means of (11) and the well-known inequality

$$\frac{1}{2d(u_1,u_2)} \leqq \rho(D_1,(u_1,u_2)) \leqq \frac{1}{d(u_1,u_2)}$$

for the convex domain D_1 (cf. [3]) we may easily show that

$$\limsup h(u_1,u_2) \leqq \log 2a/b$$

as $(u_1,u_2) \to (0,0)$ in S. Hence by the maximum principle, $h(u_1,u_2) \leqq \log 2a/b$ in S.

Finally, we may appeal to the Phragmén-Lindelöf Principle (cf. [3]) to conclude that $h(u_1,u_2) \leqq \log a/b$ for all $(u_1,u_2) \in S$. Moreover, since h is subharmonic in S, an argument based on Theorem II.12 of [5] shows that this inequality is strict. Hence the right inequality in (5) also holds. Thus (5) is established, and $L_3/\ell_3 = K(F_2)$ at each point of $B^3 \setminus \{(0,0,0)\}$. Hence F_3 is a differentiable quasiconformal mapping of $B^3 \setminus \{(0,0,0)\}$ onto $B^3 \setminus \{(-r,0,0)\}$ with $K(F_3) = K(F_2)$. By removing the singularity at the origin we see that F_3 is a (generalized) quasiconformal self-mapping of B^3 with the same dilatation.

3. Proof of theorem for general n.

We proceed by induction for $n \geqq 4$. Suppose that $F_{n-1} : B^{n-1} \to B^{n-1}$ has been shown to be an extremal displacement mapping of B^{n-1} with

$$\ell_{n-2}(x) < \frac{\operatorname{dist}(F_{n-1}(x), R^{n-2})}{\operatorname{dist}(x, R^{n-2})} < L_{n-2}(x) \tag{12}$$

for $x \in B^{n-1} \setminus (B^{n-2} \times \{0\})$, so that $K(F_{n-1}) = K(F_{n-2}) = \cdots = K(F_2)$. Following Gehring [2], for $x = (x_1,\ldots,x_n) \in B^n$ let $s = (x_{n-1}^2 + x_n^2)^{1/2}$ and set

$$F_n(x_1,\ldots,x_n) = \begin{cases} (y_1,\ldots,y_{n-2},0,0) & \text{if } s = 0\,, \\ (y_1,\ldots,y_{n-2},x_{n-1}|y_{n-1}|/s, x_n|y_{n-1}|/s) & \text{if } s > 0\,, \end{cases}$$

where $(y_1,\ldots,y_{n-1}) = F_{n-1}(x_1,\ldots,x_{n-2},s)$. We call F_n the rotation of F_{n-1} about R^{n-2} in R^n. Then F_n is a self-homeomorphism of B^n which takes each B^k onto itself and agrees with F_k on $B^k \times \{0_{n-k}\}$ for $1 \leqq k \leqq n-1$. Here 0_k denotes the k-tuple $(0,0,\ldots,0)$.

In particular, F_n has properties a), b), and c). Moreover,

$$\frac{\operatorname{dist}(F_n(x),R^{n-1})}{\operatorname{dist}(x,R^{n-1})} = \frac{x_n|y_{n-1}|/s}{x_n} = \frac{|y_{n-1}|}{s} = \frac{\operatorname{dist}(F_{n-1}(x'),R^{n-2})}{\operatorname{dist}(x',R^{n-2})}$$

in $B^n \setminus (B^{n-1} \times \{0\})$, where x' is the projection of x into $R^{n-1} \times \{0\}$. Thus $K(F_n) = K(F_2)$, d) holds, and F_n is an extremal displacement mapping in n-space.

4. Remarks.

The methods used above also show that, for each $n \geq 3$, $K_0(F_n) = K_I(F_n) = K(F_2)^{n-1}$, where K_0 and K_I denote outer and inner dilatation, respectively (cf. [6]).

Next, for $j = 1,2,3,4$, let a_j and b_j denote a pair of positively ordered quadruples of points on the real axis in $\bar{R}^2$. There is a unique extremal self-mapping G_2 of $\bar{R}^2$, called the extremal distortion mapping such that $G_2(a_j) = b_j$, $j = 1,2,3,4$. In an earlier paper [1] the present authors rotated the mapping G_2 about R^1 in R^3 to obtain an extremal distortion mapping G_3 in 3-space, subject to a condition similar to c) in §1.1 above. If we use induction to extend G_3 to n-space, $n \geq 4$, by rotating G_{n-1} about R^{n-2} in R^n as in §3 above, we obtain the following result (cf. Theorem 1 in [1]).

Theorem 2. For $j = 1,2,3,4$, let a_j and b_j be a pair of positively ordered quadruples of points on the x^1-axis in $\bar{R}^2$. Then the self-mapping G_n of $\bar{R}^n$ described above is a solution of the following extremal problem.

a) $G_n(a_j) = b_j$, $j = 1,2,3,4$,

b) G_n maps a 2-dimensional plane section through the origin onto another such, and

c) G_n is quasiconformal with minimum linear dilatation $K(G_n) = \operatorname{ess\,sup}_{x\in\bar{R}^n} (L_n(x))/\ell_n(x)$.

Moreover, $K(G_n) = K(G_2)$, $n \geq 2$.

Finally, for $n \geq 3$ and $0 < r < 1$, we let $m(n,r)$ denote the modulus of the Grötzsch ring in n-space. As a consequence of Theorem 2 we may prove, as in Corollary 1 of [1], the following

Corollary. For $n \geq 3$, $m(n,r)/m(2,r)$ is an increasing function of r, $0 < r < 1$.

References

[1] Anderson, G. D., Vamanamurthy, M. K.: Rotation of plane quasiconformal mappings. Tôhoku math. J., II. Ser. 23 (1971), 605 - 620.

[2] Gehring, F. W.: Absolute continuity properties of quasiconformal mappings. Symp. Math. 18 (1976), 551 - 559.

[3] Gehring, F. W.: Topics in geometric function theory. Unpublished lecture notes, Harvard University (1964).

[4] Teichmüller, O.: Ein Verschiebungssatz der quasikonformen Abbildung. Deutsche Math. 7 (1944), 336 - 343.

[5] Tsuji, M.: Potential theory in modern function theory. Maruzen, Tokyo (1959).

[6] Väisälä, J.: Lectures on n-dimensional quasiconformal mappings. Lecture Notes in Mathematics 229, Springer-Verlag, Berlin - Heidelberg - New York (1971).

Michigan State University
East Lansing, MI 48824
U.S.A.

University of Auckland
Auckland, New Zealand

ON THE GRÖTZSCH AND RENGEL INEQUALITIES

Cabiria Andreian Cazacu

§1. Introduction

In the frame of the vast literature on the definitions of the quasiconformality, there are many papers dedicated to the reduction as much as possible of the conditions imposed on the homeomorphisms in order that they be quasiconformal. For example we quote only the joint paper of F. W. Gehring and J. Väisälä [10], or in the n-dimensional case the papers of J. Väisälä [13] and of P. Caraman [8]. (See also [9] Part 2, and [14] Sections 13, 34, 36). We shall present here other results in the same direction. Starting from our paper [3] we established in [5] connections between Grötzsch's and Rengel's inequalities in the n-dimensional case. Now we shall develop these connections and obtain consequences for the quasiconformality coefficient and a quasiconformality criterium assuming that Grötzsch's inequality is verified only for certain families of q-surfaces, which define by their module the q-module of a topological n-cube. As usually $q = 1,2,\dots,n-1$ and the case $q = 1$ corresponds to the curves.

In this preliminary paragraph we remind some definitions and results proved in [5] which will be basic facts for this paper. We shall work with the classical module of families of q-surfaces, using the following

Convention. For each q-surface family Σ we consider the subfamily Σ_o of all the q-surfaces of Σ for which the Lebesgue measure and integral are defined and we put for the module of Σ

$$M(\Sigma) = \begin{cases} M(\Sigma_o) & \text{if } \Sigma_o \neq \emptyset \text{ and} \\ 0 & \text{if } \Sigma_o = \emptyset. \end{cases}$$

Σ_o will be called the regular subfamily of Σ. (See [4] p. 88, [5] §1).

1.1 The q-modules of a topological n-cube in R^n

Every n-segment I_n in R^n may be considered for each $q = 1,2,\dots,n-1$ in $\binom{n}{q}$ ways as the direct product of a q-segment I_q by an (n-q)-segment I_{n-q} which is orthogonal to I_q. Let us choose one of these possibilities, and take the corresponding I_q as a "base" and

I_{n-q} as a "height", which we call the "associated face with respect to I_q". Let us further denote by $\sum_{n-q} = \sum(I_n, I_q)$ the family of all the (n-q)-segments in I_n which are parallel to I_{n-q} and by $\mathcal{S}_{n-q} = \mathcal{S}(I_n, I_q)$ the family of all the (n-q)-surfaces which have in I_n the same topological position as the (n-q)-segments of $\sum_{n-q}$.

Definition 1. The ((n-q)-)module of I_n with respect to the base I_q is given by

$$M(I_n, I_q) = M[\sum(I_n, I_q)].$$

Briefly we denote $M(I_n, I_q)$ by M. One knows that

$$M = v_n h_{n-q}^{-n/(n-q)} = a_q h_{n-q}^{-q/(n-q)}, \tag{1}$$

where v_n is the volume of I_n, a_q the q-area of I_q and h_{n-q} the (n-q)-area of I_{n-q}. Further one has

$$M = M[\mathcal{S}(I_n, I_q)].$$

Let now P_n be a topological cube defined by a homeomorphism $\varphi : I_n \to P_n$ which gives also the vertices and the faces of P_n. (All the homeomorphisms we shall consider in this paper will be sense-preserving). We take $P_q = \varphi(I_q)$ as the "base" of P_n and $P_{n-q} = \varphi(I_{n-q})$ as the "associated face" of P_n with respect to P_q, and we put $\mathcal{S}(P_n, P_q) = \varphi[\mathcal{S}(I_n, I_q)]$.

Definition 2. The module of P_n with respect to P_q is defined by

$$M(P_n, P_q) = M[\mathcal{S}(P_n, P_q)].$$

Evidently the use of the classical module based on the Convention from above arises difficulties as soon as one considers a K-quasiconformal mapping $f : P_n \to P_n^* = f(P_n)$ and one wants to obtain the Grötzsch inequalities for $M(P_n, P_q)$ and $M(P_n^*, P_q^*)$, $P_q^* = f(P_q)$, since the relation between the image of the regular family $f[\mathcal{S}(P_n, P_q)_o]$ and the regular family of the image $\mathcal{S}(P_n^*, P_q^*)_o$ is generally unknown. However for what follows it will be sufficient to work with this classical module since we shall always start with an n-segment. For instance if P_n is a Q-quasiconformal cube, i.e. φ is a Q-quasiconformal mapping, then

$$M(I_n, I_q) \leqq QM(P_n, P_q).$$

1.2. Rengel's inequalities

Let us suppose that $\varphi : I_n \to P_n$ is an affine transformation. We denote in this case φ by T and call the affine n-cube P_n an n-parallelotope or n-parallelipiped. Further we denote by $\mathcal{O}_{n-q}$ the orthogonal section in P_n with respect to P_q, by $\mathcal{O}_q$ the orthogonal section in P_n with respect to $P_{n-q} = T(I_{n-q})$, by a_q and H_q the q-areas of P_q and $\mathcal{O}_q$ respectively, and by A_{n-q} and h_{n-q} the (n-q)-areas of P_{n-q} and $\mathcal{O}_{n-q}$ respectively. By usual devices one easily obtains the Rengel inequalities:

$$H_q A_{n-q}^{-q/(n-q)} \leq M(P_n, P_q) \leq a_q h_{n-q}^{-q/(n-q)} . \tag{2}$$

Equality occurs in both sides if P_q and P_{n-q} are orthogonal, in which case P_n will be called a quasirectangular parallelotope. In particular, this is the case when P_n is a rectangular parallelotope, in accordance with (1).

Obviously (2) can be also expressed by means of the volume v_n of P_n:

$$(A_{n-q}^{n/(n-q)} v_n^{-1})^{-1} \leq M(P_n, P_q) \leq (a_q^{n/q} v_n^{-1})^{q/(n-q)} . \tag{2'}$$

1.3. Connection with the oriented q-dimensional dilatation d.

Let G and G^* be domains in R^n, $f : G \to G^*$ a homeomorphism, x_o an A-point (point of regularity) for f, i.e. a point where f is differentiable with non-vanishing Jacobian, $T = A(f, x_o)$ the affine mapping associated to f at x_o, $T(x) = f(x_o) + f'(x_o)(x - x_o)$, J_n the Jacobian of f or T at x_o, Π_q a q-plane through x_o and J_q the Jacobian of $f|\Pi_q$ or $T|\Pi_q$ at x_o.

Definition 3. The q-dilatation of f at the point x_o with respect to the q-plane Π_q is equal to

$$d_{f,\Pi_q}(x_o) = \frac{J_q^{n/q}}{J_n} \tag{3}$$

(We wrote J_n, J_q instead of $J_n(x_o)$, $J_q(x_o)$ and similar abbreviation will be made for the function d too.)

We shall not insist here on the simple geometric interpretation of d (which appears even in [11], see [4] §2,4) nor on its importance as a weight in Ohtsuka's sense in the module (extremal length) theory ([4] §2, 1 - 3; [5] §1) but using this dilatation we shall give another form to the Rengel inequalities which thus become inequalities of Grötzsch's type.

Let Π_{n-q} be the (n-q)-plane through x_0 which is orthogonal to Π_q and $d_{f,\Pi_{n-q}}(x_0)$ the (n-q)-dilatation of f at x_0 with respect to Π_{n-q}.

We shall denote by Ω_n the rectangular n-parallelotope which is obtained from

$$I_n = \{x \in R^n : x = (x_1,\ldots,x_n),\ 0 \leqq x_j \leqq \ell_j,\ j = 1,\ldots,n\}$$

by a translation: $0 \mapsto x_0$ and a rotation such that

$$I_q = \{x \in I_n : x = (x_1,\ldots,x_q,\ 0,\ldots,0)\}$$

be transformed in the base Ω_q of Ω_n lying in Π_q, and

$$I_{n-q} = \{x \in I_n : x = (0,\ldots,0,\ x_{q+1},\ldots,x_n)\}$$

in the associated face Ω_{n-q} lying in Π_{n-q}.

Further we shall denote by $\tilde{\Omega}_k = T(\Omega_k)$, k = n, q and n - q.

Applying the Rengel inequalities (2') to the module $\tilde{M} = M(\tilde{\Omega}_n, \tilde{\Omega}_q)$ and the formula (1) to the module $M = M(\Omega_n, \Omega_q)$ one obtains that

$$d^{-1}_{f,\Pi_{n-q}} \leqq \frac{\tilde{M}}{M} \leqq d^{q/(n-q)}_{f,\Pi_q}. \tag{4}$$

Moreover, if we convene to write $I_q \to 0$ when the length of each side of I_q tends to 0, and analogously for I_{n-q}, then it follows that

$$\lim_{I_{n-q}\to 0} \frac{\tilde{M}}{M} = d^{q/(n-q)}_{f,\Pi_q} \tag{5}$$

and

$$\lim_{I_q\to 0} \frac{\tilde{M}}{M} = d^{-1}_{f,\Pi_{n-q}}. \tag{5'}$$

We shall call (4) the <u>Grötzsch-Rengel inequalities</u>. Similar results hold for the module of an arbitrary Fuglede's order too. ([5] §2, 3).

<u>1.4. Consequences of the Grötzsch-Rengel inequalities</u>

Let $f : G \to G^*$ be a homeomorphism, x_0 a point in the domain G and Π_q a q-plane through x_0.

<u>Definition 4.</u> We say that the <u>Grötzsch inequality is verified by f at x_0 for the q-plane</u> Π_q, if for every rectangular n-parallelotope Ω_n with the base Ω_q in Π_q (as before) and for $\Omega^*_k = f(\Omega_k)$, k = n and q, the modules $M = M(\Omega_n, \Omega_q)$ and $M^* = M(\Omega^*_n, \Omega^*_q)$ satisfy the inequality

$$M \leqq QM^* \qquad (6)$$

or

$$M^* \leqq QM, \qquad (6')$$

where $Q > 0$ is a constant.

If x_o is an A-point for f it is easy to see (by using a device of Pfluger [12]) that (6) or (6') for f implies the same property for $T = A(f,x_o)$. From the Grötzsch-Rengel inequalities (4) it follows

Theorem 1. ([5] §2, 4) Suppose that x_o is an A-point for f and Π_q a q-plane through x_o. Then
1º Grötzsch's inequality (6) for f implies

$$d_{f,\Pi_{n-q}} \leqq Q, \qquad (7)$$

i.e.

$$J^n_{n-q} \leqq Q^{n-q} J^{n-q}_n \qquad (8)$$

at the point x_o, and
2º Grötzsch's inequality (6') for f implies

$$d_{f,\Pi_q} \leqq Q^{(n-q)/q}, \qquad (7')$$

i.e.

$$J^n_q \leqq Q^{n-q} J^q_n \qquad (8')$$

at the point x_o.

For $f = T$, (6) is equivalent to (7) and (6') to (7').

If x_o is a differentiability point of f where $J_n = 0$ then the inequality (6) implies (8), i.e. $J_{n-q} = 0$, and the inequality (6') implies (8'), i.e. $J_q = 0$.

§2. Consequences of Grötzsch's inequalities for the quasiconformality coefficient

Theorem 1 permits us to deduce from Grötzsch's inequalities for a q-plane Π_q an upper bound for $d_{f,\Pi_{n-q}}$ or d_{f,Π_q} and from here to obtain an upper bound for the coefficient of quasiconformality of f. Let us remark that an upper bound for d_{f,Π_q} can be derived from two Grötzsch's inequalities:

$$M(\Omega_n^*,\Omega_q^*) \leqq QM(\Omega_n,\Omega_q) \tag{6'}$$

implies

$$d_{f,\Pi_q} \leqq Q^{(n-q)/q} \tag{7'}$$

and

$$M(\Omega_n,\Omega_{n-q}) \leqq QM(\Omega_n^*,\Omega_{n-q}^*) \tag{9}$$

implies

$$d_{f,\Pi_q} \leqq Q. \tag{10}$$

In the following we shall work with the inequality (9), the other case being similar.

2.1. Suppose that $x_o \in G$ is an A-point for the homeomorphism $f : G \to G^*$ and that Π_q is a q-plane through x_o whose position with respect to the axes of the characteristic ellipsoid, which is transformed by $T = A(f,x_o)$ in the unit ball, is known. If the axes are not uniquely determined we choose one of these possibilities. For the sake of simplicity we take $x_o = f(x_o) = 0$ and assume that T has the form

$$\tilde{x}_j = p_j x_j, \quad j = 1,\dots,n, \tag{11}$$

with $p_1 \geqq \cdots \geqq p_n = 1$. The general case may be reduced to this one by conformal mappings which have no influence on the quasiconformality coefficient. Thus the q-plane Π_q will pass through 0 and will be given by q linearly independent unit vectors $V_k = (V_{1k},\dots,V_{nk})$, $k = 1,\dots,q$. We have for an arbitrary point $x \in \Pi_q$

$$x = \sum_{k=1}^{q} u_k V_k, \tag{12}$$

i.e.

$$x_j = \sum_{k=1}^{q} u_k V_{jk}, \quad j = 1,\dots,n \tag{12'}$$

and for $\tilde{x} = T(x)$

$$\tilde{x}_j = \sum_{k=1}^{q} p_j u_k V_{jk}, \quad j = 1,\dots,n. \tag{13}$$

Using the classical method to calculate areas so clearly presented in [2] let us denote by $C(n,q)$ the set

$$C(n,q) = \{(\nu_1,\dots,\nu_q) \subset (1,\dots,n) : \nu_1 < \cdots < \nu_q\}$$

and for each $\nu = (\nu_1,\dots,\nu_q) \in C(n,q)$ put

$$A_{\nu_1\cdots\nu_q} = \begin{vmatrix} V_{\nu_1 1} & \cdots & V_{\nu_1 q} \\ \cdots & \cdots & \cdots \\ V_{\nu_q 1} & \cdots & V_{\nu_q q} \end{vmatrix} . \tag{14}$$

In our case the surface S from [2] is Π_q, hence

$$J_S^2 = \sum_{\nu\in C(n,q)} A^2_{\nu_1\cdots\nu_q} . \tag{15}$$

We write

$$B_{\nu_1\cdots\nu_q} = A^2_{\nu_1\cdots\nu_q} J_S^{-2} . \tag{16}$$

and

$$\alpha = B_{1\cdots q} . \tag{17}$$

<u>Proposition 1.</u> Assume that f verifies the inequality (9) for the $(n-q)$-plane Π_{n-q} which is orthogonal to Π_q, and that $\alpha \neq 0$. If $Q \geqq 1$, then

$$p_1 \leqq p_1\cdots p_q \leqq Q^q\alpha^{-n/2} . \tag{18}$$

If $Q < 1$, under certain conditions on α, given below, (18) remains valid.

<u>Proof.</u> The proof will even provide a better (but not explicit) upper bound than that from (18).

Indeed, if we suppose that the inequality (9) is true for Π_{n-q} then we have (10) for Π_q i.e.

$$J_q^n \leqq Q^q J_n^q . \tag{19}$$

But taking into account (11), we have

$$J_n = p_1\cdots p_{n-1} \tag{20}$$

and

$$J_q^2 = \sum_{\nu\in C(n,q)} B_{\nu_1\cdots\nu_q} p^2_{\nu_1}\cdots p^2_{\nu_q} . \tag{21}$$

Indeed, $J_q = J_{\tilde{S}} J_S^{-1}$ if we put $\tilde{S} = T(\Pi_q)$, and J_S is given by (15) while

$$J_{\tilde{S}}^2 = \sum_{\nu\in C(n,q)} A^2_{\nu_1\cdots\nu_q} p^2_{\nu_1}\cdots p^2_{\nu_q} .$$

Therefore (19) may be written

$$\Big(\sum_{\nu\in C(n,q)} B_{\nu_1\cdots\nu_q} p^2_{\nu_1}\cdots p^2_{\nu_q}\Big)^n \leqq Q^{2q}(p_1\cdots p_{n-1})^{2q} . \tag{22}$$

The greatest product $p_{\nu_1}\cdots p_{\nu_q}$ being $p_1\cdots p_q$ we majorize in the right side of (22) by $J_n^{2q} \leq (p_1 \cdots p_q)^{2(n-1)}$. Since $\alpha \neq 0$ we keep in the left side the term $B_{1\cdots q} p_1^2 \cdots p_q^2$ unchanged and replace the other products $p_{\nu_1}^2 \cdots p_{\nu_q}^2$ by 1. Further we write $p_1^2 \cdots p_q^2 = v^n$ and $Q^{2q/n} = g$. Thus we deduce from (22) the inequality

$$\alpha v^n - g v^{n-1} + (1 - \alpha) \leq 0 \tag{23}$$

which constitutes a necessary condition for (9).

Taking into account that $0 < \alpha \leq 1$, we shall now study the polynomial

$$F(v) = \alpha v^n - g v^{n-1} + (1 - \alpha).$$

1) Case $Q \geq 1$.

Since $\alpha^{n-1}(1 - \alpha) \leq n^{-1}[(n-1)/n]^{n-1}$ for $\alpha > 0$, the greatest positive zero $v_1 = v_1(\alpha)$ of $F(v)$ verifies the inequalities

$$[(n-1)/n](g/\alpha) < v_1 \leq g/\alpha.$$

Moreover, since $F(1) \leq 0$ and $F(v) > 0$ for $v > v_1$, it follows $1 \leq v_1$. Hence (23) implies

$$1 \leq p_1 \leq p_1 \cdots p_q \leq v_1(\alpha)^{n/2} \leq (g/\alpha)^{n/2} = Q^q/\alpha^{n/2}. \tag{24}$$

If $Q > 1$ and $\alpha = 1$, then $v_1 = g$; if $Q = 1$ and $(n-1)/n \leq \alpha$, then $v_1 = 1$.

2) Case $Q < 1$.

Denote by α' and α'' the two roots of the equation

$$\alpha^n - \alpha^{n-1} + (g^n/n)[(n - 1)/n]^{n-1} = 0 \tag{25}$$

which verify the inequalities $0 < \alpha' < (n-1)/n < \alpha'' < 1$. For α in the intervals $(0,\alpha')$ and $(\alpha'',1)$ it holds $F((n-1)g/n\alpha) < 0$. Hence $F(v)$ has two positive zeros $v_1 = v_1(\alpha)$ and $v_2 = v_2(\alpha)$, $v_2 < v_1$, and $F(v) < 0$ in (v_2,v_1). Since $F(1) > 0$, the single possibility to have simultaneously $F(v) \leq 0$ and $v \geq 1$ is given by the case

$$\alpha \in (0,\alpha'] \tag{26}$$

as $\alpha' < (n-1)g/n$, which implies

$$1 < v_2^{n/2} \leq p_1 \cdots p_q \leq v_1^{n/2} \leq (g/\alpha)^{n/2} = Q^q \alpha^{-n/2}. \tag{27}$$

The case $\alpha = \alpha'$ implies $v_1 = v_2 = (n - 1)g/n\alpha'$ and $p_1 \cdots p_q = v_1^{n/2}$. From (27) it follows that in the hypothesis (26) one has again

$$p_1 \leq p_1 \cdots p_q \leq v_1^{n/2} \leq Q^q \alpha^{-n/2}.$$

If the condition (26) is not fulfilled, f cannot satisfy (9).

Let us also remark that the upper bound $v_1(\alpha)$, as well as $Q^{q}\alpha^{-n/2}$, is not increasing when α increases from 0 to 1. More exactly, for $Q > 1$ the function $v_1(\alpha)$ decreases when α increases from 0 to 1, for $Q = 1$ the function $v_1(\alpha)$ decreases when α increases from 0 to $(n-1)/n$ and after that remains equal to 1, and for $Q < 1$ it decreases when α increases from 0 to α'.

We shall not discuss further the case $\alpha = 0$, when the q-plane Π_q is contained in at least one $(n-1)$-plane which contains also the axes $Ox_{q+1},\dots,Ox_n$, nor consider the case $A_{1\nu_2\cdots\nu_q} = 0$ for all $(1,\nu_2,\dots,\nu_q) \in C(n,q)$, when Π_q lies in the $(n-1)$-plane $x_1 = 0$ and no upper bound for p_1 results from (9), or the case $A_{\nu_1\cdots\nu_q} = 0$ for all $(\nu_1,\dots,\nu_q) \in C(n,q) \setminus \{(n-q+1,\dots,n)\}$ when $\Pi_q = \{x \in R^n : x_1 = \cdots = x_{n-q} = 0\}$ and no upper bound results for any p_j, but we concentrate in the following on the case $q = 1$.

2.2. In order to apply Proposition 1 for $q = 1$ we begin with some remarks. The direction Π_1 may be given by its unit vector $V_1 = OP$, where P has the coordinates $\xi_1,\dots,\xi_n$ with $\Sigma_{j=1}^n \xi_j^2 = 1$ and from symmetry reasons it is sufficient to suppose all $\xi_j \geqq 0$. Since $J_1^2 = \Sigma_{j=1}^n p_j^2\xi_j^2$ the inequality (22) will take the form

$$\left(\sum_{j=1}^n p_j^2\xi_j^2\right)^n \leqq Q^2(p_1\cdots p_{n-1})^2, \tag{22'}$$

and (17) $\alpha = \xi_1^2$ will be equal to the square of the distance d of P to the $(n-1)$-plane $x_1 = 0$.

The smaller α is, the greater becomes $v_1(\alpha)$, as well as $Q\alpha^{-n/2}$, the upper bound found for p_1. If $\alpha = 1$, i.e. Π_1 coincides with Ox_1 one has $p_1 \leqq Q$ (evidently this implies $Q \geqq 1$); if $\alpha = 0$, i.e. Π_1 lies in the $(n-1)$-plane $x_1 = 0$, one obtains no upper bound for p_1; if further Π_1 coincides with Ox_n no upper bound results for any p_j.

2.3. Let us now consider n linearly independent directions Π_1^k, $k = 1,\dots,n$, through the point $x_o = 0$ and denote by Π_{n-1}^k the $(n-1)$-plane orthogonal to Π_1^k, by OP^k the unit vector of Π_1^k, $P^k = (\xi_1^k,\dots,\xi_n^k)$, and by $\alpha_k = (\xi_1^k)^2 = d_k^2$, where $d_k = \xi_1^k$ is the distance of P^k to the $(n-1)$-plane $x_1 = 0$. (Again from symmetry reasons we may suppose $\xi_j^k \geq 0$ for $j,k = 1,\dots,n$.)

Preserve the hypothesis that the homeomorphism f has at 0 an A-point and its affine associated transformation at 0 has the form (11).

<u>Consequence 1.</u> If f verifies Grötzsch's inequality (9) for the

$(n-1)$-planes Π^k_{n-1}, $k = 1,\dots,n$, hence if (10) holds for the n directions Π^k_1, then (at least one of these directions being not contained in the $(n-1)$-plane $x_1 = 0$) p_1 is bounded from above:

1) Case $Q \geqq 1$.

$$p_1 \leqq v_1(\max_k \alpha_k)^{n/2} \leqq Q(\max_k \alpha_k)^{-n/2}. \tag{24'}$$

If $Q = 1$ and $\max_k \alpha_k \geqq (n-1)/n$, then $p_1 = 1$.

2) Case $Q < 1$.

The problem is possible only if

$$\max_k \alpha_k \leqq \alpha' \leqq (n-1)n^{-1}Q^{2/n} \tag{28}$$

and one has again (24').

2.4. However in the applications one does not know the position of the directions Π^k_1 with respect to the axes of the characteristic ellipsoid of $T = A(f,x_0)$ which in our case have been chosen as the coordinate axes. One knows, for instance, the angles (Π^k_1,Π^j_1), $k,j = 1,\dots,n$, between these directions, and one may suppose that $0 < (\Pi^k_1,\Pi^j_1) \leqq \pi/2$.

<u>Consequence 2.</u> Let f be a homeomorphism, x_0 an A-point for f, Π^k_1 n linearly independent directions through x_0 which form in x_0 the angles (Π^k_1,Π^j_1). If f verifies (10) for these directions (or (9) for the corresponding orthogonal $(n-1)$-planes Π^k_{n-1}) with a constant $Q \geqq 1$, denoting by

$$\alpha_o = \min(\max_k \alpha_k), \tag{29}$$

where min is taken over all the systems of n directions Π^k_1 which form the given angles (Π^k_1,Π^j_1) it follows that

$$p_1 \leqq v_1(\alpha_o)^{n/2} \leqq Q\,\alpha_o^{-n/2}. \tag{24"}$$

For $Q = 1$, if $\alpha_o \geqq (n-1)/n$, then again $p_1 = 1$.

In the case $Q < 1$ the homeomorphism f has the property (10) only for the systems Π^k_1, $k = 1,\dots,n$, which verify (28). Taking this time the minimum in (29) only for these systems we obtain again (24").

2.5. The problem to determine α_o can be set also in an equivalent form:

We consider $T = A(f,0)$, $T(0) = 0$, as unknown and fix the position of the system Π^k_1, $k = 1,\dots,n$, with respect to the coordinate axes, which are now independent of T. We look for the $(n-1)$-plane $\tilde{\Pi}_{n-1}$ through 0, for which the maximum of the distances from the points P^k, $k = 1,\dots,n$, to $\tilde{\Pi}_{n-1}$ becomes minimum. This minimum denoted by d_o will give α_o i: (29) by the relation $\alpha_o = d_o^2$.

As an example we solve this problem for the case when the system Π_1^k, $k = 1,\ldots,n$, is orthogonal.

In this respect we choose the coordinate axes such that $P^k = (\delta_1^k,\ldots,\delta$ and determine an $(n-1)$-plane in order to have all the distances from th points P^k to this plane equal. We find for such an $(n-1)$-plane the equation $\Sigma_{k-1}^n \varepsilon_k x_k = 0$ with $\varepsilon_k = \pm 1$ and for the corresponding distance $1/\sqrt{n}$. Further we verify that for every other $(n-1)$-plane the maximum of the distances from P^k to it, taken over all k, is $\geq 1/\sqrt{n}$, so that

$$p_1 \leq v_1^{n/2}(n^{-1}) \leq Qn^{n/2}. \tag{30}$$

Evidently for $Q < 1$ the problem is possible only if $\alpha_o = 1/n$ fulfils the necessary condition (28), for instance if $F[(n-1)g] \leq 0$ or equivalently $Q \geq (n-1)^{-(n-2)/2}$. It is clear that n must be ≥ 3.

In order to solve the problem for an arbitrary system Π_1^k, $k = 1,\ldots,n$ one transforms it in an orthogonal system by an affine mapping and then apply the result from above.

In the case $n = 3$ it is simple to make the direct calculation. For instance, it the directions Π_1^k, $k = 1,2,3$, determine on the unit sphere a spherical triangle having the sides $a \leq b \leq c \leq \pi/2$ and the angles $0 < A,B,C \leq \pi/2$, one finds

$$\alpha_o = d_o^2 = \left(1 + \frac{\cos^2(a/2)}{\cos^2(c/2)\sin^2(b/2)\sin^2 A}\right)^{-1} \tag{31}$$

and

$$p_1 \leq v_1^{3/2}(\alpha_o) \leq Q\alpha_o^{-3/2}. \tag{32}$$

(Details are given in [6].)

In the previous assertions we supposed the Grötzsch inequality (9) or the inequality (10) fulfilled at a single A-point of the homeomorphism f. Obviously assuming that f is a quasiconformal homeomorphism and tha the corresponding inequality is verified n-a.e. in G, one obtains by standard devices an upper bound for the quasiconformality coefficient of f.

§ 3. A quasiconformality criterium.

The results in §2 may be combined with other classical facts in order to obtain a simplified geometric criterium for the quasiconformality. We deal again with the case $q = 1$.

3.1. Definition 5. We say that the homeomorphism $f: G \to G^*$ verifies Grötzsch's inequality (9) $M \leqq QM^*$ for an (n-1)-plane Π_{n-1}, if (9) holds for every rectangular n-parallelotope $\Omega_n \subset G$ with the base Ω_{n-1} situated in an (n-1)-plane parallel to Π_{n-1}, where $M = M(\Omega_n, \Omega_{n-1})$ and $M^* = M(\Omega_n^*, \Omega_{n-1}^*)$, $\Omega_k^* = f(\Omega_k)$, $k = n-1$ and n.

Our first purpose is to deduce from the inequality (9) analytic properties for f.

Theorem 2. Let f be a homeomorphism which verifies Grötzsch's inequality (9) for an (n-1)-plane Π_{n-1}, and Π_1 the orthogonal direction to Π_{n-1}.

1°. f is AC on the direction Π_1, i.e. if Ω_n is an arbitrary rectangular parallelotope as in Definition 5, the mapping f is AC on (n-1)-a.e. segment in $\sum(\Omega_n, \Omega_{n-1})$, where (n-1)-a.e. refers to the Lebesgue measure in Ω_{n-1}.

2°. The derivative of f on the direction Π_1, which exists n-a.e. in Ω_n and is measurable, is L^n-integrable on Ω_n for every Ω_n as before. Its norm will be equal to J_1, where J_1 is the Jacobian of $f|\Pi_1$.

3°. At each differentiability point x_o of f it holds

$$J_1^n \leqq QJ_n \tag{33}$$

with J_1 as at nr. 2° and J_n the Jacobian of f, both taken at the point x_o.

The assertions 1° and 2° result by the method of Pfluger [12] and Väisälä (Lemma 2 in [13]), and by the method of Agard (Lemma 4.3 in [1]) respectively. The assertion 3° is contained in Theorem 1. (For 1° and 2°, see [7] and [14], 34.8.6.)

3.2. From Theorem 2 and §2, nr. 2.5 we deduce the following geometric criterium for the quasiconformality:

Theorem 3. A homeomorphism $f: G \to G^*$ which verifies Grötzsch's inequality (9) $M \leqq QM^*$ for all the coordinate (n-1)-planes with a constant $Q \geqq (n-1)^{-(n-2)/2}$ is $Q^{n-1} n^{n(n-1)/2}$-quasiconformal.

Indeed Theorem 2 shows that f is ACL_n in G. According to Väisälä's Lemma 3 in [13] f is n-a.e. differentiable in G.

If x_o is a differentiability point of f, Theorem 2 implies (33), i.e.

$$\left|\frac{\partial f}{\partial x_k}(x_o)\right|^n \leqq QJ_n$$

for each $k = 1,\dots,n$. Therefore it holds

$$|f'(x)|^n \leqq Qn^{n/2}J_n. \tag{34}$$

From these properties it follows that f is a $Q^{n-1}n^{n(n-1)/2}$-quasiconformal mapping in the sense of Väisälä's analytic definition ([14] 32.3, 34.3, 34.5, 13.1 and 34.6).

Obviously the same result may be obtained from (30) too, since at every A-point x_0 of f the linear dilatation $H(f'(x_0)) = p_1$ and one applies 13.1, 14.3 and 34.4 in [14].

By means of an affine mapping which transforms an arbitrary system of n linearly independent directions in the system of the orthogonal coordinate axes, Theorem 3 takes the following form:

Theorem 3'. If the homeomorphism $f : G \to G^*$ verifies Grötzsch's inequality (9) for n linearly independent $(n-1)$-planes and $Q \geqq 1$, then f is a quasiconformal mapping. The coefficient of quasiconformality may be also majorized, using the affine mapping mentioned above or even directly, for instance in the case $n = 3$ by (31) - (32).

Theorem 3' remains also true for a constant $Q < 1$ if the angles between the directions which are othogonal to the $(n-1)$-planes are convenient.

3.3. Suppose that at each point x_0 of the domain G there are given n linearly independent directions $\Pi_1^k(x_0)$, $k = 1,\dots,n$, such that the angles $(\Pi_1^k(x_0),\Pi_1^j(x_0))$, $k,j = 1,\dots,n$, are minorized independent of x_0 by the positive constants C_{kj} respectively. (Such a situation could be obtained starting from the directions of the coordinate axes by a quasiconformal mapping.) As usually we designate by $\Pi_{n-1}^k(x_0)$ the $(n-1)$-plane through x_0 which is orthogonal to $\Pi_1^k(x_0)$.

An ACL_n (or an ACL and n-a.e. differentiable) homeomorphism f which verifies (9) or (10) at x_0 with respect to every $(n-1)$-plane $\Pi_{n-1}^k(x_0)$, $k = 1,\dots,n$, for n-a.e. point $x_0 \in G$ is quasiconformal, the quasiconformality coefficient being bounded from above in terms of the constants C_{kj}. Here again $Q \geqq 1$, or even $Q < 1$ but in this latter case Q must satisfy a certain condition depending again on C_{kj}.

As a conclusion we remark that the criteria in this §3 contains Väisälä's criterium on the right cylinders ([14], 34.8.7).

References

[1] Agard, S.: Angles and quasiconformal mappings in space. J. Analyse math. 22 (1969), 177 - 200.

[2] Agard, S.: Quasiconformal mappings and the moduli of p-dimensional surface families, in "Proceedings of the Romanian-Finnish Seminar on Teichmüller spaces and quasiconformal mappings, Braşov, Romania 1969". Publishing House of the Academy of RSR, Bucharest (1971), 9 - 48.

[3] Andreian Cazacu, C.: Sur les inégalités de Rengel et la définition géométrique des représentations quasi-conformes. Revue Roumaine Math. pur. appl. 9 (1964), 141 - 155.

[4] Andreian Cazacu, C.: Some formulae on the extremal length in n-dimensional case, in "Proceedings of the Romanian-Finnish Seminar on Teichmüller spaces and quasiconformal mappings, Braşov, Romania 1969". Publishing House of the Academy of RSR, Bucharest (1971), 87 - 102.

[5] Andreian Cazacu, C.: Some problems in quasiconformality, in "Proceedings of the III Romanian-Finnish Seminar on Complex Analysis 1976". In print.

[6] Andreian Cazacu, C.: Affine properties of the quasiconformal mappings. Lucrările Simpozionului National Gh. Ţiţeica, 1978. In print.

[7] Andreian Cazacu, C.: On the geometric definition of the quasiconformality. To appear.

[8] Caraman, P.: About the characterization of the quasiconformality (QCf) by means of the moduli of q-dimensional surface families. Revue Roumaine Math. pur. appl. 16 (1971), 1329 - 1348.

[9] Caraman, P.: n-dimensional quasiconformal mappings. Editura Academiei Republicii Socialiste România, Bucureşti and Abacus Press, Tunbridge Wells, Kent (1974).

[10] Gehring, F. W., Väisälä, J.: On the geometric definition for quasiconformal mappings. Commentarii math. Helvet. 36 (1961), 19 - 32.

[11] Nevanlinna, R.: A remark on differentiable mappings. Michigan math. J. 3 (1955), 53 - 57.

[12] Pfluger, A.: Über die Äquivalenz der geometrischen und der analytischen Definition quasikonformer Abbildungen. Commentarii math. Helvet. 33 (1959), 23 - 33.

[13] Väisälä, J.: Two new characterizations for quasiconformality. Ann. Acad. Sci. Fenn., Ser. A I 362 (1965).

[14] Väisälä, J.: Lectures on n-dimensional quasiconformal mappings. Lecture Notes in Mathematics 229, Springer-Verlag, Berlin - Heidelberg - New York (1971).

Facultatea de Matematică
Universitatea din Bucureşti
Str. Academiei 14
70109 Bucureşti
Romania

ON INTERTWINING DILATIONS. VII

Gr. Arsene, Zoia Ceauşescu, C. Foiaş

Introduction. Contractive intertwining dilations constitute an interesting object in operator theory occurring implicitly or explicitly in several branches of the analysis. Thus, several classical extrapolation problems (e.g. the Carathéodory-Fejér [7], [8], Nevanlinna-Pick [13], [14] and Nehari [12] ones) are particular cases of a representation theorem for the commutant of certain contractions [15], which in its turn is a particular case of the theorem of existence and the indexing of contractive intertwining dilations. The existence of a Schur type description, extending the Adamjan-Arov-Kreĭn one ([1], [2], [3]) to the case of an arbitrary contraction, was established in [10], Proposition 4.1. The classical cases of extrapolation problems suggest the existence of explicit formulas for this description (see, for example [11]).

In this note we will study again some facts from [10], in order to describe the one-to-one correspondence between the contractive intertwining dilations and choice sequences (see [10], or Definition 1.1. below). All this stuff will be used in [6] for giving the explicit Schur-type labelling of all contractive intertwining dilations and for giving an algorithm for computing the choice sequence of any contractive intertwining dilation.

1. We start by recalling notations and simple facts concerning contractive intertwining dilations.

Let H and H' be some (complex) Hilbert spaces and let $L(H,H')$ be the algebra of all (linear, bounded) operators from H into H'. The space $L(H,H)$ will be denoted simply by $L(H)$. In the sequel T (resp. T') will be a contraction on H (resp. H') and we fix $U \in L(K)$ (resp. $U' \in L(K')$) to be its minimal isometric dilation. We will use freely the results from [16] concerning minimal isometric (and unitary) dilation of a given contraction.

Let $P = P_o = P_H^{K}$(*), $P' = P'_o = P_{H'}^{K'}$ and for $n \geq 1$, $P_n = P_{H_n}^{K}$, and $P'_n = P_{H'_n}^{K'}$, for $H_n = H + L + UL + \cdots + U^{n-1}L$ and $H'_n = H' + L' + U'L' + \cdots + U'^{n-1}L$ where $L = (U - T)(H)^-$ and $L' = (U' - T')(H')^-$. We de-

---

* For G a subspace (linear, closed) of H, the notation $P = P_G^H$ means the orthogonal projection of H on G; in this case $1 - P$ will be the orthogonal projection of H on $H \ominus G$.

note by T_n the operator $P_nU \mid \mathcal{H}_n$ on $\mathcal{H}_n$ (resp. $T'_n = P'_nU' \mid \mathcal{H}'_n$) for every $n \geq 0$; from the properties of isometric dilation we have that $T_o = T$ and $T'_o = T'_o$ (of course, $\mathcal{H}_o = \mathcal{H}$ and $\mathcal{H}'_o = \mathcal{H}'$). Denote by $\hat{U} \in L(\hat{\mathcal{K}})$ the minimal unitary dilation of T containing U; we have $\hat{\mathcal{K}} = \cdots + \hat{U}^{*2}\mathcal{L}^* + \hat{U}^*\mathcal{L}^* + \mathcal{L}^* + \mathcal{K}$, where $\mathcal{L}^* = (\hat{U}^* - T^*)(\mathcal{H})^-$. We define $\mathcal{K}_* = \cdots + \hat{U}^*\mathcal{L}^* + \mathcal{L}^* + \mathcal{H}$ and $U_* = \hat{U}^* \mid \mathcal{K}_*$; then $U_* \in L(\mathcal{K}_*)$ is a minimal isometric dilation of T^*. It is clear that $\hat{U}$ contains a minimal isometric dilation of T^*_n, denoted in the sequel by $U_{*n} \in L(\mathcal{K}_{*n})$, $(n \geq 1)$, where $\mathcal{K}_{*n} = \mathcal{K}_* + \mathcal{L} + \cdots + U^{n-1}\mathcal{L}$, and $U_{*n} = \hat{U}^* \mid \mathcal{K}_{*n}$. The notations $\hat{U}'$, $\hat{\mathcal{K}}'$, $\mathcal{K}'_*$, U'_*, $\mathcal{K}'_{*n}$, U'_{*n}, $(n \geq 1)$ are now clear.

By $\mathcal{I}(T',T)$ we denote the set of all operators A in $L(\mathcal{H},\mathcal{H}')$ intertwining T' and T (i.e. $T'A = AT$). In the sequel A will be a fixed <u>contraction</u> in $\mathcal{I}(T',T)$. A <u>contractive intertwining dilation</u> (CID), respectively a <u>n-partial contractive intertwining dilation</u> (n-PCID) of A is a contraction $\hat{A} \in \mathcal{I}(U',U)$, respectively $A_n \in \mathcal{I}(T'_n,T_n)$, such that $P'\hat{A} = AP$, respectively $P'A_n = AP \mid \mathcal{H}_n$, $(n \geq 0)$. It is clear that $A_o = A$. A <u>chain of PCID</u> of A is a sequence $\{A_n\}_{n=0}^{\infty}$, such that for every $n \geq 0$, A_n is a n-PCID of A and $P'_nA_{n+1} = A_nP_n \mid \mathcal{H}_{n+1}$. The applications

$$\hat{A} \to \{P'_n\hat{A} \mid \mathcal{H}_n\}_{n=0}^{\infty}, \quad \{A_n\}_{n=0}^{\infty} \to (s)\text{-}\lim_{n\to\infty} A_nP_n$$

establish a one-to-one correspondence between all CID's of A and all chains of PCID of A.

As it was pointed out in [4], [5] and [10], the following spaces

$$\begin{cases} \mathcal{F}_A(T) = \mathcal{F}_A = \{D_ATh + (U - T)h : h \in \mathcal{H}\}^- \\ \mathcal{R}_A(T) = \mathcal{R}_A = (\mathcal{D}_A + \mathcal{L}) \ominus \mathcal{F}_A \end{cases} \quad (*) \qquad (1.1)$$

$$\begin{cases} \mathcal{F}^A(T') = \mathcal{F}^A = \{D_Ah \oplus (U' - T')Ah : h \in \mathcal{H}\}^- \\ \mathcal{R}^A(T') = \mathcal{R}^A = (\mathcal{D}_A \oplus \mathcal{L}') \ominus \mathcal{F}^A \end{cases} \qquad (1.1)'$$

are very important for the structure of all CID's of A. We use also the following notations:

$$\begin{cases} p_A(T) = p_A = P^{\mathcal{D}_A+\mathcal{L}}_{\mathcal{F}_A} & p^A(T') = p^A = P^{\mathcal{D}_A\oplus\mathcal{L}'}_{\mathcal{F}^A} \\ q_A(T) = q_A = P^{\mathcal{D}_A+\mathcal{L}}_{\mathcal{L}} & q^A(T') = q^A = P^{\mathcal{D}_A\oplus\mathcal{L}'}_{\{0\}\oplus\mathcal{L}'} \end{cases} \qquad (1.2)$$

---

* For a contraction A in $L(\mathcal{H},\mathcal{H}')$ we put $D_A = (I - A^*A)^{1/2}$ and $\mathcal{D}_A = (D_A(\mathcal{H}))^-$, I being the identity operator (on any Hilbert space).

$$\begin{cases} \sigma(A;T',T) = \sigma_A : F_A \to F^A \\ \sigma_A(D_A Th + (U - T)h) = D_A h \oplus (U' - T')Ah, \qquad h \in H. \end{cases} \tag{1.3}$$

All the definitions from (1.1) to (1.3) make sense for $(A_n;T_n',T_n)$ instead of $(A;T',T)$, so the notations F_{A_n}, R_{A_n}, F^{A_n}, R^{A_n}, p_{A_n}, q_{A_n}, p^{A_n}, q^{A_n}, σ_{A_n} are clear for all $n \geq 0$.

For the sake of completeness we prove here the following facts, which result from [16], Sec. II.1 and [10], Lemma 4.4.

Lemma 1.1. (a) σ_A is unitary.

(b) $((1 - p_A)D_A)^- = R_A$.

(c) $((1 - p^A)(\{0\} \oplus L'))^- = R^A$.

Proof. (a) is obvious from the fact that:

$$\|D_A Th + (U - T)h\|^2 = \|Th\|^2 - \|ATh\|^2 + \|D_T h\|^2 = \|h\|^2 - \|T'Ah\|^2$$

$$= \|D_A h\|^2 + \|(U' - T')Ah\|^2 = \|D_A h \oplus (U' - T')Ah\|^2, \qquad (h \in H).$$

(b) Let $r = d + l$, where $r \in R_A$, $d \in D_A$ and $l \in L$; suppose that $\langle r,(1 - p_A)\tilde{d}\rangle = 0$, for every $\tilde{d} \in D_A$. This implies that $\langle r,\tilde{d}\rangle = 0$, so $\langle d,\tilde{d}\rangle = 0$ for every $\tilde{d} \in D_A$, which means that $d = 0$. Because r is in R_A, we have that $\langle r, D_A Th + (U - T)h\rangle = 0$, for every $h \in H$. But $d = 0$, so $\langle l, (U - T)h\rangle = 0$, for every $h \in H$, which implies that $l = 0$, so $r = 0$.

(c) The proof is analogous to (b).

Let us recall the following definition from [10], Definition 3.1.

Definition 1.1. A sequence $\{\Gamma_k\}_{k=1}^{\infty}$ (resp. a string $\{\Gamma_k\}_{k=1}^{n}$, $n \geq 1$) of contractions will be called an A-choice sequence (resp. an A-choice string) if $\Gamma_1 \in L(R_A, R^A)$ and $\Gamma_k \in L(D_{\Gamma_{k-1}}, D_{\Gamma_{k-1}^*})$ for every $k \geq 2$ (resp. for every $2 \leq k \leq n$).

One of the main result of [10] (see Propositions 2.1, 2.2 and 3.1) is the following:

Theorem 1.1. There exists a one-to-one correspondence between all CID's (resp., n-PCID's, $n \geq 1$) of A and all A-choice sequences (resp., A-choice strings of length n).

The aim of this paper is to give an alternate proof to this theorem, and to study the objects involved in the one-to-one correspondence claimed by the theorem. The connections with the adjoint operation is also presented.

2. In this section we will study only the so-called "first step" which means the structure of all 1-PCID's of A.

Let Γ_1 be an arbitrary contraction in $L(\mathcal{R}_A, \mathcal{R}^A)$ and define

$$A_1 = AP \mid \mathcal{H}_1 + q^A(\sigma_A p_A + \Gamma_1(1 - p_A))(D_A P + I - P) \mid \mathcal{H}_1. \tag{2.1}$$

We will prove that $A_1 \in L(\mathcal{H}_1, \mathcal{H}_1')$ is an 1-PCID of A. Indeed, from (2.1) it is clear that $P'A_1 = AP \mid \mathcal{H}_1$. Moreover

$$A_1T_1 = ATP \mid \mathcal{H}_1 + q^A(\sigma_A p_A + \Gamma_1(1 - p_A)) \cdot (D_A TP + (U - T)P) \mid \mathcal{H}_1$$

$$= T'AP \mid \mathcal{H}_1 + (U' - T')AP \mid \mathcal{H}_1 = T_1'A_1$$

so A_1 is in $\mathcal{I}(T_1', T_1)$. Finally, A_1 is a contraction because

$$\|A_1(h + l)\|^2 \leqq \|Ah\|^2 + \|\sigma_A p_A(D_A h + l)\|^2 + \|\Gamma_1(1 - p_A)(D_A h + l)\|^2$$

$$\leqq \|Ah\|^2 + \|D_A h + l\|^2 = \|h + l\|^2, \quad h \in \mathcal{H}, \quad l \in L,$$

so A_1 is an 1-PCID of A.

In particular we shall denote in the sequel by A_1^o the 1-PCID of A, associated by (2.1) to $\Gamma_1 = 0$; in other words

$$A_1^o = AP \mid \mathcal{H}_1 + q^A \sigma_A p_A(D_A P + I - P) \mid \mathcal{H}_1. \tag{2.2}$$

Conversely, let A_1 be an arbitrary 1-PCID of A. Because $P'A_1 = AP \mid \mathcal{H}_1$, we have

$$\|h\|^2 + \|l\|^2 = \|h + l\|^2 \geqq \|A_1(h+l)\|^2 = \|Ah\|^2 + \|(I - P')A_1(h+l)\|$$

which means that

$$\|(I - P')A_1(h + l)\| \leqq \|D_A h + l\| \quad h \in \mathcal{H}, \quad l \in L.$$

This relation implies the existence of a contraction $B : \mathcal{D}_A + L \to L'$, such that $(I - P')A_1 = B(D_A P + I - P) \mid \mathcal{H}_1$. This contraction verifies

$$Bp_A = q^A \sigma_A p_A. \tag{2.3}$$

Indeed

$$B(D_A Th + (U - T)h) = (I - P')A_1(Th + (U - T)h) = (1 - P')A_1T_1h$$

$$= (U' - T')Ah = q^A \sigma_A(D_A Th + (U - T)h),$$

for every $h \in \mathcal{H}$. From (2.3) we infer that $p_A B^* = \sigma_A^* p^A \mid (\{0\} \oplus L')$, which implies that

$$\|(1 - p_A)B^*1'\|^2 = \|B^*1'\|^2 - \|\sigma_A^* p^A(0 \oplus 1')\|^2$$

$$\leqq \|1'\|^2 - \|p^A(0 \oplus 1')\|^2 = \|(1 - p^A)(0 \oplus 1')\|^2, \qquad 1' \in L'.$$

This relation and Lemma 1.1 (c) imply that there exists a contraction Γ_1^* in $L(R^A, R_A)$, such that

$$\Gamma_1^*(1 - p^A)(0 \oplus 1') = (1 - p_A)B^*1', \qquad 1' \in L'. \tag{2.4}$$

We have

$$q^A \Gamma_1 (1 - p_A)(D_A P + I - P) \mid H_1 = A_1 - A_1^o; \tag{2.5}$$

indeed, because $(A_1 - A_1^o)(H') \subset L'$, ($A_1$ and A_1^o being both 1-PCID of A), we infer that

$$\langle q^A \Gamma_1 (1 - p_A)(D_A h + 1), 1' \rangle = \langle (1 - p_A)(D_A h + 1), \Gamma_1^*(1 - p^A)(0 \oplus 1') \rangle$$

$$= \langle B(1 - p_A)(D_A h + 1), 1' \rangle = \langle (I - P')(A_1 - A_1^o)(h + 1), 1' \rangle$$

$$= \langle (A_1 - A_1^o)(h + 1), 1' \rangle$$

for every $h \in H$, $1 \in L$, $1' \in L'$. We use here in order (2.4), the definition of B, (2.3) and (2.2). From (2.5) and (2.2), we have that A_1 and Γ_1 verify also (2.1). So we proved the following

Lemma 2.1. The formulas (2.1) and (2.5) establish a one-to-one correspondence between all A-choice strings of length one and all 1-PCID's of A.

In order to emphasize that Γ_1 corresponds to A_1 by Lemma 2.1, we will write that $\Gamma_1 = \Gamma(A, A_1)$. We will prove now other useful facts concerning this correspondence. Let A_1 be an 1-PCID of A and $\{\Gamma_1\}$ its A-choice string. Firstly we note that

$$R_{A_1} = D_{A_1} \ominus (D_{A_1} U(H))^-. \tag{2.6}$$

Indeed, from (1.1) it follows:

$$F_{A_1} = \{D_{A_1} T_1 (h + 1) + (U - T_1)(h + 1) : h \in H,\ 1 \in L\}^- = (D_{A_1} U(H))^- + UL,$$

so

$$R_{A_1} = (D_{A_1} + UL) \ominus F_{A_1} = D_{A_1} \ominus (D_{A_1} U(H))^-.$$

Consider the operator

$$\begin{cases} \tilde{\omega}(A_1;T_1',T_1) = \tilde{\omega}_{A_1} : \mathcal{D}_{A_1} \to \mathcal{D}_A \oplus \mathcal{D}_{\Gamma_1} \\ \tilde{\omega}_{A_1} D_{A_1} = [(1-q^A)(\sigma_A p_A + \Gamma_1(1-p_A)) \oplus D_{\Gamma_1}(1-p_A)] \cdot (D_A P + I - P) \mid \mathcal{H}_1 \end{cases} \tag{2.7}$$

<u>Lemma 2.2.</u> (a) $\tilde{\omega}_{A_1}$ is unitary.

(b) $\tilde{\omega}_{A_1} p_{A_1} \mid \mathcal{D}_{A_1} = P^{\mathcal{D}_A \oplus \mathcal{D}_{\Gamma_1}}_{\mathcal{D}_A \oplus \{0\}} \tilde{\omega}_{A_1}$.

(c) $\tilde{\omega}_{A_1} (1 - p_{A_1}) \mid \mathcal{D}_{A_1} = P^{\mathcal{D}_A \oplus \mathcal{D}_{\Gamma_1}}_{\{0\} \oplus \mathcal{D}_{\Gamma_1}} \tilde{\omega}_{A_1}$.

(d) $\tilde{\omega}_{A_1} (1 - p_{A_1}) D_{A_1} = 0 \oplus D_{\Gamma_1} (1 - p_A)(D_A P + I - P) \mid \mathcal{H}_1$.

<u>Proof.</u> Using (2.1) we have

$$||D_{A_1}(h+l)||^2 = ||h||^2 + ||l||^2 - ||Ah||^2 - ||q^A(\sigma_A p_A + \Gamma_1(1-p_A))(D_A h + l)||^2$$

$$= ||D_A h + l||^2 - ||(\sigma_A p_A + \Gamma_1(1-p_A))(D_A h + l)||^2$$

$$+ ||(1-q^A)(\sigma_A p_A + \Gamma_1(1-p_A))(D_A h + l)||^2 = ||D_{\Gamma_1}(1-p_A)(D_A h + l)||^2$$

$$+ ||(1-q^A)(\sigma_A p_A + \Gamma_1(1-p_A))(D_A h + l)||^2 = ||\tilde{\omega}_{A_1} D_{A_1}(h+l)||^2, \quad h \in \mathcal{H},\ l \in \mathcal{L}.$$

So $\tilde{\omega}_{A_1}$ is isometric. Using (2.7), (1.1) and (1.3), we have

$$\tilde{\omega}_{A_1}(D_{A_1} Uh) = \tilde{\omega}_{A_1} D_{A_1}(Th + (U-T)h) = (1-q^A)\sigma_A(D_A Th + (U-T)h) \oplus 0$$

$$= (1-q^A)(D_A h \oplus (U'-T')Ah) \oplus 0 = D_A h \oplus 0, \quad h \in \mathcal{H}. \tag{2.8}$$

From (2.8) we infer that

$$\tilde{\omega}_{A_1}(D_{A_1} U(\mathcal{H})^-) = \mathcal{D}_A \oplus \{0\}. \tag{2.9}$$

The relation (2.9) and the fact that $\tilde{\omega}_{A_1}$ is isometric imply (b) and (c). Now (d) results from (c) and (2.7). By (c), we have

$$\tilde{\omega}_{A_1}(\mathcal{R}_{A_1}) = \{0\} \oplus \mathcal{D}_{\Gamma_1}, \tag{2.9'}$$

so $\tilde{\omega}_{A_1}$ is unitary (see (2.9) and (2.9)').

Denote by $\omega(A_1;T_1',T_1) = \omega_{A_1}$ the <u>unitary operator</u> $\tilde{\omega}_{A_1} \mid \mathcal{R}_{A_1}$, considered as an operator from $\mathcal{R}_{A_1}$ onto $\mathcal{D}_{\Gamma_1}$. Explicitly:

$$\begin{cases} \omega_{A_1} : \mathcal{R}_{A_1} \to \mathcal{D}_{\Gamma_1} \\ \omega_{A_1}(1-p_{A_1})D_{A_1} = D_{\Gamma_1}(1-p_A)(D_A P + I - P) \mid \mathcal{H}_1. \end{cases} \tag{2.10}$$

3. This section and the next one are devoted to the connection between "the first step" of a contractive intertwining dilation and the adjoint operation. The general form of this connection will be given in Proposition 5.1.

For the beginning, let Z be a contraction in $L(H,H')$. Then

$$\begin{cases} H = \ker D_Z \oplus \mathcal{D}_Z , \\ H' = \ker D_{Z^*} \oplus \mathcal{D}_{Z^*} \end{cases} \tag{3.1}$$

and the matrix of Z with respect to these decompositions is

$$Z = \begin{pmatrix} Z_u & 0 \\ 0 & Z_c \end{pmatrix}. \tag{3.1)'$$

The operator $Z_u : \ker D_Z \to \ker D_{Z^*}$ is unitary and it will be called the <u>unitary core</u> of Z; the operator $Z_c : \mathcal{D}_Z \to \mathcal{D}_{Z^*}$ is a pure contraction (i.e., $h \in \mathcal{D}_Z$, $h \neq 0$ implies $\|Z_c h\| < \|h\|$) and it will be called the <u>pure contractive core</u> of Z.

We will have now a closer look to the proof of Proposition 3.2 (a), ch. VII, of [16], in order to connect the factorization of a contraction with the correspondent factorization of its adjoint. For this, let $H"$ be another Hilbert space, $B" = B' \cdot B$ a factorization of the contraction $B"$ in $L(H,H")$ by the contractions B in $L(H,H')$ and B' in $L(H',H")$. Define the spaces

$$\begin{cases} F(B' \cdot B) = \{D_B h \oplus D_{B'} Bh : h \in H\} \subset \mathcal{D}_B \oplus \mathcal{D}_{B'} \\ R(B' \cdot B) = \mathcal{D}_B \oplus \mathcal{D}_{B'} \ominus F(B' \cdot B). \end{cases} \tag{3.2}$$

Recall that by [16], Sec. VII. 3, the factorization $B' \cdot B$ is called <u>regular</u> if $R(B' \cdot B) = \{0\}$.

In order to connect the factorization $B" = B' \cdot B$ and the factorization $B"^* = B^* \cdot B'^*$, we define a contraction from $\mathcal{D}_B \oplus \mathcal{D}_{B'}$ into $\mathcal{D}_{B'^*} \oplus \mathcal{D}_{B^*}$. Because $B(\mathcal{D}_B) \subset \mathcal{D}_{B^*}$ and $B'(\mathcal{D}_{B'}) \subset \mathcal{D}_{B'^*}$, we choose this contraction to be of the form

$$J \circ \begin{pmatrix} B & Y \\ 0 & B' \end{pmatrix}, \tag{3.3}$$

where $Y : \mathcal{D}_{B'} \to \mathcal{D}_{B^*}$ and J is the operator which intertwines the terms in a direct sum. By [17], Théorème 1, the operator (3.3) is a contraction if and only if $Y = D_{B^*} X D_{B'}$, where $X : \mathcal{D}_{B'} \to \mathcal{D}_{B^*}$ is an arbitrary contraction. Define $Z(B' \cdot B) = Z$, the contraction obtained by (3.3) with $X = -P^{H'}_{\mathcal{D}_{B*}} | \mathcal{D}_{B'}$. Explicitly, we have (see (3.1) that

$$Z(B' \cdot B) = Z = J \circ \begin{pmatrix} B & -D_{B*}D_{B'} \\ 0 & B' \end{pmatrix} \tag{3.4}$$

or

$$\begin{cases} Z(B' \cdot B) = Z : \mathcal{D}_B \oplus \mathcal{D}_{B'} \to \mathcal{D}_{B'*} \oplus \mathcal{D}_{B*}\,, \\ Z(b \oplus b') = B'b' \oplus (Bd - D_{B*}D_{B'}b')\,, \quad b \in \mathcal{D}_{B'}, \quad b' \in \mathcal{D}_{B'}. \end{cases} \tag{3.4)'}$$

From (3.4) it is easy to infer that

$$\begin{cases} Z(B' \cdot B)^* = Z^* : \mathcal{D}_{B'*} \oplus \mathcal{D}_{B*} \to \mathcal{D}_B \oplus \mathcal{D}_{B'}, \\ Z^*(b'_* \oplus b_*) = B^*b_* \oplus (B'^*b'_* - D_{B'}D_{B*}b_*)\,, \quad b_* \in \mathcal{D}_{B*}, \quad b'_* \in \mathcal{D}_{B'*}, \end{cases} \tag{3.4)''}$$

which means that

$$Z(B' \cdot B)^* = Z(B^* \cdot B'^*). \tag{3.4)'''}$$

Lemma 3.1. The unitary core of Z acts between $R(B' \cdot B)$ and $R(B^* \cdot B'^*)$.

Proof. Let $b \in \mathcal{D}_B$ and $b' \in \mathcal{D}_{B'}$; then

$$\|Z(b \oplus b')\|^2 = \|B'b'\|^2 + \|Bb\|^2 + \|D_{B*}D_{B'}b'\|^2 - 2\mathrm{Re}\langle Bb, D_{B*}D_{B'}b'\rangle$$

$$= \|B'b'\|^2 + \|b\|^2 - \|D_Bb\|^2 + \|D_{B'}b'\|^2 - \|B^*D_{B'}b'\|^2$$

$$- 2\mathrm{Re}\langle D_{B*}Bb, D_{B'}b'\rangle = \|b \oplus b'\|^2 - \|D_Bb + B^*D_{B'}b'\|^2.$$

This implies that $\|D_Z(b \oplus b')\| = \|D_Bb + B^*D_{B'}b'\|$, which means that $b \oplus b' \in \ker D_Z$ if and only if $D_Bb + B^*D_{B'}b' = 0$, therefore if and only if $b \oplus b'$ is orthogonal on $F(B' \cdot B)$. As $\ker D_Z = R(B' \cdot B)$ and (see (3.4)''') $\ker D_{Z*} = R(B^* \cdot B'^*)$, the lemma follows from (3.1)'.

Corollary 3.2. (a) The factorization $B' \cdot B$ is regular if and only if $Z(B' \cdot B)$ is a pure contraction. (b) $Z(B' \cdot B) = 0$ if and only if B and B' are partial isometries such that the final space of B includes the orthogonal of the initial space of B'.

Proof. (a) is an easy consequence of Lemma 3.1. Suppose now that $Z(B' \cdot B) = 0$; using (3.4)' it follows that $B' | \mathcal{D}_{B'} = 0$ which means that B' is a partial isometry. Taking $b' = 0$ in (3.4)', one obtains that $B | \mathcal{D}_B = 0$, so B is also a partial isometry. The affirmation (b) results now from (3.4)', taking there $b = 0$.

We return to the situation considered in the first section, namely

$T \in L(H)$, $T' \in L(H')$, $A \in I(T',T)$ are contractions. We have that $A^* \in I(T^*,T'^*)$. Define the unitary operator

$$\begin{cases} \alpha_A(T) = \alpha_A : D_T \oplus D_A \to D_A + L \\ \alpha_A = (I \oplus \varphi) \circ J, \end{cases} \tag{3.5}$$

where $\varphi : D_T \to L$ is the unitary operator defined by $\varphi(D_T h) = (U - T)h$, $h \in H$ (see [16], Ch. II). Consider also the unitary operator

$$\begin{cases} \alpha^A(T') = \alpha^A : D_A \oplus D_{T'} \to D_A \oplus L' \\ \alpha^A = I \oplus \varphi', \end{cases} \tag{3.5'}$$

where φ' is analogous to φ.

We will need also the unitary operators $\alpha_{A^*} = \alpha_{A^*}(T'^*)$ and $\alpha^{A^*} = \alpha^{A^*}(T^*)$. Define the contractions:

$$\begin{cases} Z_A(T',T) = Z_A : D_A + L \to D_{A^*} \oplus L^* \\ Z_A = \alpha^{A^*} Z(A \cdot T)\alpha_A^* , \end{cases} \tag{3.6}$$

and

$$\begin{cases} Z^A(T',T) = Z^A : D_A \oplus L' \to D_{A^*} + L'^* \\ Z^A = \alpha_{A^*} Z(T' \cdot A)(\alpha^A)^* \end{cases} \tag{3.6'}$$

By virtue of (3.6), (3.5), (3.5)' and (3.4)', the explicit formulas for Z_A and $(Z_A)^*$ are

$$Z_A(a + (U - T)h) = Aa \oplus (U_* - T^*)(Th - D_A a), \quad a \in D_A, \quad h \in H, \tag{3.7}$$

$$(Z_A)^*(a^* \oplus (U_* - T^*)h) = (A^* a_* - D_A D_T^{2*} h) + (U - T)T^* h, \; a_* \in D_{A^*}, \; h \in H. \tag{3.7'}$$

The formulas for Z^A and $(Z^A)^*$ are similar, because (3.4)''' implies that

$$Z^A = (Z_{A^*})^* \tag{3.8}$$

Note also that $F_A = \alpha_A(F(A \cdot T))$, $R_A = \alpha_A(R(A \cdot T))$, $F^A = \alpha^A(F(T' \cdot A))$, $R^A = \alpha^A(R(T' \cdot A))$, so we infer from Lemma 3.1 that

<u>Corollary 3.3.</u> (a) The unitary core of Z_A acts between R_A and R^{A^*}.

(b) The unitary core of Z^A acts between R^A and R_{A^*}.

Lemma 3.2. The pure contractive cores of Z_A and Z^A verify the following relation

$$Z_A \mid F_A = \sigma_{A^*}(Z^A \mid F^A)\sigma_A. \tag{3.9}$$

Proof. The lemma follows from the equalities $(h \in H)$

$$\sigma_{A^*}Z^A\sigma_A(D_ATh + (U - T)h) = \sigma_{A^*}(AD_Ah - D_{A^*}D^2_{T'}Ah + (U'_* - T'^*)T'Ah)$$

$$= \sigma_{A^*}(D_{A^*}T'^*(T'Ah) + (U'_* - T'^*)T'Ah) = D_{A^*}(T'Ah) \oplus (U_* - T^*)A^*T'Ah)$$

$$= AD_ATh \oplus (U_* - T^*)(Th - D^2_ATh) = Z_A(D_ATh + (U - T)h).$$

Let now $\{\Gamma_n\}_{n=1}^{\infty}$ be a choice sequence for A and define the sequence $\{\Gamma_{*n}\}_{n=1}^{\infty}$ by

$$\Gamma_{*n}Z^A \mid D_{\Gamma_{n-1}} = Z_A\Gamma_n^*. \tag{3.10$_n$}$$

Proposition 3.1. The formulas $(3.10)_n$, $n \geq 1$ give an explicit one-to-one correspondence between the choice sequences of A and the choice sequences of A^*.

Proof. We have to verify that $\{\Gamma_{*n}\}_{n=1}^{\infty}$ defined by $(3.10)_n$, $n \geq 1$ is a choice sequence for A^*. Note that Γ_{*1} is a contraction from R_{A^*} into R^{A^*} (because $Z_A(R_A) = R^{A^*}$); moreover $D_{\Gamma_{*1}} = Z^A D_{\Gamma_1^*}$ and $D_{\Gamma_{*1}^*} = Z_A D_{\Gamma_1}$. The operator Γ_2^* is a contraction from $D_{\Gamma_1^*}$ into D_{Γ_1}, so Γ_{*2} is a contraction from $D_{\Gamma_{*1}}$ into $D_{\Gamma_{*1}^*}$. The proposition follows now easily by induction.

4. Proposition 3.1. and Lemma 2.1. rise the problem of finding the explicit bijection between 1-PCID's of A and 1-PCID's of A^* given by them. To this aim, define

$$(A^*)_1 = \hat{U}^*A_1^*\hat{U}'(H' + L'^*), \tag{4.1}$$

for A_1 an 1-PCID of A with $\Gamma(A,A_1) = \Gamma_1$.

Proposition 4.1. (a) $(A^*)_1$ is an 1-PCID of A^*.

(b) $\Gamma(A^*,(A^*)_1) = \Gamma_{*1}$.

Proof. Let $\tilde{B}$ the 1-PCID of A^* defined by (2.1) with $\Gamma(A^*,\tilde{B}) = \Gamma_{*1}$. We have to prove that $\tilde{B} = (A^*)_1$, or, if we take $B = \hat{U}\tilde{B}\hat{U}'^* \mid H_1$, that $B = A_1^*$.

For this, we use that $H_1 = L_* + U(H)$, where $L_* = \hat{U}L^* = (I - UT^*)(H)^-$. Therefore $B = A_1^*$ is equivalent to

$$P^{H_1}_{UH} B = P^{H_1}_{UH} A_1^* \tag{4.2}$$

and

$$P^{H_1}_{L_*} B = P^{H_1}_{L_*} A_1^*. \tag{4.3}$$

From (2.1) we have that

$$\widetilde{B}(h' + l'^*) = A^*h' + q^{A^*}(\sigma_{A^*}p_{A^*} + \Gamma_{*1}(1 - p_{A^*}))(D_{A^*}h' + l'^*), \tag{4.4}$$

for every $h' \in H'$, $l'^* \in L'^*$.
Recall that T_1 is a partial isometry from H onto UH, so $UT_1^* = T_1T_1^* = P^{H_1}_{UH}$; analogously $U'T_1'^* = T_1'T_1'^* = P^{H_1'}_{U'H'}$. Now

$$P^{H_1}_{UH} A_1^* P^{H_1'}_{U'H'} = T_1T_1^*A_1^*T_1'T_1'^* = T_1A_1^*T_1'^*T_1'T_1'^* = T_1A_1^*T_1'^* = T_1T_1^*A_1^* = P^{H_1}_{UH} A_1^*.$$

Using this, and the fact that $A_1^* \mid H = A^*$, we infer that

$$P^{H_1}_{UH} A_1 (l_*' + U'h') = UT_1^*A_1^*T_1'h' = UA_1^*h' = UA^*h', \tag{4.5}$$

for every $h' \in H'$, $l' \in L'$.
On the other hand, by (4.4) we deduce

$$P^{H_1}_{UH} B (l_*' + U'h') - P^{H_1}_{UH} \hat{U}\widetilde{B} (\hat{U}'^*l_*' + h') = UP^{\hat{K}}_{H}\widetilde{B}(h' + \hat{U}'^*l_*') = UA^*h', \tag{4.6}$$

$$h' \in H', \quad l_*' \in L_*'.$$

The relations (4.5) and (4.6) prove (4.2).

In order to obtain (4.3), we will prove

$$P^{H_1}_{L_*} Bh' = P^{H_1}_{L_*} A_1^*h' \qquad h' \in H' \tag{4.3'}$$

and

$$P^{H_1}_{L_*} B (U' - T')h' = P^{H_1}_{L_*} A_1^* (U' - T')h' \qquad h' \in H'. \tag{4.3''}$$

For (4.3)', we have:

$$P^{H_1}_{L_*}Bh' = P^{H_1}_{L_*}\hat{U}\widetilde{B}\hat{U}'^*h' = \hat{U}q^{A^*}\widetilde{B}(T'^*h' + (U_*' - T'^*)h')$$

$$= \hat{U}q^{A^*}(\sigma_{A^*}p_{A^*} + \Gamma_{*1}(1 - p_{A^*}))(D_{A^*}T'^*h' + (U_*' - T'^*)h')$$

$$= \hat{U}q^{A^*}\sigma_{A^*}(D_{A^*}T'^*h' + (U_*' - T'^*)h') = \hat{U}(U_* - T^*)A^*h'$$

$$= (I - UT^*)A^*h' \qquad h' \in H',$$

and

$$<P^{H_1}_{L_*} A_1^* h', (I - UT^*)h> = <h', A_1(D^2_{T^*}h - (U - T)T^*h>$$

$$= <h', AD^2_{T^*}h> = <(I - UT^*)A^*h', (I - UT^*)h>,$$

for every $h \in H$, $h' \in H'$. The last two relations prove (4.3)'.

The proof of (4.3)" will involve the whole construction preceding Proposition 3.1. First, we notice that

$$<P^{H_1}_{L^*} A_1^*(U' - T')h', (I - UT^*)h> = <(U' - T')h', A_1(D^2_{T^*}h - (U - T)T^*h>$$

$$= <0 \oplus (U' - T')h', (\sigma_A p_A + \Gamma_1(1 - p_A))(D_A h - (D_A TT^*h + (U - T)T^*h))>$$

$$= <0 \oplus (U' - T')h', \sigma_A p_A(D_A D^2_{T^*}h - (U - T)T^*h)>$$

$$+ <0 \oplus (U' - T')h', \Gamma_1(1 - p_A)D_A h> \qquad h \in H, \qquad h' \in H',$$

and

$$<P^{H_1}_{L_*} B(U' - T')h', (I - UT^*)h> = <\hat{U}\tilde{B}(I - U'_*T')h', (I - UT^*)h>$$

$$= <\tilde{B}(D^2_{T'}h' - (U'_* - T'^*)T'h'), (U_* - T^*)h>$$

$$= <(\sigma_{A^*}p_{A^*} + \Gamma_{*1}(1 - p_{A^*}))(D_{A^*}h' - (D_{A^*}T'^*T'h' + (U'_* - T'^*)T'h'), 0 \oplus (U_* - T^*)h>$$

$$= <\sigma_{A^*}p_{A^*}(D_{A^*}D^2_{T'}h' - (U'_* - T'^*)T'h'), 0 \oplus (U_* - T^*)h>$$

$$+ <\Gamma_{*1}(1 - p_{A^*})D_{A^*}h', 0 \oplus (U_* - T^*)h>, \qquad h \in H, \qquad h' \in H'.$$

These imply that (4.3)" is equivalent to

$$<0 \oplus (U' - T')h', \sigma_A p_A(D_A D^2_{T^*}h - (U - T)T^*h)>$$
$$= <\sigma_{A^*}p_{A^*}(D_{A^*}D^2_{T'}h' - (U'_* - T'^*)T'h'), 0 \oplus (U_* - T^*)h>, \tag{4.7}$$

and

$$<0 \oplus (U' - T')h', \Gamma_1(1 - p_A)D_A h> = <\Gamma_{*1}(1 - p_{A^*})D_{A^*}h', 0 \oplus (U_* - T^*)h>, \tag{4.7'}$$

for every $h \in H$, $h' \in H'$.

For (4.7), the point is that from (3.7)' we have that

$$D_A D^2_{T^*}h - (U - T)T^*h = -Z^*_A(0 \oplus (U_* - T^*)h), \qquad h \in H; \tag{4.8}$$

analogously

$$D_{A^*}D^2_{T'}h' - (U'_* - T'^*)T'h' = -Z^A(0 \oplus (U' - T')h'), \qquad h' \in H'. \tag{4.8'}$$

This means that (4.7) is equivalent to

$$\langle 0 \oplus (U' - T')h', \sigma_A p_A Z_A^*(0 \oplus (U_* - T^*)h)\rangle =$$

$$\langle \sigma_{A^*} p_{A^*} Z^A(0 \oplus (U' - T')h'), 0 \oplus (U_* - T^*)h\rangle, \quad h \in H,\ h' \in H'.$$

But this follows easily from (3.9), so (4.7) is proved.

For (4.7)', we have (using (3.7) and (3.7)'),

$$\langle \Gamma_{*1}(1 - p_{A^*})D_{A^*}h', 0 \oplus (U_* - T^*)h\rangle = \langle Z_A \Gamma_1^*(Z^A)^*(1 - p_{A^*})D_{A^*}h',$$

$$0 \oplus (U_* - T^*)h\rangle = \langle \Gamma_1^* Z_{A^*}(1 - p_{A^*})D_{A^*}h', (-D_A D_{T^*}^2 h + (U - T)T^*h\rangle =$$

$$\langle \Gamma_1^*(1 - p^A)Z_{A^*}D_{A^*}h', (1 - p_A)(-D_A h + D_A T(T^*h) + (U - T)T^*h)\rangle =$$

$$\langle (1 - p^A)(A^*D_{A^*}h' \oplus (U' - T')D_{A^*}^2 h'), \Gamma_1(1 - p_A)D_A h\rangle =$$

$$\langle 0 \oplus (U' - T')h', \Gamma_1(1 - p_A)D_A h\rangle, \quad h \in H,\ h' \in H'.$$

We have now (4.7) and (4.7)', so (4.3)", which completes the proof of the proposition.

Note that, in particular,

$$(A^*)_1^\circ = \hat{U}^*(A_1^\circ)^*(\hat{U}'|H' + L'^*).$$

We will define now the unitary operator between R^A and D_{Γ^*}, analogous to ω_{A_1}. First, note that from (4.1) it follows:

$$\hat{U}'(R_{(A^*)_1}) = R_{A_1^*}. \tag{4.9}$$

Consider now the operator $\omega_{(A^*)_1}$; explicitly (for $h' \in H'$, $1'^* \in L'^*$),

$$\begin{cases} \omega_{(A^*)_1} : R_{(A^*)_1} \to D_{\Gamma_{*1}} \\ \omega_{(A^*)_1}(1 - p_{(A^*)_1})D_{(A^*)_1}(h' + 1'^*) = D_{\Gamma_1^*}(1 - p_{A^*})(D_{A^*}h' + 1'^*). \end{cases} \tag{4.10}$$

Finally, we define the operator ω^{A_1} by

$$\begin{cases} \omega^{A_1}(T_1';T_1) = \omega^{A_1} : R^{A_1} \to D_{\Gamma_1^*} \\ \omega^{A_1} = (Z^A)^*\omega_{(A^*)_1}\hat{U}'Z^{A_1} \mid R^{A_1}. \end{cases} \tag{4.11}$$

<u>Lemma 4.1.</u> The operator ω^{A_1} defined by (4.11) is unitary from R^{A_1} onto $D_{\Gamma_1^*}$ and

$$\omega^{A_1}(1 - p^{A_1})(0 \oplus U'1') = D_{\Gamma_1^*}(1 - p^A)(0 \oplus 1'), \quad 1' \in L'. \tag{4.12}$$

<u>Proof.</u> From (4.11) and Corollary 3.3 it follows that ω^{A_1} is a unitary operator from R^{A_1} onto $D_{\Gamma_1^*}$. Now:

$$\omega^{A_1}(1 - p^{A_1})(0 \oplus U'l') = (Z^A)^* \omega_{(A^*)_1} \hat{U}'^* Z^{A_1}(1 - p^{A_1})(0 \oplus U'l') =$$
$$Z_{A^*}\omega_{(A^*)_1} \hat{U}'^*(1 - p_{A_1^*}) Z^{A_1}(0 \oplus (U' - T_1')l') =$$
$$Z_{A^*}\omega_{(A^*)_1}(1 - p_{(A^*)_1})\hat{U}'^*(-D_{A_1^*}D^2_{T_1'}l' + (U_{*1} - T_1')T_1'l') = \tag{4.13}$$
$$Z_{A^*}\omega_{(A^*)_1}(1 - p_{(A^*)_1})\hat{U}'^*(-D_{A_1^*}l') =$$
$$Z_{A^*}\omega_{(A^*)_1}(1 - p_{(A^*)_1})D_{(A^*)_1}(-\hat{U}'^*l'), \qquad l' \in \mathcal{L}',$$

where we used (4.11), Corollary 3.3 (for A_1 instead of A), (3.8) (for A_1), (3.7)' (for A_1^*), the facts that $D^2_{T_1'} = P^{\mathcal{H}_1'}_{\mathcal{L}'}$, that $T_1'|\mathcal{L}' = 0$ and (4.1). Suppose now that $l' = (U' - T')h'$, where $h' \in \mathcal{H}'$. From (4.13) it follows that

$$\omega^{A_1}(1 - p_{A_1})(0 \oplus U'(U' - T')h') = Z_{A^*}\omega_{(A^*)_1}(1 - p_{(A^*)_1})D_{(A^*)_1} \cdot$$
$$(-D^2_{T'}h' + (U'^* - T'^*)T'h') = Z_{A^*}D_{\Gamma_{*1}}(1 - p_{A^*})(-D_{A^*}D^2_{T'}h' + (U'^* - T'^*)T'h')$$
$$= D_{\Gamma_1^*}Z_{A^*}(1 - p_{A^*})(Z_{A^*})^*(0 \oplus (U' - T')h') =$$
$$= D_{\Gamma_1^*}(1 - p^A)(0 \oplus (U' - T')h'), \qquad h' \in \mathcal{H}',$$

(where we used (4.10), $(3.10)_1$, (3.7)' (for A^*) and Corollary 3.3 (for A^*)), which is exactly (4.12).

As an application of the construction involved in the last two sections we will give the following corollary, which clarifies the connections between the choice operator Γ_1 and the unitary operators ω_{A_1} and ω^{A_1}. As we know (see (2.6)), the space $\mathcal{R}_{A_1}$ is included in $\mathcal{D}_{A_1}$, so it is possible to consider the operator $(1 - p^{A_1})|\mathcal{R}_{A_1} : \mathcal{R}_{A_1} \to \mathcal{R}^{A_1}$.

Corollary 4.1. The diagram

$$\begin{array}{ccc}
\mathcal{R}_{A_1} & \xrightarrow{\;(1 - p^{A_1})|\mathcal{R}_{A_1}\;} & \mathcal{R}^{A_1} \\
\Big\downarrow \omega_{A_1} & & \Big\downarrow \omega^{A_1} \\
\Gamma_1 & \xrightarrow[\;-\Gamma_1\;]{} & \mathcal{D}_{\Gamma_1^*}
\end{array}$$

is commutative; that is

$$\Gamma_1\omega_{A_1} = -\omega^{A_1}(1 - p^{A_1}) \mid \mathcal{R}_{A_1}. \tag{4.15}$$

Proof. From lemma 1.1 (b) (for A_1 instead of A), we have that (4.15) is equivalent to

$$\Gamma_1\omega_{A_1}(1 - p_{A_1})D_{A_1}h_1 = -\omega^{A_1}(1 - p^{A_1})(1 - p_{A_1})D_{A_1}h_1, \tag{4.15)'}$$

for every $h_1 \in H_1$. Because $(1 - p_{A_1})D_{A_1}h_1$ is in $\mathcal{D}_{A_1}$, there exists a sequence $\{D_{A_1}h_1^n\}_{n=1}^{\infty}$, where h_1^n is in H_1 for every $n \geq 1$, such that

$$D_{A_1}h_1^n \to (1 - p_{A_1})D_{A_1}h_1. \tag{4.16}$$

From (4.16) we have that

$$\omega_{A_1}D_{A_1}h_1^n \to 0 \oplus \omega_{A_1}(1 - p_{A_1})D_{A_1}h_1,$$

which implies by (2.7) that

$$(1 - q^A)(\sigma_A p_A + \Gamma_1(1 - p_A))(D_A P + I - P)h_1^n \to 0. \tag{4.17}$$

Now, from (2.10) it follows

$$\Gamma_1\omega_{A_1}(1 - p_{A_1})D_{A_1}h_1^n = \Gamma_1 D_{\Gamma_1}(1 - p_A)(D_A P + I - P)h_1^n, \tag{4.18}$$

for every $n \geq 1$. On the other hand, the relation (1.1)' for A_1 gives that

$$F^{A_1} = \{D_{A_1}\tilde{h}_1 \oplus U'q^A A_1\tilde{h}_1 : \tilde{h}_1 \in H_1\}^-,$$

so

$$(1 - p^{A_1})D_{A_1}h_1^n = -(1 - p^{A_1})U'q^A A_1 h_1^n, \quad n \geq 1.$$

This implies by (4.12) that

$$\begin{aligned} \omega^{A_1}(1 - p^{A_1})D_{A_1}h_1^n &= -\omega^{A_1}(1 - p^{A_1})U'q^A A_1 h_1^n \\ &= -D_{\Gamma_1^*}(1 - p^A)q^A(\sigma_A p_A + \Gamma_1(1 - p_A))(D_A P + I - P)h_1^n, \quad n \geq 1. \end{aligned} \tag{4.19}$$

Because

$$\omega^{A_1}(1 - p^{A_1})D_{A_1}h_1^n \to \omega^{A_1}(1 - p^{A_1})(1 - p_{A_1})D_{A_1}h_1,$$

from (4.19), (4.17) and (4.18) it follows that

$$\begin{aligned} \omega^{A_1}(1 - p^{A_1})(1 - p_{A_1})D_{A_1}h_1 &= \lim_{n\to\infty} \omega^{A_1}(1 - p^{A_1})D_{A_1}h_1^n \\ &= -\lim_{n\to\infty} D_{\Gamma_1^*}(1 - p^A)q^A(\sigma_A p_A + \Gamma_1(1 - p_A))(D_A P + I - P)h_1^n \end{aligned}$$

$$= -\lim_{n\to\infty} D_{\Gamma_1^*}(1 - p^A)(\sigma_A p_A + \Gamma_1(1 - p_A))(D_A P + I - P)h_1^n$$

$$= -\lim_{n\to\infty} \Gamma_1\omega_{A_1}(1 - p_{A_1})D_{A_1}h_1^n = -\Gamma_1\omega_{A_1}(1 - p_{A_1})D_{A_1}h_1 .$$

5. In this section we shall prove Theorem 1.1. and the general form of Proposition 4.1. We will fix $A \in \mathcal{I}(T',T)$ a contraction, and $\hat{A}$ a CID of A; denote by $\{A_n\}_{n=0}^{\infty}$ the chain of PCID associated to $\hat{A}$. The basic way to use Sections 3 and 4 in this situation is that, for every $n \geq 1$, A_n is an 1-PCID of A_{n-1}. Therefore it is possible to apply Lemma 2.1. to the pair (A_{n-1},A_n), $(n \geq 1)$, in order to obtain an A_{n-1}-choice sequence of length one, namely a contraction $\Gamma(A_{n-1},A_n)$: $\mathcal{R}_{A_{n-1}} \to \mathcal{R}^{A_{n-1}}$, such that

$$A_n(h_{n-1} + l_{n-1})$$

$$= A_{n-1}h_{n-1}+q^{A_{n-1}}[\sigma_{A_{n-1}}p_{A_{n-1}}+\Gamma_1(A_{n-1},A_n)\cdot(1-p_{A_{n-1}})](D_{A_{n-1}}h_{n-1}+l_{n-1}),$$

$$h_{n-1} \in \mathcal{H}_{n-1}, \qquad l_{n-1} \in U^{n-1}\mathcal{L}. \qquad (5.1)_n$$

In the same way we obtain the unitary operators $\omega_{A_n} = \omega_{A_n}(T_n',T_n)$ and $\omega^{A_n} = \omega^{A_n}(T_n',T_n)$ such that:

$$\begin{cases} \omega_{A_n} : \mathcal{R}_{A_n} \to \mathcal{D}_{\Gamma_1(A_{n-1},A_n)} \\ \omega_{A_n}(1-p_{A_n})D_{A_n} = D_{\Gamma_1(A_{n-1},A_n)}(1-p_{A_{n-1}})(D_{A_{n-1}}P_{n-1}+I-P_{n-1})|\mathcal{H}_n \end{cases} \qquad (5.2)_n$$

$$\omega^{A_n} : \mathcal{R}^{A_n} \to \mathcal{D}_{\Gamma_1^*(A_{n-1},A_n)}$$

$$\omega^{A_n}(1-p^{A_n})(0 \oplus U'l_{n-1}') = D_{\Gamma_1^*(A_{n-1},A_n)}(1-p^{A_{n-1}})l_{n-1}', \; l_{n-1}' \in U'^{n-1}\mathcal{L}'. \qquad (5.2)_n'$$

We will define by induction a sequence of contractions $\{\Gamma_n\}_{n=1}^{\infty}$ and two sequences of unitary operators $\{\Omega_{A_n}\}_{n=1}^{\infty}$ and $\{\Omega^{A_n}\}_{n=1}^{\infty}$ as follows:

$$\Gamma_1 = \Gamma_1(A,A_1), \qquad \Omega_{A_1} = \omega_{A_1}, \qquad \Omega^{A_1} = \omega^{A_1} \qquad (5.3)_1$$

and for $n \geq 1$:

$$\Gamma_n = \Omega^{A_{n-1}}\Gamma_1(A_{n-1},A_n)\Omega_{A_{n-1}}, \quad \Omega^{A_n} = \Omega^{A_{n-1}} \circ \omega^{A_n}, \quad \Omega_{A_n} = \Omega_{A_{n-1}} \circ \omega_{A_n}. \qquad (5.3)_n$$

Lemma 5.1. The sequence $\{\Gamma_n\}_{n=1}^{\infty}$ is an A-choice sequence; for every $n \geq 1$ the operator Ω_{A_n} (resp. Ω^{A_n}) is a unitary from $\mathcal{R}_{A_n}$ (resp. $\mathcal{R}^{A_n}$) onto $\mathcal{D}_{\Gamma_n}$ (resp. $\mathcal{D}_{\Gamma_n^*}$).

Proof. For $n = 1$, the assertions of the lemma follow from the definitions (see $(5.3)_1$). Suppose now that $n > 1$ and that $\{\Gamma_k\}_{k=1}^{n-1}$ is an A-choice string of length $n - 1$, Ω_{A_k} (and Ω^{A_k}) are unitary operators $(1 \leq k \leq n-1)$ from R_{A_k} (resp. R^{A_k}) onto D_{Γ_k} (resp. $D_{\Gamma_k^*}$) such that

$$\Gamma_k = \Omega^{A_{k-1}}\Gamma_1(A_{k-1},A_k)\Omega^*_{A_{k-1}}, \quad \text{for every } 2 \leq k \leq n-1.$$

Define

$$\Gamma_n = \Omega^{A_{n-1}}\Gamma_1(A_{n-1},A_n)\Omega^*_{A_{n-1}};$$

it follows that Γ_n is a contraction from $D_{\Gamma_{n-1}}$ into $D_{\Gamma^*_{n-1}}$, which means that $\{\Gamma_k\}_{k=1}^n$ is an A-choice string of length n. Take now

$$\Omega_{A_n} = \Omega_{A_{n-1}} \circ \omega_{A_n} \quad \text{and} \quad \Omega^{A_n} = \Omega_{A_{n-1}} \circ \omega^{A_n}.$$

The previous definitions make sense because

$$\omega_{A_n}(R_{A_n}) = D_{\Gamma_1(A_{n-1},A_n)} \subset R_{A_{n-1}}$$

and

$$\omega^{A_n}(R^{A_n}) = D_{\Gamma_1^*(A_{n-1},A_n)} \subset R^{A_{n-1}};$$

moreover

$$\Omega_{A_n}(R_{A_n}) = \Omega_{A_{n-1}}(D_{\Gamma_1(A_{n-1},A_n)}) = D_{\Gamma_n}$$

and

$$\Omega^{A_n}(R^{A_n}) = \Omega^{A_{n-1}}(D_{\Gamma_1^*(A_{n-1},A_n)}) = D_{\Gamma_n^*},$$

just because $\Gamma_1(A_{n-1},A_n) = \Omega_{A_{n-1}}\Gamma_n(\Omega^{A_{n-1}})^*$. The lemma follows by induction.

<u>Definition 5.1.</u> The A-choice sequence $\{\Gamma_n\}_{n=1}^\infty$ and the sequences $\{\Omega_{A_n}\}_{n=1}^\infty$, $\{\Omega^{A_n}\}_{n=1}^\infty$ will be called <u>the A-choice sequences of</u> $\hat{A}$, resp. <u>the sequences of identificators</u> of $\hat{A}$. For $n \geq 1$, the A-choice string $\{\Gamma_k\}_{k=1}^n$ and the strings $\{\Omega_{A_k}\}_{k=1}^n$, $\{\Omega^{A_k}\}_{k=1}^n$ will be called <u>the A-choice string of</u> A_n, resp. <u>the string of identificators of</u> A_n.

The unitary operator ω_{A_1}, defined by (2.10), was obtained from a larger one, namely $\tilde{\omega}_{A_1}$, which maps D_{A_1} onto $D_A \oplus D_{\Gamma_1}$ (see 2.7).

It is clear that if we define recurrently

$$\tilde{\Omega}_{A_n} : \mathcal{D}_{A_n} \to \mathcal{D}_A \oplus \mathcal{D}_{\Gamma_1} \oplus \cdots \oplus \mathcal{D}_{\Gamma_n}, \quad \tilde{\Omega}_{A_n} = (\tilde{\Omega}_{A_{n-1}} \oplus \Omega_{A_{n-1}}) \circ \tilde{\omega}_{A_n} \quad n \geq 2 \qquad (5.4)_n$$

where $\tilde{\Omega}_{A_1} = \tilde{\omega}_{A_1}$, then the operators $\{\tilde{\Omega}_{A_n}\}_{n=1}^{\infty}$ are unitary; moreover we have, for every $n \geq 1$,

$$\tilde{\Omega}_{A_n}(\mathcal{R}_{A_n}) = \mathcal{D}_{\Gamma_n}, \quad P_{\{0\}\oplus\{0\}\oplus\cdots\oplus\mathcal{D}_{\Gamma_n}}^{\mathcal{D}_A\oplus\mathcal{D}_{\Gamma_1}\oplus\cdots\oplus\mathcal{D}_{\Gamma_n}} \tilde{\Omega}_{A_n} \mid \mathcal{R}_{A_n} = \Omega_{A_n}, \qquad (5.5)_n$$

$$\tilde{\Omega}^*_{A_n}(\mathcal{D}_A \oplus \mathcal{D}_{\Gamma_1} \oplus \cdots \oplus \mathcal{D}_{\Gamma_{n-1}} \oplus \{0\}) + U^n\mathcal{L} = \mathcal{F}_{A_n}. \qquad (5.6)_n$$

<u>Proof of Theorem 1.1.</u> Taking into account Lemma 5.1, we have to prove only that if $\{\Gamma_n\}_{n=1}^{\infty}$ is an A-choice sequence, then there exists a CID of A, $\hat{A}$, such that the A-choice sequence of $\hat{A}$ (see Definition 5.1) is exactly $\{\Gamma_n\}_{n=1}^{\infty}$. For this, we construct by induction a chain $\{A_n\}_{n=1}^{\infty}$ of PCID, of A, such that the A-choice string of A_n is $\{\Gamma_k\}_{k=1}^{n}$. If $n = 1$, starting with Γ_1 we define (by (2.1)) an 1-PCID, A_1, of A, such that $\Gamma_1(A,A_1) = \Gamma_1$. Suppose now that for $n \geq 1$ we have A_n, an n-PCID of A, such that the A-choice string of A_n is $\{\Gamma_k\}_{k=1}^{n}$; consider also the strings of identificators of A_n, namely $\{\Omega_{A_k}\}_{k=1}^{n}$ and $\{\Omega^{A_k}\}_{k=1}^{n}$. Define

$$\tilde{\Gamma}_n = (\Omega^{A_n})^*\Gamma_{n+1}\Omega_{A_n} \qquad (5.7)$$

It is clear that $\{\tilde{\Gamma}_n\}$ is an A_n-choice string of length one and so it defines (by (2.1)) an 1-PCID of A_n, A_{n+1}, such that $\tilde{\Gamma}_n = \Gamma_1(A_n,A_{n+1})$. Taking into account $(5.7)_n$ and $(5.3)_n$, it follows that the A-choice string of A_{n+1} is $\{\Gamma_k\}_{k=1}^{n+1}$. Since moreover it is now plain that this A_{n+1} is uniquely determined, the theorem is completely proved.

<u>Remark 5.1.</u> In [9] it is proved that the one-to-one correspondence described in Theorem 1.1 is exactly the same with that which results from Propositions 2.1, 2.2 and 3.1 of [10].

From Proposition 4.2 it follows immediately that

<u>Corollary 5.1.</u> For every $n \geq 1$,

$$\Gamma_1(A_{n-1},A_n)\omega_{A_n} = -\omega^{A_n}(1 - p^{A_n}) \mid \mathcal{R}_{A_n}. \qquad (5.8)_n$$

We will prove now the general form of Proposition 4.1. Let $\{\Gamma_n\}_{n=1}^{\infty}$ be an A-choice sequence and let $\{\Gamma_{*n}\}_{n=1}^{\infty}$ be the A*-choice sequence defined by $(3.10)_n$, $(n \geq 1)$. Consider the CID of A, $\hat{A}$, defined by $\{\Gamma_n\}_{n=1}^{\infty}$ (see Theorem 1.1) and $\widehat{A^*}$ the CID of A* defined by the chain of PCID of A^*, $\{(A^*)_n\}_{n=1}^{\infty}$, where

$$(A^*)_n = \hat{U}^{*n} A_n^* \hat{U}'^{n} \mid (\mathcal{H}' + \mathcal{L}'^* + \cdots + U_*'^{n-1}\mathcal{L}'^*), \qquad (5.9)_n$$

$(n \geq 1)$. What we have to prove is that the A^*-choice sequence of $\widehat{A^*}$ is $\{\Gamma_{*n}\}_{n=1}^{\infty}$. For this, we give firstly the following result

Lemma 5.2. The diagram

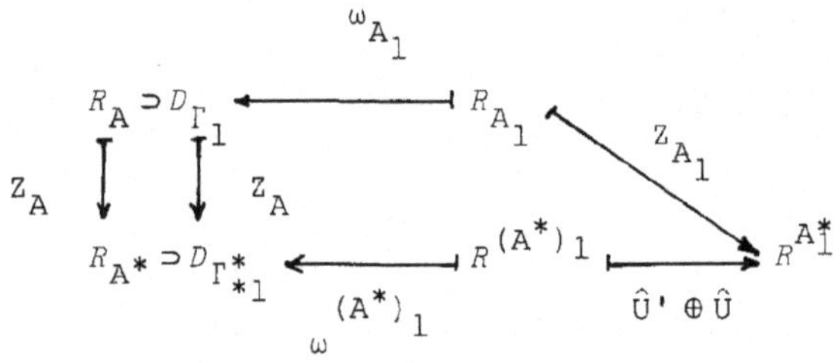

is commutative; that is

$$Z_A \omega_{A_1} = \omega^{(A^*)_1} (\hat{U}' \oplus \hat{U})^* Z_{A_1}. \qquad (5.10)$$

Proof. Because $\mathcal{R}_{A_1} \subset \mathcal{D}_{A_1}$, the relation (5.10) will be proved if

$$Z_A \omega_{A_1} (1-p_{A_1}) D_{A_1} (h+(U-T)\tilde{h}) = \omega^{(A^*)_1} (\hat{U}' \oplus \hat{U}) Z_{A_1} (1-p_{A_1}) D_{A_1} (h+(U-T)\tilde{h}), \qquad (5.10$$

for every $h, \tilde{h}$ in $\mathcal{H}$. First, we have

$$Z_A \omega_{A_1} (1-p_{A_1}) D_{A_1} (h+(U-T)\tilde{h}) = Z_A D_{\Gamma_1} (1-p_A)(D_A h + (U-T)\tilde{h})$$
$$= D_{\Gamma_{*1}^*} (1-p^{A^*}) Z_A (D_A h + (U-T)\tilde{h}) = D_{\Gamma_{*1}^*} (1-p^{A^*})(D_{A^*} A h \oplus (U_* - T^*)(T\tilde{h} - D_A^2 h)$$
$$= D_{\Gamma_{*1}^*} (1-p^{A^*})(0 \oplus (U_* - T^*)(T\tilde{h} - h)) \qquad (h, \tilde{h} \in \mathcal{H}),$$

where we use, in order, (2.10), $(3.10)_1$, Corollary 3.3, (3.7) and the structure of $\mathcal{F}^{A^*}$ (see (1.1)'). On the other hand, we have

$$\omega^{(A^*)_1} (\hat{U}' \oplus \hat{U})^* Z_{A_1} (1-p_{A_1}) D_{A_1} (h + (U-T)\tilde{h})$$
$$= \omega^{(A^*)_1} (\hat{U}' \oplus \hat{U})^* (1-p^{A_1^*})(D_{A_1^*} A_1 (h + (U-T)\tilde{h}) \oplus (U_{*1} - T_1^*)(-D_{A_1}^2 (h + (U-T)\tilde{h})$$
$$= -\omega^{(A^*)_1} (1-p^{(A^*)_1})(\hat{U}' \oplus \hat{U})^* (0 \oplus (U_{*1} - T_1^*)(h + (U-T)\tilde{h}))$$
$$= -D_{\Gamma_{*1}^*} (1-p^{A^*})(0 \oplus (U_{*1} - T_1^*)(h + (U-T)\tilde{h}))$$
$$= D_{\Gamma_{*1}^*} (1-p^{A})(0 \oplus (U_* - T^*)(T\tilde{h} - h)) \qquad (h, \tilde{h} \in \mathcal{H}),$$

where we used Corollary 3.3, (3.7), the structure of $\mathcal{F}^{A_1^*}$, $(3.10)_1$, (4.12) and the properties of isometric dilation.

Proposition 5.1. The choice sequence of $\widehat{A^*}$ is $\{\Gamma_{*n}\}_{n=1}^{\infty}$ and

$$Z_{A_n}|\mathcal{R}_{A_n} = (\hat{U}'\oplus\hat{U})^n(\Omega^{(A^*)_n})^* Z_A \Omega_{A_n}, Z^{A_n}|\mathcal{R}^{A_n} = \hat{U}'^n(\Omega_{(A^*)_n})^* Z^A \Omega^{A_n} \qquad (5.11)_n$$

for every $n \geqq 1$, where $\{\Omega_{(A^*)_n}\}_{n=1}^{\infty}$ and $\{\Omega^{(A^*)_n}\}_{n=1}^{\infty}$ are the sequences of identificators of $\widehat{A^*}$.

Proof. We proceed by induction for proving that the choice string of $(A^*)_n$ is $\{\Gamma_{*k}\}_{k=1}^{n}$ and that the formulas $(5.11)_n$ are true, for every $n \geqq 1$. For $n = 1$, the results follows from Proposition 4.1, Lemma 5.2 and (4.11). Suppose now that $n > 1$ and that the A^*-choice string of $(A^*)_n$ is $\{\Gamma_{*k}\}_{k=1}^{n}$, the formulas $(5.11)_n$ being true with $\{\Omega_{(A^*)_k}\}_{k=1}^{n}$ and $\{\Omega^{(A^*)_k}\}_{k=1}^{n}$, the strings of identificators of $(A^*)_n$. Consider A_{n+1} as a 1-PCID of A_n with the A_n-choice string $\Gamma_1(A_n,A_{n+1}) = (\Omega^{A_n})^*\Gamma_{n+1}(\Omega_{A_n})$. From Proposition 4.1 it follows that the A_n^*-choice string of $(A_n^*)_1 = \hat{U}^*A_{n+1}^*U'|(H_n' + L'^*)$ is $\Gamma_1(A_n^*,(A_n^*)_1) = Z_{A_n}\Gamma_1^*(A_n,A_{n+1})(Z^{A_n})^*$. From $(5.11)_n$ and the definition of $\Gamma_1(A_n,A_{n+1})$ it follows that

$$\begin{aligned}
&\Gamma_1(A_n^*,(A_n^*)_1)\\
&= (\hat{U}'\oplus\hat{U})^n(\Omega^{(A^*)_n})^* Z_A\Omega_{A_n}(\Omega_{A_n})^*\Gamma_{n+1}^*\Omega^{A_n}(\Omega^{A_n})^*(Z^A)^*\Omega_{(A^*)_n}\hat{U}'^{*n}|\mathcal{R}_{A_n^*}\\
&= (\hat{U}'\oplus\hat{U})^n(\Omega^{(A^*)_n})^*\Gamma_{*,n+1}\Omega_{(A^*)_n}\hat{U}'^{*n}|\mathcal{R}_{A_n^*}.
\end{aligned} \qquad (5.12)$$

But from $(5.9)_{n+1}$, it follows that

$$\begin{aligned}
(A^*)_{n+1} &= \hat{U}^{*n+1}A_{n+1}^*\hat{U}'^{n+1}|(H' + L'^* + \cdots + \hat{U}_*'^n L'^*)\\
&= \hat{U}^{*n}(A_n^*)_1\hat{U}'^n|(H' + L'^* + \cdots + U_*'^n L'^*),
\end{aligned} \qquad (5.13)$$

which implies (by (5.12)) that the $(A^*)_n$-choice string of $(A^*)_{n+1}$ is $(\Omega^{(A^*)_n})^*\Gamma_{*,n+1}\Omega_{(A^*)_n}$. This means that the A^*-choice string of $(A^*)_{n+1}$ is $\{\Gamma_{*,k}\}_{k=1}^{n+1}$. Using Lemma 5.2 for the pair (A_n,A_{n+1}), we have $Z_{A_{n+1}} = (\hat{U}'\oplus\hat{U})(\omega^{(A_n^*)_1})^*Z_{A_n}\omega_{A_n}$, and then $(5.11)_{n+1}$ follows from $(5.11)_n$, (5.13), Lemma 5.1 and (4.11) for the pair (A_n,A_{n+1}).

As in Proposition 4.1 from [10], let θ be the analytic function corresponding to $\hat{A} \in CID(A)$ (with the choice sequence $\{\Gamma_n\}_{n=1}^{\infty}$) and θ_* the analytic function corresponding to $\widehat{A^*} \in CID(A^*)$ (with the choice sequence $\{\Gamma_{*n}\}_{n=1}^{\infty}$).

Corollary 5.2. With the previous notations,

$$\theta_*(\lambda) = (Z_A \mid R_A)\widetilde{\theta}(\lambda)(Z_{A*} \mid R_{A*}),$$

where

$$\widetilde{\theta}(\lambda) = \theta(\bar{\lambda})^*,$$

for every $|\lambda| < 1$.

References

[1] Adamjan, V.M., Arov, D.Z., Kreĭn, M.G.: Bounded operators that commute with a contraction of class C_{00} of unit rank of nonunitarity (Russian). Funkcional'. Analiz Priložhenija 3.3 (1969), 86 - 87.

[2] Adamjan, V.M., Arov, D.Z., Kreĭn, M.G.: Analytic properties of Schmidt pairs for a Hankel operator and the generalized Schur-Takaji problem (Russian). Mat. Sbornik, n. Ser. 15 (1971), 31 - 73.

[3] Adamjan, V.M., Arov. D.Z., Kreĭn, M.G.: Infinite Hankel block-matrices and related continuation problems (Russian). Izvestija Akad. Nauk Armjan. SSR, Mat. 6 (1971), 87 - 112.

[4] Ando, T., Ceauşescu, Z., Foiaş, C.: On intertwining dilations. II. Acta Sci. math. 39 (1977), 3 - 14.

[5] Arsene, Gr., Ceauşescu, Z.: On intertwining dilations. IV. Tôhoku math. J., II. Ser. 30 (1978), 423 - 438.

[6] Arsene, Gr., Ceauşescu, Z., Foiaş, C.: On intertwining dilations. VIII. To appear.

[7] Carathéodory, C., Über den Variabilitetsbereich der Koeffizienten von Potenzreihen, die gegebene Werte nicht annehmen. Math. Ann. 64 (1907), 93 - 115.

[8] Carathéodory, C., Fejér, L.: Über den Zusammenhang der Extremen von harmonischen Funktionen mit ihrer Koeffizienten und über den Picard-Landauschen Satz. Rend. Circ. mat. Palermo, II. Ser. 32 (1911), 218 - 239.

[9] Ceauşescu, Z.: Operatorial extrapolations. Thesis, Bucharest (1978).

[10] Ceauşescu, Z., Foiaş, C.: On intertwining dilations. V. Acta Sci. math. 40 (1978), 9 - 32.

[11] Kreĭn, M.G., Nudel'man, A.A.: The Markov problem of moments and extremal problems (Russian). Izd. Nauka, Moscow (1973).

[12] Nehari, Z.: On bounded bilinear forms. Ann. of Math., II. Ser. 65 (1957), 153 - 162.

[13] Nevanlinna, R.: Über beschränkte Funktionen, die in gegebenen Punkten vorgeschriebene Werte annehmen. Ann. Acad. Sci. Fenn., Ser. A 13:1 (1919).

[14] Pick, G.: Über die Beschränkungen analytischer Funktionen, welche durch vorgegebene Funktionswerte bewirkt werden. Math. Ann. 77 (1915), 7 - 23.

[15] Sarason, D.: Generalized interpolation in H_∞. Trans. Amer. math. Soc. 127 (1967), 179 - 203.

[16] Sz.-Nagy, B., Foiaş, C.: Analyse harmonique des opérateurs de l'espace de Hilbert. Masson et Cie, Paris and Akadémiai Kiadó, Budapest (1967).

[17] Sz.-Nagy, B., Foiaş, C.: Forme triangulaire d'un contraction et factorization de la fonction caractéristique. Acta Sci. math. 28 (1967), 201 - 212.

INCREST
Bd. Păcii 220
77538 Bucureşti
Romania

THE STIELTJES CONE IS LOGARITHMICALLY CONVEX

Christian Berg

The Stieltjes transforms have turned out to play an important role in potential theory, in semigroups of operators and in the theory of infinitely divisible probability distributions, cf. [2-5], [9-10].

This is partly due to the fact that the cone $\mathcal{S}$ of Stieltjes transforms has some very nice stability properties, cf. below. The purpose of the present paper is to prove the stability property of $\mathcal{S}$ referred to in the title, namely that $\mathcal{S}$ is logarithmically convex: For $f_1,\dots,f_n \in \mathcal{S}$ and $\alpha_1,\dots,\alpha_n \geq 0$ with $\alpha_1 + \dots + \alpha_n \leq 1$ we have $f_1^{\alpha_1} \cdots f_n^{\alpha_n} \in \mathcal{S}$.

We give some applications of this result and remark that it does not extend to the cone $\mathcal{B}$ of Bernstein functions.

1. Introduction.

A function $f :]0,\infty[\to \mathbb{R}$ is called a Stieltjes transform, if it has the form

$$f(s) = a + \int_0^\infty \frac{d\mu(x)}{s+x}, \tag{1}$$

where $a \geq 0$ and μ is a positive measure on $[0,\infty[$.

The pair (a,μ) is uniquely determined by f. The set of Stieltjes transforms is a convex cone $\mathcal{S}$, which is closed in the topology of pointwise convergence. This and other results on $\mathcal{S}$ can be found in [3]. The Stieltjes transforms were introduced in [8].

The Stieltjes transform (1), is the Laplace transform of the measure $\tau = a\varepsilon_0 + g(x)dx$, where

$$g(x) = \begin{cases} \mathcal{L}\mu(x) & \text{for } x > 0, \\ 0 & \text{for } x \leq 0. \end{cases}$$

Here $\mathcal{L}\mu$ denotes the Laplace transform of μ. If follows that $\mathcal{S}$ is a subset of the cone of completely monotone functions. The measure τ is a potential kernel (cf. [2]), and if $\tau(\mathbb{R}) = 1$ then τ is an infinitely divisible probability distribution.

The cone $\mathcal{S}$ has the following well-known stability properties:

$$f \in \mathcal{S} \setminus \{0\} \Rightarrow \frac{1}{f(1/s)} \in \mathcal{S}. \tag{2}$$

$$f \in \mathcal{S},\ \lambda > 0 \Rightarrow \frac{f}{\lambda f + 1} \in \mathcal{S}. \tag{3}$$

$$f,g \in \mathcal{S} \setminus \{0\} \Rightarrow f \circ \frac{1}{g} \in \mathcal{S}. \tag{4}$$

$$f,g \in \mathcal{S} \setminus \{0\} \Rightarrow \frac{1}{f \circ g} \in \mathcal{S}. \tag{5}$$

$$f \in \mathcal{S} \setminus \{0\} \Rightarrow \frac{1}{sf(s)} \in \mathcal{S}. \tag{6}$$

The properties (2) - (4) are due to Hirsch [3] and property (6) was proved by Reuter [6] and rediscovered by Itô [5]. The properties (2) and (6) are easily deducible from one another. Property (5) follows from (2) and (4).

Using the Stieltjes transform $s^{-\alpha}$, $0 < \alpha \leq 1$, we get by specialization of (4) and (5):

Proposition 1. For $f \in \mathcal{S}$ and $0 < \alpha \leq 1$ we have $f^{\alpha} \in \mathcal{S}$ and $f(s^{\alpha}) \in \mathcal{S}$.

2. The log-convexity of $\mathcal{S}$.

Our main result is the following:

Theorem 2. Let $f_1,\dots,f_n \in \mathcal{S}$ and let $\alpha_1,\dots,\alpha_n$ be non-negative numbers with sum $\alpha_1 + \dots + \alpha_n \leq 1$. Then $f_1^{\alpha_1} \cdots f_n^{\alpha_n} \in \mathcal{S}$.

Remark. The theorem can be stated that the set $\log \mathcal{S} \setminus \{0\} = \{\log f \mid f \in \mathcal{S} \setminus \{0\}\}$ is a convex set. It is clearly enough to prove that if $f,g \in \mathcal{S} \setminus \{0\}$ and $0 < \alpha < 1$, then $f^{\alpha} g^{1-\alpha} \in \mathcal{S}$. Note that the first statement in Proposition 1 is a special case of Theorem 2. Another reformulation of the Theorem is the following: Let $\kappa_1,\dots,\kappa_n$ be potential kernels on $[0,\infty[$ such that $\kappa_i |]0,\infty[$ is completely monotone for $i = 1,\dots n$, and let $\alpha_1,\dots,\alpha_n$ be as above. Then $\kappa = \kappa_1^{\alpha_1} * \dots * \kappa_n^{\alpha_n}$ is a potential kernel on $[0,\infty[$ such that $\kappa |]0,\infty[$ is completely monotone. Here $\kappa_i^{\alpha_i}$ denotes the fractional power of κ_i of order α_i, $i = 1,\dots,n$.

In the following we shall need the Poisson kernel for the half-plane and the Hilbert transform. We have taken the information from [7], where the notion of non-tangential limits also can be found.

By $\mathbb{C} \setminus \mathbb{R}_-$ we denote the complex plane with the closed negative half-axis removed.

Lemma 3. Suppose $\varphi : \mathbb{R} \to [0,\infty[$ is a non-zero C^{∞}-function with compact support in $]0,\infty[$ and let $a \geq 0$. Then

$$f(z) = a + \int_0^\infty \frac{\varphi(x)}{z+x}\,dx$$

defines a holomorphic function in $\mathbb{C} \setminus \mathbb{R}_-$ satisfying $f(\bar{z}) = \overline{f(z)}$ and $\operatorname{Im} f(t+i\varepsilon) < 0$ for $t \in \mathbb{R}$, $\varepsilon > 0$. The limit function

$$f^+(t) := \lim_{\varepsilon\to 0^+} f(t+i\varepsilon)$$

exists uniformly for $t \in \mathbb{R}$ and f^+ is a bounded C^∞-function. Furthermore, $f(t+i\varepsilon)$ tends to $f^+(t_o)$ when $t+i\varepsilon$ tends non-tangentially to $t_o \in \mathbb{R}$.

Proof. The Poisson kernel and the conjugate Poisson kernel for the half-plane $\operatorname{Im} z > 0$ are denoted P and Q respectively, so we have (cf. [7])

$$P_\varepsilon(t) = \frac{1}{\pi}\,\frac{\varepsilon}{t^2+\varepsilon^2}\,,\quad Q_\varepsilon(t) = \frac{1}{\pi}\,\frac{t}{t^2+\varepsilon^2} \qquad \text{for } t \in \mathbb{R},\ \varepsilon > 0\,.$$

For $t \in \mathbb{R}$ and $\varepsilon > 0$ we find

$$f(-t+i\varepsilon) = a + \int_0^\infty \frac{\varphi(x)}{-t+i\varepsilon+x}\,dx = a - \pi Q_\varepsilon * \varphi(t) - i\pi P_\varepsilon * \varphi(t)$$

$$= a - \pi P_\varepsilon * (\mathcal{H}\varphi + i\varphi)(t)\,,$$

where

$$\mathcal{H}\varphi(t) = \lim_{\delta\to 0^+} \frac{1}{\pi} \int_{\delta\leq|u|} \frac{\varphi(t-u)}{u}\,du$$

denotes the Hilbert transform of φ. It follows that $-(1/\pi)\operatorname{Im} f(-t+i\varepsilon) = P_\varepsilon * \varphi(t)$ is the Poisson integral of φ, hence a positive harmonic function in the upper half-plane.

The Hilbert transform $\mathcal{H}\varphi$ of a C^∞-function φ with compact support is a C^∞-function tending to zero at infinity, so in particular $\mathcal{H}\varphi + i\varphi$ is a continuous function tending to zero at infinity. Therefore $P_\varepsilon * (\mathcal{H}\varphi + i\varphi)$ converges uniformly to $\mathcal{H}\varphi + i\varphi$ as $\varepsilon \to 0$, and also non-tangentially at every point, cf. [7]. We get

$$f^+(-t) = a - \pi(\mathcal{H}\varphi(t) + i\varphi(t)) \qquad \text{for } t \in \mathbb{R}\,.$$

Lemma 4. Let $\varphi,\psi : \mathbb{R} \to [0,\infty[$ be non-zero C^∞-functions with compact support in $]0,\infty[$, let $a,b \geq 0$ and let

$$f(z) = a + \int_0^\infty \frac{\varphi(x)}{z+x}\,dx,\quad g(z) = b + \int_0^\infty \frac{\psi(x)}{z+x}\,dx,\quad z \in \mathbb{C}\setminus\mathbb{R}_-\,.$$

For $0 < \alpha < 1$ the function $h(s) = (f(s))^{\alpha}(g(s))^{1-\alpha}$, $s > 0$, is a Stieltjes transform with representation

$$h(s) = a^{\alpha}b^{1-\alpha} + \int_0^{\infty} \frac{\gamma(t)}{s + t}\, dt , \tag{7}$$

where

$$\gamma(t) = -(1/\pi) \lim_{\varepsilon\to 0^+} \operatorname{Im} h(-t + i\varepsilon) \geq 0 .$$

Proof. From Lemma 3 we know that f and g are zero-free holomorphic functions in $\mathbb{C} \setminus \mathbb{R}_-$ and therefore $h(z) = (f(z))^{\alpha}(g(z))^{1-\alpha}$ is a well defined holomorphic function in $\mathbb{C} \setminus \mathbb{R}_-$ given as

$$h(z) = |f(z)|^{\alpha}|g(z)|^{1-\alpha}\exp\{i(\alpha \operatorname{Arg} f(z) + (1 - \alpha)\operatorname{Arg} g(z))\} ,$$

where $\operatorname{Arg} f(z)$ and $\operatorname{Arg} g(z)$ are continuous functions on $\mathbb{C} \setminus \mathbb{R}_-$ with values in $]-\pi,\pi[$. Furthermore $\operatorname{Arg} f(t + i\varepsilon)$ and $\operatorname{Arg} g(t + i\varepsilon)$ belong to the interval $]-\pi,0[$ for $t \in \mathbb{R}$ and $\varepsilon > 0$.

Using the uniform continuity of the function $x \to x^{\alpha}$ for $x \in [0,\infty[$ we also get from Lemma 3 that

$$\lim_{\varepsilon\to 0^+} |h(t + i\varepsilon)| = |f^+(t)|^{\alpha}|g^+(t)|^{1-\alpha}$$

uniformly for $t \in \mathbb{R}$.

We next claim that $h^+(t) := \lim_{\varepsilon\to 0^+} h(t + i\varepsilon)$ exists for all $t \in \mathbb{R}$, and that

$$\operatorname{Im} h^+(t) = \lim_{\varepsilon\to 0^+} \operatorname{Im} h(t + i\varepsilon) \leq 0 \quad \text{for} \quad t \in \mathbb{R} . \tag{8}$$

If $f^+(t) = 0$ or $g^+(t) = 0$ we get $h^+(t) = 0$, so (8) is clear. If $f^+(t) \neq 0$ we get that

$$\lim_{\varepsilon\to 0^+} \exp(i \operatorname{Arg} f(t + i\varepsilon)) = f^+(t)/|f^+(t)| ,$$

but since $\operatorname{Arg} f(t + i\varepsilon) \in]-\pi,0[$ for $\varepsilon > 0$, it follows that $\lim_{\varepsilon\to 0^+} \operatorname{Arg} f(t + i\varepsilon)$ exists and belongs to the interval $[-\pi,0]$, so that $\operatorname{Arg} f^+(t) = \lim_{\varepsilon\to 0^+} \operatorname{Arg} f(t + i\varepsilon)$. Similarly, if $g^+(t) \neq 0$ we get

$$\operatorname{Arg} g^+(t) = \lim_{\varepsilon\to 0^+} \operatorname{Arg} g(t + i\varepsilon) \in [-\pi,0] .$$

Combining these results (assuming $f^+(t) \neq 0, g^+(t) \neq 0$) we get

$$\lim_{\varepsilon\to 0^+} h(t + i\varepsilon) = |f^+(t)|^{\alpha}|g^+(t)|^{1-\alpha}\exp\{i(\alpha \operatorname{Arg} f^+(t) + (1 - \alpha)\operatorname{Arg} g^+(t))\},$$

and in particular

$$\text{Im}\, h^+(t) = |h^+(t)|\sin(\alpha \text{Arg}\, f^+(t) + (1-\alpha)\text{Arg}\, g^+(t)) \leq 0 ,$$

because

$$\alpha \;\text{Arg}\, f^+(t) + (1-\alpha)\text{Arg}\, g^+(t) \in [-\pi, 0] .$$

It follows that

$$\gamma(t) = -(1/\pi)\text{Im}\, h^+(-t) \geq 0 \quad \text{for} \quad t \in \mathbb{R} .$$

In order to prove the formula (7) we fix $s > 0$ and choose numbers ε and R such that $0 < \varepsilon < s < R$. We next consider the contour C given by the four curves

(a) $\varepsilon e^{i\theta}$, $\theta \in [-\frac{\pi}{2}, \frac{\pi}{2}]$,

(b) $Re^{i\theta}$, $\theta \in [-\pi + \tau, \pi - \tau]$, where $\tau = \text{Arcsin}(\varepsilon/R)$,

(c) $t + i\varepsilon$, $t \in [-R\cos\tau, 0]$,

(d) $t - i\varepsilon$, $t \in [-R\cos\tau, 0]$.

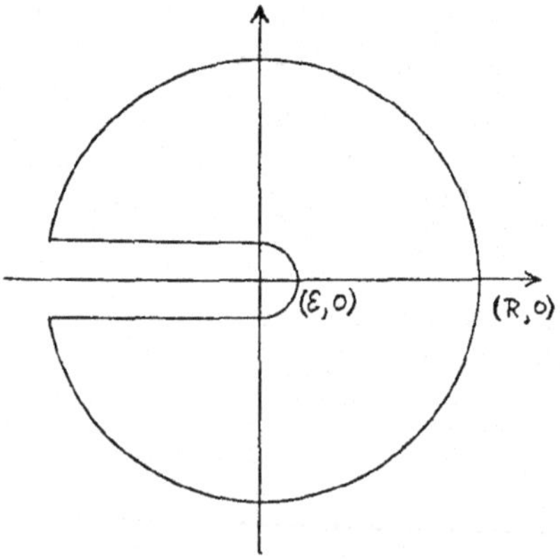

Applying Cauchy's integral formula to h and the contour C we get

$$h(s) = -\frac{1}{2\pi i}\int_{-\pi/2}^{\pi/2} \frac{h(\varepsilon e^{i\theta})i\varepsilon e^{i\theta}}{\varepsilon e^{i\theta} - s}\, d\theta + \frac{1}{2\pi i}\int_{-\pi+\tau}^{\pi-\tau} \frac{h(Re^{i\theta})iRe^{i\theta}}{Re^{i\theta} - s}\, d\theta$$

$$+ \frac{1}{\pi}\int_{-R\cos\tau}^{0} \text{Im}\left\{\frac{h(t+i\varepsilon)}{t+i\varepsilon-s}\right\} dt .$$

Let R be fixed and let $\varepsilon \to 0$. Then the first term tends to zero because $\lim_{\varepsilon\to 0} h(\varepsilon e^{i\theta})$ exists uniformly for $\theta \in [-\frac{\pi}{2},\frac{\pi}{2}]$. Since f and g have non-tangential limits for every $t \in \mathbb{R}$ we have

$$\lim_{\theta\to\pi^-} h(Re^{i\theta}) = h^+(-R), \quad \lim_{\theta\to-\pi^+} h(Re^{i\theta}) = \overline{h^+(-R)},$$

so the second term tends to

$$\frac{1}{2\pi}\int_{-\pi}^{\pi} \frac{h(Re^{i\theta})Re^{i\theta}}{Re^{i\theta}-s}\, d\theta \,.$$

By Lebesgue's theorem of dominated convergence the third term tends to

$$\frac{1}{\pi}\int_{-R}^{0} \frac{\operatorname{Im} h^+(t)}{t-s}\, dt = \int_{0}^{R} \frac{\gamma(t)}{s+t}\, dt \,,$$

and we get

$$h(s) = \frac{1}{2\pi}\int_{-\pi}^{\pi} \frac{h(Re^{i\theta})Re^{i\theta}}{Re^{i\theta}-s}\, d\theta + \int_{0}^{R} \frac{\gamma(t)}{s+t}\, dt \,.$$

Putting $A = \sup(\operatorname{supp}(\varphi))$ we have for $R > A$ and $\theta \in]-\pi,\pi[$ that

$$|f(Re^{i\theta}) - a| = \left|\int_{0}^{A} \frac{\varphi(x)}{Re^{i\theta}+x}\, dx\right| \leq \int_{0}^{A} \frac{\varphi(x)}{R-A}\, dx \,,$$

which shows that $\lim_{R\to\infty} f(Re^{i\theta}) = a$, uniformly for $\theta \in]-\pi,\pi[$.

It follows that $\lim_{R\to\infty} h(Re^{i\theta}) = a^\alpha b^{1-\alpha}$, uniformly for $\theta \in]-\pi,\pi[$, so that

$$\lim_{R\to\infty} \frac{1}{2\pi}\int_{-\pi}^{\pi} \frac{h(Re^{i\theta})Re^{i\theta}}{Re^{i\theta}-s}\, d\theta = a^\alpha b^{1-\alpha} \,.$$

Since $\gamma(t) \geq 0$ for all t we finally get

$$h(s) = a^\alpha b^{1-\alpha} + \lim_{R\to\infty}\int_{0}^{R} \frac{\gamma(t)}{s+t}\, dt = a^\alpha b^{1-\alpha} + \int_{0}^{\infty} \frac{\gamma(t)}{s+t}\, dt \,,$$

which proves that $h \in \mathcal{S}$.

<u>Proof of Theorem 2.</u> Let $f,g \in \mathcal{S} \setminus \{0\}$ and let $0 < \alpha < 1$. We shall see that $f^\alpha g^{1-\alpha} \in \mathcal{S}$.

Let f be defined as in (1) by a pair (a,μ) and g similarly by (b,ν). Using that $\mathcal{S}$ is closed under point-wise convergence we shall proceed in several steps:

(i) Suppose that $\mathrm{supp}(\mu)$ and $\mathrm{supp}(\nu)$ are compact and contained in $]0,\infty[$.

Let $(\varphi_n)_{n\in\mathbb{N}}$ be an approximate unit of C^∞-functions with compact supports shrinking to zero. Then $\mu_n = \mu * \varphi_n$ and $\nu_n = \nu * \varphi_n$ are C^∞-functions with compact supports which we may assume contained in $]0,\infty[$. If f_n and g_n are the Stieltjes transforms defined by (a,μ_n) and (b,ν_n) respectively, we know from Lemma 4 that $f_n^\alpha g_n^{1-\alpha} \in \mathcal{S}$ and hence $f^\alpha g^{1-\alpha} = \lim_{n\to\infty} f_n^\alpha g_n^{1-\alpha} \in \mathcal{S}$.

(ii) Suppose that $\mu(\{0\}) = \nu(\{0\}) = 0$.

Let μ_n and ν_n be the restrictions of μ and ν to $[\frac{1}{n},n]$ and f_n and g_n the Stieltjes transforms defined by (a,μ_n) and (b,ν_n). By (i) we know that $f_n^\alpha g_n^{1-\alpha} \in \mathcal{S}$ and hence $f^\alpha g^{1-\alpha} = \lim_{n\to\infty} f_n^\alpha g_n^{1-\alpha} \in \mathcal{S}$.

(iii) Suppose that $\mu(\{0\}) = \alpha > 0$.

The Stieltjes transform f_n defined by the pair (a,μ_n), where $\mu_n = \mu - \alpha\varepsilon_0 + \alpha\varepsilon_{1/n}$, converges pointwise to f. This shows that we can reduce the case (iii) to the case (ii).

Remark. J. Karlsson has informed the author about a result in Akhiezer's book [1] p. 127 which in our terminology may be formulated as follows:

Let $\tilde{\mathcal{S}}$ denote the set of holomorphic functions $F : \mathbb{C} \setminus \mathbb{R}_- \to \mathbb{C}$ satisfying $F(x) \geqq 0$ for $x > 0$ and $\operatorname{Im} F(z) \leqq 0$ for $\operatorname{Im} z > 0$. Then $\mathcal{S} = \{F|]0,\infty[\mid F \in \tilde{\mathcal{S}}\}$.

Theorem 2 can be derived from this result using the same convexity argument as in the proof of Lemma 4.

3. Examples. For $a > 0$ and $0 < p \leqq 1$ the function $f(s) = (1 + as)^{-p}$ belongs to $\mathcal{S}$. It is the Laplace transform of the Γ-distribution on $]0,\infty[$ with density

$$\Gamma_{a,p}(x) = \begin{cases} \frac{1}{\Gamma(p)} a^{-p} x^{p-1} e^{-\frac{x}{a}} & \text{for } x > 0 \\ 0 & \text{for } x \leqq 0 , \end{cases}$$

which is a completely monotone function on $]0,\infty[$.

For $a_1,\dots,a_n > 0$ and $p_1 > 0,\dots,p_n > 0$ such that $p_1 + \dots + p_n \leqq 1$ the function

$$f(s) = (1 + a_1 s)^{-p_1} \cdots (1 + a_n s)^{-p_n} \tag{9}$$

belongs to $\mathcal{S}$ by Theorem 2. It is the Laplace transform of the probability density $\Gamma_{a_1,p_1} * \dots * \Gamma_{a_n,p_n}$ which consequently is a completely monotone function on $]0,\infty[$. This last fact may also be verified by carrying out the convolution.

Thorin introduced the concept of generalized Γ-convolutions in [9], [10]. A probability measure μ on $[0,\infty[$ is called a <u>generalized Γ-convolution</u> if the Laplace transform $\varphi = \mathcal{L}\mu$ of μ satisfies $-\frac{\varphi'}{\varphi} \in \mathcal{S}$. This is equivalent to $\mathcal{L}\mu = e^{-f}$, where f has the integral representation

$$f(s) = bs + \int_0^\infty \log(1 + sx)d\sigma(x) ,$$

where $b \geq 0$ and σ is a positive measure on $]0,\infty[$ such that $\int_0^\infty \log(1 + x)d\sigma(x) < \infty$. Alternatively μ is a generalized Γ-convolution if and only if μ is the vague limit of translates of convolutions of finitely many Γ-distributions $\Gamma_{a,p}$ with $a,p > 0$. A generalized Γ-convolution μ is infinitely divisible and self-decomposable.

<u>Proposition 5</u> (Thorin [10]). Let μ be a generalized Γ-convolution such that $\mathcal{L}\mu = e^{-f}$ with

$$f(s) = \int_0^\infty \log(1 + sx)d\sigma(x), \; s > 0 ,$$

where σ is a positive measure on $]0,\infty[$ such that $\int_0^\infty d\sigma(x) \leq 1$.

Then $\mathcal{L}\mu = e^{-f} \in \mathcal{S}$ and μ has a completely monotone density on $]0,\infty[$.

<u>Proof.</u> The measure σ can be approximated vaguely by discrete measures $\sum_{i=1}^n p_i\varepsilon_{a_i}$ where $a_1,\dots,a_n > 0$ and $\sum_1^n p_i \leq 1$, and the Laplace transform of the corresponding generalized Γ-convolution is the Stieltjes transform (9).

4. Bernstein functions.

We recall that a function $f :]0,\infty[\to [0,\infty[$ is called a <u>Bernstein function</u>, if it is C^∞ and if f' is completely monotone, cf. [2]. The set of Bernstein functions form a convex cone $\mathcal{B}$, which is important in potential theory because $\mathcal{P} = \{1/f \mid f \in \mathcal{B} \setminus \{0\}\}$ is the set of Laplace transforms of potential kernels on $[0,\infty[$. It is known that $\mathcal{S} \setminus \{0\} \subset \mathcal{P}$, cf. [4] or [2].

Since $\mathcal{B}$ is stable under fractional powers (cf. [2]), i.e. for $g \in \mathcal{B}$ and $0 < \alpha < 1$ then $g^\alpha \in \mathcal{B}$, it is natural to examine if the log-convexity of $\mathcal{S} \setminus \{0\}$ holds for $\mathcal{B}$. That this is not the case is shown by the following example:

$$1 - e^{-s},\ 1 - e^{-2s} \in \mathcal{B} \text{ but } \sqrt{(1 - e^{-s})(1 - e^{-2s})} \notin \mathcal{B}.$$

This follows for instance from the expansion

$$\varphi(s) = 1 - \sqrt{(1 - e^{-s})(1 - e^{-2s})} = \frac{1}{2}e^{-s} + \frac{5}{8}e^{-2s} - \frac{3}{16}e^{-3s} + \cdots ,$$

because the occurrence of the negative coefficient $-\frac{3}{16}$ shows that $\varphi(s)$ is not completely monotone.

References

[1] Akhiezer, N. I.: The classical moment problem. Oliver and Boyd, Edinburg (1965).

[2] Berg, C., Forst, G.: Potential theory on locally compact abelian groups. Springer-Verlag, Berlin - Heidelberg - New York (1975).

[3] Hirsch, F.: Intégrales de résolventes et calcul symbolique. Ann. Inst. Fourier 22, 4 (1972), 239 - 264.

[4] Hirsch, F.: Transformation de Stieltjes et fonctions opérant sur les potentiels abstraits. Lecture Notes in Mathematics 404, Springer-Verlag, Berlin - Heidelberg - New York (1974), 149 - 163.

[5] Itô, M.: Sur les cônes convexes de Riesz et les noyaux de convolution complètement sous-harmoniques. Nagoya math. J. 55 (1974), 111 - 144.

[6] Reuter, G. E. H.: Über eine Volterrasche Integralgleichung mit totalmonotonem Kern. Arch. der Math. 7 (1956), 59 - 66.

[7] Stein, E. M., Weiss, G.: Introduction to Fourier analysis on euclidean spaces. Princeton University Press, Princeton, N.J. (1971).

[8] Stieltjes, T. J.: Recherches sur les fractions continues. Ann. Fac. Sci. Univ. Toulouse, 8 (1894), 1 - 122. (Oeuvres Complètes vol II, 402 - 567. Groningen (1918)).

[9] Thorin, O.: On the infinite divisibility of the Pareto distribution. Scand. Actuarial J. (1977), 31 - 40.

[10] Thorin, O.: On the infinite divisibility of the lognormal distribution. Scand. Actuarial J. (1977), 121 - 148.

København Universitet
Matematiske Institut
Universitetsparken 5
DK-2100 København Ø
Danmark

CHARACTERIZATIONS OF NORMAL MEROMORPHIC FUNCTIONS

Douglas M. Campbell and Gene Wickes

Many different and seemingly unrelated characterizations of normal meromorphic functions have been developed over the years. This paper provides a survey of twenty-two characterizations scattered throughout the literature as well as those of our own research. A historical development of normal meromorphic functions which explains why there is such a variety of characterizations can be found in [3].

In section one we give the standard characterizations in terms of normal families, uniform $\rho - \chi$ continuity, the spherical derivative condition and Lappan's derivative condition (see [12], [13], [15], [17], [21]). In section two we give Brown and Gauthier's characterization in terms of cluster sets (see [1], [8]). In section four we give characterizations for non-normality which parallel Milloux's cercles de remplissage for entire functions (see [4], [5], [6], [10], [11], [12], [20], [25]). In section six we give characterizations due to Lohwater, Pommerenke, Lappan and Yosida (see [14], [19], [25]) and connect these ideas in section seven to P, ρ, and W points (see [23]). The historical survey [3] and figure 1 provide the necessary connections between the various sections of the paper.

0. The following ideas and notations will be used in all sections of the paper. The (pseudohyperbolic) distance $\rho(z,z')$ between two points z and z' of $D = \{z : |z| < 1\}$ is $|z' - z|/|1 - \bar{z}z'|$. The chordal distance between two points w and w' of the extended complex plane $\hat{\mathbb{C}}$ is denoted by $\chi(w,w')$. The spherical derivative of a meromorphic function is the quantity $|f'(z)|/(1 + |f(z)|^2)$ and is denoted by $f^{\#}(z)$. The set $\{z : \rho(z,z_n) = \rho_n\}$ is not only a non-Euclidean circle of center z_n and pseudohyperbolic radius ρ_n, but it is also a Euclidean circle with radius $r_n = \rho_n(1 - |z_n|^2)/(1 - |z_n|^2 \rho_n^2)$ and center $z_n(1 - \rho_n^2)/(1 - |z_n|^2 \rho_n^2)$. This immediately yields $r_n/(1 - |z_n|^2) \to 0$ if and only if $\rho_n \to 0$.

1. We begin with the normal function loop which, except for the Lehto-Virtanen-Noshiro contribution, is due to Lappan.

Definition. A family F of functions meromorphic in D is normal if every sequence $\{f_n(z)\} \subset F$ contains a subsequence which converges

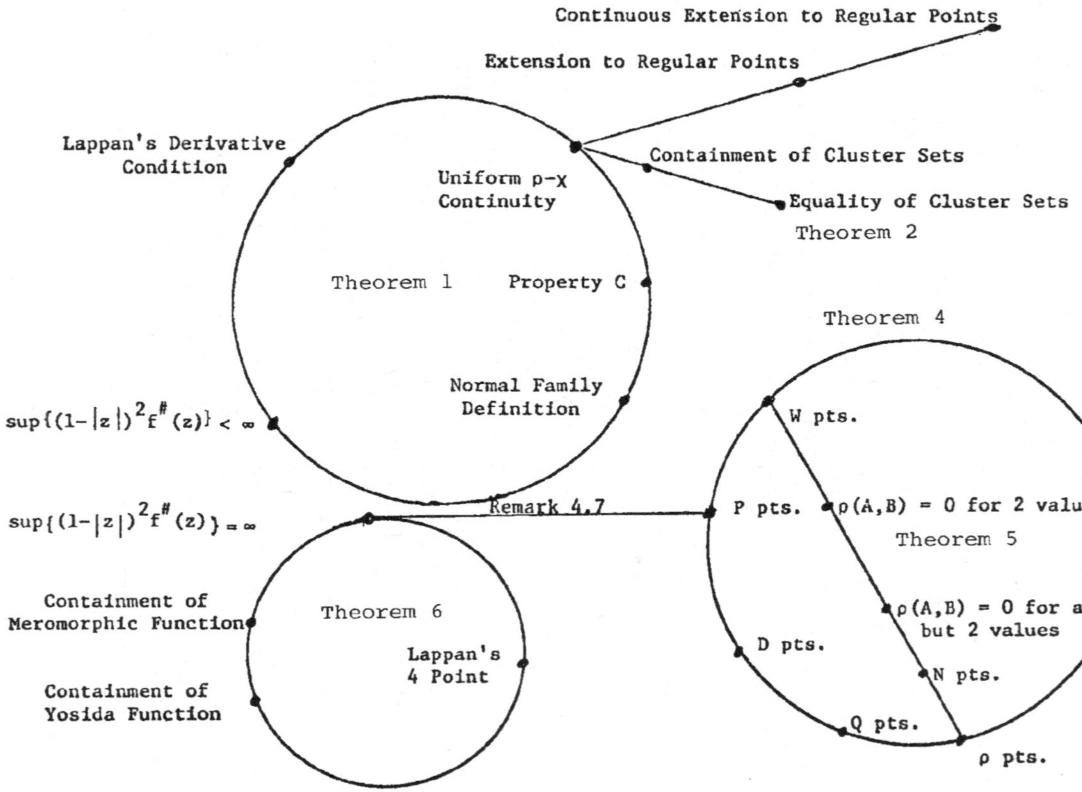
Continuous Extension to Regular Points
Extension to Regular Points
Lappan's Derivative Condition
Uniform ρ-χ Continuity
Containment of Cluster Sets
Equality of Cluster Sets
Theorem 2
Theorem 1
Property C
Theorem 4
Normal Family Definition
$\sup\{(1-|z|)^2 f^{\#}(z)\} < \infty$
W pts.
A pts.
Remark 4.7
$\sup\{(1-|z|)^2 f^{\#}(z)\} = \infty$
P pts.
ρ(A,B) = 0 for 2 values
Theorem 5
Containment of Meromorphic Function
Theorem 6
ρ(A,B) = 0 for all but 2 values
A* pts.
D pts.
Lappan's 4 Point
N pts.
Containment of Yosida Function
Q pts.
ρ pts.

FIGURE 1

uniformly in the chordal metric on every compact subset of D to a function meromorphic on D (the function identically ∞ is considered to be meromorphic on D).

Lemma 1.1. (Marty) A family F of functions meromorphic in D is normal if and only if the family of spherical derivatives $F^{\#}$ is bounded on $|z| \leq r$ for every $r < 1$.

Theorem 1. The following are equivalent for a meromorphic function $f(z)$ in $|z| < 1$:

Property C: If $\{z_n\}$ and $\{z_n'\}$ are sequences in D with $\rho(z_n,z_n') \to 0$, then $\chi(f(z_n),f(z_n')) \to 0$.

Uniform $\rho - \chi$ continuity: The function f is uniformly (ρ,χ) continuous, that is, for all $\varepsilon > 0$ there is a $\delta > 0$ such that $\rho(z,z') < \delta$ implies $\chi(f(z),f(z')) < \varepsilon$.

Lappan's derivative condition: For any $K > 0$ and any $n \geq 1$ $\sup\{(1 - |z|^2)^n|f^{(n)}(z)| : |f(z)| \leq K\} < \infty$.

Spherical derivative condition: $\sup\{(1 - |z|^2)f^{\#}(z) : z \in D\} < \infty$.

Normal function definition: The family $\{f(G)\}$ is normal, where G is the set of all Möbius maps of D onto D.

Proof. Property C implies uniform $\rho - \chi$ continuity. If f is not uniformly $\rho - \chi$ continuous, then there is an $\varepsilon > 0$ and sequences $\{z_n\}$, $\{z_n'\}$ such that $\rho(z_n,z_n') < 1/n$ but $\chi(f(z_n),f(z_n')) \geq \varepsilon$, which shows that property C fails.

Uniform $\rho - \chi$ continuity implies Lappan's derivative condition. We must show for $K > 0$ and $n \geq 1$ there is a finite constant $E(n,K,f)$ such that $\sup\{(1 - |z|^2)^n|f^{(n)}(z)| : |f(z)| \leq K\} \leq E(n,K,f)$. The uniform $\rho - \chi$ continuity of f implies there is a $\delta > 0$ such that $\chi(f(z), f(z')) < \chi(K,2K)$ whenever $\rho(z,z') < \delta$. Let z_0 be such that $|f(z_0)| \leq K$. For all z in the disc $|z - z_0| < \delta(1 - |z_0|)$ we have $\rho(z,z_0) < \delta$. Elementary geometry on the sphere and the above $\rho - \chi$ continuity yield $|f(z)| \leq 2K$ for all z in the disc $|z - z_0| \leq \delta(1 - |z_0|)$. Applying this bound on $f(z)$ to Cauchy's integral formula for $f^{(n)}(z_0)$ on the circle $|z - z_0| = \delta(1 - |z_0|)$ yields $|f^{(n)}(z_0)| \leq 2Kn!\,\delta^{-n}(1 - |z_0|)^{-n}$. Thus $E(n,K,f) \leq 2^{n+1}Kn!\delta^{-n}$.

Lappan's derivative condition implies the spherical derivative condition. If f satisfies Lappan's derivative condition, then so does $F(z) = 1/f(z)$. The proof is by induction using the fact that $\sum_{j=0}^{n} f^{(j)}(z)F^{(n-j)}(z) = 0$. Let A be the set of points where $|f(z)| \leq 1$. By Lappan's derivative condition for f we have for all z in A that $(1 - |z|^2)f^{\#}(z) \leq (1 - |z|^2)|f'(z)| \leq E(1,1,f)$, while by Lappan's derivative condition for $F = 1/f$ we have for all z not in A that $(1 - |z|^2)f^{\#}(z) \equiv (1 - |z|^2)F^{\#}(z) \leq (1 - |z|^2)|F'(z)| \leq E(1,1,F)$. Therefore, $(1 - |z|^2)f^{\#}(z) \leq E(1,1,f) + E(1,1,F) < \infty$.

The spherical derivative condition implies the normal family definition. Let $\sup (1 - |z|^2)|f'(z)|/(1 + |f(z)|^2) = M$. By Marty's criterion (lemma 1.1) the family $\{f(G)\}$ will be normal if $\{f^{\#}(G)\}$ is uniformly bounded on $|z| \leq r$. A computation shows that for every Möbius map $g(z)$ the max of $|g'(z)|/(1 - |g(z)|^2)$ on $|z| \leq r$ is $(1 - r^2)^{-1}$. Therefore for all $|z| \leq r$

$$f^{\#}(g(z)) = \frac{|f'(g(z))||g'(z)|}{1 + |f(g(z))|^2} = \frac{|g'(z)|}{1 - |g(z)|^2} \cdot \frac{(1 - |g(z)|^2)|f'(g(z))|}{1 + |f(g(z))|^2}$$

$$\leq \frac{M}{1 - r^2} .$$

Normal function definition implies property C. Indeed, if property C does not hold then f cannot be normal. For let $\{z_n\}$ and $\{z_n'\}$ be sequences in D with $\rho(z_n,z_n') \to 0$, $f(z_n) \to A$, $f(z_n') \to B$, $\chi(A,B) > 0$. The family $\{g_n(\zeta)\} = \{f((\zeta + z_n)/(1 + \bar{z}_n\zeta))$ cannot converge to a continuous function on any compacta containing the origin since $\zeta_n = (z_n - z_n')/(1 - z_n'\bar{z}_n)$ satisfies $\rho(0,\zeta_n) \to 0$ while $g_n(0) \to A$, $g_n(\zeta_n) \to B$, $A \neq B$. This completes the proof of Theorem 1.

Remarks. A normal function is uniformly uniformly continuous, that is, $\delta \leq \varepsilon/M$, where $M = \sup (1 - |z|^2)f^{\#}(z)$. The constants $E(n,K,f)$ need not be bounded with n. In fact for $f(z) = (1 - z)^{-1}$, and $K = 1$, $E(n,1,f)$ is $\geq n!$. Since f is normal if and only if $1/f$ is normal we see from Lappan's derivative condition that normality is also equivalent to $\sup\{(1 - |z|^2)^n \cdot |f^{(n)}(z)|/|f(z)| : f(z) \geq K\} < \infty$ for every $K > 0$ and $n \geq 1$.

2. We now turn to Brown and Gauthier's characterization of normality in terms of cluster sets.

Definition. Let S and T be subsets of D. A set S is (hyperbolically) close to a set T if for every $\varepsilon > 0$ there is an r, $1 - \varepsilon < r < 1$, such that for all s in S with $|s| > r$ there is a t in T with $\rho(t,s) < \varepsilon$. Two sets S and T in D are (hyperbolically) equivalent if S is close to T and T is close to S.

Definition. If S is a subset of D, then $C(S,f)$, the cluster set of f on S, is $\{w \in \hat{\mathbb{C}} : f(z_n) \to w,\ z_n \in S,\ |z_n| \to 1\}$.

Theorem 2. The following are equivalent for a function $f(z)$ meromorphic in $|z| < 1$:

Uniform $\rho - \chi$ continuity. For all $\varepsilon > 0$ there is a $\delta > 0$ such that $\rho(z,z') < \delta$ implies $\chi(f(z),f(z')) < \varepsilon$.

Containment of cluster sets. For every set S close to T, $C(S,f) \subset C(T,f)$.

Equality of cluster sets. For every set S equivalent to T, $C(S,f) = C(T,f)$.

Proof. Uniform $\rho - \chi$ continuity implies containment of cluster sets. If $C(S,f) = \emptyset$, then we are done. Otherwise, let A belong to $C(S,f)$ and $\{z_n\}$ be a sequence in S, $f(z_n) \to A$, $|z_n| \to 1$. To show that A is in $C(T,f)$ it suffices to prove for each integer m that there is a point t_m in T with $\chi(f,(t_m),A) < 1/m$, $|t_m| \to 1$. Since $f(z)$ is uniformly $\rho - \chi$ continuous there is a $\delta > 0$ such that $\rho(a,b) < \delta$ implies $\chi(f(a),f(b)) < 1/(2m)$. The convergence of $f(z_n)$ to A guarantees an n_0 such that for all $n \geq n_0$, $\chi(f(z_n),A) < 1/(2m)$. Since S is close to T there is an $r > 1 - \varepsilon$, such that for all s in S, $|s| > r$, there is a t in T with $\rho(s,t) < \delta$. Choose any z_n, $n \geq n_0$, $|z_n| > r$. Then there is a t_m in T with $\rho(z_n,t_m) < \delta$. Hence $\chi(f(z_n),f(t_m)) < 1/(2m)$ and $\chi(f(t_m),A) < 1/m$ which concludes the proof.

Containment of cluster sets is equivalent to equality of cluster sets. Clear.

Containment of cluster sets implies uniform $\rho - \chi$ continuity. Indeed, if f is not uniformly $\rho - \chi$ continuous then, by passing to a subsequence if necessary, there are sequences $\{z_n\}$, $\{z_n'\}$, $\rho(z_n,z_n') \to 0$, with $f(z_n) \to A$, $f(z_n') \to B$, $\chi(A,B) \neq 0$. Clearly $\{z_n\}$ is close to $\{z_n'\}$ but $C(\{z_n\},f) \not\subset C(\{z_n'\},f)$. This completes the proof of Theorem 2.

Remark 2.1. Given a nonnormal function it is easy to find equivalent sets on which f will have unequal cluster sets. One need only find close sequences which do not go to close sequences. If $f(z) = \exp(i/(1 - z))$, $z_n = 1 - (2\pi n)^{-1}$, $z_n' = 1 - (2\pi n)^{-1} + in^{-2}$, then $\rho(z_n,z_n') \to 0$ and $\{z_n\}$ and $\{z_n'\}$ are equivalent. Clearly, $C(\{z_n\},f) = 1$ while $C(\{z_n'\},f) = \infty$.

Remark 2.2. These tools simplify a number of proofs in the literature. For example, let $Z_n \to 1$ (possibly tangentially) with $\rho(Z_n,Z_{n+1}) \to 0$. Let us prove that for any Blaschke product B which vanishes on $\{Z_n\}$, $B(Z)$ must go to zero in any Stolz angle at 1. Since the path γ formed by connecting the points Z_n to Z_{n+1} is equivalent to $\{Z_n\}$ and

$C(\{z_n\},B) = 0$, the above theorem quarantees that $B(z)$ has the asymptotic value 0 on γ. Normal functions with an asymptotic value have the same limiting value in any Stolz angle. One should compare this with more direct attacks even for special cases such as [22, p. 341, 13], or [24].

4. The loop of this section contains contributions from Lange, Gavrilov, and Lappan.

Lemma 4.1. (Montel) If for each meromorphic function $f_\alpha(z)$ there are three values a_α, b_α, c_α, which depend on f_α and which f_α omits in $|z| < r$, then the family $\{f_\alpha(z)\}$ is normal in $|z| < r$.

Theorem 4. Let $\{z_n\}$ be a sequence of points of $|z| < 1$ and $f(z)$ a meromorphic function in $|z| < 1$. The following are equivalent for $f(z)$:

$\{z_n\}$ is a sequence of P points, that is, for each $r > 0$ and each subsequence $\{z_{n(k)}\}$ the function $f(z)$ assumes every value of the extended complex plane, with at most two exceptions, infinitely often in the union of the discs

$$D_k = \{z : \rho(z, z_{n(k)}) < r\} .$$

$\{z_n\}$ is a sequence of D points, that is, for every c in $\hat{\mathbb{C}}$ there exists a sequence of points $\{z_n'\}$, $\rho(z_n, z_n') \to 0$, such that $\{f(z_n')\}$ converges to c.

$\{z_n\}$ is a sequence of Q points, that is, for each $r > 0$ there are sequences of sets $\{E_n(r)\}$ and $\{G_n(r)\}$ whose chordal diameters do not exceed r and an integer $N(r)$ such that for all $n \geq N(r)$ the function $f(z)$ assumes all values of $\hat{\mathbb{C}} - \{E_n(r) \cup G_n(r)\}$ in $\{\rho(z_n, z) < r\}$.

$\{z_n\}$ is a sequence of ρ points, that is, there are monotonic nonincreasing sequences $\{L_n\}$ and $\{r_n\}$ both tending to zero, such that in each disc $\{z : \rho(z_n, z) < r_n\}$ the function $f(z)$ assumes all values of $\hat{\mathbb{C}}$ with the possible exception of two sets E_n and G_n whose chordal diameters do not exceed L_n

$\{z_n\}$ is a sequence of A^* points, that is, there is a sequence $\{r_n\} \to 0$, such that the spherical area of f in $\{z : \rho(z_n, z) < r_n\}$ tends to π.

$\{z_n\}$ is a sequence of A points, that is there is a sequence $\{r_n\} \to 0$, such that the spherical area of f in $\{z : \rho(z_n, z) < r_n\}$ does not tend to zero.

$\{z_n\}$ is a sequence of W points, that is, there is a sequence of points $\{z_n'\}$ such that $\rho(z_n,z_n') \to 0$ and $\lim_n \sup \chi(f(z_n),f(z_n')) > 0$.

Proof. P points are D points. Let $\{z_n\}$ be a sequence of P points. We must show for any point w in $\hat{\mathbb{C}}$ that there is a sequence of points $\{z_n'\}$ such that $\rho(z_n',z_n) \to 0$, and $f(z_n') \to w$. Since $\{z_n\}$ is a sequence of P points, then for each positive integer m there are at most a finite number of points z_n for which the image of the disc $\{z : \rho(z_n,z) < 1/m\}$ is spherically bounded away from w by $1/m$, that is, for each m there is an integer n_m such that for each $n > n_m$ there is a point z_n', $\rho(z_n,z_n') < 1/m$, with $\chi(f(z_n'),w) < 1/m$. For $n \leq n_1$, let $z_n' = z_n$, and for $n_m < n \leq n_{m+1}$, let z_n' be as above. The sequence $\{z_n'\}$ is then close to $\{z_n\}$ and $\{f(z_n')\}$ converges to w. Hence, $\{z_n\}$ is a sequence of D points.

D points are Q points. Indeed suppose that $\{z_n\}$ is not a sequence of Q points. Then there is an $r > 0$ such that on the disc $\{z : \rho(z,z_n) < r\}$ f does not assume three values a_n, b_n, and c_n and $\chi(a_n,b_n) > r/2$, $\chi(a_n,c_n) > r/2$, and $\chi(b_n,c_n) > r/2$, $n = 1,2,3,\ldots$ If $f_n(z) = f((z + z_n)/(1 + \bar{z}_n z))$, then for each n, $f_n(z)$ omits a_n, b_n, and c_n in the disc $\{z : \rho(0,z) < r\}$. By Montel's theorem, $\{f_n(z)\}$ is a normal family of functions. Therefore there is a subsequence, which we continue to denote by $\{f_n(z)\}$, which converges spherically uniformly on $\{z : \rho(0,z) \leq r/2\}$ to a function $g(z)$ which is either meromorphic of identically infinite. In either case $g(z)$ is continuous. The continuity of $g(z)$ together with $f_n(z) \to g(z)$ spherically uniformly implies the existence of a δ and an $N(\delta)$ such that

$$\chi(f_n(0),g(0)) + \chi(g(0),g(z')) + \chi(g(z'),f_n(z')) < \varepsilon$$

whenever $n > N(\delta)$ and $\rho(0,z') < \delta$. If z satisfies $\rho(z_n,z) < \delta$, then $z' = (z - z_n)/(1 - \bar{z}_n z)$ satisfies $\rho(0,z') < \delta$. Hence, for all z satisfying $\rho(z_n,z) < \delta$ and all $n > N(\delta)$

$$\chi(f(z_n),f(z)) = \chi(f_n(0),f_n(z'))$$
$$\leq \chi(f_n(0),g(0)) + \chi(g(0),g(z')) + \chi(g(z'),f_n(z')) < 3\varepsilon ,$$

which implies that $\{z_n\}$ is not a sequence of D points.

Q points imply ρ points. Let $\{z_n\}$ be a sequence of Q points. Then for each $r > 0$, there are sequences of sets $\{E_n(r)\}$ and $\{G_n(r)\}$ whose chordal diameters do not exceed r, and an integer $N(r)$ such that in each disc $\{z : \rho(z_n,z) < r\}$, $n > N(r)$, the function $f(z)$ assumes all values of the Riemann sphere with the possible exception of the sets $E_n(r)$ and $G_n(r)$. Letting $r = 1/m$ we obtain sets

$\{E_n(1/m)\}_{n=1}^{\infty}$ and $\{G_n(1/m)\}_{n=1}^{\infty}$. Choose the integers $N(1/m)$ so that $N(1/(m+1)) > N(1/m)$. For $1 \leq n < n(1/1)$ let $r_n = 1$, $L_n = 1$. For $N(1/m) < n \leq N(1/(m+1))$ let $r_n = 1/m$, $L_n = 1/m$. Clearly $\{r_n\}$ and $\{L_n\}$ are monotonic nonincreasing with $r_n \to 0$, $L_n \to 0$. With $\{r_n\}$ and $\{L_n\}$ defined in this manner it is easily seen that in each of the discs $D_n = \{z : \rho(z_n,z) < r_n\}$, the function f assumes all values of the Riemann sphere with the possible exception of two sets E_n and G_n whose chordal diameters do not exceed L_n. Therefore, $\{z_n\}$ is a sequence of ρ points.

<u>ρ points imply A^* points.</u> If $\{z_n\}$ is a sequence of ρ points, then under $f(z)$ the image of $\{z : \rho(z_n,z) < r_n\}$ must cover the sphere with the possible exception of two sets whose chordal diameters are less than L_n. The area on the unit sphere bounded by a circle of chordal radius r is πr^2. Therefore, the area of $\{z : \rho(z_n,z) < r_n\}$ under f must be at least $\pi - 2(\pi(L_n/2)^2)$ which, since $L_n \to 0$, approaches π.

<u>A^* points imply A points.</u> Since the spherical area tends to π on the discs $\{z : \rho(z_n,z) < r_n\}$ it certainly does not tend to zero.

<u>A points imply W points.</u> By passing to a subsequence we may assume that the sequence of points $\{z_n\}$ and the sequence of radii $r_n \to 0$ are such that the spherical area of the image of $f(z)$ on $\rho(z,z_n) < r_n$ is always greater than some $\pi\delta^2 > 0$. In each disc $\{z : \rho(z_n,z) < r_n\}$, there must then be at least one point z_n' with $\chi(f(z_n),f(z_n')) > \delta/2$ (the area on the unit sphere which is bounded by a circle of chordal radius δ is $\pi\delta^2$). Thus $\rho(z_n,z_n') \to 0$, and $\chi(f(z_n),f(z_n')) \geq \delta/2 > 0$ together imply that $\{z_n\}$ is a sequence of W points.

<u>W points imply P points.</u> Let $\{z_n\}$ and $\{z_n'\}$ be such that $\rho(z_n,z_n') \to 0$ and $\chi(f(z_n),f(w_n)) > \varepsilon > 0$. Let $g_n(z) = f((z+z_n)/(1+\bar{z}_n z))$. The point $z_n'' = (z_n' - z_n)/(1 - z_n'\bar{z}_n)$ satisfies $\rho(0,z_n'') = \rho(z_n,z_n')$ and $(z_n'' + z_n)/(1 + \bar{z}_n z_n'') = z_n'$. Since $\rho(z_n,z_n') \to 0$, then for any $r > 0$ we can find an N such that for all $n \geq N, \rho(0,z_n'') < r$ and $\chi(g_n(0),g_n(z_n'')) \geq \varepsilon$. Therefore no subsequence of $\{g_n\}$ can ever converge to a continuous let alone meromorphic function at 0. Consequently $\{g_n\}$ is not a normal family in any disc $\{z : \rho(0,z) < r\}$. Applying Montel's theorem we can conclude that for every $r > 0$ the family $\{g_n\}$ must assume each value of the Riemann sphere, with at most two exceptions, infinitely often on the union of the discs $\rho(z_n,z) < r$, that is, for any $r > 0$, $f(z)$ assumes each value of the Riemann sphere, with at most two exceptions, infinitely often on the union of the discs $\rho(z_n;z) < r$. The same argument holds for any sub-

sequence of z_n, which proves that $\{z_n\}$ is a sequence of P points. This concludes the proof of Theorem 4.

Remark 4.1. If $\{z_n\}$ is a sequence of P, ρ, A, etc. points and $\{z'_n\}$ satisfies $\rho(z_n, z'_n) \to 0$, then $\{z'_n\}$ is also a sequence of P, ρ, A, etc. points. Note that if there exists discs $\rho(z_n, z) < r_n$, $r_n \to 0$, whose image under f have spherical areas bounded away from zero, then there are discs $\rho(z_n, z) < r'_n$, $r'_n \to 0$, whose images under f have spherical areas which go to π.

Remark 4.2. Any sequence $\{z_n\}$, $0 < z_n < 1$, $z_n \to 1$, is a sequence of P, W and D points for the nonnormal function $\exp(i/(1-z))$.

Remark 4.3. If $\{z_n\}$ is a sequence of P points, then $(1 - |z_n|^2) \cdot f^{\#}(z_n)$ need not diverge as $n \to \infty$ as we see for $f(z) = \exp(i/(1-z))$, $z_n = n^2/(1+n^2) - i/(n+n^3)$. Since $z^*_n = n^2/(1+n^2)$ is a sequence of P points by remark 4.2 and $\rho(z_n, z^*_n) \to 0$, then by remark 4.1 we know that $\{z_n\}$ is a sequence of P points. A simple computation shows that $(1 - |z_n|^2) f^{\#}(z_n) \to 0$.

Remark 4.4. The following theorem was suggested by a result of Gavrilov [6] for functions meromorphic in the neighborhood of an essential singularity . It relates P points to the places where the spherical derivative diverges and connects loops 4 and 6 of figure 1.

Theorem. Let $f(z)$ be meromorphic in $|z| < 1$. Then $\{z_n\}$ is a sequence of P points for $f(z)$ if and only if there is a sequence of positive numbers $\varepsilon_n \to 0$ such that

$$\lim_{n\to\infty} \{ \sup_{\rho(z,z_n)<\varepsilon_n} (1 - |z|^2) f^{\#}(z) \} = \infty . \tag{4.1}$$

Proof. Since P points are W points we may choose sequences $\{a_n\}$ and $\{b_n\}$ with $\rho(a_n, z_n) \to 0$, $\rho(b_n, z_n) \to 0$ which satisfy $f(a_n) \to 0$, $f(b_n) \to 2$. Let $\varepsilon_n = \max\{|a_n - z_n|, |b_n - z_n|\}$. As we noted at the first of the paper there are simple relationships among the Euclidean diameter of a non-Euclidean disc, the non-Euclidean diameter, and the distance of the non-Euclidean center to $|z| = 1$. In fact since c_n and b_n lie in a disc whose non-Euclidean radius goes to zero, then $\varepsilon_n/(1 - |z_n|)$ must also go to zero. Let L denote the line segment from a_n to b_n and let c_n denote the first point from a_n to b_n at which $|f(z)| = 1$. Then

$$|a_n - c_n||f'(w_n)| \geq \int_L |f'(z)||dz| \geq |\int_L f'(z)dz| = |f(a_n) - f(c_n)| \to 2$$

for some w_n on L. Therefore

$$\frac{(1-|w_n|^2)|f'(w_n)|}{1+|f(w_n)|^2} \geq \frac{1-|w_n|^2}{|c_n - a_n|} \geq \frac{1-(|z_n|+\varepsilon_n)^2}{2\varepsilon_n} \geq \frac{1}{2}(\frac{1-|z_n|}{\varepsilon_n}) - \frac{3}{2} \to \infty ,$$

which proves (4.1) as claimed.

Conversely suppose there is a sequence of positive numbers $\varepsilon_n \to 0$ such that (4.1) holds. Let $\{g_n(\zeta)\} = \{f((\zeta + z_n)/(1 + \bar{z}_n\zeta))\}$ and z_n^* be such that $\rho(z_n^*, z_n) < \varepsilon_n$ and $(1 - |z_n^*|^2)f^{\#}(z_n^*) \to \infty$. Let $\zeta_n = (z_n^* - z_n)/(1 - \bar{z}_n z_n^*)$. A direct computation yields $g_n^{\#}(\zeta_n) = |1 - \bar{z}_n z_n^*|^2(1 - |z_n^*|^2)^{-1}$ $(1 - |z_n|^2)^{-1}(1 - |z_n^*|^2)f^{\#}(z_n^*)$. Since $|1 - z_n z_n^*|^2(1 - |z_n^*|^2)(1 - |z_n^*|^2)^{-1} \cdot (1 - |z_n|^2)^{-1} \geq 1$, we see that $g_n^{\#}$ is unbounded on every compacta $|\zeta| \leq r$. Thus no subsequence of g_n is normal on any $|\zeta| \leq r$. It follows that $\{z_n\}$ is a sequence of P points.

5. The following theorem shows that the nesting property displayed by $\exp(i/(1 - z))$ near $z = 1$ is characteristic of non-normal functions. The notion of N points seems to be a natural, simple, and pleasing geometric way to introduce non-normal functions.

Theorem 5. If f is a meromorphic function in $|z| < 1$, then the following are equivalent

f has a sequence of ρ points.

f has a sequence of N points, that is, there are monotonic sequences L_n and r_n, $L_n \to 0$, $r_n \to 0$, such that in each disc $\{z : \rho(z_n, z) < r_n\}$ the function $f(z)$ assumes all values of $\hat{\mathbb{C}}$ with the possible exception of two discs D_n' and D_n'' whose chordal diameters do not exceed L_n and $D_1' \supset D_2' \supset \cdots$, $D_1'' \supset D_2'' \supset \cdots$.

If $A = \{z : f(z) = a\}$ and $B = \{z : f(z) = b\}$, then for all but at most two values of a and b in $\hat{C}$, $\rho(A,B) = 0$.

If $A = \{z : f(z) = a\}$ and $B = \{z : f(z) = b\}$, then for some a, b in $\hat{C}$, $a \neq b$, $\rho(A,B) = 0$.

f has a sequence of W points.

Proof. ρ points imply N points. Let $\{z_n\}$ be a sequence of ρ points and $\{E_n\}$ and $\{G_n\}$ be the associated sequences of exceptional sets. Passing to a subsequence we may choose points e_n in E_n converging to e and g_n in G_n converging to g (e and g may coincide) such that $\chi(e_n, e)$ and $\chi(g_n, g)$ go monotonically to 0. Let D_n' be the disc $\{w : \chi(e,w) < \chi(e_n,e) + L_n\}$ and D_n'' be the disc $\{w : \chi(g,w) < \chi(g_n,g) + L_n\}$. The other implications are clear.

Remark 5.1. A function has N points if and only if f is non-normal. Although a sequence of P points need not be a sequence of N points, it must contain a subsequence of N points. Notice that the existence of a single pair of numbers for which the root set is close implies the root set is close for all but at most two numbers. This cannot be improved as $\exp(i/(1 - z))$ shows. Furthermore, for $f(z) = \exp(i/(1 - z))$ and any $a, b \in \mathbb{C} - \{0\}$, we see that $z_n = 1 - i(\log a + 2n\pi i)^{-1}$, $z_n' = 1 - i(\log b + 2n\pi i)^{-1}$ satisfy $f(z_n) = a$, $f(z_n') = b$, $\rho(z_n, z_n') \to 0$.

6. Our final characterization of nonnormal functions contains contributions by Lappan, Lohwater, and Pommerenke. Lohwater and Pommerenke expressed their part of this theorem in the following somewhat picturesque language, 'a function is nonnormal if and only if its Riemann image surface contains asymptotically a Riemann surface of parabolic type.'

Definition. A function meromorphic in $\mathbb{C}$ is Yosida if and only if for any sequence of complex numbers $\{a_j\}$ the family $\{f(z + a_j)\}$ is normal.

Lemma 6.1. A function meromorphic in $\mathbb{C}$ is Yosida if and only if there is a finite constant M such that $f^{\#}(z) \leq M$ for all z in $\mathbb{C}$.

For example, the function $\exp(z)$ is Yosida while $\exp(\exp(z))$ and $\exp(z) - \exp(iz)$ are non-Yosida (consider $z_n = \log(1 + ni)$ as $n \to \infty$ and $z_n = \pi n(1 - i)$ as $n \to -\infty$ respectively).

Lemma 6.2. (Hurwitz) [9, p. 205]) Let $\{g_n\}$ and g be functions meromorphic in $\mathbb{C}$, $\chi(g_n, g) \to 0$ on compacta of $\mathbb{C}$, and t_0 an arbitrary element of $\mathbb{C}$. Then all but a finite number of the functions g_n assume the value $g(t_0)$ in each neighborhood of t_0.

The spherical derivative provides a clean way of discussing multiple roots (or poles) of a meromorphic function, In fact, $f^{\#}(z_0) = 0$ if and only if z_0 is a multiple root (pole) of $f(z)$. A value a in $\mathbb{C}$ is said to be completely ramified with respect to $f(z)$ if all the roots of the equation $f(z) = a$ are multiple roots. Nevanlinna theory guarantees that a meromorphic function can have at most four completely ramified values [9, p. 231]. We summarize this as follows:

Lemma 6.3. If $g(z)$ is a nonconstant meromorphic function in $\mathbb{C}$, then for all but at most four choices of β in $\hat{\mathbb{C}}$, there is a finite complex number t_0 with $g(t_0) = \beta$, $g^{\#}(t_0) \neq 0$.

Theorem 6. Let $f(z)$ be meromorphic in $|z| < 1$. The following are equivalent:

Spherical derivative condition. $\sup_{|z|<1} (1 - |z|^2) f^{\#}(z) = \infty$.

f contains a Yosida function, that is, there exists sequences $\{z_n\}$, $\{\rho_n\}$, with z_n in D, $\rho_n > 0$, $\rho_n/(1 - |z_n|) \to 0$ such that $\{f(z_n + \rho_n\zeta)\}$ converges locally uniformly to a nonconstant Yosida function.

f contains a meromorphic function, that is, there exists sequences $\{z_n\}, \{\rho_n\}$, with z_n in D, $\rho_n > 0$, $\rho_n/(1 - |z_n|) \to 0$, such that $\{f(z_n + \rho_n\zeta)\}$ converges locally uniformly to a nonconstant meromorphic function in C.

f satisfies the four point condition, that is, for all but at most four values β of $\hat{\mathbb{C}}$, $\sup\{(1 - |z|^2) f^{\#}(z) : z \in f^{-1}(\beta)\} = \infty$.

Proof. Spherical derivative condition implies f contains a Yosida function. Since $\sup(1 - |z|^2) f^{\#}(z) = \infty$, there exists a sequence of points $\{z_n^*\}$ such that $(1 - |z_n^*|^2) f^{\#}(z_n^*) \to \infty$, $|z_n^*| \to 1$. Let $r_n^2 = (2|z_n^*|^2)/(1 + |z_n^*|^2)$ and note that

$$(1 - \frac{|z_n^*|^2}{r_n^2}) f^{\#}(z_n^*) = \frac{(1 - |z_n^*|^2)}{2} f^{\#}(z_n^*) \quad . \tag{6.1}$$

Choose $\{z_n\}$ such that

$$M_n = \max_{|z| \leq r_n} (1 - \frac{|z|^2}{r_n^2}) f^{\#}(z) = (1 - \frac{|z_n|^2}{r_n^2}) f^{\#}(z_n) \; ; \tag{6.2}$$

the maximum exists since $f^{\#}(z)$ is continuous in $\{z : |z| \leq r_n\}$. Since z_n^* is itself a point within $|z| \leq r_n$, (6.1) implies that $M_n \to \infty$. Setting

$$\rho_n \equiv \frac{1}{M_n} (1 - \frac{|z_n|^2}{r_n^2}) = \frac{1}{f^{\#}(z_n)} , \tag{6.3}$$

we have

$$\frac{\rho_n}{r_n - |z_n|} = \frac{r_n + |z_n|}{r_n^2 M_n} \leq \frac{2}{r_n^2 M_n} \to 0 \; . \tag{6.4}$$

Therefore the function $g_n(\zeta) = f(z_n + \rho_n\zeta)$ is defined for $|\zeta| < R_n \equiv (r_n - |z_n|)/\zeta_n$. It follows from (6.3) that

$$g_n^{\#}(0) = \rho_n f^{\#}(z_n) = 1. \tag{6.5}$$

We now show that $\{g_n(\zeta)\}$ is a normal family by proving that $g_n^{\#}(\zeta)$ is uniformly bounded on $|\zeta| \leq R$, R arbitrary. By (6.4) we have that $R \leq R_n$ for n sufficiently large. Hence for sufficiently large n, all points of $|\zeta| \leq R$ satisfy $|z_n + \rho_n\zeta| < r_n$. Therefore by (6.2), (6.3), and the triangle inequality

$$g_n^{\#}(\zeta) = \rho_n f^{\#}(z_n + \rho_n\zeta) \leq \frac{\rho_n M_n}{1 - (|z_n + \rho_n\zeta|/r_n)^2} \tag{6.6}$$

$$\leq \frac{(r_n + |z_n|)(r_n - |z_n|)}{(r_n + |z_n| + \rho_n R)(r_n - |z_n| - \rho_n R)} \leq (1 - \frac{R\rho_n}{r_n - |z_n|})^{-1}$$

Since R is fixed this last term tends to 1 by (6.4). Hence, $\{g_n(\zeta)\}$ is a normal family. By passing to a subsequence we can therefore assume that there are sequences $\{z_n\}$ and $\{\rho_n\}$, $|z_n| \to 1$, $\rho_n/(1 - |z_n|) \to 0$, such that $\{f(z_n + \rho_n\zeta)\}$ converges locally uniformly to a function $g(\zeta)$ which is meromorphic in $\mathbb{C}$. By (6.5), $g^{\#}(0) = 1$ so g is non-constant and by (6.6), $g^{\#}(\zeta) \leq 1$ for all complex ζ so that $g(\zeta)$ is a Yosida function.

<u>f contains a Yosida function implies f contains a meromorphic function.</u> Clear.

<u>f contains a meromorphic function implies f satisfies the four point condition.</u> By hypothesis there exist sequences $\{z_n\}$ and $\{\rho_n\}$ with z_n in D, $|z_n| \to 1$, $\rho_n > 0$, $\rho_n/(1 - |z_n|) \to 0$, and a non-constant function g meromorphic in the complex plane such that the sequence of functions $\{g_n(\zeta)\} = \{f(z_n + \rho_n\zeta)\}$ converges locally uniformly to g. By lemma 6.3 there are at most four points β for which $g(z) = \beta$ either fails to have roots or has only multiple roots. Let β be a nonexceptional point and t_0 a point satisfying $g(t_0) = \beta$, $g^{\#}(t_0) \neq 0$. For any sequence $t_n \to t_0$, the local uniform convergence of g_n to g implies by lemma 6.2 that all but a finite number of the functions g_n assume the value β in each neighborhood of t_0. Therefore there is a sequence of points $\{t_n\}$, $t_n \to t_0$, such that for sufficiently large n we have $g_n(t_n) = \beta$, that is, $f(z_n + \rho_n t_n) = \beta$. Letting $s_n = z_n + \rho_n t_n$, we note that

$$\left|1 - \frac{1 - |s_n|}{1 - |z_n|}\right| = \left|\frac{|z_n + \rho_n t_n| - |z_n|}{1 - |z_n|}\right| \leq \frac{|\rho_n t_n|}{1 - |z_n|} \to 0$$

and therefore $(1 - |s_n|)/(1 - |z_n|) \to 1$. Also, $g_n^{\#}(t_n) = \rho_n f^{\#}(s_n)$

so that $f^{\#}(s_n)(1-|s_n|) = g_n^{\#}(t_n)(1-|s_n|)/\rho_n = g_n^{\#}(t_n)\ (1-|z_n|)/\rho_n \cdot (1-s_n)/(1-|z_n|)$. Letting $n \to \infty$, we have $g_n^{\#}(t_n) \to g^{\#}(t_0)$, $(1-|z_n|)/\rho_n \to \infty$, and $(1-|s_n|)/(1-|z_n|) \to 1$. Therefore $f^{\#}(s_n)(1-|s_n|) \to \infty$. Thus, for all but at most four values of β, $\sup\{(1-|z|^2) f^{\#}(z) : z \in f^{-1}(\beta)\} = \infty$.

The four point condition implies the spherical derivative condition. If $\sup\{(1-|z|^2)f^{\#}(z) : z \in f^{-1}(\beta)\} = \infty$ for each complex number β, with at most four exceptions, then clearly $\sup\{(1-|z|^2)f^{\#}(z) : |z| < 1\} = \infty$. This concludes the proof of Theorem 6.

Remark 6.1. Lohwater and Pommerenke originally stated their theorem with no restriction on the speed at which $\rho_n \to 0$. In proving their theorem they asserted, "if f is normal and $f(z_n + \rho_n\zeta) \to g(\zeta)$ locally uniformly, then $\rho_n/(1-|z_n|) \to 0$". The statement in quotes is false as one can see from $f(z) \equiv z$, $z_n = 1 - n^{-1}$, $\rho_n = n^{1/2}$, $g(\zeta) \equiv 1$.

Remark 6.2. We give two examples of sequences z_n and ρ_n such that $f(z_n + \rho_n\zeta)$ tends locally uniformly to a non-constant, meromorphic Yosida function in $\mathbb{C}$.

Example 1. Let a be a nonzero real number. If $f(z) = \exp(i/(1-z))$ $z_n = 1 - 1/(na)$, and $\rho_n = 1/n^2$, then $f(z_n + \rho_n\zeta)$ converges uniformly on compacta to $c\exp(a^2 i\zeta)$, where c is some point with modulus one. Since $|e^{ani}| = 1$, choose a subsequence so that e^{ani} approaches some c, $|c| = 1$. Let $|\zeta| \leq R$, R arbitrary. Then

$$|c\exp(a^2 i\zeta) - f(z_n + \rho_n\zeta)| = |c\exp(a^2 i\zeta) - \exp(\frac{ani}{1 - a\zeta/n})|$$

$$\leq |c\exp(a^2 i\zeta) - \exp(ani)\cdot\exp(a^2 i\zeta)| + |\exp(ani)\cdot\exp(a^2 i\zeta)$$

$$- \exp(\frac{ani}{1 - a\zeta/n})| = |\exp(a^2 i\zeta)||c - \exp(ani)| + |\exp(ani)\cdot\exp(a^2 i\zeta)|$$

$$\cdot |1 - \exp(\frac{a^3 i\zeta^2}{n - \zeta a})| \leq \exp(a^2 R)|c - \exp(ani)| + \exp(a^2 R)$$

$$\cdot |1 - \exp(\frac{a^3 i\zeta^2}{n - \zeta a})| .$$

Since e^{ani} converges to c the first term goes to zero. Using the continuity of the exponential ($|\zeta| \leq R$, R fixed but arbitrary), the second term also tends to zero and the result is established.

Example 2. If $f(z) = (1-z)^{-\alpha}\exp-(\frac{1+z}{1-z})$, $z_n = (n - 1 + ie^{n/\alpha})/(n + 1 + ie^{n/\alpha})$ and $\rho_n = 2Ae^{-2n/\alpha}$, $\alpha > 0$, $A > 0$, then $f(z_n + \rho_n\zeta)$ converges uniformly on compacta to $c\exp(A\zeta)$, where c is some point with $|c| = 2^{-\alpha}$.

Remark 6.3. There are nonnormal functions with sequences z_n, ρ_n such that $f(z_n + \rho_n\zeta)$ converges to a non-Yosida function. In fact, using example 1 we see that $F(z) = \exp(\exp(i/(1 - z)))$, $z_n = 1 - 1/(2\pi n)$, $\rho_n = 1/n^2$ satisfies $F(z_n + \rho_n\zeta) \to \exp(\exp(4\pi^2 i\zeta))$ which is non-Yosida.

7. It is difficult to use Lohwater and Pommerenke's proof to exhibit a meromorphic function $g(\zeta)$ which is asymptotically contained in a nonnormal function $f(z)$ since one must find a point z_n at which the max of $(1 - |z|^2/r_n^2)f^{\#}(z)$ occurs for $|z| \leqq r_n$. Furthermore, there are no a priori clues to the size of the required ρ_n. The following theorems connect Lohwater and Pommerenke's theorem to P, ρ, W, etc. points and to the divergence of the spherical derivative.

Theorem 7.1. Let $f(z)$ be meromorphic in $|z| < 1$. If $\{z_n\}$, and $\{\rho_n\}$ satisfy $\rho_n/(1 - |z_n|) \to 0$, and $f(z_n + \rho_n\zeta)$ converges to a non-constant meromorphic function in $\mathbb{C}$, then $\{z_n\}$ is a sequence of W points.

Theorem 7.2. If $(1 - |z_n^*|^2)f^{\#}(z_n^*) \to \infty$, then there is a sequence of points z_n in D and a sequence of points $\rho_n > 0$ such that $\rho_n/(1 - |z_n|) \to 0$, $\rho(z_n, z_n^*) \to 0$, $f(z_n + \rho_n\zeta)$ converges to a non-constant Yosida function. The points ρ_n satisfy $\rho_n \leqq f^{\#}(z_n^*)^{-1}$.

Proof of Theorem 7.1. If $\rho_n/(1 - |z_n|) \to 0$ and $f(z_n + \rho_n\zeta) \to g(\zeta)$, g non-constant, then choose any ζ_1 such that $g(\zeta_1) \neq g(0)$. Then $\rho(z_n, z_n + \rho_n\zeta_1) \leqq \rho_n|\zeta_1|/(1 - |z_n|) \to 0$ and $\chi(f(z_n), f(z_n + \rho_n\zeta_1)) \to \chi(g(0), g(\zeta_1)) > 0$ which proves that $\{z_n\}$ is a sequence of W points.

Proof of Theorem 7.2. Let $p_n = (1 - |z_n^*|^2)^{1/2} f^{\#}(z_n^*)^{-1/2}$ and z_n be a point at which $M_n = \max_{|z-z_n|\leqq p_n} (p_n - |z_n - z_n^*|)f^{\#}(z)$ is attained. Since z_n^* is itself a contender we find that $M_n \geqq p_n f^{\#}(z_n^*) = ((1 - |z_n^*|^2)f^{\#}(z_n))^{1/2}$ so $M_n \to \infty$. We note that $f^{\#}(z_n) \geqq f^{\#}(z_n^*)$ since

$$p_n f^{\#}(z_n) \geqq (p_n - |z_n - z_n^*|)f^{\#}(z_n) \geqq p_n f^{\#}(z_n^*) .$$

Therefore, upon letting

$$\rho_n \equiv f^{\#}(z_n)^{-1} = (p_n - |z_n - z_n^*|)/M_n$$

we have

$$\rho_n/(p_n - |z_n - z_n^*|) = M_n^{-1} \to 0 . \tag{7.1}$$

Define $g_n(\zeta) = f(z_n + \rho_n\zeta)$ and let ζ lie in the arbitrary compacta $|\zeta| \leq R$. Then for sufficiently large n we obtain from (7.1) that $z_n + \rho_n\zeta$ is in $|z - z_n^*| \leq p_n$. By the definition of ρ_n we note

$$g^{\#}_n(0) = \rho_n f^{\#}(z_n) = 1 \ . \tag{7.2}$$

As in Lohwater and Pommerenke's proof we have from (7.1)

$$\begin{aligned} g_n^{\#}(\zeta) &= \rho_n |f^{\#}(z_n + \rho_n\zeta)| \leq \rho_n M_n/(p_n - |z_n + \rho_n\zeta - z_n^*|) \\ &\leq (p_n - |z_n - z_n^*|)/(p_n - |z_n - z_n^*| - \rho_n R) \\ &\leq [1 - R(\rho_n/(\rho_n - |z_n - z_n^*|))]^{-1} \to 1 \ . \end{aligned}$$

This proves that $g_n(\zeta)$ forms a normal family and so contains a subsequence which converges locally uniformly in $\mathbb{C}$ to a non-constant Yosida function. We see that

$$\begin{aligned} \rho(z_n, z_n^*) &= |z_n - z_n^*|/|1 - \bar{z}_n z_n^*| \leq p_n/(1 - |z_n^*|) \\ &= [(1 - |z_n^*|^2) f^{\#}(z_n^*)]^{-1} \to 0 \ , \end{aligned}$$

and similarly

$$\left|1 - \frac{1 - |z_n|}{1 - |z_n^*|}\right| \leq \left|\frac{|z_n| - |z_n^*|}{1 - |z_n^*|}\right| \leq \frac{|z_n - z_n^*|}{1 - |z_n^*|} \leq \frac{p_n}{1 - |z_n^*|} \to 0 \ .$$

Consequently,

$$\frac{\rho_n}{1 - |z_n|} = \frac{1}{(1 - |z_n|) f^{\#}(z_n)} \leq \frac{1 - |z_n^*|}{1 - |z_n|} \cdot \frac{1 + |z_n^*|}{(1 - |z_n^*|^2) f^{\#}(z_n^*)} \to 0 \ .$$

8. We conclude with three open problems.

1) Must every non-normal meromorphic function contain a Yosida function of the form $A \exp(B\zeta)$?

2) Given a non-normal function predict the form of an associated limit function.

3) Characterize the set of all limit functions of a given non-normal function.

References

[1] Brown, L., Gauthier, P.: Cluster sets in a Banach algebra of non-tangential curves. Ann. Acad. Sci. Fenn., Ser. A I 460 (1969), 1 - 4.

[2] Brown, L., Gauthier, P.: Behavior of normal meromorphic functions on the maximal ideal space of H^∞, Michigan math. J. 18 (1971),

365 - 371.

[3] Campbell, D., Wickes, G.: A historical survey of normal meromorphic functions. To appear.

[4] Gavrilov, V. I.: Boundary properties of functions meromorphic in the unit disc. Doklady Akad. Nauk SSSR, 151 (1963), 19 - 22.

[5] Gavrilov, V. I.: On the distribution of values of functions meromorphic in the unit circle which are not normal. Mat. Sbornik 67 (109) (1965), 408 - 427.

[6] Gavrilov, V. I.: The behavior of a meromorphic function in the neighborhood of an essentially singular point. Izvestija Akad. Nauk. SSSR, 30 (1966), 767 - 788.

[7] Gauthier, P. M.: The non-Plessner points for the Schwarz triangle functions. Ann. Acad. Sci. Fenn., Ser. A I 422 (1968), 1 - 6.

[8] Gauthier, P. M.: A criterion for normalacy. Nagoya math. J. 32 (1968), 277 - 282.

[9] Hille, E.: Analytic function theory, Vol. II. Ginn, Lexington, Mass. (1962).

[10] Lange, L.: The existence of non-euclidean cercles de remplissage in certain subsets of the unit disc. Nagoya math. J. 19 (1961), 41 - 47.

[11] Lange, L.: Sur les cercles de remplissage non-euclideans. Ann. sci. École norm. sup., III. Sér. 77 (1960), 257 - 280.

[12] Lappan, P.: Some sequential properties of normal and non-normal functions with applications to automorphic functions. Commentarii math. Univ. St. Pauli 12 (1964), 41 - 47.

[13] Lappan, P.: Some results on harmonic normal functions. Math. Z. 90 (1965), 155 - 159.

[14] Lappan, P.: A criterion for a meromorphic function to be normal. Commentarii math. Helvet. 49 (1974), 492 - 495.

[15] Lappan, P.: The spherical derivative and normal functions. Ann. Acad. Sci. Fenn., Ser. A I 3 (1977), 301 - 310.

[16] Lehto, O.: The spherical derivative of meromorphic functions in the neighborhood of an isolated singularity. Commentarii math. Helvet. 33 (1959), 196 - 205.

[17] Lehto, O., Virtanen, K. I.: Boundary behavior and normal meromorphic functions. Acta. math. 97 (1957), 47 - 65.

[18] Lehto, O., Virtanen, K. I.: On the behavior of meromorphic functions in the neighborhood of an isolated singularity. Ann. Acad. Sci. Fenn., Ser. A I 240 (1957).

[19] Lohwater, A. J., Pommerenke, C.: On normal meromorphic functions. Ann. Acad. Sci. Fenn., Ser. A I 550 (1973).

[20] Milloux, H.: Le théorème de M. Picard, suites de fonctions holomorphes, fonctions méromorphes et fonctions entieres. J. Math. pur. appl., IX. Sér. 3 (1924), 345 - 401.

[21] Noshiro, K.: Contributions to the theory of meromorphic functions in the unit circle. J. Fac. Sci., Hokkaido Univ. 7 (1938), 149 - 159.

[22] Rudin, W.: Real and complex Analysis. McGraw-Hill, New York (1974).

[23] Rung, D. C.: A local form of Lappan's five point theorem for normal functions. Michigan math. J. 23 (1976), 141 - 145.

[24] Somadosa, H.: Blaschke products with zero tangential limits. J. London math. Soc., II. Ser. 41 (1966), 293 - 303.

[25] Yosida, K.: On a class of meromorphic functions. Proc. phys. math. Soc. Japan, 16 (1934), 227 - 235.

Brigham Young University
Provo, UT 84602
U.S.A.

ABOUT CAPACITIES AND MODULI IN INFINITE-DIMENSIONAL SPACES

Petru Caraman

1. Introduction.

In the main part of our communication at the Colloquium on Complex Analysis and the IV Romanian-Finnish Seminar on Complex Analysis, we established some inclusion relations between conformal capacity, Bessel capacities, Φ-capacities and Hausdorff h-measures in R^n, showing that most of these results are best possible. The last part of the talk contained some remarks on conformal capacity in infinite-dimensional spaces. Since the results presented in the first part of our lecture will appear elsewhere [3], we shall give here only the results connected with the second part.

We shall consider first the different capacities in R^n. Some of them may be used in an infinite-dimensional space without any change. For the definition of another category, Lebesgue integral and then Lebesgue measure is involved. We propose to consider them in a Banach space and to take abstract Wiener measure instead of the Lebesgue one. However, it is possible to maintain Lebesgue measure as a starting point for restricted products of measures in the sense of Moore [9]. There is also a third category of capacities in R^n, where the dimension n appears explicitly in their expression, as for instance in the case of the conformal capacity. In this case we propose a change of the original definition (change letting invariant the class of sets in R^n of capacity zero) so that, letting $n \to \infty$, we obtain a well defined concept.

2. Concepts of capacity in R^n which may be used unchanged in normed spaces

Let X be a normed space and $|\cdot|$ its norm.

In general, by capacity we mean a non-negative, extended real-valued set function C, whose domain is a class $\mathcal{A}$ of subsets of X (usually the field of Borel sets), which contains the compact sets and is closed under countable unions. Furthermore, C will satisfy

(i) $C(\emptyset) = 0$, $\emptyset$ the empty set.

(ii) $E_1, E_2 \in \mathcal{A}$ and $E_1 \subset E_2 \Rightarrow C(E_1) \leq C(E_2)$.

(iii) $E_i \in \mathcal{A}$, $i = 1,2,\ldots \Rightarrow C(\bigcup_i E_i) \leq \sum_i C(E_i)$.

Now, let us define the Φ-capacity of a Borel set $E \subset X$ as

$$C_\Phi(E) = [\inf \iint \Phi(|x - y|)d\mu(x)d\mu(y)]^{-1},$$

where the kernel $\Phi : R^+ \to R^+$ is continuous, strictly decreasing, with $\lim_{r\to\infty}\Phi(r) = +\infty$, and the infimum is taken over all measures $\mu \geqq 0$ such that $\mu(X) = 1$ and the support $S_\mu \subset E$; E is supposed to be contained in a ball of radius r_o such that $\Phi(r_o) = 0$. If E does not satisfy this condition, we shall consider $C_\Phi(E) = 0$ if and only if $C_\Phi(B(x,r)\cap E) = 0$ for every ball $B(x,r)$ with $0 < r < r_o$.

Next, let us define

$$c_\Phi(E) = \sup_\mu \mu(X),$$

where the supremum is taken over all measures $\mu \geqq 0$ with $S_\mu \subset E$ and

$$\int \Phi(|x - y|)d\mu(y) \leqq 1 \quad \text{for every} \quad x \in E.$$

In a similar way, we have

$$c'_\Phi(E) = \inf \mu(X),$$

where the infimum is taken over all measures $\mu \geqq 0$ with $S_\mu \subset E$ and

$$\int \Phi(|x - y|)d\mu(y) \geqq 1 \quad \text{for every} \quad x \in E.$$

Now, let us suppose X is a normed space with a countable basis. Then we define

$$M(E) = [\lim_{m\to\infty} \sup_{\{x_k\}} \inf_{x\in E} \frac{1}{m} \sum_{k=1}^{m} \Phi(|x - x_k|)]^{-1}.$$

Finally, let us give a generalization of the <u>transfinite diameter</u>

$$D(E) = [\lim_{m\to\infty} \inf_{x_p, x_q \in E} \binom{m}{2}^{-1} \sum_{1\leqq p,q\leqq m} \Phi(|x_p - x_q|)]^{-1}.$$

3. Capacities and moduli in abstract Wiener spaces.

We shall consider in this paragraph the moduli and some capacities, which can not be used in the infinite-dimensional case without changes.

Let $\Gamma \subset R^n$ be an arc family and $F(\Gamma)$ the class of functions such that $\rho(x) \geqq 0$ is Borel measurable and $\int_\gamma \rho(x)ds \geqq 1$ for every $\gamma \in \Gamma$. Then the p-modulus of Γ is defined as

$$M_p(\Gamma) = \inf_{\rho\in F(\Gamma)} \int \rho(x)^p dx,$$

where the integration is taken with respect to Lebesgue measure.

The <u>p-extremal length</u> of $\Gamma \subset R^n$ is given by

$$\lambda_p(\Gamma) = \frac{1}{M_p(\Gamma)}.$$

The p-capacity of a compact set $F \subset R^n$ is

$$\text{cap}_p F = \inf_u \int |\nabla u(x)|^p dx,$$

where $\nabla u = (\frac{\partial u}{\partial x_1},\ldots,\frac{\partial u}{\partial x_n})$ is the gradient of u and the infimum is taken over all real functions $u \in C_0^1$ (i.e. continuously differentiable and with compact support) and with the restriction $u_{|F} = 1$. In the case $p = n$, S_u is supposed to be contained in a fixed ball.

Then,

$$C_{\Phi,p}(E) = \inf_f \|f\|_p^p,$$

where $\|f\|_p = (\int f(x)^p dx)^{1/p}$, $f \in L_p^+$, i.e. $f(x) \geq 0$ is L^p-integrable and, by definition,

$$\Phi * f(x) = \int \Phi(|x - y|)f(y)dy \geq 1 \quad \text{for every} \quad x \in E \subset R^n.$$

Also

$$c_{\Phi,p}(E) = \sup_\mu \mu(E),$$

where $\mu \in L_1^+(E)$, i.e. μ is a non-negative Radon measure with $S_\mu \subset E$ and $\mu(E) < \infty$. E is measurable with respect to all such measures and

$$\|\Phi * \mu\|_{p'} = \left(\int \left(\int \Phi(|x - y|)d\mu(y)\right)^{p'} dx\right)^{1/p'} \leq 1, \quad \frac{1}{p} + \frac{1}{p'} = 1.$$

Finally,

$$\tilde{c}_{\Phi,p}(E) = \sup_\mu \mu(E),$$

where $\mu \in L_1^+(E)$. E is measurable with respect to all measures and

$$\Phi * (\Phi * \mu)^{1/(p-1)}(x) \leq 1 \quad \text{for every} \quad x \in R^n.$$

In all the above definitions for $p < n$ (except that of $c_{\Phi,p}$, which for $p = 1$ comes to c_Φ of the preceding paragraph) integration with respect to Lebesgue measure is involved. (The case $p = n$ is even more complicated and will be considered separately.) We propose to use in all these definitions, in the infinite-dimensional case, instead of Lebesgue measure, a Wiener abstract measure, which may be constructed by means of Gauss measure, which is equivalent, in the n-dimensional case, to Lebesgue measure and is considered by some authors (as for instance L. Gross) the most natural generalization of Lebesgue measure in the infinite-dimensional case. As it is well known (see L. Gross [5]), a Wiener abstract space is obtained as follows: given a real separable Hilbert space H, one completes it with respect to a measurable norm getting a Banach space B. Then, one defines Gauss measure on the cylinder sets $Z \subset H$

and the same formula yields its σ-additive (countably additive) extension (the abstract Wiener measure) to the field of all Borel sets of B, the topology of B being induced by the measurable norm allowing us to transform H in B. All this procedure is explained in detail in L. Gross' paper quoted above and in H. H. Kuo's monograph [8]. Let us remind the basic concepts:

A cylinder set $Z \subset H$ is

$$Z = \{x \in H : (<x,y_1>,\dots,<x,y_n>) \in E\},$$

where $<x_1,x_2>$ is the scalar product in H, $|x| = \sqrt{<x,x>}$ its norm and E a Borel set in R^n.

A cylinder set $Z \subset B$ is

$$Z = \{x \in B : (<x,y_1>,\dots,<x,y_n>) \in E\},$$

where this time $<x,y_k> = y_k(x)$ and $y_k \in B^*$ (the topological dual of B), $k = 1,\dots,n$. If a subspace $K \subset B^*$ contains $\{y_1,\dots,y_n\}$, then Z is said to be based on K.

Let R_B be the set of all cylinders in B and S_K the subset of all cylinders based on K.

A set function $\mu \geq 0$ on R_B is called a cylinder set measure if 1) $\mu(B) = 1$, 2) for each finite-dimensional subspace $K \subset B^*$, μ is σ-additive when restricted to the σ-ring S_K. Clearly a cylinder set measure is finitely additive on R_B.

Gauss' measure in a Hilbert space H is a cylinder set measure μ from the ring R_H to $(0,\infty)$ such that if $P_n : H \to R^n$ is an orthogonal projection and $Z = \{x \in H : P_n x \in E\}$, $E \subset R^n$ a Borel set, then

$$\mu(Z) = (\sqrt{2\pi})^{-n} \int_E \exp(-|x|^2/2)dx ,$$

where the integration is with respect to Lebesgue measure.

A norm $\|\cdot\|$ on H is called measurable if for every $\varepsilon > 0$, there is a finite-dimensional projection P_o on H such that

$$\mu\{x \in H : \|Px\| > \varepsilon\} < \varepsilon , \qquad (1)$$

whenever P is a finite-dimensional projection orthogonal to P_o (since $\{x \in H : \|Px\| > \varepsilon\}$ is a cylinder set based on the range of P, (1) makes sense).

Let us select a few results from L. Gross' paper and H. H. Kuo's monograph (quoted above) in order to justify the procedure indicated above.

The σ-algebra generated by R_B coincides with the algebra of Borel sets (H. H. Kuo [8]). Since Gauss' measure μ has not a σ-additive extension to the σ-ring generated by R_H, we have to consider a corresponding measure m on B induced by μ as follows:

$$m\{x \in B : (\langle x,y_1\rangle,\ldots,\langle x,y_n\rangle) \in E\} = \mu\{x \in H : (\langle x,y_1\rangle,\ldots,\langle x,y_n\rangle) \in E\}.$$

Then we have

Proposition 1. Let $\|\cdot\|$ be a measurable norm on a real separable Hilbert space H. Denote by B the completion of H with respect to $\|\cdot\|$ and by m the cylinder set measure on R_B induced by Gauss' measure on H. Then m is σ-additive on R_B (L. Gross [5]).

Hence we deduce that the cylinder set measure m on R_B may be extended to a σ-additive measure π on the σ-ring generated by R_B, i.e. to the algebra of all Borel sets. The probability measure π defined on the algebra of all Borel sets of B is called the abstract Wiener measure on B.

A triplet (i,H,B), where $i : H \to B$ is a continuous injection of the real Hilbert space into the real Banach space B with dense range is called an abstract Wiener space if the B-norm pulled back to H is a measurable norm.

It is important to observe that any Banach space B can arise as the third element of some abstract Wiener space. Another important result is also

Proposition 2. If (i,H,B) is an abstract Wiener space and $\|\cdot\|$ is the measurable norm with respect to which H was completed (the B-norm), then for every $\varepsilon > 0$, $\pi\{x \in B : \|x\| \leqq \varepsilon\} > 0$ (H. H. Kuo [8], Corollary 1.1(b)).

Hence, we deduce that open sets of B have strictly positive abstract Wiener measure.

If $\dim H = \infty$, then $|\cdot|$ is not a measurable norm and then the completion of H with respect to this norm will not be an abstract Wiener space. There are different ways of obtaining measurable norms. Thus, for instance, any injective Hilbert-Schmidt operator A yields a measurable norm $\|\cdot\|$ by means of the relation $\|x\| = |Ax|$, where $|\cdot| = \sqrt{\langle\cdot,\cdot\rangle}$.

We remind that a Hilbert-Schmidt operator of H is a linear operator A such that $\sum_{n=1}^{\infty}|Ae_n|^2 < \infty$, where $\{e_n\}$ is an orthonormal basis of H. Its norm is given by $\|A\|_2 = (\sum_{n=1}^{\infty}|Ae_n|^2)^{1/2}$.

Clearly, Gauss' measure μ, and then also m and the abstract Wiener measure π are cylinder measures and if for instance $Z \subset H$ is a cylinder of the form $Z = \prod_{k=1}^{n} E_k \times (\prod_{k=n+1}^{\infty} X_k) \subset H$, where $X_k = R$ $(k = 1,2,\ldots)$, then

$$\mu(Z) = \prod_{k=1}^{n} \mu_k(E_k),$$

where μ_k is Gauss' measure on R.

Now, let us remind that a triple $(X,\mathcal{A},\mu)$ is said to be a measure space if X is a set, $\mathcal{A}$ is an algebra of subsets of X and $\mu \geq 0$ is a σ-additive and σ-finite measure defined on $\mathcal{A}$. If μ is a measure defined on $\mathcal{A}$ such that $B \in \mathcal{A}$ whenever $E \in \mathcal{A}$, $\mu(E) = 0$ and $B \subset E$ i.e. all subsets of measure zero are measurable, then μ is said to be a complete measure and $(X,\mathcal{A},\mu)$ is called a complete measure space. A function $f : X \to R$ is called measurable (some authors use to say μ-measurable or $\mathcal{A}$-measurable) on X if, for every open set $D \subset R$, the set $\{x \in X : f(x) \in D\} \in \mathcal{A}$.

Proposition 3. Let $J_n = \{1,...,n\}$, and $(X_k,\mathcal{A}_k,\mu_k)$ be a measure space for every $k \in J_n$ and $\mathcal{A}_{J_n}$ the family of all cartesian products $E_{J_n} = E_1 \times \cdots \times E_n$ with $E_k \in \mathcal{A}_k$. Then there is a unique measure μ_{J_n} on $\mathcal{A}_{J_n}$ such that

$$\mu_{J_n}\left(\prod_{k=1}^{n} E_k\right) = \prod_{k=1}^{n} \mu_k(E_k)$$

for all $\prod_{k=1}^{n} E_k \in \mathcal{A}_{J_n}$ (E. Hewitt and K. Stromberg [6], Lemma 22.5).

Next, let us introduce (according to [6], § 12) the concept of abstrac Lebesgue integral.

We remind that a measurable dissection of X is any finite, pairwise disjoint family $\{E_1,...,E_n\} \subset \mathcal{A}$ such that $\bigcup_{k=1}^{n} E_k = X$. Let $(X,\mathcal{A},\mu)$ be a measure space and f be any function from X into $[0,\infty]$. Define

$$L(f) = \sup\{\sum_{k=1}^{n} \inf_{x \in E_k} f(x)\, \mu(E_k) : E_1,...,E_n \text{ a measurable dissection of } X$$

Here $\inf \emptyset = 0$. For an extended real-valued function f, we define $f^+ =$ $\max(f,0)$, $f^- = -\min(f,0)$, $f = f^+ - f^-$.

The abstract Lebesgue integral (or simply the integral) of f is $L(f) = L(f^+) - L(f^-)$ provided that at least one of the numbers $L(f^+)$ and $L(f^-)$ is finite. If $L(f^+) = L(f^-) = \infty$, then we do not define $L(f)$ The functional L is ordinarily written in integral notation:

$$L(f) = \int_X f(x)d\mu(x) = \int_X f\, d\mu = \int f\, d\mu .$$

Then, let $L_1^r(X,\mathcal{A},\mu)$ be the set of all μ-measurable real-valued functions f defined μ-a.e. (μ-almost everywhere) on X (i.e. everywhere except in a set of μ-measure zero) such that $\int f\, d\mu$ exists and is finite (when confusion seems impossible, we will write simply L_1^r).

Let $\{(X_n,\mathcal{A}_n,\mu_n)\}$ denote a sequence of measure spaces such that

$\mu_n(X_n) = 1$ for every $n \in N$, $X = \prod_{n\in N} X_n$, $\mathcal{N}$ be the smallest algebra (not σ-algebra) of subsets of X that contains all sets of the form $E_{J_n} \times X_{J_n'}$, where $J_n = \{1,\dots,n\}$, $J_n' = \{n+1,n+2,\dots\} = N - J_n$, $E_k \in \mathcal{A}_k$, $E_{J_n} = \prod_{k=1}^{n} E_k$, $X_{J_n'} = \prod_{k=n+1}^{\infty} X_k$ and finally, $\mathcal{M}$ be the smallest σ-algebra of subsets of X such that it contains the algebra $\mathcal{N}$. With these notations, we have

Proposition 4. There is a unique finitely additive measure μ on the algebra of sets $\mathcal{N}$ such that

$$\mu(E_{J_n} \times X_{J_n'}) = \mu_{J_n}(E_{J_n})$$

for all $n \in N$ and all $E_{J_n} \in \mathcal{A}_{J_n}$ (E. Hewitt and K. Stromberg [6], Theorem 22.7).

Proposition 5. The finitely additive measure μ on $\mathcal{N}$ admits a unique extension over $\mathcal{M}$ that is σ-additive (E. Hewitt and K. Stromberg [6], Theorem 22.8).

Proposition 6. Let $(X,\mathcal{M},\mu)$ be the measure space from above and $f \in L_1^r(X,\mathcal{M},\mu)$. Then

$$\int f(x)\,d\mu(x) = \lim_{n\to\infty} \int f(x_{J_n}, x_{J_n'})\,d\mu_{J_n}(x_{J_n}) \tag{2}$$

holds μ-a.e. in X, where $x_{J_n} = (x_1,\dots,x_n)$, $x_{J_n'} = (\tilde{x}_{n+1},\tilde{x}_{n+2},\dots)$ and $d\mu_{J_n}(x_{J_n}) = d\mu_1(x_1)\cdots d\mu_n(x_n)$ (E. Hewitt and K. Stromberg [6], Theorem 22.22).

Lemma 1. Let $(X,\mathcal{M},\mu)$ be the measure space from above. Suppose it is a normed space such that $E_o \subset X$ with $\mu(E_o) = 0$ implies $\overline{X - E_o} = X$ and $f \in L_1^r(X)$ is also uniformly continuous, then the relation (2) holds everywhere in X.

Proof. Let us denote $\int f(x)\,d\mu(x) = \alpha$ and let $E_\alpha \subset X$ be the set of points where (2) holds. Suppose, to prove it is false, that there exists a point $x^o \in CE_\alpha$, where $\lim_{n\to\infty} \int f(x_{J_n}, x^o_{J_n'})\,d\mu_{J_n}(x_{J_n})$ does not exist or is different of α. Then there is a sequence $\{n_k\}$, $n_k \in N$ such that

$$\lim_{k\to\infty} \int f(x_{J_{n_k}}, x^o_{J'_{n_k}})\,d\mu_{J_{n_k}}(x_{J_{n_k}}) = \alpha' \neq \alpha ,$$

where $x^o_{J'_{n_k}} = (x^o_{n_k+1}, x^o_{n_k+2},\dots)$. Hence, since f is uniformly continuous and $\overline{X - E_\alpha} = X$, we deduce that given $\varepsilon > 0$, there exists a $\delta = \delta(\varepsilon)$ such that for $x \in E_\alpha$ with $|x - x^o| < \delta$, we can find a $k_o = k_o(\tilde{x},x^o,\varepsilon)$ so that, for $k > k_o$, we have

$$|\alpha - \alpha'| = |\alpha - \int f(x_{J_{n_k}}, x_{J'_{n_k}}) d\mu_{J_{n_k}}(x_{J_{n_k}}) + \int f(x_{J_{n_k}}, x_{J'_{n_k}}) d\mu_{J_{n_k}}(x_{J_{n_k}}) -$$

$$\int f(x_{J_{n_k}}, x^{o}_{J'_{n_k}}) d\mu_{J_{n_k}}(x_{J_{n_k}}) + \int f(x_{J_{n_k}}, x^{o}_{J'_{n_k}}) d\mu_{J_{n_k}}(x_{J_{n_k}}) - \alpha'|$$

$$< 2\varepsilon + \int |f(x_{J_{n_k}}, x_{J'_{n_k}}) - f(x_{J_{n_k}}, x^{o}_{J'_{n_k}})| d\mu_{J_{n_k}}(x_{J_{n_k}})$$

$$\leq 2\varepsilon + \varepsilon\mu_{J_{n_k}}(X_{J_{n_k}}) < \varepsilon(2 + \mu_{J_{n_k}}(X_{J_{n_k}})) = 3\varepsilon,$$

and letting $\varepsilon \to 0$, we get $\alpha' = \alpha$. This contradiction allows us to conclude that

$$\Phi(x) = \lim_{n\to\infty} \int f(x_{J_n}, x_{J'_n}) d\mu_{J_n}(x_{J_n}) = \alpha = \int f(x) d\mu(x)$$

everywhere in X, as desired.

Corollary 1. In the hypotheses of the preceding lemma,

$$\left(\int |f(x)|^p d\mu(x)\right)^{1/p} = \lim_{n\to\infty} \left(\int |f(x_{J_n}, x_{J'_n})|^p d\mu_{J_n}(x_{J_n})\right)^{1/p}.$$

Corollary 2. The preceding lemma and corollary hold if X is a real separable Banach space and π is the corresponding abstract Wiener measure.

From the above considerations about abstract Wiener spaces, and the preceding corollary, it follows that we may deduce the definition of some capacities in a Banach space if, in the conditions of the corresponding definition in R^n, expressed by means of Lebesgue measure, we change it in a Gaussian measure and then let $n \to \infty$.

Theorem 1. The following definitions may be obtained from the corresponding ones in R^n by taking a Wiener abstract measure π instead of the Lebesgue one:

$$c_{\Phi,p}(E) = \sup_{\mu} \mu(X),$$

where X is a real separable Banach space, $\mu \in L_1^+(E)$ and

$$\left(\int\left(\int \Phi(|x - y|) d\mu(y)\right)^{p'} d\pi(x)\right)^{1/p'} \leq 1 ;$$

$$\tilde{c}_{\Phi,p}(E) = \sup_{\mu} \mu(X),$$

where X is a real separable Banach space, $\mu \in L_1^+(E)$ and

$$\int \Phi(|x - y|)\left(\int \Phi(|y - z|) d\mu(z)\right)^{1/(p-1)} d\pi(y) \leq 1 \quad \text{for every } x \in X .$$

Now, let us remind some other results and prove another lemma in order to be able to extend also the concepts of p-moduli and p-extremal length.

Proposition 7. Let F be a compact set in a topological space, let E be an arbitrary set and let Φ be a real-valued function defined on $F\times E$ and continuous on F for every $y\in E$. Then the following statements are equivalent:

(a) For every $\beta\in R$ and $y_1,\dots,y_n\in E$ such that $\beta > \max_{x\in F}\min_{1\leq i\leq m}\Phi(x,y_i)$, there is an $y_o\in E$ such that $\beta > \max_{x\in F}\Phi(x,y_o)$;

(b) Φ satisfies the relation

$$\max_{x\in F}\inf_{y\in E}\Phi(x,y) = \inf_{y\in E}\max_{x\in F}\Phi(x,y).$$

(See for instance the monograph of V. Barbu and T. Precupanu [1], Ch. 2, Theorem 3.4.)

Lemma 2. In the hypotheses of proposition 6, if $\mathcal{C}$ is the class of functions $f(x)\geq 0$, $f\in L_1^r(X,\mathcal{M},\mu)$, then

$$\lim_{n\to\infty}\inf_f\int_{X_{J_n'}}\inf f(x_{J_n},x_{J_n'})d\mu_{J_n}(x_{J_n}) = \inf_f\int f(x)d\mu(x). \tag{3}$$

Proof. First, let us prove that

$$\inf_f\int f(x)d\mu(x) = \inf_f\lim_{n\to\infty}\int_{X_{J_n'}}\inf f(x_{J_n},x_{J_n'})d\mu_{J_n}(x_{J_n}). \tag{4}$$

Indeed, on account of Proposition 6, we have

$$\lim_{n\to\infty}\int_{X_{J_k'}}\inf f(x_{J_k},x_{J_{kn}},x_{J_n'})d\mu_{J_n}(x_{J_n}) \leq \lim_{n\to\infty}\int_{X_{J_n'}}\inf f(x_{J_n},x_{J_n'})d\mu_{J_n}(x_{J_n})$$
$$\leq \lim_{n\to\infty}\int f(x_{J_n},x_{J_n'})d\mu_{J_n}(x_{J_n}) = \int f(x)d\mu(x)$$

μ-a.e., where $x_{J_{kn}} = (x_{k+1},\dots,x_n)$. Hence, since $\inf_{X_{J_k'}} f = f_k\in\mathcal{C}$, we deduce that

$$\inf_f\int f\,d\mu = \inf_f\lim_{n\to\infty}\int f(x_{J_n},x_{J_n'})d\mu_{J_n}(x_{J_n}) \leq$$
$$\inf_f\lim_{n\to\infty}\int_{X_{J_k'}}\inf f(x_{J_k},x_{J_{kn}},x_{J_n'})d\mu_{J_n}(x_{J_n}) \leq \inf_f\lim_{n\to\infty}\int_{X_{J_n'}}\inf f(x_{J_n},x_{J_n})d\mu_{J_n}(x_{J_n'})$$
$$\leq \inf_f\int f\,d\mu$$

μ-a.e., so that, since the extreme parts of the inequality do not depend on x and coincide, we get (4), and then (3) may be written also as

$$\lim_{n\to\infty}\inf_f\int_{X_{J_n'}}\inf f(x_{J_n},x_{J_n'})d\mu_{J_n}(x_{J_n}) = \inf_f\lim_{n\to\infty}\int_{X_{J_n'}}\inf f(x_{J_n},x_{J_n'})d\mu_{J_n}(x_{J_n}).$$

Next, let us show that

$$\lim_{n\to\infty}\int \inf_{X_{J'_n}} f(x_{J_n},x_{J'_n})d\mu_{J_n}(x_{J_n}) = \sup_n \int \inf_{X_{J'_n}} f(x_{J_n},x_{J'_n})d\mu_{J_n}(x_{J_n}). \tag{5}$$

Indeed,

$$\int \inf_{X_{J'_n}} f(x_{J_n},x_{J'_n})d\mu_{J_n}(x_{J_n}) \leq \int \inf_{X_{J'_{n+1}}} f(x_{J_n},x_{n+1},x_{J'_{n+1}})d\mu_{J_n}(x_{J_n}),$$

hence, integrating with respect to x_{n+1} and taking into account that $\mu_{n+1}(X_{n+1}) = 1$, we obtain

$$\int \inf_{X_{J'_n}} f(x_{J_n},x_{J'_n})d\mu_{J_n}(x_{J_n}) \leq \int \inf_{X_{J'_{n+1}}} f(x_{J_{n+1}},x_{J_{n+1}})d\mu_{J_{n+1}}(x_{J_{n+1}})$$

and since this happens for every $n \in N$, we deduce (5).

We are able to obtain even more, i.e.

$$\lim_{k\to\infty}\int \inf_{X_{J'_{n_k}}} f(x_{J_{n_k}},x_{J'_{n_k}})d\mu_{J_{n_k}}(x_{J_{n_k}}) = \sup_k \int \inf_{X_{J'_{n_k}}} f(x_{J_{n_k}},x_{J'_{n_k}})d\mu_{J_{n_k}}(x_{J_{n_k}}) \tag{6}$$

for any sequence $\{n_k\}$. Indeed, since $\{\int \inf_{X_{J'_n}} f(x_{J_n},x_{J'_n})d\mu_{J_n}(x_{J_n})\}$ is a non-decreasing infinite sequence of non-negative numbers, any infinite subsequence $\{\int \inf_{X_{J'_{n_k}}} f(x_{J_{n_k}},x_{J'_{n_k}})d\mu_{J_{n_k}}(x_{J_{n_k}})\}$ has the same limit. Hence arguing as above, it follows that also (6) holds, so that, on account of (4) and of the equality of the limit of the above sequence and of its subsequence, we get

$$\inf_f \sup_k \int \inf_{X_{J'_{n_k}}} f(x_{J_{n_k}},x_{J'_{n_k}})d\mu_{J_{n_k}}(x_{J_{n_k}}) = \inf_f \int f(x)d\mu(x) \tag{7}$$

for every infinite sequence $\{n_k\}$ $(n_k \in N)$.

Next, by means of the preceding proposition, let us derive

$$\inf_f \sup_k \int \inf_{X_{J'_{n_k}}} f(x_{J_{n_k}},x_{J'_{n_k}})d\mu_{J_{n_k}}(x_{J_{n_k}}) = \sup_k \inf_f \int \inf_{X_{J'_{n_k}}} f(x_{J_{n_k}},x_{J'_{n_k}})d\mu_{J_{n_k}}(x_{J_{n_k}}). \tag{8}$$

Let us show first that we are in the hypotheses of the preceding proposition. Let us denote $\alpha_n = 1/n$ and let us consider the compact set $F = \{0\} \cup \{\alpha_n : n = 1,2,\ldots\}$. This is evidently a compact set. Now, let us introduce the function

$$\Phi(\alpha,f) = \begin{cases} \int \inf_{X_{J'_n}} f(x_{J_n},x_{J'_n})d\mu_{J_n}(x_{J_n}) & \text{for } \alpha = \alpha_n, \\ \lim_{n\to\infty}\int \inf_{X_{J'_n}} f(x_{J_n},x_{J'_n})d\mu_{J_n}(x_{J_n}) & \text{for } \alpha = 0. \end{cases}$$

Then, let us identify $\mathcal{C}$ with E and let us consider a number $\beta > 0$ and $f_1,\dots,f_m \in \mathcal{C}$ so that $\beta > \max_{\alpha\in F} \min_{1\le i\le m} \Phi(\alpha,f_i)$. Let us prove now that there exists a function $f_o \in \mathcal{C}$ with the property that $\beta > \max_{\alpha\in F} \Phi(\alpha,f_o)$. From (7), we deduce that

$$\Phi(\alpha,f_i) \leqq \Phi(0,f_i) = \max_{\alpha\in F} \Phi(\alpha,f_i) = \lim_{n\to\infty} \Phi(\alpha_n,f_i),$$

hence

$$\max_{\alpha\in F} \min_{1\leqq i\leqq m} \Phi(\alpha,f_i) = \lim_{n\to\infty} \min_{1\leqq i\leqq m} \Phi(\alpha_n,f_i) = \min_{1\leqq i\leqq m} \Phi(0,f_i).$$

Indeed, let us suppose that the maximum is attained for $\alpha = \alpha_p$, i.e.

$$\max_{\alpha\in F} \min_{1\leqq i\leqq m} \Phi(\alpha,f_i) \leqq \min_{1\leqq i\leqq m} \Phi(\alpha_p,f_i) = \Phi(\alpha_p,f_{i_o}),$$

but

$$\Phi(\alpha_p,f_i) \leqq \Phi(0,f_i), \quad i = 1,\dots,m,$$

hence

$$\Phi(\alpha_p,f_{i_o}) \leqq \min_{1\leqq i\leqq m} \Phi(0,f_i),$$

so that the maximum is attained for $\alpha = 0$.

Now, for every $n \in N$, let us define f_n by means of the relation

$$\min_{1\leqq i\leqq m} \Phi(\alpha_n,f_i) = \Phi(\alpha_n,f_n).$$

Clearly,

$$\{\Phi(\alpha_n,f_n)\} = \{\Phi(\alpha_{1_k},f_1)\} \cup\cdots\cup \{\Phi(\alpha_{m_k},f_m)\},$$

where $\{\alpha_n\} = \{\alpha_{1_k}\} \cup\cdots\cup\{\alpha_{m_k}\}$, in other words, $\{\Phi(\alpha_{i_k},f_i)\}$ for instance contains those elements of $\{\Phi(\alpha_n,f_n)\}$ for which $n = i_k$ implies $f_n = f_i$. Since $\{\Phi(\alpha_n,f_n)\}$ is an infinite sequence, it follows that at least one of the m subsequences, let us say $\{\Phi(\alpha_{i_k},f_i\}$, will be infinite and the existence of the limit of $\{\Phi(\alpha_n,f_n)\}$ will imply the existence of the limit for any of its infinite subsequences and the equality of all these limits, so that

$$\beta > \max_{\alpha\in F} \min_{1\leqq i\leqq m} \Phi(\alpha,f_i) = \lim_{n\to\infty} \Phi(\alpha_n,f_n) = \lim_{k\to\infty} \Phi(\alpha_{i_k},f_i) = \max_{\alpha\in F} \Phi(\alpha,f_i),$$

where the last equality is a consequence of the existence of the limit of the sequence $\{\Phi(\alpha_n,f_i)\}$; thus, condition (a) of the preceding proposition is verified for $f_o = f_i$, and then relation (8) is true. But this implies

$$\lim_{n\to\infty} \inf_f \int_{X_{J_n'}} \inf f(x_{J_n},x_{J_n'})d\mu_{J_n}(x_{J_n}) = \inf_f \int f(x)d\mu(x) = C_o.$$

Indeed, let us suppose, to prove it is false, that

$$\lim_{n\to\infty} \inf_{f} \int_{X_{J'_n}} \inf f(x_{J_n}, x_{J'_n}) d\mu_{J_n}(x_{J_n}) = C_o$$

does not hold; then it would exist an $\varepsilon > 0$ and a sequence $\{n_k\}$ so that, for every k,

$$|C_o - \inf_{f} \int_{X_{J'_{n_k}}} \inf f(x_{J_{n_k}}, x_{J'_{n_k}}) d\mu_{J_{n_k}}(x_{J_{n_k}})| \geqq \varepsilon,$$

which would contradict (8), which, on account of (7), implies for every subsequence $\{n_k\}$,

$$\sup_{k} \inf_{f} \int_{X_{J'_{n_k}}} \inf f(x_{J_{n_k}}, x_{J'_{n_k}}) d\mu_{J_{n_k}}(x_{J_{n_k}}) = C_o$$

and our lemma is completely proved.

Corollary. In the hypotheses of the preceding lemma, if F is a compact set of X and $\mathcal{U}$ is a class of admissible functions for F, i.e. $u \in C_o^1$ such that $u_{|F} = 1$ and $|\nabla u| \in L_\mu^p$, then

$$\inf_{u} \int |\nabla u(x)|^p d\mu(x) = \lim_{n\to\infty} \inf_{u} \int_{X_{J'_n}} \inf |\nabla u(x_{J_n}, x_{J'_n})|^p d\mu_{J_n}(x_{J_n}),$$

where $\nabla u(x) = (\frac{\partial u(x)}{\partial x_1}, \frac{\partial u(x)}{\partial x_2}, \ldots.)$ is the gradient of u.

This corollary (taking into account also the considerations on abstract Wiener spaces) yields (in a similar way as theorem 1)

Theorem 2. The following definition of p-capacity of a compact set F of a real separable Banach space X may be deduced from the corresponding definitions in R^n by taking Wiener abstract measure π instead of the Lebesgue one, i.e.

$$\mathrm{cap}_p F = \inf_{u \in \mathcal{U}} \int |\nabla u(x)|^p d\pi(x),$$

where $\mathcal{U}$ is the class of admissible functions for F.

All the concepts of capacity given in this paragraph clearly satisfy conditions (i), (ii), (iii) characterizing the capacity and mentioned at the beginning of the paragraph.

Now, if in the definition of the class $F(\Gamma)$ of functions ρ, involved in the definition of the p-modulus and of the p-extremal length, we suppose ρ continuous (not only Borel measurable) and denote the corresponding p-modulus and p-extremal length by $\tilde{M}_p(\Gamma)$ and $\tilde{\lambda}_p(\Gamma)$, respectively, then, we deduce as above (using Wiener measure instead of the Lebesgue one and then letting $n \to \infty$)

Theorem 3. By the same procedure as in the preceding two theorems, we obtain the following definition of p-modulus and of p-extremal length of an arc family Γ contained in a real separable Banach space X:

The p-modulus is

$$\tilde{M}_p(\Gamma) = \inf_{\rho \in \tilde{F}(\Gamma)} \int \rho(x)^p d\pi(x)$$

where π is an abstract Wiener measure and $\tilde{F}(\Gamma)$ is the class of continuous functions $\rho(x) \geqq 0$ such that $\int \rho(s)ds \geqq 1$ for every $\gamma \in \Gamma$, the integral being taken with respect to linear Hausdorff measure.

The p-extremal length is

$$\tilde{\lambda}_p(\Gamma) = (\tilde{M}_p(\Gamma))^{-1}.$$

The restriction of supposing ρ continuous is, at least in some cases, only apparent as it follows from our previous result in [2]:

Proposition 8. If the domain $D \subset R^n$ is bounded, $C_o, C_1 \subset \bar{D}$ are two closed disjoint sets and Γ is the family of the arcs joining C_o and C_1 in $D - C_o - C_1$, then $\tilde{M}_p(\Gamma) = M_p(\Gamma)$.

The arc family Γ, involved in the preceding proposition, contains as particular cases the arc family Γ_A joining the boundary components of a ring A and the arc family Γ_Z joining the bases of a topological cylinder Z.

In order to generalize also the capacity $C_{\Phi,p}$ in the way described above, it would be necessary to ask the more restrictive condition of the continuity of f instead of $f \in L_p^+$.

And now, let us find a method of extending the concept of conformal capacity, modulus and extremal length, i.e. of p-capacity, p-modulus and p-extremal length, respectively in the case $p = n$.

Let us remind first that a μ-measurable function f is called essentially bounded if there exists a constant $\alpha \geqq 0$ such that $|f(x)| \leqq \alpha$ μ-a.e. The infimum of all these α is said to be the essential supremum of $|f(x)|$ and is denoted $\operatorname{ess\,sup}_{x\in X}|f(x)|$ or $\operatorname{vrai\,max}_{x\in X}|f(x)|$.

The space $L^\infty(X,\mathcal{M},\mu)$ or simply $L^\infty(X)$ is the set of all μ-measurable essentially bounded functions given μ-a.e. in X. In this space one uses the norm $\|f\|_\infty = \operatorname{ess\,sup}_{x\in X}|f(x)|$.

Proposition 9. If X is a topological space, $\mu(X) < \infty$ and $f \in L^\infty(X)$, then

$$\lim_{p\to\infty} \left(\int |f(x)|^p d\mu(x)\right)^{1/p} = \operatorname*{ess\,sup}_{x\in X} |f(x)|.$$

(For the proof, see for instance K. Yosida [11], Ch. I, §3, Theorem 1.)

Corollary. In the hypotheses of the preceding proposition, if $f \in C(X;R)$ and μ is such that $E_o \subset X$ with $\mu(E_o) = 0$ implies $\overline{X - E_o} = X$, then

$$\lim_{p\to\infty} \left(\int |f(x)|^p d\mu(x)\right)^{1/p} = \sup_X |f(x)|.$$

Lemma 3. In the hypotheses of the preceding corollary

$$\lim_{p\to\infty}\lim_{n\to\infty} \left(\int_{X_{J'_n}} \sup |f(x_{J_n}, x_{J'_n})|^p d\mu_{J_n}(x_{J_n})\right)^{1/p} =$$

$$\lim_{n\to\infty}\lim_{p\to\infty} \left(\int_{X_{J'_n}} \sup |f(x_{J_n}, x_{J'_n})|^p d\mu_{J_n}(x_{J_n})\right)^{1/p} = \sup_X |f(x)|.$$

Proof. Indeed, proposition 6 and the preceding corollary imply

$$\sup_X |f(x)| = \lim_{p\to\infty}\left(\int |f(x)|^p d\mu(x)\right)^{1/p} = \lim_{p\to\infty}\lim_{n\to\infty} \left(\int |f(x_{J_n}, x_{J'_n})|^p d\mu_{J_n}(x_{J_n})\right)^{1/p}$$

$$\leq \lim_{p\to\infty}\lim_{n\to\infty} \left(\int_{X_{J'_n}} \sup |f(x_{J_n}, x_{J'_n})|^p d\mu_{J_n}(x_{J_n})\right)^{1/p}$$

$$\leq \lim_{p\to\infty}\lim_{n\to\infty} \left(\int_X \sup |f(x)|^p d\mu_{J_n}(x_{J_n})\right)^{1/p}$$

$$\leq \sup_X |f(x)| \lim_{p\to\infty}\lim_{n\to\infty} \left(\int d\mu_{J_n}(x_{J_n})\right)^{1/p} = \sup_X |f(x)|;$$

from the preceding corollary we also deduce

$$\lim_{n\to\infty}\lim_{p\to\infty} \left(\int_{X_{J'_n}} \sup |f(x_{J_n}, x_{J'_n})|^p d\mu_{J_n}(x_{J_n})\right)^{1/p} = \lim_{n\to\infty} \sup_{X_{J_n}} \sup_{X_{J'_n}} |f(x_{J_n}, x_{J'_n})|$$

$$= \lim_{n\to\infty} \sup_X |f(x)| = \sup_X |f(x)|.$$

Corollary 1. In the hypotheses of the preceding lemma,

$$\lim_{n,p\to\infty} \left(\int_{X_{J'_n}} \sup |f(x_{J_n}, x_{J'_n})|^p d\mu_{J_n}(x_{J_n})\right)^{1/p} = \sup_X |f(x)|.$$

Corollary 2. In the hypotheses of the preceding lemma,

$$\lim_{n\to\infty} \left(\int_{X_{J'_n}} \sup |f(x_{J_n}, x_{J'_n})|^n d\mu_{J_n}(x_{J_n})\right)^{1/n} = \sup_X |f(x)|.$$

Lemma 4. In the hypotheses of lemma 3, if $f \geq 0$,

$$\inf_f \sup_X f(x) = \lim_{n\to\infty} \inf_f \left(\int_{X_{J'_n}} \sup f(x_{J_n}, x_{J'_n})^n d\mu_{J_n}(x_{J_n})\right)^{1/n}.$$

Since the argument is similar to that of lemma 2, we shall point out only the part involving some new difficulties. Thus, for instance, this time, in order to prove that

$$\lim_{n\to\infty}\Big(\int_{X_{J'_n}}\sup f(x_{J_n},x_{J'_n})^n d\mu_{J_n}(x_{J_n})\Big)^{1/n} = \sup_k\Big(\int_{X_{J'_{n_k}}}\sup f(x_{J_{n_k}},x_{J'_{n_k}})^{n_k} d\mu_{J_{n_k}}(x_{J_{n_k}})\Big)^{\frac{1}{n_k}},$$

we observe that, since $\mu_n(X_n) = 1$,

$$\Big(\int_{X_{J'_n}}\sup f(x_{J_n},x_{J'_n})^n d\mu_{J_n}(x_{J_n})\Big)^{1/n} \leqq \Big(\int \sup_X f(x_{J_n},x_{J'_n})^n d\mu_{J_n}(x_{J_n})\Big)^{1/n} =$$

$$\sup_X f(x)\Big(\int d\mu_{J_n}(x_{J_n})\Big)^{1/n} = \sup_X f(x) = \lim_{n\to\infty}\Big(\int_{X_{J'_n}}\sup f(x_{J_n},x_{J'_n})^n d\mu_{J_n}(x_{J_n})\Big)^{1/n}$$

and then

$$\Big(\int_{X_{J'_n}}\sup f(x_{J_n},x_{J'_n})^n d\mu_{J_n}(x_{J_n})\Big)^{1/n} \leqq \sup_n\Big(\int_{X_{J'_n}}\sup f(x_{J_n},x_{J'_n})^n d\mu_{J_n}(x_{J_n})\Big)^{1/n}$$

$$\leqq \lim_{n\to\infty}\Big(\int_{X_{J'_n}}\sup f(x_{J_n},x_{J'_n})^n d\mu_{J_n}(x_{J_n})\Big)^{1/n},$$

and letting $n \to \infty$, the extremal members of the inequality become equal so that the last two members (which do not depend on n) have to be equal too. But hence, we may deduce even more, i.e. that for any sequence $\{n_k\}$ with $n_k \in N$,

$$\sup_n\Big(\int_{X_{J'_n}}\sup f(x_{J_n},x_{J'_n})^n d\mu_{J_n}(x_{J_n})\Big)^{1/n} = \sup_k\Big(\int_{X_{J'_{n_k}}}\sup f(x_{J_{n_k}},x_{J'_{n_k}})^{n_k} d\mu_{J_{n_k}}(x_{J_{n_k}})\Big)^{\frac{1}{n_k}}$$

$$= \lim_{n\to\infty}\Big(\int_{X_{J'_n}}\sup f(x_{J_n},x_{J'_n})^n d\mu_{J_n}(x_{J_n})\Big)^{1/n},$$

and the rest of the argument is as in lemma 2.

Corollary. In the hypotheses of the preceding lemma, if F is a compact set in X and u are admissible for F with the additional condition that, for every $n \in N$, the n-dimensional orthogonal projection P_n of the support S_u is contained in a fixed n-dimensional ball $B_n(R)$, then

$$\inf_u \sup_X |\nabla u(x)| = \lim_{n\to\infty}\inf_u\Big(\int_{X_{J'_n}}\sup |\nabla u(x_{J_n},x_{J'_n})|^n d\mu_{J_n}(x_{J_n})\Big)^{1/n}.$$

Arguing as for the other theorems, we have

Theorem 4. By the same procedure as in the other theorems, we obtain the following definition of the conformal capacity, the modulus and the extremal length in a real separable Banach space X:

The conformal capacity of a compact set $F \subset X$ is of the form

$$\text{cap } F = \inf_u \sup_X |\nabla u(x)|,$$

where the infimum is taken over all admissible functions u with $P_n(S_u)$ $\subset B_n(R)$.

The modulus $\tilde{M}(\Gamma)$ of an arc family of X is given by

$$\tilde{M}(\Gamma) = \inf_{\rho\in\tilde{F}(\Gamma)} \sup_X \rho(x).$$

The extremal length of Γ is defined as

$$\tilde{\lambda}(\Gamma) = (\tilde{M}(\Gamma))^{-1}.$$

Of course, in these definitions, we did not take as starting point the corresponding definitions of cap, $\tilde{M}$ and $\tilde{\lambda}$ in R^n (with Gauss' measure instead of the Lebesgue one), but $(\text{cap})^{1/(n-1)}$, $\tilde{M}^{1/(n-1)}$ and $\tilde{\lambda}^{1/(n-1)}$, or $(\text{cap})^{1/n}$, $\tilde{M}^{1/n}$ and $\tilde{\lambda}^{1/n}$ (which asymptotically comes to the same) in order to eliminate the unpleasant effect of the fact that, for instance, $\rho(x)^n$ becomes 0, 1 or ∞ as $\rho(x)$ is < 1, $= 1$ or > 1. This additional modification is not essential for two reasons: first because it is not so important to know the exact value of the conformal capacity of a set, or of the modulus or the extremal length of an arc family, as it is to know if a set is exceptional or not (i.e. has conformal capacity 0 or not), or if an arc family is exceptional or not (i.e. has modulus 0 or not), and the class of exceptional sets or arc families does not change if we take $(\text{cap})^{1/n}$ or $\tilde{M}^{1/n}$ instead of cap or $\tilde{M}$; the second reason is that the new expression $\tilde{M}^{1/(n-1)}$ (calculated with respect to Lebesgue measure) in the case of the arc family Γ_A, corresponds to the modulus of a ring introduced by F. Gehring [4] and has even in R^n some advantages compared to $\tilde{M}$. Finally, the fact that we have to use in R^n Gauss' measure instead of Lebesgue one does not matter since in R^n these two measures are equivalent and then, to exceptional sets or arc families in one case, correspond again exceptional sets or arc families in the other case.

Finally, in the more restrictive case of $f \in C(X;R)$, we have the following generalization of $C_{\Phi,p}$ $(1 < p \leqq \infty)$ and $c_{\Phi,\infty}$:

Theorem 5. Arguing as for theorems 1 and 4, we have in a real separable Banach space X:

$$C_{\Phi,p}(E) = \inf_f \left(\int f(x)^p d\pi(x)\right)^{1/p} \quad (1 < p < \infty), \quad C_{\Phi,\infty}(E) = \inf_f \sup_X f(x),$$

where the infimum is taken over all $f \in C(X;R)$, $f(x) \geqq 0$ in X and such that

$$\int \Phi(|x-y|)f(y)d\pi(y) \geqq 1 \quad \text{for every} \quad x \in X$$

and

$$c_{\Phi,\infty}(E) = \sup_{\mu} \mu(X) ,$$

where the supremum is taken over all $\mu \in L_1^+(E)$ such that

$$\iint \Phi(|x - y|)\, d\pi(x)\, d\mu(y) \leq 1 .$$

4. Capacities and moduli by means of restricted products of measures.

Another way of generalization of the capacities and the moduli is by means of "restricted products" of measures (in the sense of C. Moore [9]). By this method, it is possible to use Lebesgue measure so that we have not to make changes any more in the analytic aspect of the definitions, except for the conformal capacity and the modulus.

Let us remind the concept of restricted product of measures (the notations belong to D. Hill [7]).

Let $\{(X_n, \mathcal{M}_n, \mu_n)\}$ be a sequence of measure spaces, where μ_n are probability measures, and let us denote $(X, \mathcal{M}, \mu) = \prod_{n \in N} (X_n, \mathcal{M}_n, \mu_n)$. It follows $\mu(X) = 1$. Next, let ν_n be a σ-finite measure on $(X_n, \mathcal{M}_n)$ and let $Y_n \in \mathcal{M}_n$ be such that $0 < \nu_n(Y_n) < \infty$. Let us denote by $\mathcal{F}$ the family of finite subsets of N, ordered by inclusion $Y = \prod_{n \in N} Y_n$, $Y_F = \prod_{n \in F} Y_n$, $Y_F^* = \prod_{n \in CF} Y_n$, $X_F = \prod_{n \in F} X_n$, where $F \in \mathcal{F}$ and also we denote $S^{(Y,F)} = X_F \times Y_F^*$, $S^{(Y)} = \bigcup_{F \in \mathcal{F}} S^{(Y,F)}$. This last set is measurable since it is countable. Now, normalizing ν_n, we get $\bar{\nu}_n = \nu_n / \nu_n(Y_n)$ and let λ_n be the restriction of $\bar{\nu}_n$ to Y_n, i.e. $\lambda_n(B_n) = \nu_n(Y_n \cap B_n)/\nu_n(Y_n)$ for every $B_n \in \mathcal{M}_n$. $\bar{\nu}_n$ is a σ-finite measure, while λ_n is a probability measure on $(X_n, \mathcal{M}_n)$ and $\bar{\nu}_n(Y_n) = \lambda_n(X_n) = 1$. Finally, let us define the following infinite products of measures:

$$\nu^{(Y,F)} = (\prod_{n \in F} \bar{\nu}_n) \times (\prod_{n \in CF} \lambda_n) = \bar{\nu}_F \times \lambda_F^* .$$

Such a product is well-defined since all its factors, except a finite number of them are probability measures. We observe that $S^{(Y,F)}$ is the support of $\nu^{(Y,F)}$ and that $\nu^{(Y,F)}(Y) = 1$. Let us remind also

<u>Proposition 10.</u> For every $B \in \mathcal{M}$, $\{\nu^{(Y,F)}(B) : F \in \mathcal{F}\}$ is an increasing net so that $\lim_F \nu^{(Y,F)}(B)$ exists (possibly infinite). This limit, which is denoted by $\nu^{(Y)}(B)$ is a σ-finite σ-additive measure on $(X, \mathcal{M})$ with the support $S^{(Y)}$ and $\nu^{(Y)}(Y) = 1$ (D. Hill [7]).

The measure $\nu^{(Y)}$ is said to be the <u>restricted product of the sequence of measures</u> $\{\nu_n\}$ with respect to Y.

D. Vandev [10] considers the particular case in which $X_n = R$, $\nu_n = m_1$ (the linear Lebesgue measure) and $Y_n = [-1/2, 1/2]$. Denoting by m the restricted product of Lebesgue linear measures with respect to the infinite product K of segments [-1/2,1/2] (abusively called sometimes Hilbert cube, but which is only homeomorphic to what is usually called so), we have the following measure space:

$$(R^{\infty},m) = \prod_{n\in N} (R,m_1,[-\tfrac{1}{2},\tfrac{1}{2}]) \tag{9}$$

$$= \lim_{n\to\infty}\{ \prod_{k=1}^{n} (R,\mathcal{M}_k,m_1) \times \prod_{k=n+1}^{\infty} ([-\tfrac{1}{2},\tfrac{1}{2}],\mathcal{M}_k \cap [-\tfrac{1}{2},\tfrac{1}{2}],m_1)\},$$

where R^{∞} is the set of all sequences of real numbers. D. Vandev shows that this measure has a kind of invariance with respect to the translations. Let us denote $S_x(E) = E + x$ and $S_x\mu(E) = \mu(E + x) = \mu(S_x(E))$.

A translation S_x is said to be <u>quasi-invariant with respect to a measure</u> μ if the measure $S_x\mu$ is equivalent to μ $(S_x\mu \sim \mu)$. We remind that a measure μ is AC (absolutely continuous) with respect to a measure ν if $\nu(E) = 0$ implies $\mu(E) = 0$ and two measures μ, ν are called equivalent if $\mu(E) = 0$ if and only if $\nu(E) = 0$.

D. Vandev [10] establishes

<u>Proposition 11.</u> Let M_μ be the set of translations quasi-invariant with respect to a measure μ of R^{∞}. A translation $S_x \in M_\mu$ if and only if $x \in \ell_1$ (i.e. if and only if $\sum_{n\in N}|x_n| < \infty$). In this case, if m is the restricted product from above (corresponding to (9)), it follows that $S_x m = m$ and also that

$$C_2 \|x\|_{\ell_1} \leqq m(K \Delta S_x K) \leqq C_3 \|x\|_{\ell_1},$$

where $\|x\|_{1} = \sum_{n\in N}|x_n|$ and $E_1 \Delta E_2 = (E_1 - E_2) \cup (E_2 - E_1)$ is the symmetrical difference of E_1, E_2. For the invariance of m it is sufficient that at least one of the following conditions hold

$$\|x\|_{\ell_1} \leq \frac{1}{2}, \quad m(K \Delta S_x K) \leqq \frac{1}{4}.$$

<u>References</u>

[1] Barbu, V., Precupanu, T.: Convexity and optimisation in Banach spaces. Editura Academiei Republicii Socialiste România, Bucureşti and Noordhoff International Publishing, Leyden (1978).

[2] Caraman, P.: p-capacity and p-modulus, in "Symposia Mathematica 18". Academic Press, New York - San Francisco - London (1976), 455 - 484.

[3] Caraman, P.: Relations between capacities, Hausdorff h-measures and p-modules. Math.-Rev. Anal. numér. Théor. Approximation, Math. 7 (1978), 13 - 49.

[4] Gehring, F.: Rings and quasiconformal mappings in space. Trans.

Amer. math. Soc. 103 (1962), 353 - 393.

[5] Gross, L.: Abstract Wiener measure and infinite dimensional potential theory. Lecture Notes in Mathematics 140, Springer-Verlag, Berlin - Heidelberg - New York (1970), 84 - 116.

[6] Hewitt, E., Stromberg, K.: Real and abstract analysis. Springer-Verlag, Berlin - Heidelberg - New York (1965).

[7] Hill, D. G. B.: σ-finite invariant measures on infinite product spaces. Trans. Amer. math. Soc. 153 (1971), 347 - 370.

[8] Kuo, H.-H.: Gaussian measures in Banach spaces. Lecture Notes in Mathematics 463, Springer-Verlag, Berlin - Heidelberg - New York (1975).

[9] Moore, C. C.: Invariant measures on product spaces, in "Proc. Fifth Berkeley Sympos. Math. Statist. and Probability (Berkeley 1965/66), Vol. II". Univ. California Press, Berkeley and Los Angeles (1967), 447 - 459.

[10] Vandev, D.: Lebesgue measure in the space R^{∞} (Bulgarian), in "Proc. 3rd Spring Conf. Bulg. Math. Soc., Burgas 1974". Publishing House of the Bulgarian Academy of Sciences, Sofia (1976).

[11] Yosida, K.: Functional analysis. Springer-Verlag, Berlin - Heidelberg - New York (1971).

Facultatea de Matematică
Universitatea din Iaşi
Str. 23 August nr 11
Iaşi
Romania

AN APPLICATION OF QUASI-CONFORMAL METHODS TO A PROBLEM IN VALUE-DISTRIBUTION THEORY

David Drasin*

Introduction

The theory of quasi-conformal mappings [1] is about fifty years old, and plays an important role in many areas of analysis. I will apply these methods to a problem in R. Nevanlinna's theory of value-distribution [8], [11]. Quasi-conformal methods have been used in this subject for about forty years, beginning with work of Teichmüller, but heretofore exclusively to construct examples (cf. [7, Ch. 7], [3] and [4]). The use here is different.

We shall study functions $f(z)$ of order $\lambda < \infty$ which are extremal for Nevanlinna's deficiency relation so that

$$\sum \delta(a) = \sum \delta(a,f) = 2. \tag{1}$$

The solution to this problem for entire functions (or <u>nearly entire</u> functions: those with $\delta(\infty,f) = 1$) is well-understood ([5], [6], [12]) but much less is known for general meromorphic functions. Thus if f is nearly entire and (1) holds with $\lambda < \infty$, then λ is a positive integer and all deficiencies are integral multiples of λ^{-1}. In fact, $f(z)$ is almost an integral of

$$f_p(z) = Ae^{Bz^p} \quad (A \cdot B \neq 0). \tag{2}$$

However, F. Nevanlinna [10] showed that for general meromorphic functions, λ could be of the form $n/2$ ($n \geqq 2$ integer) with all deficiencies integral multiples of λ^{-1} and conjectured this is true for all meromorphic solutions of (1). Thus, the nearly entire functions are a small subclass of functions extremal in the sense of (1).

We shall outline a method which, given a meromorphic function $f(z)$ of finite order satisfying (1), produces a deformation by means of quasi-conformal mappings of small dilatation into a nearly entire function. This latter function, must then behave like (2), and so we obtain the following

---

* Supported in part by (U.S.) National Science Foundation.

Theorem. Let f be meromorphic in the plane of order $\lambda < \infty$ such that (1) holds. Then $\delta(a)$ is rational for all a. More precisely, there is a nonnegative integer q,

$$q \leqq [2\lambda] \tag{3}$$

($[\ ]$ = greatest integer function) such that if $\delta(a) > 0$, then

$$\delta(a) = \frac{p(a)}{q}$$

where $p(a)$ is an even number.

Corollary [13]. Under our hypotheses, the number of deficient values is $\leqq [2\lambda]$.

Here is the principle of our method. Let f satisfy the hypotheses of the Theorem. By applying a nearly-analytic change of variables ω on the Riemannian image of f, we form

$$F(z) = \omega(f(z^2)). \tag{4}$$

Now F is not meromorphic, but has (in a natural sense)

$$\delta(0,F) = \delta(\infty,F) = 1; \tag{5}$$

thus, ω deforms the deficient values of F to all lie over 0 and ∞.

It is possible to estimate the partial derivatives F_z, $F_{\bar{z}}$ and discover that the dilatation

$$|\mu_F(z)| \equiv \left|\frac{F_{\bar{z}}(z)}{F_z(z)}\right| \tag{6}$$

is small. We now apply two known principles from the theory of quasi-conformal mappings. The first is that the factorization on the right side of (4) commutes in the sense that also

$$F(z) = H_1(\phi^{-1}(z)) \tag{7}$$

where H_1 is a (genuinely) meromorphic function and ϕ a quasi-conformal homeomorphism of the sphere with $\mu_\phi = \mu_F$ ([1, Ch. 5]). Next, we are able to study F by studying H_1, since after normalization, ϕ is nearly the identity map (see [3, Ch. 2]). Since H_1 is nearly entire, our theorem comes from re-examining standard proofs for these functions. Thus the additional range of λ in the meromorphic case arises from the need to introduce z^2 in (4).

This approach has one serious weakness, in that the deformation (4) cannot be written for all z, but only for z in a sequence of annuli

$$\mathcal{A}_n = \{R_n < |z| < A_n^4 R_n\}, \tag{8}$$

where $A_n \to \infty$, $R_n \to \infty$, and (3)-(5) must be interpreted relative to the $\mathcal{A}_n$. In these $\mathcal{A}_n$, f behaves like one of the functions of F. Nevanlinna, but we are unable to connect these annuli together. If $\{\rho_n\}$ is a sequence of Pólya peaks of order λ (cf. [8], Ch. 4) so that in particular

$$\frac{T(r)}{T(\rho_n)} \leq \{1 + o(1)\}\left(\frac{r}{\rho_n}\right)^{\lambda} \qquad (K_n^{-1} \leqq \frac{r}{\rho_n} \leqq K_n) \tag{9}$$

where $K_n \to \infty$, we only know that

$$\mathcal{A}_n \subset \{\rho_n \leqq |z| \leqq K_n \rho_n\} \tag{10}$$

for each n. (We write A_n^4 in (9) for reasons that will be clear in §3).

Full details will appear elsewhere.

1. The geometry of functions having maximal deficiency sum

An important property of the functions $f_p(z)$ of (2) is that f_p partitions every annulus $\mathcal{A}_n$ (cf. (8)) with A_n, R_n large, into 2p sectors in a natural way. Each of these has equal angular measure πp^{-1}. In p of these, f_p is small, and these sectors alternate with p in which f_p is large. The value-distribution occurs near the boundary of the sectors, where all numbers $a \neq 0, \infty$ are taken on.

Our first result captures a weak form of this, but again only holds for a sparse set of annuli $\mathcal{A}_n$. Some notation is required: if D is any open set, let

$$m(r,a,D) = \frac{1}{2\pi} \int_{re^{i\theta} \in D} \log \frac{1}{|f(re^{i\theta}) - a|} d\theta,$$

$$n(r,a,D) = \sum_{\substack{f(z)=a \\ |z| \leqq r \\ z \in D}} 1,$$

$$N(r,a,D) = \int_1^r \frac{n(t,a,D)}{t} dt.$$

Proposition 1. Let f(z) satisfy (1) with $\lambda < \infty$. Then we may find $A_n \to \infty$, $R_n \to \infty$ with

$$T(A_n^4 R_n) \lesssim C A_n^{4\lambda} T(R_n) \tag{1.1}$$

such that the corresponding annuli $\mathcal{A} = \mathcal{A}_n$, given by (8)*, may be partitioned

$$\mathcal{A} = \bigcup_{j=1}^{N} E_j \quad \bigcup_{j=1}^{N} F_j ,$$

where

$$N(= N(n)) \leq [2\lambda], \tag{1.2}$$

so that the following hold:

(i) Let $\{a_i\} = \{a;\ \delta(a) > 0\}$. Then to each $j \leq N$ may be associated $b_j \in \{a_i\}$ such that

$$b_j \neq b_{j+1} \tag{1.3}$$

and

$$\sum_{k,j} n(4A^4R, a_k, \cup E_j) + n(4A^4R, b_j, F_j) + n(4A^4R, b_{j+1}, F_j) = o(1)T(R), \quad (n \to \infty) \tag{1.4}$$

(in (1.3), (1.4) and henceforth, all indexing is made mod N = mod $N(n)$);

(ii) Each set E_j, F_j is connected and simply-connected relative to $\mathcal{A}$, and intersects both components of $\partial\mathcal{A}$. As we go about $\mathcal{A}$, these sets are confronted in the order $E_1, F_1, E_2, F_2, E_3, \ldots, E_N, F_N, E_1, \ldots$.

Inequality (1.4) generalizes our discussion of the f_p; now as we cross from E_j to E_j via F_j, both b_j and b_{j+1} are (asymptotically) local Picard values.

The proof of this proposition is based on the ideas of [2], along with some simple potential theory, as used in [14]. Start with a sequence $\{\rho_n\}$ of Pólya peaks as in (9). To each n will be associated sequences ε $(= \varepsilon_n) \to 0$, M $(= M(n)) \to \infty$, and γ $(= \gamma(n)) \to 0$, all limits taken very slowly, so that (as $n \to \infty$)

$$\varepsilon^{-1} = o(M), \quad M^2 = o(\gamma T(\rho)). \tag{1.5}$$

According to [2], the components $\bigcup_m E_{km}$ of $f^{-1}\{|z - a_k| < \varepsilon\} \cap \{\rho \leq |z| \leq K\rho\}$ which contribute significantly to $m(r, a_k)$ are (asymptotically) noncompact coverings of the punctured disc $\{0 < |z - a_k| < \varepsilon\}$. The classical Ahlfors method, introduced to prove Denjoy's conjecture, gives (1.2) as an upper bound for the number of these components, at least for $r = o(K_n)\rho_n$. We recall the sequence $\gamma_n \to 0$ which, from (1.5), satisfies $\varepsilon_n^{-1} = o(1)\gamma_n T(\rho_n)$, and only consider those components such that

---

* To streamline notation, we often drop the subscript n.

$$m(r,a_k,E) > \gamma T(\rho) \tag{1.6}$$

for some such r, and then choose $\mathcal{A}$ as in (8) and (10) so that each such E connects the inner and outer boundary of $\mathcal{A}$.

The E_j are, after reindexing, these components, plus all components of $\mathcal{A} - \cup E$ whose boundary in $\mathcal{A}$ is contained in the closure of components E_{km} which belong to a single k. This essentially estimates the first sum on the left side of (1.4). Finally, F_j is the connected subset of $\mathcal{A}$ which separates E_j from E_{j+1} in $\mathcal{A}$; that b_j and b_{j+1} are asymptotically omitted in F_j comes from the argument principle and a study of Ahlfors's theory of covering surfaces [11, Ch. 13] with J. Miles's useful estimate of the length of the relative boundary [9].

It is also not too hard to see that if (1.6) holds for $E = E_j$ and $r = r_j$, then E_j must contain a component D_j' of the set

$$\mathcal{A}_n \cap \{z \in E_j;\ \log|f(z) - b_j| < -\tfrac{1}{8\pi}\gamma_n T(\rho_n)\} \tag{1.7}$$

which connects $\{|z| = R_n\}$ to $\{|z| = A_n^4 R_n\}$ in $\mathcal{A}_n$. We take D_j to be the union of D_j' to all components of $E_j - D_j'$ whose boundary in $\mathcal{A}$ is contained in $\overline{D}_j'$.

2. A nearly analytic change of variables

Let $\mathcal{A}$ be as in (8), and let

$$\mathcal{B}(=\mathcal{B}_n) = \{z : z^2 \in \mathcal{A}\}; \tag{2.1}$$

also we set $s_0^2 = R$, $s_1^2 = A^4R$, $s^2 = A^2 s_0$. The preimages of the E_j and F_j divide $\mathcal{B}$ into $2N$ (now an <u>even</u> number) of subsets, each of which joins $\{|z| = s_0\}$ to $\{|z| = s_1\}$. To limit notation, we also call these E_j, F_j*, and index mod $2N$; thus $b_{N+i} = b_i$. Proposition 1 transfers in an obvious manner to $f(z^2)$ $(z \in \mathcal{B})$.

We now construct our $F(z)$ $(= F_n(z))$, as in (4), so that the key estimate (5) holds. The reader is referred to Chapter 2 of [3] for a discussion and references to the notions from quasiconformal mappings used here. Our function F will satisfy

$$\max_{z\in\mathcal{B}} |\mu_F(z)| = o(1) \qquad (n \to \infty) \tag{2.2}$$

(where μ_F is defined in (6)) and

---

* These definitions require that the number of E_j and F_j be even.

$$n(0,F,\mathcal{B}) + n(\infty,F,\mathcal{B}) = o(1)T(A_n^2R,f) \qquad (n \to \infty); \tag{2.3}$$

(2.3) is a precise form of (5).

It is easy to define F in $\cup F_j$: use (4) with

$$\begin{cases} \omega(W) = \dfrac{W - b_j}{W - b_{j+1}} & (j \text{ even}) \\ \omega(W) = \dfrac{W - b_{j+1}}{W - b_j} & (j \text{ odd}). \end{cases} \tag{2.4}$$

Also, for $j = 1,\ldots,N$, we recall $D_j \subset E_j$ as described in (1.7) and set

$$\begin{cases} \omega(W) = W - b_j & (z \in D_j,\ j \text{ even}) \\ \omega(W) = \dfrac{1}{W - b_j} & (z \in D_j,\ j \text{ odd}); \end{cases} \tag{2.5}$$

thus

$$\mu_F(z) \equiv 0 \qquad (z \in \mathcal{B} \cap \{F_j \cup D_j\}). \tag{2.6}$$

The delicate part is to extend (4) to $\cup(E_j - D_j)$; it is here that it is essential that D_j unite both components of $\partial\mathcal{B}$.

For $1 \leq j \leq 2N$, let $h_j = b_j - b_{j+1}$; according to (1.3), $h_j \neq 0$. An elementary lemma (which is geometrically obvious, and proved in [3], pp. 106-7) shows that given $\eta > 0$, we may choose $M = M(\eta) > 1$ and a q.c. homeomorphism ω_j on the Riemann sphere with

$$\omega_j(W) = \begin{cases} W & (|W| > M^2) \\ 1 + h_jW & (|W| < M) \end{cases} \tag{2.7}$$

and

$$\|\mu_{\omega_j}(W)\|_\infty < \eta. \tag{2.8}$$

Since $1 + h_j(W - b_j)^{-1} = (W - b_{j+1})(W - b_j)^{-1}$, we observe that

$$\omega_j\left(\frac{1}{W - b_j}\right) = \begin{cases} \dfrac{1}{W - b_j} & \left(\dfrac{1}{|W - b_j|} > M^2\right) \\ \dfrac{W - b_{j+1}}{W - b_j} & \left(\dfrac{1}{|W - b_j|} < M\right). \end{cases} \tag{2.9}$$

Sequences ε_n, and γ_n have already been discussed in §1 (cf. (1.5), (1.6)), and we now take $M = M_n \to \infty$ so slowly that (1.5) holds. Mapping ω^j may also be constructed as in (2.7) - (2.9), with b_{j-1} in place of b_{j+1}.

Recall from (1.5) and (1.7) that for each j, $\{E_j - D_j\} \cap \mathcal{B}$ consists of two components, each of which meets $\{|z| = S_0\}$, $\{|z| = S_1\}$. Let E_j^+ be that which connects D_j' to F_j and E_j^- that which connects D_j' to F_{j+1}. Then (4) is completed by

$$F(z) = \begin{cases} \omega_j\left(\dfrac{1}{f(z^2) - b_j}\right) & (z \in F_j^+,\ i \text{ odd}) \\ \omega^j\left(\dfrac{1}{f(z^2) - b_j}\right) & (z \in E_j^-,\ i \text{ odd}), \end{cases}$$

and

$$F(z) = \begin{cases} \dfrac{1}{\omega_j\left(\dfrac{1}{f(z^2) - b_j}\right)} & (z \in E_j^+,\ i \text{ even}) \\ \dfrac{1}{\omega^j\left(\dfrac{1}{f(z^2) - b_j}\right)} & (z \in E_j^-,\ i \text{ even}). \end{cases}$$

Conditions (1.5) and (2.9) show that F is continuous. Assertion (2.2) follows from (2.6) and (2.8), since $M \to \infty$, $\eta \to 0$ and also ([1, p. 9]) $\mu_{\omega\circ g}(z) = \mu_\omega(f(z))$ if f is meromorphic. Finally, the key estimate (2.3) follows from (1.4) and the particular Möbius transformations used above.

3. Back to meromorphic functions

For each n, a function $\mu = \mu_n$ may be defined on the plane by

$$\begin{cases} \mu = \mu_F & (z \in \mathcal{B}) \\ \mu = 0 & (z \notin \mathcal{B}) \end{cases} \tag{3.1}$$

so that, by (2.2), $\|\mu\|_\infty = o(1)$ $(n \to \infty)$. We now recall from (7) that there exists a homeomorphism ϕ of the sphere which fixes $z = 0$, S and ∞, such that $\mu_\phi = \mu$, and

$$F(z) = H_1(\phi^{-1}(z)) \qquad (z \in \phi(\mathcal{B})), \tag{3.2}$$

where $H_1 = H_{1,n}$ is a meromorphic function. A standard normal family argument (cf. [3, Lemma 2, p. 91]) also shows that there exist constants $k = k_n \to \infty$ such that

$$\left|\frac{\phi(z)}{S}\right| = \{1 + o(1)\}\left|\frac{z}{S}\right| \qquad \left(k_n^{-1} < \left|\frac{z}{S}\right| < k_n\right). \tag{3.3}$$

If the rate at which the $\{A_n\} \to \infty$ in (3) is sufficiently restricted, we can be sure that $\{k_n^{-1}S < |z| < k_nS\} \supset \mathcal{B}$. Finally we set

$$H(z) = H_1(z) \frac{\pi(1 - a/b_\nu)}{\pi(1 - z/a_\mu)}$$

where the $\{a_\mu\}$, $\{b_\nu\}$ are the zeros and poles of H_1 in $\phi(\mathcal{B})$. Once we have H, we are essentially back in the classical Pfluger-Edrei-Fuchs situation, and we find (as $n \to \infty$) that

$$H(z) \sim cz^{2\lambda}$$

where 2λ is an integer ≥ 2. To go back from H to f is quite easy, since the changes of variables are rather explicit. Thus

$$m(R_n, a_i) \sim 2\{ j \leq N;\ b_j = a_i\} T(R_n)$$

and since (1) implies that $m(r,a) \sim \delta(a)T(r)$ for all a, the theorem is proved.

Remark. A. Baernstein has observed that one can replace our use of quasi-meromorphic functions with quasi-harmonic functions.

References

[1] Ahlfors, L. V.: Lectures on quasiconformal mappings. Van Nostrand, Princeton, N.J. (1966).

[2] Drasin, D.: A note on functions with deficiency sum two. Ann. Acad. Sci. Fenn., Ser. A I 2 (1976), 59 - 66.

[3] Drasin, D.: The inverse problem of the Nevanlinna theory. Acta math. 138 (1977), 83 - 151.

[4] Drasin, D., Weitsman, A.: Meromorphic functions with large sums of deficiencies. Advances Math. 15 (1974), 93 - 126.

[5] Edrei, A., Fuchs, W. H. J.: On the growth of meromorphic functions with several deficient values. Trans. Amer. math. Soc. 93 (1959), 292 - 328.

[6] Edrei, A., Fuchs, W. H. J.: Valeurs déficientes et valeurs asymptotiques des fonctions méromorphes. Commentarii math. Helvet. 33 (1959), 258 - 295.

[7] Gol'dberg, A. A., Ostrovskii, I. V.: The distribution of values of meromorphic functions (Russian). Izd. Nauka, Moscow (1970).

[8] Hayman, W. K.: Meromorphic functions. Oxford University Press, Oxford (1964).

[9] Miles, J.: A note on Ahlfors' theory of covering surfaces. Proc. Amer. math. Soc. 21 (1969), 30 - 32.

[10] Nevanlinna, F.: Über eine Klass meromorpher Funktionen in "Comptes rendus de septième congrès des mathématiciens scandinaves tenu a Oslo, 19 - 22 août 1929". A. W. Brøggers boktrykkeri A/S, Oslo (1930).

[11] Nevanlinna, R.: Analytic functions. Springer-Verlag, Berlin - Heidelberg - New York (1970).

[12] Pfluger, A.: Zur Defktrelation ganzer Funktionen endlicher Ordnung. Commentarii math. Helvet. 19 (1946), 91 - 104.

[13] Weitsman, A.: Meromorphic functions with maximal deficiency sum and a conjecture of F. Nevanlinna. Acta math. 123 (1969), 115 - 139.

[14] Weitsman, A.: A theorem on Nevanlinna deficiencies. Acta math. 128 (1972), 41 - 52.

Purdue University
Mathematics Department
West Lafayette, IN 47907
U.S.A.

SOME JACOBIAN VARIETIES WHICH SPLIT

Clifford J. Earle*

1. Introduction.

It is well known that the period matrix of a closed Riemann surface with respect to a canonical homology basis never splits into a direct sum. In other words, the canonically polarized Jacobian variety of a closed Riemann surface is never isomorphic (as a polarized variety) to a nontrivial product. However, a Jacobian variety may be isomorphic as a complex torus to a nontrivial product. Many examples in genus two were given by Hayashida and Nishi [2] and by Hayashida [1]. H. H. Martens [3] has cited an unpublished example of A. Weil in genus four and has raised the question whether examples exist in all genera.

In this paper we give simple examples of closed Riemann surfaces of arbitrary even genus whose Jacobian varieties are isomorphic to nontrivial products. Our cxamples have a number of special properties. For instance each surface we construct belongs to a one parameter family of closed surfaces whose Jacobian varieties split. We have no idea whether that is a general phenomenon, nor do we have any examples with odd genus.

In genus two the situation can be analysed rather completely. It is easy to prove that each surface of genus two whose Jacobian variety splits belongs to a one parameter family of such surfaces. Moreover, each such family contains surfaces of the form studied by Hayashida [1]. We will discuss these matters in a forthcoming paper.

2. The examples.

Let X be a closed Riemann surface of genus $g = 2k$, $k \geq 1$. Suppose that Y_1 and Y_2 are closed Riemann surfaces of genus k and that $f_1 : X \to Y_1$ and $f_2 : X \to Y_2$ are holomorphic maps. There are induced homomorphisms

$$J(f_j) : J(X) \to J(Y_j), \quad j = 1,2,$$

---

* This research was partly supported by a grant from the National Science Foundation.

of the Jacobian varieties, and a product homomorphism

$$J(f_1) \times J(f_2) : J(X) \to J(Y_1) \times J(Y_2) .$$

Both $J(X)$ and $J(Y_1) \times J(Y_2)$ are complex tori of dimension g. The homomorphism $J(f_1) \times J(f_2)$ will be an isomorphism of complex tori if and only if the induced map from $H_1(J(X),Z)$ to $H_1(J(Y_1) \times J(Y_2),Z)$ is an isomorphism. Since the canonical embedding of a closed Riemann surface in its Jacobian variety induces an isomorphism of first homology groups, we conclude that $J(f_1) \times J(f_2)$ is an isomorphism of complex tori if and only if

$$f_1 \times f_2 : X \to Y_1 \times Y_2 \tag{2.1}$$

induces an isomorphism of $H_1(X,Z)$ onto $H_1(Y_1,Z) \times H_1(Y_2,Z)$.

One way to obtain a surface X of genus $g = 2k$ whose Jacobian variety splits is therefore to find a surface X and holomorphic maps $f_j : X \to Y_j$, $j = 1,2$, so that the product map (2.1) induces an isomorphism of integral first homology groups. Our examples are of that form.

<u>Theorem.</u> For any complex number $t \neq 0,1$, let X be the hyperelliptic Riemann surface of genus $g = 2k$, $k \geq 1$, defined by the equation

$$y^2 = (x^n - 1)(x^n - t), \; n = 2k + 1 . \tag{2.2}$$

There are Riemann surfaces Y_1, Y_2 of genus k such that $J(X)$ is isomorphic to $J(Y_1) \times J(Y_2)$.

The theorem will be proved by finding holomorphic maps $f_j : X \to Y_j$, $j = 1,2$, so that the map (2.1) induces the required isomorphism of homology groups. The surfaces Y_j will be quotients of X by certain involutions, and the maps f_j will be obtained from the natural quotient maps.

<u>3. Some automorphisms of X.</u>

3.1. Choose a complex number λ so that $t = \lambda^{2n}$, and define involutions h_1 and h_2 on X by

$$h_1(x,y) = (\lambda^2 x^{-1}, \lambda^n x^{-n} y), \; h_2(x,y) = (\lambda^2 x^{-1}, -\lambda^n x^{-n} y) . \tag{3.1}$$

Each of these maps fixes exactly two points of X, h_1 the points where $x = \lambda$ and h_2 those where $x = -\lambda$. The quotient surfaces

$$Y_j = X_j/\{id, h_j\}, \; j = 1,2, \tag{3.2}$$

therefore have genus k.

3.2. We shall want to know the action of h_1 and h_2 on $H_1(X,Z)$, computed with respect to an appropriate homology basis. To simplify visualization we shall assume that λ is positive and greater than one. Put $\alpha = e^{2\pi i/n}$. The Riemann surface X is obtained from two copies of the extended complex plane, slit along the line segments joining α^j to $\lambda^2\alpha^j$, $1 \leq j \leq n$, by pasting the two sheets together along the slits in the usual way. Figure 1 shows the top sheet, with the homology basis we shall use. The B-curves are closed by returning on the lower sheet.

Since h_2 maps the top sheet (shown in Figure 1) onto itself, the matrix of the induced map on $H^1(X,Z)$ is easy to compute. That matrix (with respect to the homology basis $A_1,\ldots,A_g$, $B_1,\ldots,B_g$ in Figure 1) is

$$M = \begin{pmatrix} L & O \\ O & L \end{pmatrix},$$

where L is the $2k$ by $2k$ matrix

$$L = \begin{pmatrix} O & I \\ I & O \end{pmatrix},$$

and I is the $k \times k$ identity matrix.

The hyperelliptic involution $(x,y) \mapsto (x,-y)$ induces minus the identity on $H_1(X,Z)$, so the homology matrix for h_1 is $-M$.

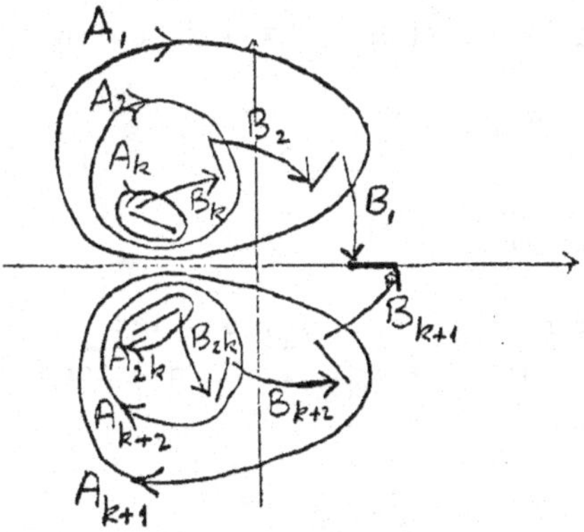

Figure 1 (k = 3)

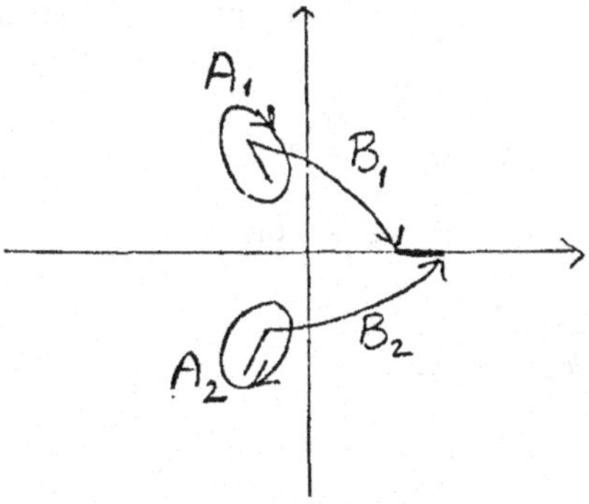

Figure 2 (k = 1)

3.3 Let $\pi_j : X \to Y_j$, $j = 1,2$, be the quotient map. The loops $\pi_j(A_1)$, $\ldots,\pi_j(A_k)$, $\pi_j(B_1),\ldots,\pi_j(B_k)$ on Y_j form a canonical homology basis. With respect to that basis on Y_j and the original basis on X, the maps on homology induced by π_1 and π_2 are given by the matrices

$$\begin{pmatrix} I & -I & O & O \\ O & O & I & -I \end{pmatrix} \tag{3.3}$$

and

$$\begin{pmatrix} I & I & O & O \\ O & O & I & I \end{pmatrix} \tag{3.4}$$

respectively. Here I is the $k \times k$ identity matrix and O is the $k \times k$ zero matrix.

3.4. From equations (3.3) and (3.4) we see that the holomorphic map $\pi_1 \times \pi_2 : X \to Y_1 \times Y_2$ induces a map on first homology given by the matrix

$$\begin{pmatrix} I & -I & O & O \\ O & O & I & I \\ I & I & O & O \\ O & O & I & I \end{pmatrix} \tag{3.5}$$

The matrix (3.5) is non-singular, so $\pi_1 \times \pi_2$ induces an isogeny from $J(X)$ to $J(Y_1) \times J(Y_2)$. That isogeny is not an isomorphism because the determinant of the matrix (3.5) is not ± 1.

3.5. To obtain an isomorphism between $J(X)$ and $J(Y_1) \times J(Y_2)$ we shall define a different holomorphic map $f_1 \times f_2 : X \to Y_1 \times Y_2$. A glance at equation (2.2) shows that X has an automorphism $r : X \to X$ defined by

$$r(x,y) = (\alpha x, y), \quad \alpha = e^{2\pi i/n} . \tag{3.6}$$

Our theorem follows at once from the following.

<u>Lemma 1.</u> Let $f_1 : X \to Y_1$ be the quotient map $f_1 = \pi_1$, and let $f_2 : X \to Y_2$ be the map $f_2 = \pi_2 \circ r$, where $\pi_2 : X \to Y_2$ is the quotient map. Then

$$f_1 \times f_2 : X \to Y_1 \times Y_2$$

induces an isomorphism of $H_1(X,Z)$ onto $H_1(Y_1,Z) \times H_1(Y_2,Z)$.

Lemma 1 will be proved in §4 for genus two $(k = 1)$ and in §5 for higher genera.

4. Genus two.

Let X have genus two. We shall prove Lemma 1 by finding the matrix of the map on first homology induced by $f_1 \times f_2$ and computing its determinant. Since $f_1 = \pi_1$, the homology map induced by $f_1 : X \to Y_1$ has the matrix (3.3). In genus two, with $k = 1$, that matrix is simply

$$\begin{pmatrix} 1 & -1 & 0 & 0 \\ 0 & 0 & 1 & -1 \end{pmatrix} \tag{4.1}$$

To find the homology matrix for $f_2 = \pi_2 \circ r$, we shall compute the action of $r : X \to X$ on $H_1(X,\mathbb{Z})$. In genus two Figure 1 takes the simpler form shown in Figure 2. The automorphism r maps the top sheet of X (shown in Figure 2) onto itself, carrying A_1 to A_2 and B_2 to $-B_1$. The loop $r(A_2)$ crosses both B_1 and B_2 once from right to left and crosses neither A_1 nor A_2, so

$$r(A_2) = -A_1 - A_2 .$$

Computing $r(B_1)$ similarly we find that the homology map induced by r has the matrix

$$\begin{pmatrix} 0 & -1 & 0 & 0 \\ 1 & -1 & 0 & 0 \\ 0 & 0 & -1 & -1 \\ 0 & 0 & 1 & 0 \end{pmatrix} . \tag{4.2}$$

Multiplying the matrices (3.4) and (4.2) we find that $f_2 : X \to Y_2$ has the homology matrix

$$\begin{pmatrix} 1 & -2 & 0 & 0 \\ 0 & 0 & 0 & -1 \end{pmatrix} . \tag{4.3}$$

It follows from (4.1) and (4.3) that the map on first homology induced by $f_1 \times f_2 : X \to Y_1 \times Y_2$ has the matrix

$$\begin{pmatrix} 1 & -1 & 0 & 0 \\ 0 & 0 & 1 & -1 \\ 1 & -2 & 0 & 0 \\ 0 & 0 & 0 & -1 \end{pmatrix} .$$

Since the determinant of that matrix is -1, the lemma is proved for genus two.

5. Higher genera.

5.1. Now let X have genus $2k$, $k > 1$. Lemma 1 will be proved by the same method as in §4, but more effort is required because the matrices are larger. Again $f_1 = \pi_1 : X \to Y_1$ has the homology matrix (3.3), and we shall obtain the matrix of $f_2 = \pi_2 \circ r$ by computing the action of $r : X \to X$ on $H_1(X,Z)$.

5.2. Looking at intersection numbers we find that

$$r(A_j) = A_{j+1} + A_{2k}, \qquad r(B_j) = B_{j+1}, \qquad 1 \leqq j \leqq k-1,$$

$$r(A_k) = A_{2k}, \qquad r(B_k) = -(B_1 + \cdots + B_k) + B_{k+1} + \cdots + B_{2k},$$

$$r(A_{k+1}) = -A_1 - A_{2k}, \qquad r(B_{k+1}) = -B_1,$$

$$r(A_{k+j}) = A_{k+j-1} - A_{2k}, \quad r(B_{k+j}) = B_{k+j-1}, \quad 2 \leqq j \leqq k.$$

Therefore the map of $H_1(X,Z)$ induced by r has a matrix of the form

$$\begin{pmatrix} A & O \\ O & D \end{pmatrix} \tag{5.1}$$

where A and D are $2k$ by $2k$ matrices

$$A = \begin{pmatrix} P & Q \\ R & S \end{pmatrix}, \quad D = \begin{pmatrix} {}^tS & {}^tQ \\ {}^tR & {}^tP \end{pmatrix}, \tag{5.2}$$

and P, Q, R, S are the $k \times k$ matrices with entries $p_{j+1,j} = 1$, $q_{1,1} = -1$, $r_{k,j} = 1$, $s_{k,j} = -1$, $s_{i,i+1} = 1$, and all other entries zero. We are using the notation ${}^tM$ for the transpose of a matrix M.

5.3. Multiplying the matrices (3.4) and (5.1) we find that the homology map induced by f_2 has the matrix

$$\begin{pmatrix} P+R & Q+S & O & O \\ O & O & {}^tS + {}^tR & {}^tQ + {}^tP \end{pmatrix}.$$

Therefore $f_1 \times f_2 : X \to Y_1 \times Y_2$ has the homology matrix

$$M = \begin{pmatrix} I & -I & O & O \\ O & O & I & -I \\ P+R & Q+S & O & O \\ O & O & {}^tS + {}^tR & {}^tQ + {}^tP \end{pmatrix}.$$

5.4. The matrix M has the same determinant as the matrix

$$\begin{pmatrix} I & O & O & O \\ O & O & I & O \\ * & P+Q+R+S & * & O \\ * & O & * & {}^t(P+Q+R+S) \end{pmatrix}.$$

The determinant of that matrix is $(-1)^k \det(P + Q + R + S)^2$. Lemma 1 is therefore an immediate consequence of

Lemma 2. The matrix $U = P + Q + R + S$ has determinant ± 1.

5.5. All that remains is to prove Lemma 2. The definitions of P, Q, R, and S in §5.2 imply that the entries of U satisfy $u_{1,1} = -1$, $u_{i,i+1} = 1$, and $u_{j+1,j} = 1$, with all other entries zero. It is easy to verify that there is exactly one permutation σ of the set $\{1,\dots,k\}$ with the property that $u_{i,\sigma(i)} \neq 0$ for each i. Since every entry of U has absolute value zero or one, Lemma 2 follows.

References

[1] Hayashida, T.: A class number associated with a product of two elliptic curves. Natur. Sci. Rep. Ochanomizu Univ. 16 (1965), 9 - 19.

[2] Hayashida, T., Nishi, M.: Existence of curves of genus two on a product of two elliptic curves. J. math. Soc. Japan 17 (1965), 1 - 16.

[3] Martens, H. H.: Riemann matrices with many polarizations, in "Complex analysis and its applications, Vol. III". International Atomic Energy Agency, Vienna (1976), 35 - 48.

Cornell University
Ithaca, NY 14853
U.S.A.

SINGULAR POINTS OF THETA FUNCTIONS, QUADRIC RELATIONS AND HOLOMORPHIC DIFFERENTIALS WITH PRESCRIBED ZEROS

H. M. Farkas*

Introduction. In this, for the most part, expository paper we are concerned with the connection between singular points of the Riemann theta function, relations among holomorphic quadratic differentials and representations of holomorphic differentials with prescribed zeros. The main tools used are the Riemann-Roch theorem, Abel's theorem and the theory of theta functions on a Riemann surface. These are all treated in [RF].

The fact that there are connections between the objects in the title is not a new fact. The ideas are contained implicitly in many papers. It is our purpose here to make explicit many of these ideas and to highlight the role of the Riemann theta function in the solution of function theoretic problems on a Riemann surface. It is our hope to convince the reader that the subject of theta functions can still be an important tool in solving function theory problems on compact Riemann surfaces. The two main function theoretic problems we deal with here are the construction of explicit rank four (three) relations between the holomorphic quadratic differentials on a compact surface and explicit representations of holomorphic differentials with prescribed zeros. The first problem is treated in [AM] and the second problem is a generalization of a very classical idea. Some of the ideas contained in this paper are also contained in the papers [Ma I, II].

Part of the motivation for this work was also to understand a recent theorem of Mumford [Mu] on singular points for the theta function associated with the Prym variety of a Riemann surface. The Prym variety and the theta functions on it have been a useful tool in the investigations of the Schottky problem for compact Riemann surfaces [F, B, FR_1, RF]. Andreotti and Mayer [AM] characterized Jacobian varieties in terms of the singular sets of the theta functions on them. One of the by products of this investigation will be that there is probably no analogue of their result for the Prym varieties.

I. In this section we recall some of the function theory on compact

* Research partially supported by the National Science Foundation (MCS 760496A01).

Riemann surfaces which will be necessary in the sequel.

Let S denote a compact Riemann surface of genus $g \geq 2$ and let $\gamma_1,\dots,\gamma_g,\delta_1,\dots,\delta_g$ denote a canonical homology basis on S. If $\omega_1,\dots,\omega_g$ is any basis for the vector space of holomorphic differentials on S then $\{\omega_i \cdot \omega_j\}_{i,j=1,\dots,g}$ belong to the vector space of holomorphic quadratic differentials on S. The dimension of this space is $3g-3$ and thus there necessarily exists $\frac{(g-3)(g-2)}{2}$ linearly independent relations of the type $\sum_{i,j=1}^{g} a_{ij}\omega_i\omega_j \equiv 0$ on S. We call such a relation a quadric relation.

A quadric relation is thus given by a $g \times g$ symmetric matrix and as such has a rank. The possible ranks, ρ, of a quadric relation satisfy the inequalities $3 \leq \rho \leq g$. A rank ρ with $\rho < 3$ would imply a linear dependence among the holomorphic differentials ω_i, $i = 1,\dots,g$, which is ruled out since the ω_i by hypothesis form a basis.

If ω_i, $i = 1,\dots,g$, is any basis we can form the $g \times 2g$ matrix obtained by integrating the ω_i over the elements of the canonical homology basis. A basis $\{\varphi_i\}$ is called normalized with respect to a canonical homology basis when $\int_{\gamma_j} \varphi_i = \delta_{ij}$, $\begin{smallmatrix} i=1,\dots,g \\ j=1,\dots,g \end{smallmatrix}$. When this is the case $\int_{\delta_j} \varphi_i = \pi_{ij}$ and $\Pi = (\pi_{i,j})$ is a $g \times g$ complex symmetric matrix with positive definite imaginary part. If we now denote by Φ the vector whose components are the elements of the normalized basis φ_i, we can consider the map φ, of S into C^g, defined by $S \ni P \to \int_{P_o}^{P} \Phi$ where P_o is some fixed point on S. This is not a well defined map on S since the image in C^g depends on the path of integration from P_o to P. Any two images of a point P can only differ by an integral linear combination of the columns of the $g \times 2g$ matrix $(I \;\; \Pi)$ where I as usual denotes the identity matrix and Π is the matrix defined above. If we therefore define a new space $C^g/\langle I,\Pi\rangle$, which is C^g identified under the group of translations of C^g generated by the $2g$ translations $z \to z + e^i$, $i = 1,\dots,g$, and $z \to z + \pi^i$, $i = 1,\dots,g$, where e^i is the i-th column of the identity matrix and π^i is the i-th column of the matrix Π, then the map φ described above is a well defined map into this space. This space will be called the Jacobi variety of S denoted by $J(S)$.

The map φ can be extended easily to divisors on S. If $P_1^{\alpha_1} \cdots P_n^{\alpha_n}$ is any divisor on S then its image under the map φ, $\varphi(P_1^{\alpha_1} \cdots P_n^{\alpha_n})$ $= \alpha_1\varphi(P_1) + \cdots + \alpha_n\varphi(P_n)$.

The matrix Π is an element of $\mathfrak{S}_g$, the Siegel upper half plane of degree g. Given any such Π one can form the theta function

$$\theta(Z,\Pi) = \sum_{N\in Z^g} \exp 2\pi i(\tfrac{1}{2}{}^tN\Pi N + {}^tNZ)$$

where N runs over all integer vectors in C^g and Z is an element of C^g. For Π fixed in $\mathfrak{S}_g$ the theta function is an entire function in C^g and is actually a holomorphic function on $C^g \times \mathfrak{S}_g$.

A singular point for the theta function is a point $e \in C^g$ such that $\theta(e,\pi) = 0$ and $\frac{\partial}{\partial z_i}\theta(e,\pi) = 0$, $i = 1,\ldots,g$. The Riemann vanishing theorem gives another description of the singular points of the theta function when the point $\Pi \in \mathfrak{S}_g$ arises from a compact Riemann surface by the previous construction. The description is in terms of the map φ and is the following: A point $e \in C^g$ is a singular point for the theta function if and only if $e = \varphi(P_1\cdots P_{g-1}) + K$ and $i(P_1\cdots P_{g-1}) \geq 2$. In the above statement $P_1\cdots P_{g-1}$ is an integral divisor of degree $g - 1$ on S, K is the vector of Riemann constants which depends on the canonical homology basis and the base point P_o for the integrals and $i(P_1\cdots P_{g-1})$ is the dimension of the space of holomorphic differentials which vanish at the points $P_1\cdots P_{g-1}$.

As stated already in the introduction a reference for all the above concepts is the author's book with Rauch [RF] and in the book are given additional references.

II. In this section we begin to show the connections between the first two objects in the title.

Lemma 1. Let Q be a quadric relation of rank $\rho \leq 4$. Then Q gives rise to at least one and at most a $g - 3$ dimensional set of singular points.

Proof. If Q is a quadric relation of rank $\rho \leq 4$ then by a change of basis the relation can be brought into the form $\omega_1\omega_2 - \omega_3\omega_4 = 0$ or $\omega_1^2 - \omega_2\omega_3 = 0$. In the first case it is evident that either ω_1/ω_3 or ω_1/ω_4 is a meromorphic function on S of degree $h \leq g - 1$. This follows because at least half of the $2g - 2$ zeros of ω_1 must be also zeros of ω_3 or ω_4.

Let us now denote this meromorphic function by f and f is a function of degree h. We can denote the poles of f by $P_1\cdots P_h$. The Riemann-Roch theorem then yields that $r[\frac{1}{P_1\cdots P_h}] \geq 2$ and therefore for arbitrary integral divisor $Q_{h+1},\ldots,Q_{g-1}$ of degree $g - 1 - h$ it is also true that

$$r[\frac{1}{P_1\cdots P_n\; Q_{h+1}\cdots Q_{g-1}}] \geq 2.$$

It thus follows once again from the Riemann-Roch theorem that $i(P_1\cdots P_n\ Q_{h+1}\cdots Q_{g-1}) \geqq 2$. Hence if we denote by $e = \varphi(P_1\cdots P_h\ Q_{h+1}\cdots Q_{q-1}) + K$, e is singular point for the theta function. The exact same argument holds for the second case.

Our original hypothesis $g \geqq 2$ forces $h \geqq 2$ and so we obtain when $h = 2$ (and the surface is hyperelliptic) a $(g-3)$-dimensional set of singular points while if $h = g - 1$ we obtain two or possibly only one singular point. This completes the proof of the lemma.

The important point we wish to discuss is the converse situation. Suppose we start with a singular point e of the theta function. Then by the discussion in section I we know that $e = \varphi(L_1\cdots L_{g-1}) + K$ with $i(L_1\cdots L_{g-1}) \geqq 2$. We now must consider two possibilities. The first possibility is that $L_1\cdots L_{g-1}$ is the zero divisor of a meromorphic function on S and the second possibility is that it is not the zero divisor of a function on S. In this latter case the divisor class will have fixed points.

No matter which case we happen to be in we certainly have the result that there is at least one non-constant meromorphic function f in $L(\frac{1}{L_1\cdots L_{g-1}})$ where $L(\frac{1}{L_1\cdots L_{q-1}})$ denotes the vector space of meromorphic functions on S with at most poles at the points L_i, $i = 1,\ldots,g-1$. In the above a repeated index would allow a pole of higher order at that point. Let the polar divisor of f (and for the sake of definiteness choose the function of lowest degree in the space) be $L_1\cdots L_h$ and its zero divisor be $Q_1\cdots Q_h$ with $2 \leqq h \leqq g - 1$.

We now construct holomorphic differentials $\theta_1,\ldots,\theta_4$ with the following zeros:

differential	zeros of differential
θ_1	$L_1\cdots L_h L_{h+1}\cdots L_{g-1}\ S_1\cdots S_{g-1}$
θ_2	$Q_1\cdots Q_h L_{h+1}\cdots L_{g-1}\ S_1\cdots S_{g-1}$
θ_3	$Q_1\cdots Q_h L_{h+1}\cdots L_{g-1}\ T_1\cdots T_{g-1}$
θ_4	$L_1\cdots L_h L_{h+1}\cdots L_{g-1}\ T_1\cdots T_{g-1}$

The fact that $i(L_1\cdots L_{g-1}) \geqq 2$ explains the existence of θ_1 and θ_4 and the fact that they are independent. Since $Q_1\cdots Q_h$ is in the same divisor class as $L_1\cdots L_h$ we also see that the divisors of what we have called θ_2 and θ_3 are equivalent to the divisor of say θ_1 and thus that there are holomorphic differentials θ_2 and θ_3 as above. We immediately observe that these four differentials satisfy the condition $\theta_1\theta_3 - \theta_2\theta_4 = 0$. If we knew that the four differentials θ_i,

$i = 1,\ldots,4$, were independent we would have produced here a rank 4 quadrick relation.

We shall now show that either the four differentials are linearly independent or if they are dependent we shall produce a rank 3 quadric relation. To this end assume that we have a linear dependence among the θ_i. This implies that the four meromorphic functions 1, θ_2/θ_1, θ_3/θ_1, θ_4/θ_1 are also dependent. θ_2/θ_1 is precisely the function f we began our discussion with. θ_4/θ_1 is a function g whose divisor is $\frac{T_1\cdots T_{g-1}}{S_1\cdots S_{g-1}}$ after cancellation, and $\theta_3/\theta_1 = fg$. The dependence of $1, f, g, fg$ implies that g is a fractional linear transformation of f and thus that the polar divisor of g is equivalent to the polar divisor of f. We therefore have that whenever we have dependence

$$\frac{T_1\cdots T_{g-1}}{S_1\cdots S_{g-1}} = \frac{T_1\cdots T_h\ S_{h+1}\cdots S_{g-1}}{S_1\cdots S_h\ S_{h+1}\cdots S_{g-1}} = \frac{T_1\cdots T_h}{S_1\cdots S_h}.$$

Moreover, we also have $L_1\cdots L_h \sim Q_1\cdots Q_h \sim S_1\cdots S_h \sim T_1\cdots T_h$ where $\sim$ denotes linear equivalence.

In this case we construct holomorphic differentials $\tilde\theta_1$, $\tilde\theta_2$, $\tilde\theta_3$ with the following zeros.

differential	zeros of differential
$\tilde\theta_1$	$L_1\cdots L_h L_{h+1}\cdots L_{g-1}\ L_1\cdots L_h\ \ S_{h+1}\cdots S_{g-1}$
$\tilde\theta_2$	$T_1\cdots T_h L_{h+1}\cdots L_{g-1}\ T_1\cdots T_h\ \ S_{h+1}\cdots S_{g-1}$
$\tilde\theta_3$	$L_1\cdots L_h L_{h+1}\cdots L_{g-1}\ T_1\cdots T_h\ \ S_{h+1}\cdots S_{g-1}$

In this construction we have simply used the above linear equivalences and it is clear that $\tilde\theta_1\tilde\theta_2 - \tilde\theta_3^2 = 0$. It is also immediate that the three differentials are linearly independent.

<u>Lemma 2.</u> Let $e = \varphi(P_1\cdots P_{g-1}) + K$ be a singular point for the theta function. Let h be the smallest degree of a function in $L(\frac{1}{P_1\cdots P_{g-1}})$ and let $P_1\cdots P_h$ be the polar divisor of this function. Then e (in fact the divisor $P_1\cdots P_h$) determines a quadric relation of rank 3 or 4 on S depending on whether $i(P_1^2\cdots P_h^2\ P_{h+1}\cdots P_{g-1})$ is greater than zero or equal to zero. In the former case the rank is three while in the latter case the rank is four.

<u>Proof.</u> The discussion preceding the statement of the lemma proves everything except for the last statement. In the discussion we have observed that one can always get a rank four relation unless f is a fractional linear transformation of g. In this case we have seen that the zeros of $\tilde\theta_1$ are $L_1^2\cdots L_h^2\ L_{h+1}\cdots L_{g-1}\ S_{h+1}\cdots S_{g-1}$ so that

indeed $i(L_1^2\cdots L_h^2\, L_{h+1}\cdots L_{g-1}) \geqq 1$. Conversely if $e = \varphi(L_1\cdots L_{g-1}) + K$ with $L_1\cdots L_h$ the smallest polar divisor of a function then if $i(L_1^2\cdots L_h^2\, L_{h+1}\cdots L_{g-1}) \geqq 1$ we have a rank three quadric relation as follows: Choose $\tilde{\theta}_1$ with divisor of zeros $L_1^2\cdots L_h^2 L_{h+1}\cdots L_{g-1}S_{h+1}\cdots S_{g-1}$. Choose $\tilde{\theta}_2$ with divisor of zeros $Q_1^2\cdots Q_n^2\, L_{h+1}\cdots L_{g-1}\, S_{h+1}\cdots S_{g-1}$. This is clearly okay since $Q_1\cdots Q_h \sim L_1\cdots L_h$. Finally choose $\tilde{\theta}_3$ with divisor of zeros $L_1\cdots L_h\, Q_1\cdots Q_h\, L_{h+1}\cdots L_{g-1}S_{h+1}\cdots S_{g-1}$. Clearly $\tilde{\theta}_1\tilde{\theta}_2 - \tilde{\theta}_3^2 = 0$ is a rank three quadric relation.

We now wish to consider the quadric relation in another light. We begin by extending the three or four linearly independent differentials to a basis of the space of holomorphic differentials. In order to fix ideas let us assume that we are in the rank four situation and that $\theta_1,\ldots,\theta_4,\theta_5,\ldots,\theta_g$ is a basis for the space of holomorphic differentials. Let us now view $\theta_1,\ldots,\theta_g$ as coordinates on CP^{g-1} and the quadric $\theta_1\theta_3 - \theta_2\theta_4$ as a quadric on $CP^{(g-1)}$. Under the map $P \to (\theta_1(P),\ldots,\theta_g(P))$ we have the Riemann surface sitting in CP^{g-1} and the quadric vanishes on this subset of CP^{g-1}. It is now convenient to lift everything to C^g and ask what is the zero set of the quadric on C^g.

In order to answer the above question we make some simplifying assumptions. We shall assume that the singular point $e = \varphi(P_1\cdots P_{g-1}) + K$ has the property that $i(P_1\cdots P_{g-1}) = 2$. In this case let $P_1\cdots P_h$ be the polar divisor of any non-constant $f \in L(\frac{1}{P_1\cdots P_{g-1}})$ and $P_1\cdots P_h$ $2 \leqq h \leqq g-1$ is unique. Notice that in this case the quadric relation is also unique and is determined by e. When the index is larger than two more than one relation is possible.

In order to make some of the subsequent results easier to state we return for a moment to the map φ of divisors on S into $J(S)$ discussed in I. If we restrict ourselves to integral divisors $P_1^{\alpha_1}\cdots P_k^{\alpha_k}$ (integral means $\alpha_i \geqq 0$) and fix the degree $(\sum_{i=1}^{k} \alpha_i = \text{degree})$ of the divisor, then we can think of the integral divisors of this fixed degree as being elements of the symmetric product of S with itself degree of divisor times. This space is a complex analytic manifold of dimension degree of divisor and local coordinates on this manifold can be chosen as the elementary symmetric functions of the local coordinates on the Cartesian product of S with itself. These coordinates are called the Andreotti coordinates and they have the following property as shown in [Fl, AM]: Let S_r denote the symmetric product of S with itself r times and let φ be the map of $S_r \to J(S)$.

Let $\zeta = P_1^{\alpha_1}\cdots P_k^{\alpha_k}$ with $\sum_{i=1}^{k} \alpha_i = r$ be a point of S_r and let

$\varphi_1, \ldots, \varphi_r$ be the Andreotti coordinates at ζ. Then the Jacobian matrix of the map $\varphi: S_r \to J(S)$ at ζ is equal to

$$\begin{pmatrix} \theta_1(P_1)\theta_1'(P_1)\cdots\theta_1^{(\alpha_1-1)}(P_1)\cdots\theta_1(P_k)\cdots\theta_1^{(\alpha_k-1)}(P_k) \\ \vdots \quad \vdots \quad \vdots \quad \vdots \\ \theta_g(P_1)\theta_g'(P_1) \quad \theta_g^{(\alpha_1-1)}(P_1)\cdots\theta_g(P_k) \quad \theta_g^{(\alpha_k-1)}(P_k) \end{pmatrix} \cdot D$$

where D is an $r \times r$ non-singular diagonal matrix. Furthermore, the rank of the Jacobian matrix at this point is equal to $g - i(P_1^{\alpha_1}\cdots P_k^{\alpha_k})$.

We have been talking about the vector K of Riemann constants without having said much about its significance. It is clear from Abel's theorem that the divisors of abelian differentials on S are linearly equivalent and thus have the same image under φ in $J(S)$. The image is precisely $-2K$. It thus follows that if $e = \varphi(P_1\cdots P_{g-1}) + K$ and $P_1\cdots P_{g-1}\,R_1\cdots R_{g-1}$ the divisor of a holomorphic differential then $-e = \varphi(R_1\cdots R_{g-1}) + K$.

It is clear from the definition of $\theta(z,\pi)$ given in I that $\theta(z,\pi)$ is an even function of z so that if e is a singular point so is $-e$. Moreover from our comments above it is clear what the associated integral divisor of degree $g - 1$ is. It is any divisor for which $P_1\cdots P_{g-1}\,R_1\cdots R_{g-1}$ is the divisor of holomorphic differential where $e = \varphi(P_1\cdots P_{g-1}) + K$.

We can now state the following.

Theorem 1. Let $e = \varphi(P_1\cdots P_{g-1}) + K$ be a singular point with $i(P_1\cdots P_{g-1}) = 2$. Then the zero set of this quadric relation of rank three or four determined by e vanishes on the union of linear subspaces of dimension $g - 2$ on C^g. The linear spaces are precisely the span of the columns of the Jacobian matrix of the map $\varphi: S_{g-1} \to J(S)$ at the point $P_1\cdots P_{g-1}$ and the union is over all divisors equivalent to $P_1\cdots P_{g-1}$ and divisors $R_1\cdots R_{g-1}$ equivalent to the divisor of degree $g - 1$ associated with $-e$.

Proof. The fact that the matrices in question span a $g - 2$ dimensional space follows from the remarks preceding the theorem and the assumption that $i(P_1\cdots P_{g-1}) = 2$. We now show that the quadric $\theta_1\theta_3 - \theta_2\theta_4$ vanishes on each of the linear spaces in question. We treat only the case of divisors with no multiple points since obvious minor modifications of the argument show how to deal with the case of multiple points.

To this end let $Q_1\cdots Q_{g-1}$ be a divisor equivalent to $P_1\cdots P_{g-1}$.

We wish to show that

$$(\lambda_1\theta_1(Q_1) + \cdots + \lambda_{g-1}\theta_1(Q_{g-1}))\ (\lambda_1\theta_3(Q_1) + \cdots + \lambda_{g-1}\theta_3(Q_{g-1}))$$

$$- (\lambda_1\theta_2(Q_1) + \cdots + \lambda_{g-1}\theta_2(Q_{g-1}))\ (\lambda_1\theta_4(Q_1) + \cdots + \lambda_{g-1}\theta_4(Q_{g-1})) = 0$$

This is so if and only if

$$\lambda_1^2[\theta_1(Q_1)\theta_3(Q_1)-\theta_2(Q_1)\theta_4(Q_1)]+\cdots+\lambda_{g-1}^2[\theta_1(Q_{g-1})\theta_3(Q_{g-1})-\theta_2(Q_{g-1})\theta_4(Q_{g-1})]$$

$$+ \sum_{i\neq j}\lambda_i\lambda_j[\theta_1(Q_i)\theta_3(Q_j)+\theta_1(Q_j)\theta_3(Q_i)-\theta_2(Q_i)\theta_4(Q_j)-\theta_2(Q_j)\theta_4(Q_i)] = 0$$

Since the quadric vanishes on the Riemann surface by construction it is clear the coefficients of λ_i^2 all vanish. The only point at issue is what about the coefficients of $\lambda_i\lambda_j$. The claim is that each such coefficient also vanishes.

Consider the span $x_1\theta_1 + x_2\theta_2 = \psi_{12}$. As x_1, x_2 vary over all complex numbers we have ψ_{12} varies over all holomorphic differentials which vanish at $S_1\cdots S_{g-1}$ and whose remaining $g - 1$ zeros are equivalent to $P_1\cdots P_{g-1}$ say $Q_1\cdots Q_{g-1}$. Moreover each such differential is obtainable in this fashion. Thus given such a divisor $Q_1\cdots Q_{g-1}$ we have a non-zero solution to $x_1\theta_1(Q_i) + x_2\theta_2(Q_i) = 0$ $x_1\theta_1(Q_j) + x_2\theta_2(Q_j) = 0$. It is therefore the case that $\theta_1(Q_i)\theta_2(Q_j) - \theta_2(Q_i)\theta_i(Q_j) = 0$ for all pairs i,j. The same remark applied to θ_3 and θ_4 yields $\theta_3(Q_i)\theta_4(Q_j) - \theta_4(Q_i)\theta_3(Q_j) = 0$. Hence we have

$$\frac{\theta_1(Q_i)}{\theta_2(Q_i)} = \frac{\theta_1(Q_j)}{\theta_2(Q_j)}\ ;\ \frac{\theta_3(Q_i)}{\theta_4(Q_i)} = \frac{\theta_3(Q_j)}{\theta_4(Q_j)}$$

and of course for each point Q in the surface

$$\frac{\theta_1(Q)}{\theta_2(Q)} = \frac{\theta_4(Q)}{\theta_3(Q)}.$$

If we now set $Q = Q_j$ and use the first and second equations we find

$$\frac{\theta_1(Q_i)}{\theta_2(Q_i)} = \frac{\theta_4(Q_j)}{\theta_3(Q_j)} \quad\text{and}\quad \frac{\theta_1(Q_j)}{\theta_2(Q_j)} = \frac{\theta_4(Q_i)}{\theta_3(Q_i)}.$$

Finally we have

$$\theta_1(Q_i)\theta_3(Q_j) + \theta_1(Q_j)\theta_3(Q_i) - \theta_2(Q_i)\theta_4(Q_j) - \theta_2(Q_j)\theta_4(Q_i) = 0.$$

The same argument applied to θ_1,θ_4 and θ_2,θ_3 give the same result for the complementary divisor which corresponds to the singular point $-e$. The case where the divisor has multiple points can be treated in a similar manner. This completes the proof of the theorem.

The Riemann vanishing theorem mentioned at the end of I actually has much more information than we have mentioned. In particular it contains the result that the zero divisor of the theta function is exactly the set of points $e = \varphi(P_1 \cdots P_{g-1}) + K$ where $P_1 \cdots P_{g-1}$ ranges over all integral divisors of degree $g - 1$ on S. Moreover the order of vanishing of the theta function at the point is precisely equal to $i(P_1 \cdots P_{g-1})$. This allows us to draw several conclusions.

Theorem 2. Let $e = \varphi(P_1 \cdots P_{g-1}) + K$ and assume $i(P_1 \cdots P_{g-1}) = 2$. Then the quadric relation of rank three or four determined by e can be written in the form

$$\sum_{i,j=1}^{g} \frac{\partial^2\theta}{\partial z_i \partial z_j}(e)\theta_i\theta_j = 0.$$

In particular the rank of the Hessian matrix of theta at a singular point where the zero is precisely second order is either three or four.

Proof. In theorem 1 we have described the zero set of the quadric determined by e on C^g. We prove theorem 2 by showing that the quadric

$$\sum_{i,j} \frac{\partial^2\theta}{\partial z_i \partial z_j}(e)\theta_i\theta_j$$

has the same zero set.

It follows from the Riemann vanishing theorem that $\theta(\varphi(Q_1 \cdots Q_{g-1}) + K)$ vanishes identically on the $(g-1)$st symmetric product of S with itself. Hence if we expand in Taylor series all the coefficients must vanish. If we now denote by φ the Andreotti coordinates on the symmetric product and recall our discussion preceding the statement of theorem 1 we have

$$\theta(e) = 0, \quad \sum_i \frac{\partial\theta}{\partial z_i}(e)\frac{\partial z_i}{\partial\varphi_\alpha} = \sum_i \frac{\partial\theta}{\partial z_i}(e)\theta_i(P_\alpha) = 0$$

(at least in the case $P_i \neq P_j$ for $i \neq j$ and an appropriate insertion of derivatives when we have $P_i = P_j$) and finally

$$\sum_{i,j} \frac{\partial^2\theta}{\partial z_i \partial z_j}(e)\frac{\partial z_i}{\partial\varphi_\alpha}\frac{\partial z_j}{\partial\varphi_\beta} + \sum \frac{\partial\theta}{\partial z_i}(e)\frac{\partial^2 z_j}{\partial\varphi_\beta\partial\varphi_\alpha} = \sum_{i,j} \frac{\partial^2\theta}{\partial z_j \partial z_i}(e)\theta_i(P_\alpha)\theta_j(P_\beta)$$

$$+ \sum_i \frac{\partial\theta}{\partial z_i}(e)\frac{\partial^2 z_i}{\partial\varphi_\beta\partial\varphi_\alpha} = 0.$$

In the case that we are considering we have $\theta(e) = 0$ and $\frac{\partial\theta}{\partial z_i}(e) = 0$ so we have no new information from the first two identities. The third identity however yields

$$\sum_{i,j} \frac{\partial^2\theta}{\partial z_i \partial z_j}(e)\theta_i(P_\alpha)\theta_j(P_\beta) = 0$$

for any choice of α and β. In particular if we choose $\alpha = \beta$ we have

$$\sum_{i,j} \frac{\partial^2\theta}{\partial z_i \partial z_j}(e)\theta_i(P_\alpha)\theta_j(P_\alpha) = 0.$$

The important point now is that since $i(P_1 \cdots P_{g-1}) = 2$ given any point $P \in S$ there is an integral divisor of degree $g - 1$ say ζ which contains P and $e = \varphi(\zeta) + K$. It thus follows that

$$\sum_{i,j} \frac{\partial^2\theta}{\partial z_i \partial z_j}(e)\theta_i\theta_j$$

vanishes for all points $P \in S$ and thus that the above is a quadric relation on S. The fact that

$$\sum_{i,j} \frac{\partial^2\theta}{\partial z_i \partial z_j}(e)\theta_i(P_\alpha)\theta_j(P_\beta) = 0$$

for every P_α, P_β in the divisor ζ of degree $g - 1$ on S such that $e = \varphi(\zeta) + k$ yields the result that the relation vanishes on the linear span of the columns of the Jacobian matrix of the map of $S_{g-1} \to J(S)$. Observe also that the coefficients

$$\frac{\partial^2\theta}{\partial z_i \partial z_j}(e) = \frac{\partial^2\theta}{\partial z_i \partial z_j}(-e)$$

so that whatever is true for e is also true for $-e$. Since the two quadrics have the same zero set they agree up to a constant multiple.

This concludes the most important ideas in the connection between the singular points of the theta function and the quadric relations of rank less than or equal to four on a compact Riemann surface.

III. We now turn to the last object in the title namely holomorphic differentials with prescribed zeros. Since the dimension of the space of holomorphic differentials on a compact surface of genus g is equal to the genus g, prescribing $g - 1$ zeros should in general give us a unique differential up to constant multiples. Our problem here is to represent the differentials as linear combinations of a normalized basis with respect to a fixed canonical homology basis on the surface.

We begin with the simplest case of points $P_1 \cdots P_{g-1}$ on S with $i(P_1 \cdots P_{g-1}) = 1$ and ask for the representation of the holomorphic differential which vanishes at these (for the moment distinct) $g - 1$ points. To this end we let $e = \varphi(P_1 \cdots P_{g-1}) + K$ and consider the holomorphic differential $\omega = \sum_{i=1}^{g} \frac{\partial\theta}{\partial z_i}(e)\theta_i$. We claim that it vanishes at the points in question. The proof follows the same line of reasoning as in the proof of theorem 2 above. By the Riemann vanishing theorem $\theta(\varphi(P_1 \cdots P_{g-1}) + K)$ vanishes identically on the (g-1)-st symmetric product of S with itself. Expanding in power series in a neighborhood of $P_1 \cdots P_{g-1}$ we get that for each $\alpha = 1,\ldots,g-1$ $\frac{\partial\theta}{\partial\phi_\alpha}(e) = 0$ or that $\sum_{i=1}^{g} \frac{\partial\theta}{\partial z_i}(e)\frac{\partial z_i}{\partial\phi_\alpha} = 0$ or that $\sum_{i=1}^{g} \frac{\partial\theta}{\partial z_i}(e)\theta_i(P_\alpha) = 0$.

It is now clear that the condition that the $(g-1)$ points be distinct was really not necessary to the discussion at all. If the integral divisor $P_1^{\alpha_1}\cdots P_k^{\alpha_k}$ with $\sum_{i=1}^{k} \alpha_i = g-1$ is given and we let $e = \varphi(P_1^{\alpha_1} P_2^{\alpha_2}\cdots P_k^{\alpha_k}) + K$ then $\omega = \sum_{i=1}^{g} \frac{\partial\theta}{\partial z_i}(e)\theta_i$ vanishes to under α_j at P_j.

In fact there is another way to achieve the same result. We simply consider the expression

$$\det \begin{pmatrix} \theta_1(P_1)\cdots\theta_1(P_{g-1})\theta_1(P) \\ \cdot \\ \cdot \\ \theta_g(P_1)\cdots\theta_g(P_{g-1})\theta_g(P) \end{pmatrix}.$$

It is clear that this is a holomorphic differential which vanishes at the points P_i, $i = 1,\ldots,g-1$, and that this differential is not identically zero since the rank of

$$M = \begin{pmatrix} \theta_1(P_1)\cdots\theta_1(P_{g-1}) \\ \cdot \\ \cdot \\ \theta_g(P_1)\cdots\theta_g(P_{g-1}) \end{pmatrix}$$

is equal to $g-1$ in the case under consideration. Hence comparing the two results we've obtained we in fact have $\frac{\partial\theta}{\partial z_i}(e) = \pm\det$ of M_i where M_i is the matrix M with the i-th row deleted.

The situation where the points $P_1\cdots P_{g-1}$ satisfy the condition $i(P_1\cdots P_{g-1}) \geq 2$ is more complicated and we here treat only the case $i(P_1\cdots P_{g-1}) = 2$. Our point of departure is the same as it was previously. We consider $e = \varphi(P_1\cdots P_{g-1}) + K$. The condition $i(P_1\cdots P_{g-1}) = 2$ yields from the Riemann vanishing theorem that $\frac{\partial\theta}{\partial z_i}(e) = 0$, $i = 1,\ldots,g$, so that the holomorphic differential $\sum_{i=1}^{g} \frac{\partial\theta}{\partial z_i}(e)\theta_i$ is the zero differential and not very interesting. In the proof of theorem 2 however we showed that

$$\sum_{i,j=1}^{g} \frac{\partial^2\theta}{\partial z_i \partial z_j}(e)\theta_i(P_\alpha)\theta_j(P_\beta) = 0$$

for any choice of P_α and P_β in $P_1\cdots P_{g-1}$. Hence if we consider now the holomorphic differential

$$\sum_{i=1}^{g}\left(\sum_{j=1}^{g} \frac{\partial^2\theta}{\partial z_i \partial z_j}(e)\theta_j(P_\alpha)\right)\theta_i(P)$$

for some P_α in $P_1 \cdots P_{g-1}$, we immediately conclude that this differential vanishes at the points $P_1 \cdots P_{g-1}$. The only additional fact we need is to know that the differential we have written down is not identically zero.

Lemma 3. The differential

$$\varphi_\alpha = \sum_{i=1}^{g} \left(\sum_{j=1}^{g} \frac{\partial^2 \theta}{\partial z_i \partial z_j}(e) \theta_j(P_\alpha) \right) \theta_i(P)$$

cannot vanish identically for every P_α in $P_1 \cdots P_{g-1}$.

Proof. If φ_α were identically zero then for each $i = 1,\ldots,g$

$$\sum_{j=1}^{g} \frac{\partial^2 \theta}{\partial z_i \partial z_j}(e) \theta_j$$

would vanish at the point P_α. We have previously seen that the rank of the matrix $\left(\left(\frac{\partial^2 \theta}{\partial z_i \partial z_j}\right)(e)\right)$ is either three or four so that we would have at least three linearly independent differentials which vanish at P_α. Since $i(P_1 \cdots P_{g-1}) = 2$ it could not be the case that $\varphi_\alpha \equiv 0$ for each α.

We have now shown that for at least one α $\varphi_\alpha \not\equiv 0$ so the question naturally arises whether φ_α is ever identically zero. The next lemma shows that this can occur.

Lemma 4. Let $e = \varphi(P_1 \cdots P_{g-1}) + K$ with $i(P_1 \cdots P_{g-1}) = 2$. Assume that $P_1 \cdots P_{g-1}$ is not the polar divisor of a meromorphic function so that it contains a fixed divisor $P_{h+1} \cdots P_{g-1}$. Then if $\alpha = h+1,\ldots,g-1$, $\varphi_\alpha \equiv 0$.

Proof. Consider

$$\varphi_\alpha = \sum_{i,j=1}^{g} \frac{\partial^2 \theta}{\partial z_i \partial z_j}(e) \theta_j(P_\alpha) \theta_i(P)$$

with $\alpha = h+1,\ldots,g-1$. We claim that for any point Q on the Riemann surface $\varphi_\alpha(Q) = 0$. This follows from the fact that given any point Q there is a divisor $R_1 \cdots R_h$ which contains Q and $e = \varphi(R_1 \cdots R_h\, P_{h+1} \cdots P_{g-1}) + K$. Hence $\varphi_\alpha(Q) = 0$ for every Q.

We have thus far shown that if $\varphi_\alpha \not\equiv 0$ it satisfies the condition that it vanishes at the points $P_1 \cdots P_{g-1}$. Since $i(P_1 \cdots P_{g-1}) = 2$ there is a two dimensional set of holomorphic differentials with this property. Which one have we written down?

We unfortunately cannot at this time give a complete answer to the above question. What we can say is contained in the following.

Theorem 3. Let $e = \varphi(P_1 \cdots P_{g-1}) + K$ with $i(P_1 \cdots P_{g-1}) = 2$ and $P_i \neq P_j$ for $i \neq j$. Assume $\varphi_\alpha \not\equiv 0$ and that $i(P_1 \cdots P_\alpha^2 \cdots P_{g-1}) = 1$.

Then φ_α is the holomorphic differential which vanishes at $P_1 \cdots P_{g-1}$ and has a double zero at P_α.

Proof. The main ingredient of the proof is the fact that replacing e by $-e$ in φ does not change anything. We know that $-e = \varphi(Q_1 \cdots Q_{g-1}) + K$ and $i(Q_1 \cdots Q_{g-1}) = 2$. Hence there is a divisor $R_1 \cdots R_{g-1}$ equivalent to $Q_1 \cdots Q_{g-1}$ which contains P_α. The reasoning preceding lemma 3 then gives that φ_α vanishes at $P_1 \cdots P_\alpha^2 \cdots P_{g-1}$.

In the above theorem we have inserted the hypothesis $i(P_1 \cdots P_\alpha^2 \cdots P_{g-1}) = 1$ because in this case we can also express the differential as a determinant. In particular we have up to a constant multiple

$$\varphi_\alpha = \det \begin{pmatrix} \theta_1(P_1) \cdots \theta_1(P_{\alpha-1}) & \theta_1'(P_\alpha) & \theta_1(P_{\alpha+1}) \cdots \theta_1(P_{g-1}) & \theta_1(P) \\ \theta_g(P_1) \cdots \theta_g(P_{\alpha-1}) & \theta_g'(P_\alpha) & \theta_g(P_{\alpha+1}) \cdots \theta_g(P_{g-1}) & \theta_g(P) \end{pmatrix} .$$

IV. In this last section we wish to make some remarks and pose some problems. In this paper we have seen that the singular points which we were able to deal with best were those points $e = \varphi(P_1 \cdots P_{g-1}) + K$ and $i(P_1 \cdots P_{g-1}) = 2$. This was so in the case of the quadric relations and also in the case of the representation of the holomorphic differential which vanishes at $P_1 \cdots P_{g-1}$. It is natural to ask what happens when $i(P_1 \cdots P_{g-1}) > 2$. It is clear from lemma 2 that in this case there are also quadric relations of rank four (three) determined by the singular point. In fact many such are determined. The question is whether we have an analogue of theorem 2 in this case.

Consider therefore $e = \varphi(P_1 \cdots P_{g-1}) + K$ with for example $i(P_1 \cdots P_{g-1}) = 3$. The Riemann vanishing theorem guarantees $\frac{\partial \theta}{\partial z_i}(e) = 0$, $i = 1,\ldots,g$ and $\frac{\partial^2 \theta}{\partial z_j \partial z_i}(e) = 0$, $i,j = 1,\ldots,g$. Hence by the reasoning used repeatedly above

$$\sum_{i,j,k=1}^{g} \frac{\partial^3 \theta}{\partial z_k \partial z_j \partial z_i}(e)\theta_i(P_\alpha)\theta_j(P_\beta)\theta_k(P_\gamma) = 0$$

where P_α, P_β, P_γ are points in the divisor $P_1 \cdots P_{g-1}$. We recall that this equation comes from the fact that

$$\frac{\partial^3 \theta}{\partial \varphi_\alpha \partial \varphi_\beta \partial \varphi_\gamma}(e) = 0$$

where φ_α are now the Andreotti coordinates on the $(g-1)$-st symmetric product of S with itself. In particular choosing $\alpha = \beta$ we obtain

$$\frac{\partial^3 \theta}{\partial \varphi^2 \partial \varphi_\gamma}(e) = 0 = \sum_{i,j,k=1}^{g} \frac{\partial^3 \theta}{\partial z_k \partial z_j \partial z_i}(e)\theta_i(P_\alpha)\theta_j(P_\beta)\theta_k(P_\gamma)$$

and this holds for any two points P_α, P_γ on S provided

$e = \varphi(P_1 \cdots P_{g-1}) + k$ and P_α, P_γ part of $P_1 \cdots P_{g-1}$. Since $i(P_1 \cdots P_{g-1}) = 3$ given any two points on S say Q, R it is true that $e = \varphi(T_1 \cdots T_{g-1}) + k$ and Q, R are in $T_1 \cdots T_{g-1}$. From this we can conclude that the expression

$$\sum_{i,j,k=1}^{g} \frac{\partial^3 \theta}{\partial z_k \partial z_j \partial z_i}(e)\theta_i(Q)\theta_j(Q)\theta_k(R) .$$

vanishes identically on the second symmetric product of S with itself. In particular for any fixed Q is a holomorphic differential in R which vanishes identically on S. We can therefore conclude

$$\sum_{i,j=1}^{g} \frac{\partial^3 \theta}{\partial z_k \partial z_j \partial z_i}(e)\theta_i(Q)\theta_j(Q) \equiv 0$$

for each $k = 1,\ldots,g$.

Hence we have produced quadric relations on S again but do not know what the rank of the quadric is nor how many linearly independent quadrics we have in this way produced.

In a similar fashion one can produce a holomorphic differential which vanishes at $P_1 \cdots P_{g-1}$. We simply consider

$$\sum_{k=1}^{g} \Big(\sum_{i,j=1}^{g} \frac{\partial^3}{\partial z_k \partial z_j \partial z_i}(e)\theta_i(P_\alpha)\theta_j(P_\beta)\Big)\theta_k(P).$$

It is clear that this holomorphic differential vanishes at $P_1 \cdots P_{g-1}$ but we don't know how to prove that it is not identically zero.

Our final remark concerns what was the original motivation for this study and the reason for our concern whether the quadric relation is rank three or rank four. Given a compact Riemann surface S of genus $g \geq 2$ one can always construct a compact Riemann surface $\hat{S}$ of genus $2g - 1$ which is a smooth two sheeted covering of S. The analytic mapping $\pi : \hat{S} \to S$ induces a homomorphism of $J(\hat{S}) \overset{\text{onto}}{\to} J(S)$. The kernel of this homomorphism is an abelian variety of dimension $g - 1$ which is called the Prym variety of S corresponding to the given smooth cover. There is in a natural way a theta function defined on the Prym and the question raised was what is the dimension of the singular set of this theta function. In a beautiful paper [Mu] Mumford gives an answer to this question. He proves that if S is not hyperelliptic and $g \geq 5$ then the dimension of the singular set of the Prym is always bounded from above by $g - 5$ and is equal to $g - 5$ only in special cases which we shall not enumerate.

The idea of the proof is roughly the following:
Mumford shows that the singular points of the theta function on the Prym are also singular points for the Riemann theta function on $J(\hat{S})$. In fact he shows that they are all singular points of even order. In

the case when the order is two he shows that these points actually arise from singular points of the theta function on $J(S)$. In fact in this case by lemma 2 and theorem 2 the singular points arise from singular points which give rise to rank three quadric relations on S.

The ideas in [F1] can be used to show that the existence of rank three quadric relations are special in the sense of moduli. It thus would follow that the existence of theta functions on the Pryms with singular points are also special in the sense of moduli. We cannot yet draw this conclusion however since we don't know yet how to characterize those singular points (if any) on the Prym which give rise to higher order singular points on $J(\hat{S})$.

References

[AM] Andreotti, A., Mayer, A.: On period relations for abelian integrals on algebraic curves. Ann. Sc. norm. super. Pisa, Cl. Sci., IV. Ser. 21 (1967), 189 - 238.

[B] Beauville, A.: Prym varieties and the Schottky problem. Inventiones math. 41 (1977), 149 - 196.

[F1] Farkas, H.M.: Special divisors and analytic subloci of Teichmüller space. Amer. J. Math. 88 (1966), 881 - 901.

[F2] Farkas, H.M.: On the Schottky relation and its generalization to arbitrary genus. Ann. of Math., II. Ser. 92 (1970), 56 - 81.

[FR] Farkas, H.M., Rauch, H.E.: Period relations of Schottky type on Riemann surfaces. Ann. of Math., II. Ser. 92 (1970), 434 - 461.

[Ma 1] Martens, H.H.: On the varieties of special divisors on a curve I. J. reine angew. math. 227 (1967), 111 - 120.

[Ma 2] Martens, H.H.: On the varieties of special divisors on a curve II. J. reine angew. math. 233 (1968), 89 - 100.

[Mu] Mumford, D.: Prym varieties I, in "Contributions to Analysis". Academic Press, New York and London (1974), 325 - 350.

[RF] Rauch, H.E., Farkas, H.M.: Theta function with applications to Riemann surfaces. Williams and Wilkins Co., Baltimore, Md. (1974).

Hebrew University
Jerusalem
Israel

HARMONIC MORPHISMS

Bent Fuglede

The notion of a harmonic morphism (also called harmonic map) between two harmonic spaces was introduced in 1965 by Constantinescu and Cornea [2] as a natural generalization of the holomorphic mappings between Riemann surfaces. One major purpose was to extend the results earlier obtained by them, and by Doob, concerning the behaviour of a holomorphic function at an ideal boundary.

This project of extension was a very natural one in view of the creation and development of axiomatic potential theory since the fifties by Tautz, Doob, Brelot, Hervé, Bauer, Boboc, Constantinescu, Cornea, and others.

For the purpose of my talk to-day the most important (and also the simplest) version of the axiomatic theory of harmonic spaces is the one introduced by Brelot [1]. He defined a harmonic space as a locally compact, connected and locally connected Hausdorff space (let us say here with a countable base) endowed with a sheaf of vector spaces of continuous functions, called harmonic functions, and subjected to two further axioms - one about the solvability of the Dirichlet problem for certain (so-called regular) open sets, which should form a base for the topology, and the other one (nowadays called the Brelot convergence axiom) postulating the validity of what corresponds to the classical Harnack theorem about limits of increasing sequences of harmonic functions.

A very large part of classical potential theory has been extended meanwhile to this axiomatic theory of harmonic spaces. For those of you who may not be very familiar with this theory it will be quite sufficient to think of the classical case of Riemannian manifolds - or even Euclidean spaces - the harmonic functions being then of course the solutions of the Laplace-Beltrami equation. Part of my talk will even deal specifically with the manifold case.

Definition (cf. [2]). A harmonic morphism $f : X \to Y$ between Brelot harmonic spaces X and Y is a continuous mapping such that, for any open subset V of Y and any harmonic function v in V, the composition $v \circ f$ is harmonic in $f^{-1}(V)$.

A related definition was given independently at about the same time by Sibony [10].

The main result concerning harmonic morphisms as such, obtained by Constantinescu and Cornea in [2], is the following:

Theorem 1. (Constantinescu and Cornea [2]). Every non-constant harmonic morphism $f : X \to Y$ is open in the fine topologies on the two Brelot spaces X and Y, provided that the points of Y (or just of $f(X)$) are polar.

The fine topology is defined as the weakest one for which all superharmonic functions are continuous.

A point a is called polar if there exists a superharmonic function s in some neighbourhood of a which takes the value $+\infty$ at a.

When I became interested in harmonic morphisms - through discussions with Prof. Laine in Erlangen in 1972 - I found it interesting to study the following 3 questions:

I. How can we describe the harmonic morphisms more explicitly, and relate them to other types of mappings, in the case of Riemannian manifolds?

II. Does Theorem 1 have a counterpart for the usual topologies on X and Y?

III. Can we make a parallel study of finely harmonic morphisms between fine domains?

In my talk to-day I shall concentrate on the first two questions, referring to [5] as to III.

As to question I, let now X and Y be two Riemannian manifolds - say C^∞, connected, and with countable base. Their dimensions

$$\dim X = m, \dim Y = n$$

need not be the same, but we shall assume that $m \geq n$ since otherwise every harmonic morphism $f : X \to Y$ turns out to be constant.

The Laplace-Beltrami operator on X is denoted by Δ_X, and similarly on Y. Since this operator is elliptic, the harmonic functions are C^∞ like the manifolds.

The main result concerning question I is:

Theorem 2 (cf. [6]). The following are equivalent for a mapping $f : X \to Y$ between Riemannian manifolds:

1) f is a harmonic morphism.
2) f is a harmonic mapping (in the sense of differential geometry) and moreover semiconformal.
3) f is semiconformal, and the components of f in terms of

harmonic local coordinates in Y are harmonic functions (in X).

4) f is C^∞ (or equivalently just C^2), and there exist a function $\lambda : X \to [0,+\infty[$ such that

$$\Delta_X(v \circ f) = \lambda^2[(\Delta_Y v) \circ f]$$

for every C^2-function v on Y (or equivalently just in some open subset of Y).

As to the notion of a harmonic mapping in the sense of differential geometry I refer to Eells and Sampson [3]. A mapping $f : X \to Y$ is called harmonic if f makes a certain energy integral - or generalized Dirichlet integral - stationary. The corresponding Euler equations for this variational problem are not linear but only quasi-linear. There are n of them:

$$\Delta_X f^k + \sum_{\alpha,\beta=1}^{n} g_X(\nabla f^\alpha, \nabla f^\beta)(\Gamma^k_{\alpha\beta} \circ f) = 0, \ k = 1,\cdots,n . \qquad (*)$$

Here g_X denotes the metric tensor on X, and ∇ the gradient operator on X. In the special case when Y is Euclidean, with Euclidean coordinates, the Christoffel symbols $\Gamma^k_{\alpha,\beta}$ (on Y) vanish, and (*) just states that the components of f are harmonic. (The equivalence between 2) and 3) is therefore obvious in the case when Y is Euclidean.)

The harmonic mappings thus form a wider - much wider - class of mappings of X into Y than the harmonic morphisms (except in special cases such as the case $Y = \mathbb{R}$, where both types of mappings reduce to the harmonic functions on X).

Unlike the situation for harmonic morphisms, the composition of two harmonic mappings is not again such a mapping, except in particular cases (for instance if the mapping to be applied first is a harmonic morphism).

Nevertheless, the harmonic mappings play a very important role in differential geometry and topology. For example, the harmonic mappings of a circle into a compact manifold Y are precisely the closed geodesics on Y. Harmonic mappings also enter in the Plateau problem.

Now to the term semiconformal. For the lack of a better name I call a C^1-mapping $f : X \to Y$ semiconformal if, for any point $a \in X$ at which $df \neq 0$, the restriction $df|(\ker df)^\perp$ of df to the orthogonal complement of $\ker df$ within the tangent space X_a is conformal and surjective. Its coefficient of conformality equals $\lambda(a)$ from 4), if 4) is fulfilled. Note that points at which $df = 0$ (or equivalently $\lambda = 0$) are allowed.

Explicitly, the definition of semiconformality may be stated as follows in terms of local coordinates in Y:

$$g_X(\nabla f^k, \nabla f^l) = \lambda^2 \cdot (g_Y^{kl} \circ f) .$$

Again, when $Y = \mathbb{R}^n$ with Euclidean coordinates, this means that the gradients $\nabla f^1, \cdots, \nabla f^n$ of the components of f should be mutually orthogonal and of equal length $\lambda(a)$ (at each point $a \in X$).

For the proof of theorem 2 I must refer to [6]. Only the implication 4) ⇒ 1) is trivial (but the whole theorem is elementary in the case $Y = \mathbb{R}^n$). I should like to mention, though, that the possibility of choosing local coordinates in Y which are harmonic functions is a corollary of a remarkable embedding theorem due to Greene and Wu [7].

It should be added that only the case $m > n$ leads to something new. If $m = n = 2$, the harmonic morphisms are precisely the holomorphic or antiholomorphic mappings between the two Riemann surfaces X and Y. And if $m = n \neq 2$, the non-constant harmonic morphisms are just the local isometries, up to a constant change of scale.

In the rest of my talk I shall no longer consider harmonic mappings (in the sense of Eells and Sampson) but only harmonic morphisms (in the sense of Constantinescu and Cornea, see the above definition).

After the preceding discussion of the nature of these harmonic morphisms in the manifold case, let us now turn to question II - the problem of openness.

The property of openness is clearly important, for - as you know - the classical fact that every non-constant holomorphic mapping between Riemann surfaces is open is very essential. It is therefore not surprising that in the studies of harmonic morphisms between harmonic spaces, made by Sibony, Hansen, Ikegami, and others, the harmonic morphisms were often <u>assumed</u> to be open.

In the manifold case we have the following positive answer to question II:

<u>Theorem 3</u> (cf. [6]). a) Every non-constant harmonic morphism between Riemannian manifolds is open (in the usual topologies), even at points where $df = 0$.

b) Every semiconformal C^∞-mapping $f : X \to Y$ with $\dim Y \geqq 2$ is open at least at any point a of X at which f has finite order (that is, not all partial derivatives of all orders of the components of f in terms of local coordinates on Y are allowed to vanish at the point a).

Part a) follows easily from Part b) in view of the Carleman-Aronszajn-Cordes <u>uniqueness theorem</u> applied to the Laplace operator. The proof of Part b) is not quite easy.

Quite recently I obtained the following positive answer to the problem of openness in the axiomatic case of harmonic morphisms between Brelot harmonic spaces. This covers, in particular, the case of Riemannian manifolds X and Y when $\dim Y \geq 2$. Since the case $\dim Y = 1$ is easy, this new approach produces a much simpler proof of Theorem 3 a. (Theorem 3 b cannot be covered by the new method since the notion of a semiconformal mapping cannot be defined in potential theoretic terms).

Theorem 4. Every non-constant harmonic morphism $f : X \to Y$ between Brelot harmonic spaces X and Y is open (in the initial topologies on the two spaces), provided that the points of Y (or just of $f(X)$) are strongly polar.

Definition. A point a of a Brelot harmonic space is called strongly polar if every superharmonic function s in some neighbourhood U of a which is harmonic in $U \setminus \{a\}$, but not at a, takes the value $+\infty$ at a.

I don't know whether "strongly polar" can be replaced by "polar" in Theorem 4 (as it could in Theorem 1). There is a simple, unpublished example, due to Cornea, showing that both Theorem 1 and Theorem 4 may break down if Y contains non-polar points.

In the case of a Brelot harmonic space admitting a potential > 0, each of the following 3 conditions is sufficient to ensure that every polar point is strongly polar:

1) The continuity principle holds: Every potential (or superharmonic function) is continuous if its restriction to its harmonic support is finite and continuous.
   - This condition is equivalent to the strong domination property $\overline{D}$.

2) There exists a symmetric Green kernel.

3) There exists a Green kernel which is infinite on the diagonal set.

Here 2) implies 1), as shown by Forst [4] in the case of a harmonic group, and subsequently by Janssen [9] in the general case. - Condition 3) even ensures that all points are strongly polar. As shown by Hervé [8], 1) - 3) are all fulfilled in the case of a Riemannian manifold (of dimension ≥ 2 in the case of the last condition).

Proof of Theorem 4 in the case 3): Let $G : Y \times Y \to]0,+\infty]$ be a Green kernel on Y such that $G(y,y) = +\infty$ for all $y \in Y$. Let U be an open neighbourhood of a given point $x_o \in X$. We shall prove that $f(U)$ is a neighbourhood of $y_o := f(x_o)$ in Y. Since X is locally connected, we may assume that U is connected.

Proceeding by contradiction, suppose there is a sequence of points $y_n \in Y \setminus f(U)$ converging to $y_o = f(x_o)$. Then

$$v_n := G(\cdot, y_n) > 0$$

is harmonic in the open set $Y \setminus \{y_n\}$ $(\supset f(U))$ and hence

$$u_n : v_n \circ f > 0$$

is harmonic in $f^{-1}(Y \setminus \{y_n\})$, in particular in U. Since G is l.s.c. and infinite on the diagonal, G is continuous there, and so

$$u_n(x_o) = v_n(y_o) = G(y_o, y_n) \to G(y_o, y_o) = +\infty \tag{1}$$

as $n \to \infty$.

Next fix a point $a \in U \setminus f^{-1}(y_o)$. Such a point exists because the point $y_o = f(x_o)$ is polar in Y, and hence the fiber $f^{-1}(y_o)$ is a polar set in X according to [2, Theorem 3.2]. It follows that $f^{-1}(y_o)$ cannot contain the whole open neighbourhood U of x_o.

Writing $b := f(a)$, we thus have $b \neq y_o$, and so

$$u_n(a) = v_n(b) = G(b, y_n) \to G(b, y_o) < +\infty \tag{2}$$

because the Green kernel G is finite and continuous off the diagonal set.

Combining (1) and (2), we finally obtain

$$\frac{u_n(x_o)}{u_n(a)} \to +\infty ,$$

but this is impossible by the well-known Harnack property of the connected open set U - a consequence of the Brelot convergence axiom for X.

This proves the theorem in the case 3) (for Y). The general case will be treated elsewhere.

By the above method it can also be easily proved, for a Brelot harmonic space satisfying the domination axiom $\overline{D}$, that a domain U in the fine topology does not have the natural Harnack property with respect to the finely harmonic functions > 0 on U (viz. the boundedness of $u(x)/u(y)$ for such functions u and for fixed $x, y \in U$) except in the trivial case where U is a domain even in the initial topology.

In closing, may I return to the semiconformal mappings between two Riemannian manifolds X and Y with $\dim X > \dim Y$. It is easy enough to define a more general class of mappings $X \to Y$ which are related to the semiconformal mappings in the analogus way as the quasiconformal, or more precisely the quasiregular mappings are related to conformal mappings (in the case of equal dimensions). Perhaps it would be worth while to study this extension of the quasiregular mappings to the case dim X > dim Y, and among other things to try to prove that also these mappings - if not constant - are open, thus extending in a non-trivial way the theorem of Rešetnjak on the openness of quasiregular mappings.

Remarks

1) The implication 1) ⇒ 4) in Theorem 2 above may be proved much simpler than in Lemma 4 in [6] as follows: Let $f : X \to Y$ be a harmonic morphism, let $x \in X$ be given, and write $f(x) = y$. We first show that, for any C^2-function v defined in an open neighbourhood V of y in Y, we have the implication

$$(\Delta_Y v)(y) = 0 \Rightarrow [\Delta_X(v \circ f)](x) = 0.$$

As in the beginning of the proof of [6, Lemma 4] choose a C^2-function w in some open neighbourhood W of y in Y so that $\Delta_Y w > 0$ in W. For any $\varepsilon > 0$ we then have $\Delta_Y(v + \varepsilon w) > 0$ in some open neighbourhood V_ε of y in Y, and hence $\Delta_X[(v + \varepsilon w) \circ f] \geqq 0$ in $f^{-1}(V_\varepsilon)$, according to [6, §1, (1)]. Inserting the given point x and letting $\varepsilon \to 0$, we obtain $\Delta_X(v \circ f)(x) \geqq 0$, and here the sign of equality holds, as shown by replacing v by $-v$.

Next, consider any C^2-function $v : Y \to \mathbb{R}$, and write

$$t = \Delta_Y v(y)/\Delta_Y w(y), \qquad y = f(x), \qquad x \text{ fixed in } X,$$

with a fixed w as before. The implication established above applies to $v - tw$ (in W) in place of v and shows that indeed

$$\Delta_X(v \circ f)(x) = t\Delta_X(w \circ f)(x) = \lambda(x)^2 (\Delta_Y v)(f(x))$$

when $\lambda(x) \geqq 0$ is defined (independent of v) by

$$\lambda(x)^2 = \frac{\Delta_X(w \circ f)(x)}{(\Delta_Y w)(f(x))}.$$

Note that this simple proof does not use Lemma 3 in [6], which therefore can be entirely omitted, cf. the remark on p. 116 of [6].

2) I have just received a paper by Ishihara [12] (in manuscript) which overlaps with [6] in that the equivalence of 1) and 2) in the above Theorem 2 (= Theorem 7 in [6]) is obtained independently in [12] by a different approach involving a rather technical construction of

a harmonic function with prescribed first and second order partial derivatives at a given point of Y.

3) Through [12] I have become acquainted with a paper by Watson [13] in which it is proved among other things that a smooth mapping $f: X \to Y$ commutes with the Laplacian if and only if f is a harmonic, Riemannian submersion. This is precisely the case λ constant (= 1) of the bi-implication 2) $\Leftrightarrow$ 4) in Theorem 2 above. The general case of a variable λ cannot be reduced to the case $\lambda = 1$ by a local change of metric on X (except in the case $\dim X = 2$).

4) In the case $m = n$ with X a domain in $\mathbb{R}^n$ and $Y = \mathbb{R}^n$, Gehring and Haahti [11] established some of the results mentioned in this lecture, and they allowed the metric on $\mathbb{R}^n$ to be indefinite.

References

[1] Brelot, M.: Lectures on potential theory. Tata Institute of Fundamental Research, Bombay (1960).

[2] Constantinescu, C., Cornea, A.: Compactifications of harmonic spaces. Nagoya math. J. 25 (1965), 1 - 57.

[3] Eells, J., Sampson, J. H.: Harmonic mappings of Riemannian manifolds. Amer. J. Math. 86 (1964), 109 - 160.

[4] Forst, G.: Symmetric harmonic groups and translation invariant Dirichlet spaces. Inventiones math. 18 (1972), 143 - 182.

[5] Fuglede, B.: Finely harmonic mappings and finely holomorphic functions. Ann. Acad. Sci. Fenn. Ser. A I 2 (1976), 113 - 127.

[6] Fuglede, B.: Harmonic morphisms between Riemannian manifolds. Ann. Inst. Fourier 28,2 (1978), 107 - 144.

[7] Greene, R. E., Wu, H.: Embedding of open Riemannian manifolds by harmonic functions. Ann. Inst. Fourier 25,1 (1975), 215 - 235.

[8] Hervé, R.-M.: Recherches axiomatiques sur la théorie des fonctions surharmoniques et du potentiel. Ann. Inst. Fourier 12 (1962), 415 - 571.

[9] Janssen, K.: A co-fine domination principle for harmonic spaces. Math. Z. 141 (1975), 185 - 191.

[10] Sibony, D.: Allure à la frontière minimale d'une classe de transformations. Théorème de Doob généralisé. Ann. Inst. Fourier 18,2 (1968), 91 - 120.

[11] Gehring, F.W., Haahti, H.: The transformations which preserve the harmonic functions. Ann. Acad. Sci. Fenn. Ser. A I 293 (1960).

[12] Ishihara, T.: A mapping of Riemannian manifolds which preserves harmonic functions. Manuscript.

[13] Watson, B.: Manifold maps commuting with the Laplacian. J. diff. Geometry 8 (1973), 85 - 94.

Københavns Universitet
Matematiske Institut
Universitetsparken 5
DK-2100 København Ø
Danmark

NOTES ON CLUSTER SETS AT IDEAL BOUNDARY POINTS

Tatsuo Fuji'i'e

Let R be a hyperbolic Riemann surface, and R_M^* and R_W^* be its Martin and Wiener Compactification, respectively. Put $\Delta_M = R_M^* - R$ and $\Delta_W = R_W^* - R$. We call them Martin and Wiener boundary of R, respectively. We give some notes to the theory of cluster sets at Martin boundary points, especially to a theorem of Koebe type and that of Gross-Iversen type.

1. For a Martin boundary point p we define the full cluster set $C(f,p)$ of a non-constant meromorphic function f on R as

$$C(f,p) = \{\alpha \in \overline{C};\ f^{-1}(V) \cap U \neq \emptyset \text{ for all neighborhoods } V = V(\alpha) \text{ and } U = U(p)\}.$$

And for a closed set S on R we define the set γ_S as

$$\gamma_S = \{q \in \Delta_1 ;\ G \cap S \neq \emptyset \text{ for all } G \in \mathcal{G}_q\},$$

where Δ_1 is the set of minimal points of Δ_M and $\mathcal{G}_q$ is the filter basis of fine neighborhoods at q. Then we have

Proposition. If γ_S is of positive harmonic measure then $\overline{S} \cap \Gamma \neq \emptyset$, where Γ denotes the harmonic boundary of R_W^* and $\overline{S}$ is the closure of S in R_W^*.

Proof. Let q be a point of γ_S at which $\hat{1}_S(q)$ exists, where 1_S is the least superharmonic function on R which is not smaller than 1 on S, and $\hat{1}_S(q)$ is its fine limit at q. Since $G \cap S$, for arbitrary $G \in \mathcal{G}_p$, is non-polar there exists a regular point in $G \cap S$. $1_S(z) = 1$ for the regular point z and $\hat{1}_S(q) = 1$. The set of points such as q is of positive harmonic measure, so 1_S is not a Green potential and $\Gamma \cap \overline{S}$ is not empty [2].

For a special case, we consider γ_α of a cluster value $\alpha \in C(f,p)$. By taking as S $f^{-1}(\overline{V}_n)$ of a $1/n$ neighborhood V_n of α, we define γ_n and put $\gamma_\alpha = \bigcap_n \gamma_n$. To avoid a trivial case we suppose p is a regular point in the following, that is, $U \cap \Delta$ is of positive harmonic measure for all neighborhoods U of p.

Lemma 1. If there is a cluster value α for which γ_α is of harmonic measure 0 in a neighborhood of p, then γ_β is also of harmonic measure 0 for all β of $C(f,p)$.

<u>Proof.</u> If $\gamma_\alpha = \cap \gamma_n$ is of harmonic measure 0 then f has fine limit almost everywhere on $U \cap \Delta_1 - \gamma_n = \{q \in U \cap \Delta_1;\ \exists G \in \mathcal{G}_q,\ G \cap f^{-1}(\overline{V}_n) = \emptyset\}$ for all n, and on $\cup(U \cap \Delta_1 - \gamma_n) = U \cap \Delta_1 - (\cap \gamma_n)$, that is, almost everywhere on $U \cap \Delta$. Hence, if γ_β is of positive harmonic measure f must be a constant by the theorem of Lusin-Privalov type.

This fact shows that p is an F_1-point in the sense of Collingwood-Lohwater [1], that is, Fatou point set of f is of density 1 at p. And the lemma means also that if there exists a cluster value α for which γ_α is of positive harmonic measure then, for all cluster values β, γ_β is of positive harmonic measure.

<u>Lemma 2.</u> If there exists a cluster value $\alpha \in C(f,p)$ for which γ_α is of positive harmonic measure, then f is a constant or $\overline{C} - R(f,p)$ is of capacity 0, where $R(f,p)$ is the range of values of f at p. This means if f is not a constant p is a Frostman point.

<u>Proof.</u> If $\overline{C} - R(f,p)$ is of positive capacity, then there exists a neighborhood U of p in each connected component G_i of which f is a Fatou mapping. Let $\Delta_1(U) = \{p \in \Delta_1;\ U \in \mathcal{G}_p\}$, then $\Delta_1(U) = \bigcup_i \Delta_1(G_i)$, so some $\Delta_1(G_i)$ contains a part of γ_α of positive harmonic measure. Therefore, by the theorem of Lusin-Privalov type, f is a constant ($= \alpha$).

This lemma is a generalization of Koebe's theorem.

As a consequence of the above lemmas we have the following theorem which is a generalization of that of Collingwood-Lohwater [1].

<u>Theorem.</u> Every point of Δ_1 is an F_1-point or a Frostman point.

More precisely, the symmetric difference $\gamma_\alpha - \gamma_\beta$ must be of harmonic measure 0, otherwise f is a constant and $\alpha = \beta$. And in the case where γ_α is of positive harmonic measure in any neighborhood of p there exists a point $\mathcal{P}$ of Γ_p (harmonic boundary part in the fiber of Δ_W over p) for which $C(f,p) = C(f,\mathcal{P})$. Because, $\overline{f^{-1}(\overline{V})} \cap \Gamma \neq \emptyset$ by the proposition and so $\overline{f^{-1}(\overline{V})} \cap \Gamma_p \neq \emptyset$ for all neighborhoods V of α. Hence, $\bigcap_n (\Gamma_p \cap \overline{f^{-1}(\overline{V}_n)}) \neq \emptyset$ for a neighborhood system V_n of α. This means that there exists a point $\mathcal{P}$ of Γ_p which is contained in $\overline{f^{-1}(\overline{V}_n)}$ for all V_n. We conclude that $\alpha \in C(f,\mathcal{P})$.

We remark, in the above proof, that positivity of harmonic measure of γ_n for all V_n in place of γ_α is sufficient to conclude that $C(f,p) = C(f,\mathcal{P})$.

2. Now we decompose the cluster set $C(f,p)$ into two sets S_1 and S_2. S_1 is the set of $\alpha \in C(f,p)$ such that, for any neighborhood U

of p, there exists a neighborhood V of α such that at least one connected component D of $f^{-1}(V)$ is contained in U. $S_2 = C(f,p) - S_1$, that is, the set of α for which there exists a neighborhood U_o of p such that, for all neighborhoods V of α, any connected component of $f^{-1}(V)$ is not contained in U_o. Further, we consider two subsets S_1' and S_1'' of S_1. S_1' is the set of α such that among the above components D, we can find those of the class SO_{HB}. S_1' is an open set in $C(f,p)$. Each α of S_1' belongs to $R(f,p)$ except for a set of capacity 0. The exceptional set consists of asymptotic values α, and every neighborhood of p contains an asymptotic path for α [3].

S_1'' is the set of α for which we find, among the above components D, a component G which is not of the class SO_{HB}. S_1'' is a closed set. At almost every point q of $\Delta_1(G)$ f has a fine limit, so α belongs to the essential closed range $C^*(f,p)$ of fine boundary function $\hat{f}$ of f at p and $\alpha \in C(f,\mathscr{P})$ for $\mathscr{P} \in \Gamma_p$.

For each α of the set S_2 we can find a Koebe sequence of arcs converging to a continuum on Δ_M. Because, there exists a sequence of points $\{z_n\}$ tending to p such that $f(z_n)$ tends to α and z_n is contained in a component of $f^{-1}(V_n)$. Since this component contains a point of ∂U_o it contains an arc combining z_n to the point on ∂U_o. R_M^* is a compact metric space and f is non constant. Therefore, the sequence of those arcs converges to a continuum on Δ_M. We call this continuum a Koebe continuum and α a Koebe value. If γ_α for Koebe value α is of positive harmonic measure then f is a constant or $R(f,p) = \overline{C}$ - a set of capacity 0 [3].

Let $C_B^*(f,p)$ denote the essential fine boundary cluster set of f at p, that is, $C_B^*(f,p) = \bigcap_{E \in J} C_{B-E}^*(f,p)$, where $C_{B-E}^*(f,p)$ is the fine boundary cluster set modulo E and J denotes the collection of the sets of harmonic measure 0. Let α be in $C_B^*(f,p) - C^*(f,p)$ and suppose the harmonic measure of γ_n tends to 0 for $V_n \to \alpha$. Then almost every point of $U \cap \Delta_1$ is a Fatou point for a neighborhood U of p. Since $\alpha \in C_B^*(f,p)$, γ_n is of positive harmonic measure for all V_n and so, $U \cap F(f) \cap \gamma_n$ is of positive harmonic measure, where $F(f)$ denotes the set of Fatou points of f. f has fine limit at almost every point on $U \cap F(f) \cap \gamma_n$, which is contained in $\overline{V}_n$. We have $\alpha \in C^*(f,p)$. This contradicts our assumption, and γ_α must be of positive harmonic measure. Therefore, if $C_B^*(f,p) - C^*(f,p)$ is not empty, then γ_α is of positive harmonic measure for all $\alpha \in C(f,p)$ by Lemma 1 and hence $C_B^*(f,p) = C(f,p)$. And if f is not a constant p is a Frostman point and $R(f,p) = \overline{C}$ - Cap 0, and both $C(f,p)$ and $C_B^*(f,p)$ are total. Consequently, from the fact that $C(f,p) - C_B^*(f,p)$ is not empty, it follows that $p \in F_1(f)$ and $C_B^*(f,p) = C^*(f,p)$.

We have the following theorem of Gross-Iversen type.

Theorem. If $\Omega = C(f,p) - C_B^*(f,p) \neq \phi$, then $C_B^*(f,p) = C^*(f,p)$ and Ω consists of the range of values $R(f,p)$, a set of asymptotic values of capacity 0 and a set of Koebe values α for which γ_α is of harmonic measure 0. If there exists, in this case, a Koebe value α with γ_α of positive harmonic measure f is a constant.

References

[1] Collingwood, E. F., Lohwater, A. J.: The theory of cluster sets. Cambridge University Press, Cambridge (1966).

[2] Constantinescu, C., Cornea, A.: Ideale Ränder Riemannscher Flächen. Springer-Verlag, Berlin - Göttingen - Heidelberg (1963).

[3] Fuji'i'e, T.: On asymptotic behaviors of analytic mappings at Martin boundary points. J. Math. Kyoto Univ. 13 (1973), 373 - 380.

Kyoto University
Kyoto 606
Japan

A TECHNIQUE FOR EXTENDING QUASICONFORMAL EMBEDDINGS

David B. Gauld

1. Introduction

The torus furling technique of geometric topology, described in §8 of [9], has also been found to be directly useful in codimension zero extensions of topological embeddings, see [3] and [10]. This technique also respects quasiconformality, so it may be used to obtain quasiconformal extensions. Furthermore, it is easy to calculate the dilatation, cf. the special case in [12]. Although the technique itself is described in a fairly general setting, the illustrations will be given in the context of the following problem, which is related to the second goal of the paper.

Problem. Let A be a subset of $\mathbb{R}^n$ which is homeomorphic to S^{n-1}, the (n-1)-sphere. Let I denote the inside (bounded) component of $\mathbb{R}^n - A$. Is $\bar{I} \approx \bar{B}^n$?

In the statement of the problem as well as in what follows, $\bar{S}$ denotes the closure of the set S, and $\approx$ denotes either topological or quasiconformal homeomorphism.

Of course it is well-known that for $n \geq 3$, the answer to the question is in general "no". However, by successive localization of the above problem, we are able to obtain a useful criterion for an affirmative answer. More precisely, consider the following five statements.

(I) $\bar{I} \approx \bar{B}^n$.

(II) There is a neighbourhood N of S^{n-1} in $\bar{B}^n$ and an embedding of the pair (N, S^{n-1}) in $(\bar{I}, A)$.

(III) There is a homeomorphism $h : S^{n-1} \to A$ so that at each $x \in S^{n-1}$, h extends to a homeomorphism of a neighbourhood of x in $\bar{B}^n$ to a neighbourhood of $h(x)$ in $\bar{I}$.

(IV) For each $y \in A$ there is an embedding of the triple $(B^n_+, B^{n-1}, 0)$ in the triple $(\bar{I}, A, y)$.

(V) Each $y \in A$ has a neighbourhood N in $\bar{I}$ so that $N \cap I$ is homeomorphic to B^n.

In statement (IV), B^n_+ denotes the set $\mathbb{R}^n_+ \cap B^n$, where $\mathbb{R}^n_+ = \{(x_1,\dots,x_n) \in \mathbb{R}^n \mid x_n \geq 0\}$, and B^{n-1} is considered as a subset of

B^n (hence B^n_+) by identifying $(x_1,\dots,x_{n-1}) \in B^{n-1}$ with $(x_1,\dots,x_{n-1},0) \in B^n$.

The above five statements have interpretations in either the topological or the quasiconformal context. It is clear that (I) ⇒ (II) ⇒ (III) ⇒ (IV) ⇒ (V) in either context. We shall consider the reverse implications. In particular we shall find that each implication is reversible in the quasiconformal context except possibly (III) ⇒ (IV) when $n = 5$. Thus we obtain the following theorem.

Theorem. Suppose $n \neq 5$. A Jordan domain D in $\overline{\mathbb{R}}^n$ can be mapped quasiconformally onto B^n by a homeomorphism which extends homeomorphically over the boundary if and only if each point of ∂D has a neighbourhood N such that $N \cap D$ is quasiconformally equivalent to B^n.

Of course this theorem was already known for $n = 3$, see [13] and [17].

2. The extension technique

In its basic form, the technique may be described as follows. Let X be a "nice" (compact, locally connected and Hausdorff will do) space and $e : X \times (-4,4) \to (X \times \mathbb{R})$ an embedding near the inclusion. Define $\alpha,\beta,\gamma : X \times \mathbb{R} \to X \times \mathbb{R}$, homeomorphisms, as follows:

$\alpha(x,t) = (x,t+3)$;

β is the identity on $X \times [2,\infty)$, α^{-1} on $X \times (-\infty,1]$ and expands $X \times [1,2]$ linearly on the second factor to $X \times [-2,2]$;

γ is the identity on $X \times (-\infty,-2]$, α^{-1} on $X \times [2,\infty)$ and contracts $X \times [-2,2]$ linearly on the second factor to $X \times [-2,-1]$.

The composition

$$E = e\gamma e^{-1}\beta e\alpha : X \times [-3,0] \to X \times \mathbb{R},$$

represented schematically in [10, figure 1] and [11, figure 7], has the same effect on $X \times \{-3\}$ as on $X \times \{0\}$; more precisely, for any $x \in X$,

$$E(x,-3) = E(x,0) - 3,$$

where, for $y \in X \times \mathbb{R}$ and $r \in \mathbb{R}$, $y - r$ means "subtract r from the $\mathbb{R}$-coordinate of y". This nice feature of E allows one to stack such embeddings together, as in [3], to obtain a homeomorphism $\hat{E} : X \times \mathbb{R} \to X \times \mathbb{R}$. The homeomorphism $\hat{E}$ is defined as follows. Let $(x,t) \in X \times \mathbb{R}$. Choose the integer n so that $t + 3n \in [-3,0]$, and set

$$\hat{E}(x,t) = E(x,t+3n) - 3n.$$

The nice property of E eliminates ambiguity when t is an integer multiple of 3. Note that if e is quasiconformal, then so are E and $\hat{E}$. The only apparent singular set of $\hat{E}$ is $\bigcup_{n \in Z} X \times \{3n\}$, which has measure 0 : in fact even this set is not singular since in a neighbourhood of $X \times \{3n\}$, $\hat{E}$ is merely a conjugation of e by translations.

There is a variation on the above stacking, discussed in [10]. In this variation, $X \times \mathbb{R}$ is replaced by an open cone on X, say $X \times [-\infty,\infty) \mid X \times \{-\infty\}$. The procedure is the same as above, but we must crush $X \times \{-\infty\}$ to the vertex of the cone. The stacked homeomorphism preserves this vertex.

In our discussion of the extension technique above, we have not yet mentioned the problem of well-definition of E. There are two problems, each caused by the use of e^{-1}. If e is close enough to the inclusion, then it will stretch $X \times (-4,4)$ all the way from $X \times \{-3\}$ to $X \times \{3\}$, which disposes of one problem. Even so, $e(X \times (-4,4))$ might not contain all of $\beta e\alpha(X \times [-3,0])$ because e might not be locally surjective. This is where the niceness assumption relating to X comes in: we appeal to [19, 1.7].

3. First illustration

The first illustration is the major part of a proof of (II) $\Rightarrow$ (I) in the quasiconformal context. Since it is treated fully in §5 of [11], full details are not given here. In lemma 9 of [11], the extension technique is applied to the case where $X = S^{n-1}$, and the cone over X is identified with $\overline{B}^n$, with the origin as vertex and the radial lines as generators of the cone. The role of $X \times (-4,4)$ is played by the neighbourhood N. Essentially one takes an annulus within the interior of N, moves it out radially so as still to be within N, applies the embedding from (II), stretches radially inward, and so on as in §2. At the end one has an embedding which agrees with the original on one sphere (so can be the original outside this sphere as well) and is a conjugate (via expansion and contraction) on a sphere of smaller radius. One then stacks such embeddings of the annulus between these spheres together to obtain an embedding of all of $\overline{B}^n$ onto $\overline{I}$.

The remainder of §5 of [11] is devoted to simplifying the general case of (II) to the specific case considered in the previous paragraph. It might be noted, however, that lemma 8 of [11] is superfluous. In fact the procedure described in §2 above is wasteful in the sense that near $X \times \{0\}$, the composition $\gamma e^{-1}\beta e\alpha$ is the identity as is the composition $e\gamma e^{-1}$ near $\beta e\alpha(X \times \{-3\})$.

4. A sewing lemma

Although the sewing lemma can be formulated and verified in a more general setting, we will consider it only in the context of the above problem. In the lemma we use piecewise linear homeomorphisms which are assumed to preserve spheres, so we must use the following norm on $\mathbb{R}^n$: for $x = (x_1,\dots,x_n) \in \mathbb{R}^n$, let

$$|x| = \max\{|x_i| \mid i = 1,\dots,n\}.$$

For the purpose of the lemma only, the sets C_1 and C_2 are the following adjacent cubes in $\mathbb{R}^{n-1}$ with an (n-2)-face in common:

$$C_1 = \bar{B}^{n-1},\ C_2 = \{(x_1,\dots,x_{n-1}) \in \mathbb{R}^{n-1} \mid (x_1,\dots,x_{n-1} - 2) \in \bar{B}^{n-1}\}.$$

Sewing lemma. Suppose $n \neq 5$. Let M be a neighbourhood of $C_1 \cup C_2$ in $\mathbb{R}^{n-1}$ and N a neighbourhood of C_1 in $\mathbb{R}^n_+$. Let $g : M \to A$ and $h_i : N \to \bar{I}$ $(i = 1,2)$ be embeddings satisfying the following:
(i) for $i = 1,2$, h_i is quasiconformal;
(ii) $h_i(N \cap \mathbb{R}^{n-1})$ is a neighbourhood of $g(C_i)$ in A $(i = 1,2)$;
(iii) $h_1 | M \cap N = g | M \cap N$.

Then there are embeddings $g' : M \to A$ and h of a neighbourhood in $\mathbb{R}^n_+$ of $C_1 \cup C_2$ to $\bar{I}$ satisfying:
(a) h is quasiconformal;
(b) the image of h is a neighbourhood of $g(C_1 \cup C_2)$ in $\bar{I}$;
(c) h and g' agree on a neighbourhood of $C_1 \cup C_2$ in $\mathbb{R}^{n-1}$;
(d) $g'(M) = g(M)$, g' agrees with g on a neighbourhood in $\mathbb{R}^{n-1}$ of C_1 and near the frontier of M, and g' is arbitrarily near g.

Outline of proof. Choose a piecewise linear homeomorphism $f : \mathbb{R}^{n-1} \to \mathbb{R}^{n-1}$ which is the identity on a neighbourhood of C_1 and near and beyond the frontier of the domain of $h_2^{-1}g$ (which is a neighbourhood of C_2) and takes C_2 into $M \cap N$. Since f has compact support, $f \times 1 : \mathbb{R}^n \to \mathbb{R}^n$ is quasiconformal, cf. [20]. The composition $h_1(f \times 1)$ is almost the embedding h which is being sought, except that the shrinking effect of f needs to be reversed to attain (b). This reversal is carried out in the range of $h_2^{-1}g$.

Using [15] when $n - 1 \leq 3$ and [2] and [7] or [18, page 194] when $n - 1 \geq 5$, approximate $h_2^{-1}g$ sufficiently closely by a piecewise linear homeomorphism $\varphi : \mathbb{R}^{n-1} \to \mathbb{R}^{n-1}$. The required embedding h is just h_1 in a neighbourhood of C_1 and in a neighbourhood of C_2 it is the composition

$$h_2(\varphi f^{-1} \varphi^{-1} \times 1)h_2^{-1}h_1(f \times 1). \qquad \square$$

Returning to statement (IV) in the quasiconformal context, compactness of A gives finitely many quasiconformal embeddings

$$h_i : (B^n_+, B^{n-1}) \to (\bar{I}, A) \quad (i = 1,\dots,k)$$

so that $\bigcup_{i=1}^{k} h_i(B^{n-1}) = A$. Using the sewing lemma, we can reduce k to 2, at least when $n \neq 5$. In the next section we will see how to use the extension technique to deduce statement (III).

Using the generalised Schoenflies theorem, we can assume that $h_1 \mid B^{n-1}$ extends to a homeomorphism $\bar{h}_1 : \bar{\mathbb{R}}^{n-1} \to A$. We can also assume that $\bar{h}_1(\infty) \in h_k(B^{n-1})$. Let C be a cube which contains the compact set $\bar{\mathbb{R}}^{n-1} - \bar{h}_1^{-1} h_k(B^{n-1})$ in its interior. Rectilinearly subdivide C into small cubes each of which lies in $\bar{h}_1^{-1} h_i(B^{n-1})$ for some i, and ordered in some way so that the first lies in $B^{n-1} = \bar{h}_1^{-1} h_1(B^{n-1})$ and that any cube K_i is related to the union of its predecessors $\bigcup_{j<i} K_j$ by the following:

$$(\mathbb{R}^{n-1};\, C_1, C_2) \underset{PL}{\approx} (\mathbb{R}^{n-1};\, \bigcup_{j<i} K_j, K_i) \quad (i > 1)$$

where C_1 and C_2 are the cubes of the sewing lemma.

One now applies the sewing lemma inductively to the cubes $K_1, K_2, \dots$ and the embeddings $\bar{h}_1, h_i$ to obtain a homeomorphism $\tilde{h}_1 : \bar{\mathbb{R}}^{n-1} \to A$ and a quasiconformal embedding $h_1' : (B^n_+, B^{n-1}) \to (\bar{I}, A)$ so that $\tilde{h}_1$ and h_1' agree on B^{n-1} and

$$\tilde{h}_1(B^{n-1}) \cup h_k(B^{n-1}) = A.$$

Thus we have reduced the number of local collars to 2.

5. Second illustration

Consider the following situation which, on reflection, is the situation at the end of §4. Given a homeomorphism $\bar{h}_1 : \bar{\mathbb{R}}^{n-1} \to A$ and quasiconformal embeddings h_1 of a neighbourhood in $\bar{\mathbb{R}}^n_+$ of $\bar{\mathbb{R}}^{n-1} - B^{n-1}$ and h_2 of B^n_+ into $\bar{I}$, so that h_1 and $\bar{h}_1$ agree on the intersection of their domains, $h_i^{-1}(A)$ is the intersection of the domain of h_i with $\bar{\mathbb{R}}^{n-1}$ and

$$h_1(\bar{\mathbb{R}}^{n-1} - \bar{B}^{n-1}) \cup h_2(B^{n-1}) = A.$$

Now $h_2^{-1} h_1$ embeds a neighbourhood of S^{n-2} in $(\mathbb{R}^n_+, \mathbb{R}^{n-1})$ into (B^n_+, B^{n-1}) so by the extension technique of §2 we may adjust $h_2^{-1} h_1$ to a new embedding φ which agrees with the old on a neighbourhood in $\mathbb{R}^n_+$ of S^{n-2} and repeats itself on two concentric spheres as did the embedding E in §2. The domain of φ is a neighbourhood of the base of the cone with vertex the origin and base the intersection of the above neighbourhood of S^{n-2} with S^{n-1}_+.

Just as in §2, we can stack embeddings like φ together to give an embedding, call it φ also, of the whole of the cone into (B^n_+, B^{n-1}). If M denotes the union of this cone with that part of the domain of h_1 lying outside S^{n-1}, then we may define $g : M \to \overline{I}$ as follows: if $x \in M$, then set

$$g(x) = \begin{cases} h_1(x), & \text{if } |x| \geqq 1 \\ h_2\varphi(x), & \text{if } |x| \leqq 1. \end{cases}$$

Note that g is quasiconformal, takes $\overline{\mathbb{R}}^{n-1}$ homeomorphically onto A and provides a pinched collar for A in $\overline{I}$, ie. a collar at each point of A except $g(0)$. The embedding h_2 provides a collar around the point $g(0)$, but we do not yet have condition (III) as h_2 and g do not necessarily agree when restricted to $\overline{\mathbb{R}}^{n-1}$. To rectify this situation we give two separate arguments, the first is valid when $n \leqq 4$ and the second is valid when $n \neq 4$.

Firstly, consider the embedding $g^{-1}h_2 : B^{n-1} \to \mathbb{R}^{n-1}$. Thus is quasiconformal, so by [1] when $n = 3$ or [5] when $n = 4$, it extends to a quasiconformal embedding $f : B^n_+ \to \mathbb{R}^n_+$. If we replace the local collar h_2 by the local collar $h_2 f^{-1}$ then we achieve statement (III).

Secondly, we enlarge the image of the collar h_1 even before appealing to the extension technique. Since $\mathrm{Im}(h_1) \cup \mathrm{Im}(h_2)$ is a neighbourhood of A in $\overline{I}$, there is a piecewise linear embedding $\psi : \overline{B}^n \to I$ satisfying:

$$\mathrm{Im}(h_1) \cup \mathrm{Im}(h_2) \cup \mathrm{Im}(\psi) = \overline{I};\ \psi(S^{n-1}_+) \subset \mathrm{Im}(h_1) \text{ and } \psi(S^{n-1}_-) \subset \mathrm{Im}(h_2).$$

This situation requires $n \neq 4$, and is achieved by firstly using the Schoenflies theorem to obtain a topological embedding like ψ and then, since $n \neq 4$, piecewise linearly approximating. Now choose a piecewise linear homeomorphism χ of $\overline{B}^n$ which is 1 on S^{n-1} and satisfies

$$\chi\psi^{-1}\mathrm{Im}(h_1) \cup \psi^{-1}\mathrm{Im}(h_2) = \overline{B}^n.$$

If we replace h_1 by the collar which agrees with h_1 off $h_1^{-1}\psi(\overline{B}^n)$ but with $\psi\chi\psi^{-1}h_1$ on $h_1^{-1}\psi(\overline{B}^n)$ then we achieve the situation where the union of the images of the collars is all of $\overline{I}$. Repeating the above construction of g in this new situation we find that g essentially exhibits statement (I).

6. Comments and conclusions

There are two topological methods for proving (II) ⇒ (I), one in [4] and the other in [14] and [16]. Each of these has been adapted to the quasiconformal context, the second by Gehring in [13], and the first more recently in [11], as summarised in §3. As noted in [20] the proof from [4] does not readily carry over to the quasiconformal context and it is the extension technique of §2 which allows this.

There are also two topological methods for proving (III) ⇒ (II), one in [5] and the other in [8]. Again these adapt to the quasiconformal context, the first in [13] and the second in [11].

By way of contrast, the situations regarding the implications (IV) ⇒ (III) and (V) ⇒ (IV) are completely different. At the one extreme, the implication (IV) ⇒ (III) is trivial in the topological context but is distinctly non-trivial in the quasiconformal context, remaining unknown for $n = 5$. At the other extreme, the implication (V) ⇒ (IV) is false in the topological context, at least for $n = 3$, see [18, p. 68]. Its truth in the quasiconformal context is established in the proof of theorem 4.7 in [17].

Conversations with Prof. Jussi Väisälä subsequent to the Colloquium helped rectify a flaw in the original version of this paper, and recent joint work with him has resulted in a number of generalizations of the theorem.

References

[1] Ahlfors, L.: Extension of quasiconformal mappings from two to three dimensions. Proc. nat. Acad. Sci. USA 51 (1964), 768 - 771.

[2] Bing, R.H.: Radial engulfing, in "Conference on the topology of manifolds". Prindle, Weber and Schmidt, Boston - London - Sydney (1968), 1 - 18.

[3] Brakes, W.R.: An improved version of the non-compact weak canonical Schoenflies theorem. Trans. Amer. math. Soc. 213 (1975), 61 - 69.

[4] Brown, M.: A proof of the generalized Schoenflies theorem. Bull. Amer. math. Soc. 66 (1960), 74 - 76.

[5] Brown, M.: Locally flat imbeddings of topological manifolds. Ann. of math., II. Ser. 75 (1962), 331 - 341.

[6] Carleson, L.: The extension problem for quasiconformal mappings, in "Contributions to analysis". Academic Press, New York and London (1974), 39 - 47.

[7] Connell, E.H.: Approximating stable homeomorphisms by piecewise linear ones. Ann. of math., II. Ser. 78 (1963), 326 - 338.

[8] Connelly, R.: A new proof of Brown's collaring theorem. Proc. Amer. math. Soc. 27 (1971), 180 - 182.

[9] Edwards, R.D., Kirby, R.C.: Deformations of spaces of imbeddings. Ann. of math., II. Ser. 93 (1971), 63 - 88.

[10] Gauld, D.B.: Local contractibility of spaces of homeomorphisms. Compositio math. 32 (1976), 3 - 11.

[11] Gauld, D.B., Vamanamurthy, M.K.: Quasiconformal extensions of mappings in n-space. Ann. Acad. Sci. Fenn., Ser. A I 3 (1977), 229 - 246.

[12] Gauld, D.B., Vamanamurthy, M.K.: A special case of Schönflies theorem for quasiconformal mappings in n-space. Ann. Acad. Sci. Fenn., Ser. A I 3 (1977), 311 - 316.

[13] Gehring, F.W.: Extension theorems for quasiconformal mappings in n-space. J. Analyse math. 19 (1967), 149 - 169.

[14] Mazur, B.C.: On embeddings of spheres. Acta math. 105 (1961), 1 - 17.

[15] Moise, E.E.: Geometric topology in dimensions 2 and 3. Springer-Verlag, New York - Berlin - Heidelberg (1977).

[16] Morse, M.: A reduction of the Schoenflies extension problem. Bull. Amer. math. Soc. 66 (1960), 113 - 115.

[17] Näkki, R.: Boundary behaviour of quasiconformal mappings in n-space. Ann. Acad. Sci. Fenn., Ser. A I 484 (1970), 1 - 50.

[18] Rushing, T.B.: Topological embeddings. Academic Press, New York and London (1973).

[19] Siebenmann, L.C.: Deformations of homeomorphisms on stratified sets. Commentarii math. Helvet. 47 (1972), 123 - 163.

[20] Väisälä, J.: Lectures on n-dimensional quasiconformal mappings. Lecture Notes in Mathematics 229, Springer-Verlag, Berlin - Heidelberg - New York (1971).

Department of Mathematics
University of Auckland
New Zealand

UNIFORM HARMONIC APPROXIMATION ON UNBOUNDED SETS*

P. M. Gauthier and W. Hengartner

Let u be a function defined on a subset F of the plane $\mathbb{R}^2$. We wish to approximate u by a function harmonic on F. This problem has received considerable attention for the case that F is bounded. In the present report we consider the situation where F is unbounded.

Suppose for the moment that such an approximation is possible. That is, for each positive ε, there is a function h harmonic on F (this means harmonic in a neighbourhood of F) such that

$$|h(z) - u(z)| < \varepsilon, \; z \in F. \tag{1}$$

Clearly this forces u to be continuous on F and harmonic on F^o. Thus, we can and shall assume, without loss of generality, that u has these properties.

Moreover, (1) implies that the "sequence" h_ε converges on the closure of F to an extension of u. Thus, we also assume, without loss of generality, that F is closed.

Notice that the approximation (1) is required to be uniform "all the way out" and not just on compact portions. Also, there is no assumption whatsoever on the behaviour of u at infinity; it may be unbounded and oscillate. Thus, there does not appear to be any easy way of performing "unbounded approximation" as a corollary of "compact approximation". Moreover, approximation on unbounded sets is not easily amenable to linear methods, since the set of continuous functions on F, with the topology of uniform convergence, is not a linear topological space. Indeed, scalar multiplication is not continuous, since unbounded functions are allowed.

If u can be approximated by functions harmonic in a neighbourhood of F, it is preferable to do so in the largest possible neighbourhood.

In the present paper, we only seek to approximate by a function h harmonic on F. In [3] one is given a function h harmonic on F and one attempts to approximate h by a function h_o globally harmonic on all of $\mathbb{R}^2$ except for isolated logarithmic singularities. We call such a function h_o essentially harmonic. By combining the present paper with [3], we are thus equipped to approximate fairly general functions

* Research supported by the Natural Sciences and Engineering Research Council in Canada and the Ministère d'Education du Québec.

on F by essentially global harmonic functions. In case $F^{\circ} = \emptyset$, this has already been accomplished [7].

The analogous problem of holomorphic approximation has been treated completely ([5], [6]) and has yielded profound applications in function theory (see [2] for a short survey). It seems reasonable to expect that the harmonic approximation also will have important applications in potential theory.

Unbounded harmonic approximation may be of interest also in mathematical physics. For example, it is frequently convenient to consider an equilibrium temperature distribution on an unbounded region, and it is natural to ask whether this temperature distribution can be approximated by a temperature distribution on a larger region.

We introduce some terminology to describe the type of set F on which we can approximate. The interior of F is said to be <u>without holes</u> if each component of F° is simply connected. A boundary point $p \in \partial F$ is said to be a <u>Jordan point</u> if there is a neighbourhood V of p such that $V \cap \partial F$ is a simple Jordan arc. The set F is said to have <u>Jordan windows</u> if there is an exhaustion of the plane by Jordan domains G_n, $n = 1,2,\ldots$, such that for each n, the boundary ∂G_n meets ∂F in a finite set of Jordan points.

A non-negative real-valued function is said to be <u>quasibounded</u> on an open set U if it is the limit of an increasing sequence of bounded harmonic functions. A real-valued function is said to be quasibounded if it is a difference of positive quasibounded functions.

Let H_u^U denote the generalized solution of the Dirichlet problem for U with boundary values u. Then, a continuous function u on $\bar{U}$ is quasibounded on U if and only if it is the solution of its own Dirichlet problem; that is, if and only if

$$u = H_u^U, \text{ on } U.$$

We now state our principal result.

<u>Theorem.</u> Let F be a closed set of $\mathbb{R}^2$ for which $\mathbb{R}^2 \setminus F$ is thin at no point of ∂F. Suppose also that F has Jordan windows and F° is without holes. Then, each function u, continuous on F and harmonic quasibounded on F°, can be uniformly approximated by functions essentially harmonic on $\mathbb{R}^2$.

As a direct consequence of the results in [3], we have:

<u>Corollary.</u> Let F and u be as in the theorem. If moreover $(\mathbb{R}^2 \cup \{\infty\}) \setminus F$ is connected and locally connected, then u can be

uniformly approximated by functions harmonic on $\mathbb{R}^2$.

The proof of our theorem is based on the following lemmas.

Lemma. Let F and u be as in the theorem. Then, u can be uniforml approximated on F by functions continuous on F and harmonic on

$$F^o \cup \bigcup_{n=1}^{\infty} \partial F \cap \partial G_n . \qquad (2)$$

Fusion Lemma [3]. Let K_1 and K_2 be compact sets on $\mathbb{R}^2 \cup \{\infty\}$ an let V be a neighbourhood of $K_1 \cap K_2$. There is a positive constant a such that if q_1 and q_2 are essentially harmonic functions on $\mathbb{R}^2 \cup \{\infty$ satisfying for some $\varepsilon > 0$,

$$|q_1 - q_2|_V < \varepsilon,$$

then there is a function ω, essentially harmonic on $\mathbb{R}^2 \cup \{\infty\}$, such tha for $j = 1,2$,

$$|\omega - q_j|_{K_j} < a\varepsilon .$$

Suppose, for the moment, that our lemma is valid and let us prove the theorem.

Let $\varepsilon > 0$ and let u_1 be harmonic on (2) and continuous on F suc that

$$|u - u_1| < \varepsilon/2 \quad \text{on} \quad F.$$

Then, u_1 is harmonic in some closed neighbourhood $\bar{U}$ of

$$\bigcup_{n=1}^{\infty} \partial F \cap \partial G_n .$$

Set $U_n = (F \cup \bar{U}) \cap G_n$, where G_n is the exhaustion associated to the Jordan windows. We apply the Fusion Lemma, where we replace K_1, K_2, V by $\overline{G_n}$, $F \cup \{\infty\} \setminus G_n$, U_{n+1}. We may choose the $\{a_n\}$ in the Fusion Lemma so that $1 < a_n < a_{n+1}$ and select positive numbers $\varepsilon_1, \varepsilon_2, \ldots$, so that

$$\varepsilon_{n+1} < \varepsilon_n \quad \text{and} \quad \sum_{n=1}^{\infty} \varepsilon_n < \varepsilon/2.$$

By a theorem of Deny ([1], [4, p. 341]), there exist essentially harmonic functions q_n on $\mathbb{R}^2 \cup \{\infty\}$ such that, for $n = 1,2,\ldots$,

$$|q_n - u_1| < \varepsilon_n/2a_n \quad \text{on} \quad \bar{U}_{n+1}, \qquad (3)$$

and therefore

$$|q_{n+1} - q_n| < \varepsilon_n/a_n \quad \text{on} \quad \bar{U}_{n+1}, \qquad (4)$$

By the Fusion Lemma, for each $n = 1,2,\dots,$ there exists an essentially harmonic function Ω_n on $\mathbb{R}^2 \cup \{\infty\}$ such that

$$|\Omega_n - q_n| < \varepsilon_n \quad \text{on} \quad \overline{G_n}, \tag{5}$$

$$|\Omega_n - q_{n+1}| < \varepsilon_n \quad \text{on} \quad F \cup \{\infty\} \setminus G_n. \tag{6}$$

The inequalities (5) yield

$$\sum_{\nu=n}^{\infty} |\Omega_\nu - q_\nu| < \sum_{\nu=n}^{\infty} \varepsilon_\nu \quad \text{on} \quad \overline{G_n},$$

and therefore,

$$h = q_1 + \sum_{\nu=1}^{\infty} (\Omega_\nu - q_\nu)$$

is essentially harmonic on $\mathbb{R}^2 = \bigcup_{n=1}^{\infty} G_n$.

We show now that on F, h approximates u_1 uniformly and therefore u. Set $F_n = F \cap \overline{G_n}$. From (3) and (5), we have on F_1:

$$|h - u_1| \leqq |q_1 - u_1| + \sum_{\nu=1}^{\infty} |\Omega_\nu - q_\nu| < \frac{\varepsilon_1}{2a_1} + \sum_{\nu=1}^{\infty} \varepsilon_\nu < \varepsilon.$$

Now on $F_n \setminus F_{n-1}$, $n = 2,3,\dots,$ the inequalities (6), (3) and (5) imply

$$|h - u_1| \leqq \sum_{\nu=1}^{n-1} |\Omega_\nu - q_{\nu+1}| + |q_n - u_1| + \sum_{\nu=n}^{\infty} |\Omega_\nu - q_\nu| < \sum_{\nu=1}^{n-1} \varepsilon_\nu + \frac{\varepsilon_n}{2a_n} + \sum_{\nu=n}^{\infty} \varepsilon_\nu < \varepsilon.$$

This completes the proof of the theorem modulo the lemma.

We now prove the lemma. Let p be a Jordan point on ∂F and let $\varepsilon > 0$. By induction, it is sufficient to find a neighbourhood V of p and a function u_p with the following properties:

1) F_p^o is without holes, where $F_p = F \cup V$.
2) u_p is continuous on F_p, harmonic and quasibounded on F_p^o.
3) $|u - u_p| < \varepsilon$ on F.

The property that $\mathbb{R}^2 \setminus F$ be not thin at each point of ∂F is not required in the proof of the lemma.

We may assume that $u(p) = 0$. Since u is quasibounded,

$$u = H_u = H_{u^+} - H_{u^-}.$$

Thus, we may assume that $u \geqq 0$. We shall also extend u continuously to all of $\mathbb{R}^2$ and continue to denote the extension by u.

Since p is a Jordan point, there is a Jordan domain V containing p, such that $\overline{V} \cap \partial F$ is a simple Jordan arc. Thus F_p^o is without holes. We may also assume that V is so small that $u < \varepsilon/2$ on V.

We define

$$u_p = \begin{cases} u & \text{on } F \setminus F_p^o \\ H_u^{F_p^o} & \text{on } F_p^o . \end{cases}$$

Let ω denote harmonic measure for F^o, and ω^p, harmonic measure for F_p^o. We know that on ∂F^o, $u \in L(\omega)$, since $u = H_u^{F^o}$. We wish to show that u is also in $L(\omega^p)$ on ∂F_p^o. We may restrict our attention to a single component Ω_p of F_p^o. Since F_p^o is without holes, Ω is simply connected. We may also assume that $p \in \Omega_p$. Denote by Ω the component $F^o \cap \Omega_p$ of F^o. We have $p \in \partial\Omega$ and $\Omega_p = \Omega \cup V$.

Let f and f_p be conformal maps of the unit disc Δ onto Ω and Ω_p respectively. We normalize so that

$$f_p^{-1}(\partial\Omega \setminus V) = f^{-1}(\partial\Omega \setminus V). \tag{7}$$

Let Γ denote the arc (7) on $\partial\Delta$, and let I be an arc on $\partial\Delta$ containing $\partial\Delta \setminus \Gamma$ in its interior. We may assume that u is bounded by ε on $f_p(I)$ and $f(I)$.

Since harmonic measure is invariant under conformal mapping our problem is to show that $v_p = u \circ f_p$ is integrable on $\partial\Delta$ given that $v = u \circ f$ is.

We have

$$\int_0^{2\pi} v_p(\varphi)\,d\varphi = \int_{\partial\Delta\setminus I} v_p(\varphi)\,d\varphi + \int_I v_p(\varphi)\,d\varphi .$$

Since v_p is bounded on I, we need only consider the integral over $\partial\Delta \setminus I$.

Set $\varphi = f^{-1}f_p$. Then $\varphi : \Gamma \to \Gamma$ and therefore $\varphi(I)$ is again a neighbourhood of $\partial\Delta \setminus \Gamma$. Moreover, φ admits a conformal extension to a neighbourhood of $\partial\Delta \setminus I$. Note that φ' is bounded away from zero here. Therefore, we have

$$\int_{\varphi(\partial\Delta\setminus I)} v(\varphi)\,d\varphi = \int_{\partial\Delta\setminus I} v(\varphi(t))\varphi'(t)\,dt.$$

Since v is integrable, $v \circ \varphi = v_p$ is also integrable on $\partial\Delta \setminus I$ with respect to Lebesgue measure. Thus $u \in L(\omega^p)$. Therefore, u_p is finite on Ω and hence harmonic.

We now have that u_p is harmonic and quasibounded on F_p^o and since F_p^o is regular for the Dirichlet problem, u_p is continuous on F_p.

There remains to show that u_p is near u on F. Again, let Ω be the component of F^o which meets V, and let $\Omega_p = \Omega \cup V$. Since $u = u_p$ on $F \setminus \overline{\Omega}$, we restrict our attention to $\overline{\Omega}$.

Since both u and u_p are quasibounded, it is sufficient to show that u and u_p are near on $\partial\Omega$. Now u and u_p are actually equal on $\partial\Omega \setminus V$ so we need only show that u and u_p are close on $\partial\Omega \cap V$. Since $u < \varepsilon/2$ here, we have only to show that u_p is also small.

Fix $z \in \Omega$. Then

$$u_p(z) = \int_{\partial\Omega_p} u(q)\,d\omega_z^p(q) = \int_{\partial\Omega\setminus V} u(q)\,d\omega_z^p(q) + \int_{V\setminus\partial\Omega} u(q)\,d\omega_z^p(q).$$

Since $u < \varepsilon/2$ on $\partial\Omega \cap V$, the last integral is small, and we may restrict our attention to the integral over $\partial\Omega \setminus V$. We note that on this set, $\omega^p < \omega$, and so

$$\int_{\partial\Omega\setminus V} u(q)\,d\omega_z^p(q) \leq \int_{\partial\Omega\setminus V} u(q)\,d\omega_z(q) = u(z) - \int_{\partial\Omega\cap V} u(q)\,d\omega_z(q).$$

Since this last integral is again bounded by $\varepsilon/2$, we have

$$u_p(z) \leq u(z) + \varepsilon, \quad z \in \Omega.$$

Since $u \leq \varepsilon/2$ on $\partial\Omega \cap V$, it follows that $u_p \leq 3\varepsilon/2$ on $\partial\Omega \cap V$. This completes the proof.

Finally, we remark that our theorem holds for any set F for which $\mathbb{R}^2 \setminus F$ is thin at no point of ∂F, and for which the lemma holds.

References

[1] Deny, J.: Systèmes totaux de fonctions harmoniques. Ann. Inst. Fourier 1 (1949), 103 - 113.

[2] Gauthier, P. M.: Approximation complexe. To appear in Gazette Sciences Math. Québec 3 (1979).

[3] Gauthier, P. M., Goldstein, M., Ow, W. H.: Uniform approximation on unbounded sets by harmonic functions with non-essential singularities. Unpublished manuscript.

[4] Landkof, N. S.: Foundations of modern potential theory. Springer-Verlag, Berlin - Heidelberg - New York (1972).

[5] Nersesjan, A. A.: On the uniform and tangential approximation by meromorphic functions (Russian). Izvestija Akad. Nauk Armjan. SSR, Mat. 7 (1972), 405 - 412.

[6] Roth, A.: Uniform and tangential approximations by meromorphic functions on closed sets. Canadian J. Math. 28 (1976), 104 - 111.

[7] Saginjan, A. A.: Uniform and tangential harmonic approximation of continuous functions on arbitrary sets. Math. Notes 9 (1971), 78 - 84.

Université de Montréal
C.P. 6128
Montréal, P.Q., H3C 3J7
Canada

Université Laval
Cite Universitaire
Québec G1K 7P4
Canada

ASSOCIATED MEASURES AND THE QUASICONFORMALITY

D. Ghişa

To each real function f defined on a Riemann surface W and having a.e. partial derivatives which belong locally to $\mathcal{L}^2$, may be associated a Borel measure μ_f on W so that for each local map (V,h) and for each Borel set $B \subseteqq V$ the measure $\mu_f(B)$ coincides with

$$\int_{h(B)} \|\mathrm{grad}\ f \circ h^{-1}\|^2 dx\, dy$$

This integral not depending on h will be denoted by

$$\int_B \|\mathrm{grad}\ f\|^2 (x,y) dx\, dy.$$

It has a meaning even if B is not included in a parametric set V (see [1]). Such a measure μ_f is called a measure on the Riemann surface W corresponding to the function f.

Let W and W' be two Riemann surfaces and let be $\varphi : W \to W'$ a quasiconformal mapping. For the real function f defined on W and having a.e. partial derivatives which belong locally to ^{2}, we denote by $f' = f \circ \varphi^{-1}$.

The analytic characterization of the quasiconformality, [2], assures us that f' has also partial derivatives a.e. on W' which belong locally to $\mathcal{L}^2$.

Let $\mu_{f'}$ be the measure on W' corresponding to the function f'. The measures μ_f and $\mu_{f'}$ are called associated measures by means of the quasiconformal mapping φ.

We proposed to study the relationship between these measures and we found that they verify a well-known type of inequality.

Theorem: If $\mu_{f'}$ is associated to μ_f by means of the K-quasiconformal mapping φ, then for each Borel set $B \subseteqq W$ we have

$$\frac{1}{K}\mu_f(B) \leqq \mu_{f'}(\varphi(B)) \leqq K\mu_f(B)$$

Proof. We suppose that φ may be written by means of a local parameter in the form

$$\varphi(x,y) = x'(x,y) + iy'(x,y),$$

where $z = x + iy$ is a point of regularity of φ.

Let $J_\varphi(x,y)$ be the Jacobian matrix of φ. Let us denote the matrix $\bar{J}_\varphi J_\varphi$ by

$$\begin{pmatrix} a_{11} & a_{12} \\ a_{12} & a_{22} \end{pmatrix}.$$

We suppose that f' has partial derivatives at the point (x',y') and we put $\partial f'/\partial x' = X$, $\partial f'/\partial y' = Y$.

We note first that $\operatorname{grad} f = J_\varphi \operatorname{grad} f'$, hence

$$\begin{aligned} \|\operatorname{grad} f\|^2 &= (\overline{\operatorname{grad} f})(\operatorname{grad} f) = (\overline{J_\varphi \operatorname{grad} f'})(J_\varphi \operatorname{grad} f') \\ &= \overline{\operatorname{grad} f'}\, \bar{J}_\varphi J_\varphi \operatorname{grad} f' = (X,Y)\begin{pmatrix} a_{11} & a_{12} \\ a_{12} & a_{22} \end{pmatrix}\begin{pmatrix} X \\ Y \end{pmatrix} \\ &= a_{11}X^2 + 2a_{12}XY + a_{22}Y^2. \end{aligned}$$

The last expression represents a positively defined quadratic form, as follows by applying the Schwarz inequality to the

$$\left(\frac{\partial x'}{\partial x}, \frac{\partial x'}{\partial y}\right) \quad \text{and} \quad \left(\frac{\partial y'}{\partial x}, \frac{\partial y'}{\partial y}\right).$$

Hence the preceding relation represents the equation of an ellipse.

Let $a(x,y)$ and $b(x,y)$ be its semi-axes. Then:

$$b^2(X^2 + Y^2) \leqq a_{11}X^2 + 2a_{12}XY + a_{22}Y^2 \leqq a^2(X^2 + Y^2)$$

and hence

$$b^2(\|\operatorname{grad} f'\|^2 \circ \varphi) \leqq \|\operatorname{grad} f\|^2 \leqq a^2(\|\operatorname{grad} f'\|^2 \circ \varphi).$$

By integrating these inequalities over B we obtain

$$\begin{aligned} \int_B b^2(x,y)(\|\operatorname{grad} f'\|^2 \circ \varphi)(x,y)\,dx\,dy &\leqq \mu_f(B) \\ &\leqq \int_B a^2(x,y)(\|\operatorname{grad} f'\|^2 \circ \varphi)(x,y)\,dx\,dy. \end{aligned}$$

According to a classical result, the K-quasiconformality of φ may be expressed by the following inequalities which take place in W:

$$\frac{1}{K}\,|J_\varphi(x,y)| \leqq b^2(x,y) \leqq a^2(x,y) \leqq K|J_\varphi(x,y)|$$

Bearing in mind these inequalities, we obtain for example in the right-hand part of the preceding inequalities:

$$\mu_f(B) \leqq \int_B a^2(x,y)(\|\operatorname{grad} f'\|^2 \circ \varphi)(x,y)\,dx\,dy$$

$$\leq K \int_B (\|\text{grad } h'\|^2 \circ \varphi)(x,y)\,|J_\varphi(x,y)|\,dx\,dy$$

$$= K \int_{\varphi(B)} \|\text{grad } h'\|^2 dx'dy' = K\,\mu_{f'}(\varphi(B)).$$

With these, the second inequality of the theorem is proved. The proof of the first inequality is analogous.

References

[1] Ghişa, D.: Mesures sur les surfaces de Riemann. Actas del V Congreso de la Agrupación de Matemáticos de Expresión Latina, Madrid (1978).

[2] Lehto, O., Virtanen, K.I.: Quasikonforme Abbildungen. Springer-Verlag, Berlin - Heidelberg - New York (1965).

Universitatea din Timişoara
Bv. V. Pîrvan 4
1900 Timişoara
Romania

REALITY OF THE ZEROS OF DERIVATIVES OF A MEROMORPHIC FUNCTION

Simon Hellerstein* and Jack Williamson

We have recently proved [3] [4]

Theorem A. Let f be a (constant multiple of a) real entire function (i.e., z real implies $f(z)$ real) with only real zeros. Assume that f' has only real zeros. Then f'' has only real zeros if and only if f is in the Laguerre-Pólya class, i.e.

$$f(z) = z^m e^{-az^2+bz+c} \prod_n (1 - z/z_n) e^{z/z_n} ,$$

where m is a non-negative integer, $a \geqq 0$, b and the z_n are real, and $\sum_n z_n^{-2} < \infty$.

We denote the Laguerre-Pólya class by U_o. The "if" half of Theorem A was well known and is an easy consequence of the classical result of Laguerre [5] and Pólya [7] that $f \in U_o$ if and only if it can be uniformly approximated on compacts in the complex plane by polynomials with only real zeros. The "only if" assertion of Theorem A settled a question of Pólya [8] which asked whether the preservation of reality of the zeros under differentiation characterizes U_o within the class of real entire functions with only real zeros.

We have now obtained, what appears to us to be, a surprising analogue of Theorem A for reciprocals of functions of finite order satisfying the hypothesis of Theorem A.

Theorem 1. Let $F = \frac{1}{f}$ where f is a real entire function of finite order with only real zeros. Suppose F' (equivalently f') has only real zeros, then F'' has only non-real zeros if and only if $f \in U_o$.

In [3], we determined the exact number of non-real zeros of f'' in Theorem A for f of finite order. We have also determined the exact number of real zeros of F'' in Theorem 1. Indeed, in the setting of this latter theorem the number of real zeros of F'' is the same as the number of non-real zeros of f''. This strikes us as a remarkable duality. In order to state this duality more precisely, we recall the notation employed in [3]. For each integer $p \geqq 0$, we denote by U_{2p}

* Supported in part by NSF grant MCS7903017

the class of entire functions of the form

$$Kz^m \exp(-az^{2p+2} + bz^{2p+1} + cz^{2p} + Q(z))\pi(z) ,$$

where $a \geqq 0$, b and c are real, Q is a real polynomial of degree $\leqq 2p - 1$, K a complex constant, m a non-negative integer, and π is a canonical product with only real zeros of genus $\leqq 2p + 1$. If a, b, and c are 0, we require π to be of genus $2p$ or $2p + 1$. If a and b are 0, we must have either $c > 0$ or π of genus $2p$ or $2p + 1$. For $p = 0$, we have the Laguerre-Pólya class. Also, every real entire function of finite order with only real zeros is in U_{2p} for some $p \geqq 0$.

In [3] we showed

Theorem B. Let $f \in U_{2p}$. If f' has only real zeros, then f'' has exactly $2p$ non-real zeros.

The dual result for $F = 1/f$ may now be stated as

Theorem 2. Let $f \in U_{2p}$ and $F = 1/f$. If F' has only real zeros, then F'' has exactly $2p$ real zeros.

It is evident that Theorem 2 ⇒ Theorem 1. For $p = 0$, the proof of Theorem 2 is elementary. In fact, although we are unable to provide a reference, it is no doubt known, since for $f \in U_0$, it follows from the log-convexity of $|F|$, on any real interval free of zeros of f, that $|F|$ is convex and hence that F'' is zero free on the real axis. The log-convexity property of $|F|$ is a simple consequence of the fact that for $f \in U_0$ we have $(f'/f)' < 0$ on the real axis, whereever $f \neq 0$. When F is the usual gamma-function Γ, this is the standard proof of the log-convexity property of $|\Gamma|$, [cf. e.g. [1], p. 17]. Having established Theorem 2 for $p = 0$, we note that the "if" portion of Theorem 1 follows.

For $p > 0$, our proof of Theorem 2 is somewhat intricate and relies heavily on the methods employed by us in [3]. We shall indicate the main points in this proof in § 2. A detailed proof will appear elsewhere.

Theorem 2 gives the precise number of real zeros of F''. As for the non-real zeros of F'', we have

Theorem 3. Let f be a real entire function of finite order with only real zeros. Suppose $F = 1/f$. Then

(a) F'' has only real zeros if and only if either

(i) $f(z) = e^{az^2+bz+c}$ $(a \geqq 0$, b and c real constants$)$, or

(ii) $f(z) = (az + b)^n$

where $a \neq 0$ and b are real constants and n is a positive integer.

(b) F'' has infinitely many non-real zeros if f has infinitely many zeros.

We observe that if f has only finitely many zeros, then $f = Pe^Q$, P and Q polynomials. Combining Theorem 2 with a straight forward count of the total number of zeros of F'' shows that unless f is either of form (i) or (ii) F'' has some non-real zeros. To complete the proof of (a) it is sufficient then to assume that f has infinitely many zeros - in which case we may infer the rest of (a) from (b). To prove (b) we could appeal to the following.

Theorem C. Let F be meromorphic in the complex plane. Suppose F and $F^{(k)}$, for some $k \geq 2$, have only finitely many zeros. Then $F = \frac{P_1}{P_2} e^{P_3}$, where P_1, P_2, and P_3 are polynomials.

Theorem 2 coupled with Theorem C when $k = 2$ gives (b) of Theorem 3. However, Theorem C appears in [2] only for $k \geq 3$. Although in this latter work the authors promise a proof for the case $k = 2$ in a future publication, this proof is still forthcoming.

In our setting we may take advantage of the particular form of F to give a proof, as follows. If f has infinitely many real zeros, and F'' finitely many, we would have (*)$F''/F' = Q/g$ where Q is a polynomial and g a real entire function of finite order with infinitely many real zeros. On the other hand, in [3] we obtained the representation $f'/f = P\psi$, where P is a polynomial and ψ a meromorphic function with interlaced zeros and poles. (In the next section we exhibit ψ precisely.) This gives us (**)$F''/F' = P'/P + \psi'/\psi - f'/f$.

It is not difficult to show that (*) and (**) are incompatible. To see this, we may exploit known growth properties of g, ψ'/ψ, and $P\psi$. In particular if the genus of $g \geq 2$ and g is as above, there exist non-real rays along which $|g|$ decays exponentially fast, and (*) would imply that $|F''/F'|$ could grow exponentially along such rays. The relation (**) and a simple analysis of the growth of ψ'/ψ and f'/f away from the real axis show that such growth is impossible. If g is as above, and has genus 0 or 1, then one may verify that $|g(iy)| \neq O(|y|^n)$ for any n, in which case from (*) we get $|(F''/F')(iy)| = O(|y|^{-n})$ for all n. To obtain a contradiction from (**) it suffices to prove that $|(\psi'/\psi)(iy)| = O(|y|^{-1})$, while for all $p \geq 0$ and $f \in U_{2p}$ with infinitely many zeros, $|(f''/f')(iy)| \neq O(|y|^{-1})$, so that from (**) we conclude that $|(F''/F')(iy)| \neq O(|y|^{-1})$ and we have our desired contradiction.

§2. Outline of the proof of Theorem 2. It is no loss of generality to assume that $f(0) \neq 0$ and $f'(0) \neq 0$. Denote by $\{a_n\}$ the distinct zeros of f with the following enumeration

$$\cdots < a_{k-1} < a_k < a_{k+1} < \cdots (-\infty \leq \alpha \leq k \leq \omega \leq +\infty,\ k \text{ finite}).$$

Applying Rolle's Theorem choose exactly one zero of f' in each interval (a_k, a_{k+1}) and denote it by b_k so that $a_k < b_k < a_{k+1}$ and (reindex, if needed) $b_{-1} < 0 < a_1$. We put

$$\psi(z) = \begin{cases} (z - b_0)/(z - a_0) \prod_{k\neq 0} (1 - z/b_k)/(1 - z/a_k) & \text{if } \omega = +\infty \\ [(z - b_0)/(z - a_0)(a_\omega - z)] \prod_{k\neq 0,\omega} (1 - z/b_k)/(1 - z/a_k) & \text{if } \omega < +\infty, \end{cases}$$

with $\psi(z) = (a_0 - z)^{-1}$ if f has only one zero and $\psi(z) \equiv 1$ if f has no zeros. Then ψ is meromorphic and maps the upper half-plane into the upper half plane [6, p. 308 - 309]. In [3] we showed that for $f \in U_{2p}$, $f'/f = P\psi$, where P is a polynomial of degree $2p$, $2p + 1$, or $2p + 2$.

We confine ourselves here to the case where f has infinitely many positive and negative zeros. (In the contrary situation the proof follows similar lines, the semi-infinite intervals free of zeros of f being given special consideration.) In this case the degree of P is always $2p$ [3]. We term the interval between two distinct consecutive zeros of f' which contains a zero of f "typical" and one which contains no zero of f "atypical". In a typical interval given by $\gamma_n < a_k < \gamma_{n+1}$, γ_n and γ_{n+1} consecutive zeros of f' and a_k a zero of f, we observe that F''/F' has an even number of zeros on each of the subintervals (γ_n, a_k) and (a_k, γ_{n+1}). Introducing the auxiliary function H_k defined by $H_k(z) = (z - a_k)P(z)\psi(z) = -(z - a_k)(F'/F)(z)$ for $z \neq a_k$ and $H_k(z) = m_k$, the multiplicity of the zero a_k of f, when $z = a_k$; we have $(F''/F')(z) = (H_k'/H_k)(z) - 1/(z - a_k) - H_k(z)/(z - a_k)$. Imitating the proof of Lemma 7 of [3] with minor variations, we establish that $(z - a_k)(F''/F')(z)$ has at most one simple zero on the interval (γ_n, γ_{n+1}), and we conclude that F'' has no zeros on this interval. On an atypical interval (γ_n, γ_{n+1}), it is immediate that F''/F' has at least one zero. In this case we set $H = P\psi = -F'/F$ so that $F''/F' = H'/H - H$. Here we show that F''/F' has at most one simple zero on the interval (γ_n, γ_{n+1}) to conclude that F'' has exactly one simple zero on an atypical interval. The key observation in both the typical and atypical case is that $(H_k'/H_k)' < 0$ on (γ_n, γ_{n+1}). The proof is completed with the additional observation that the sum of the orders of the zeros of F'' at multiple zeros of F' plus the number of atypical intervals totals $2p$.

References.

[1] Artin, E.: The gamma function. Holt, Rinehart, and Winston, Inc., New York (1964).

[2] Frank, G., Hennekemper, W., Polloczek, G.: Über die Nullstellen meromorpher Funktionen und deren Ableitungen. Math. Ann. 225 (1977), 145 - 154.

[3] Hellerstein, S., Williamson, J.: Derivatives of entire functions and a question of Pólya. Trans. Amer. math. Soc. 227 (1977), 227 - 249.

[4] Hellerstein, S., Williamson, J.: Derivatives of entire functions and a question of Pólya II. Trans. Amer. math. Soc. 234 No. 2 (1977), 497 - 503.

[5] Laguerre, E.: Sur les fonctions du genre zéro et du geure un. C. R. Acad. Sci. 98 (1882); Ouevres 1 (1898), 174 - 177.

[6] Levin, B. Ja.: Distribution of zeros of entire functions. Transl. Math. Monographs 5, Amer. Math. Soc., Providence, R.I. (1964).

[7] Pólya, G.: Über Annäherung durch Polynome mit lauter reellen Wurzeln. Rend. Circ. mat. Palermo 36 (1913), 279 - 295.

[8] Pólya, G.: Bemerkung zur Theorie der ganzen Funktionen. J.-ber. Deutsch. Math. -Verein. 24 (1915), 392 - 400.

Department of Mathematics
University of Wisconsin
Madison, WI 53706
U.S.A.

and

Department of Mathematics
University of Hawaii
Honolulu, HI 96822
U.S.A.

SOME RESULTS ON FUNCTIONS OF BOUNDED INDEX

Wilhelm Hennekemper

Let us first recall the definition of a function of bounded index.

Definition. An entire function $f : C \to C$ is said to be of bounded index, if

$$I(f) := \left\{ \begin{matrix} n \in Z, \\ n \geqq 0 \end{matrix} \;\middle|\; \frac{|f^{(\mu)}(z)|}{\mu!} \leqq \max_{0 \leqq \nu \leqq n} \left\{ \frac{|f^{(\nu)}(z)|}{\nu!} \right\} \begin{matrix} \text{for all } \mu \in Z, \\ \mu \geqq 0,\ z \in C \end{matrix} \right\} \neq \emptyset .$$

f is said to be of bounded index N, if f is of bounded index and $N = \min I(f)$.

A survey of known results on functions of bounded index was recently given by S. M. Shah [4].

Let us now look at the sinus-function - a function of bounded index 1 - and the famous Euler-formula $e^{iz} = \cos z + i \sin z$. If we set $f := \sin z$, it can be expressed as follows: There are entire functions g_1, g_0 of exponential type such that $1 = g_1 f' + g_0 f$. This result is a special case of the following theorem.

Theorem 1. Let f be a function of bounded index N, f not identically zero. Then there exist functions $g_N, \ldots, g_0$ of exponential type such that

$$1 = g_N f^{(N)} + \cdots + g_0 f.$$

Proof. We shall show that there is a $C > 0$ such that

$$\sum_{j=1}^{N+1} |f^{(j-1)}(z)|^2 \geqq \exp(-C|z|) \qquad z \in C,\ |z| \geqq 1. \tag{1}$$

Then our result follows from [3], Theorem 1.1.

To prove (1) we choose a $C_1 > 0$ such that

$$\log\left(\sum_{j=1}^{N+1} |f^{(j-1)}(z)|^2 \right) \leqq C_1 |z| \qquad z \in C,\ |z| \geqq 1. \tag{2}$$

Such a constant exists because ([4], Theorem 2.1) f is of exponential type and so ([1], Theorem 2.4.1) are the derivatives of f.

Next, we choose a $C_2 > 0$ with the following property: If $z_0 \in C$, $|z_0| \geqq 1$ then there is a $\delta = \delta(z_0) \in (\frac{1}{2}, 1)$ such that

$$|f(z)|^2 \geqq \exp(-C_2 |z_0|) \qquad z \in C,\ |z - z_0| = \delta |z_0|. \tag{3}$$

The existence of such a C_2 is a simple consequence of [4], Theorem 2.1 and a well-known theorem ([1], Theorem 3.7.4) about the minimum modulus of an entire function of exponential type. Now we claim that (1) is fulfilled with $C = C_2 + 2C_1 + 2(N+1)^2(N+1)!$. To verify this take $z_0 \in C$, $|z_0| \geq 1$. Then we choose a $\delta \in (\frac{1}{2},1)$ such that (3) is true. If we define $u := \log(\sum_{j=1}^{N+1} |f^{(j-1)}|^2)$, it follows from (2) and (3) that $|u(z)| \leq (C_2 + 2C_1)|z_0|$ for $z \in C$, $|z - z_0| = \delta|z_0|$ and from the definition of a function of bounded index that $|u_{\bar{z}}| \leq (N + 1)^2(N + 1)!$. Thus the general Cauchy-formula yields

$$|u(z_0)| \leq \frac{1}{2\pi\delta|z_0|} \int_{|z-z_0|=\delta|z_0|} |u(z)|\,|dz| + \frac{1}{\pi} \int_{|z-z_0|\leq\delta|z_0|} \frac{|u_{\bar{z}}(z)|}{|z - z_0|} d\lambda(z)$$

$$\leq (C_2 + 2C_1)|z_0| + 2(N + 1)^2(N + 1)!\,|z_0| \leq C|z_0|.$$

Theorem 1 can be generalized to functions of bounded α-index and to solutions of certain differential equations (see [2]).

Now, in order to give an application of Theorem 1 let us deal with some results concerning functions of bounded index and linear differential equations. There are many results of this kind. One of them states ([4], Theorem 9.1) that if an entire function is a solution of a linear differential equation $f^{(n)} + a_{n-1}f^{(n-1)} + \cdots + a_0f = 0$ with constant coefficients, then f is of bounded index. We are interested in a converse of this result, i.e. we seek a class K of entire functions, such that every function of bounded index satisfies a linear differential equation $f^{(n)} + g_{n-1}f^{(n-1)} + \cdots + g_0f = 0$, $g_\nu \in K$. Corollary 6.11 of [4] and [5], V.2. show that K must contain transcendental functions. As an immediate consequence of Theorem 1 we have

<u>Theorem 2.</u> Let f be a function of bounded index $n - 1$. Then f is a solution of a linear differential equation $f^{(n)} + g_{n-1}f^{(n-1)} + \cdots + g_0f = 0$, where $g_{n-1},\ldots,g_0$ are entire functions of exponential type.

<u>Proof.</u> By Theorem 1 there are entire functions $g_{n-1},\ldots,g_0$ of exponential type such that $1 = g_{n-1}f^{(n-1)} + \cdots + g_0f$. Thus we have $f^{(n)} - (f^{(n)}g_{n-1})f^{(n-1)} - \cdots - (f^{(n)}g_0)f = 0$ and because of [4], Theorem 2.1 and [1], Theorem 2.4.1 our theorem is proved.

It is evident that there is an entire function of exponential type which is not a solution of any linear differential equation $f^{(n)} + g_{n-1}f^{(n-1)} + \cdots + g_0f = 0$ with entire coefficients. However Theorem 2 and [4], Theorem 5.5 give

<u>Theorem 3.</u> Every entire function of exponential type can be expressed

as a difference of two entire functions of exponential type which are solutions of linear differential equations $f^{(n)} + g_{n-1}f^{(n-1)} + \cdots + g_0 f = 0$, where $g_{n-1},\ldots,g_0$ are entire functions of exponential type.

References

[1] Boas, R.P.: Entire functions. Academic Press, New York, N.Y. (1954).

[2] Hennekemper, W.: Einige Ergebnisse über Ideale in Ringen ganzer Funktionen mit Wachstumsbeschränkung. Dissertation, Fernuniversität Hagen (1978)

[3] Kelleher, J.J., Taylor, B.A.: Finitely generated ideals in rings of analytic functions. Math. Ann. 193 (1971), 225 - 237.

[4] Shah, S.M.: Entire functions of bounded index. Lecture Notes in Mathematics 599, Springer-Verlag, Berlin - Heidelberg - New York (1977), 117 - 145.

[5] Wittich, H.: Neuere Untersuchungen über eindeutige analytische Funktionen. Springer-Verlag, Berlin - Heidelberg - New York (1968).

Universität Dortmund
Mathematisches Institut
Postfach 50 05 00
D-4600 Dortmund 50
BR Deutschland

THE BOUNDARY BEHAVIOR OF ANALYTIC MAPPINGS OF RIEMANN SURFACES

Teruo Ikegami

Introduction. Let φ be an analytic mapping of a hyperbolic Riemann surface R into an arbitrary Riemann surface R'. The boundary behavior of φ at the Martin boundary points was investigated by Constantinescu-Cornea [1],[2]. In their most successful work - theorem of Riesz, theorem of Fatou and theorem on mappings of type Bl - the fine limit or the fine cluster set $\hat{\varphi}(x)$ at the minimal boundary point x plays an essential role. Later they also studied the boundary behavior of φ at the Wiener boundary points in a more general setting, that is, they considered harmonic maps of harmonic spaces in the sense of Brelot [3]. The theory developed there contains, as a special case, the case of Riemann surfaces. Relations between the Wiener and the Martin boundaries and those between the fine cluster set $\hat{\varphi}(x)$ and the cluster set $\tilde{\varphi}(\tilde{x})$ at a Wiener boundary point $\tilde{x}$ were investigated by the author [4],[6]. The purpose of this paper is to establish a more detailed relation between $\hat{\varphi}(x)$ and $\tilde{\varphi}(\tilde{x})$ (Main Theorem) and, using this relation, derive theorems on the boundary behavior of φ at the Martin boundary from those at the Wiener's. A part of the results - theorem of Plessner type - was published in [6] for harmonic maps. We note finally that the proof of the theorem on mappings of type Bl in this paper was carried out in a different way in the previous paper [7].

1. Preliminaries

Let R be a hyperbolic Riemann surface and let φ be a non-constant analytic mapping of R into an arbitrary Riemann surface R'. A continuous function f on a Riemann surface S is called a Wiener function if there exists a compact subset K of S such that $S \setminus K \notin O_G$ and f is harmonizable [1] on $S \setminus K$. The set of all Wiener functions on S is denoted by $\mathcal{W}(S)$.

We consider two compactifications of R, that is, Wiener's R^W and Martin's R^M. Their boundaries are denoted by Δ^W and Δ^M respectively. The harmonic Wiener boundary Γ^W of Δ^W and the set of all minimal boundary points Δ_1^M of Δ^M play a special role. ω^W (resp. ω^M) is the harmonic measure with respect to R^W (resp. R^M). It is known that

1) For the terminologies in this paragraph we refer to [2].

there exists a continuous mapping of R^W onto R^M fixing each point of R. This mapping is called the canonical mapping and denoted by π.

For the compactification R'^* of R' we shall make the following convention: if R' is compact, $R'^* = R'$; if R' is non-compact and $R' \in O_G$, then R'^* is an arbitrary metrizable compactification of R'; if $R' \notin O_G$, then R'^* is a metrizable and resolutive compactification of R'. Then the restriction to R' of every bounded continuous function on R'^* is a Wiener function on R'.

We define

$$\tilde{\varphi}(\tilde{x}) = \cap\{\overline{\varphi(\tilde{U} \cap R)};\ \tilde{U} \text{ is a neighborhood of } \tilde{x} \text{ in } R^W\} \quad \text{for } \tilde{x} \in \Delta^W,$$

$$\hat{\varphi}(x) = \cap\{\overline{\varphi(E \cap R)};\ R \setminus E \text{ is thin at } x\} \quad \text{for } x \in \Delta_1^M,$$

where closures are taken in R'^*. We also define

$$\tilde{P} = \{\tilde{x} \in \Delta^W;\ \tilde{\varphi}(\tilde{x}) = R'^*\},$$

$$\tilde{F} = \{\tilde{x} \in \Delta^W;\ \tilde{\varphi}(\tilde{x}) \text{ consists of a single point}\}$$

and

$$\hat{P} = \{x \in \Delta_1^M;\ \hat{\varphi}(x) = R'^*\},$$

$$\hat{F} = \{x \in \Delta_1^M;\ \hat{\varphi}(x) \text{ consists of a single point}\}.$$

2. Behavior of analytic mappings on the Wiener boundary

In [3], Constantinescu and Cornea considered a harmonic mapping of Brelot spaces and obtained some interesting results about the boundary behavior of a mapping on the Wiener boundary. We state their results in our context.

1. A subset A' of R'^* is called a polar set if for every subdomain U' of R' with $U' \notin O_G$ there exists a positive superharmonic function s' on U' such that $\lim_{a' \to x'} s'(a') = +\infty$ for all $x' \in A' \cap \overline{U}'$.

Theorem of Riesz type[2]. Let A' be a polar subset of R'^*. If $\tilde{\varphi}(\tilde{x}) \in A'$ for every $\tilde{x} \in \tilde{A} \subset \Delta^W$, then $\omega^W(\tilde{A}) = 0$.

2. A mapping φ is termed a Fatou mapping if $f' \circ \varphi \in \mathscr{W}(R)$ for every bounded Wiener function f' on R'.

Theorem of Plessner type[3].

a) $\tilde{P} \cup \tilde{F} = \Delta^W$;

---

2) Cf. [3], Th. 4.10.
3) Cf. [3], Th. 6.3.

b) $\tilde{P}$ is an open and closed subset of Δ^W;

c) let $\tilde{U}$ be an open subset of R^W, if $\tilde{U} \cap \tilde{P} \neq \emptyset$, then $\tilde{U} \cap R$ contains a component on which (the restriction of) φ is not a Fatou mapping.

3. Mappings of type Bl: a mapping φ is of type Bl if each point $a' \in R'$ possesses a neighborhood $U'(a')$ such that $H_1^{\varphi^{-1}(U'(a'))} = 1$.

<u>Theorem</u>[4].

(i) if $R' \in O_G$, then φ is of type Bl if and only if $\Gamma^W = \tilde{P}$;

(ii) if $R' \notin O_G$, then φ is of type Bl if and only if $\tilde{\varphi}(\tilde{x}) \cap R' = \emptyset$ for every $\tilde{x} \in \Gamma^W$.

3. Auxiliary lemmas

In order to obtain information about the boundary behavior of analytic mappings on the Martin boundary from that on Δ^W, we prepare some lemmas which play a role like a bridge.

By making use of the relation between the Wiener and the Martin boundaries, we can derive the following results concerning the cluster sets $\tilde{\varphi}$ and $\hat{\varphi}$ (for the proof, see [6]).

<u>Lemma A.</u> Let A be a $d\omega^M$-measurable subset of Δ_1^M, G' be an open subset of R'^* and f' be a bounded continuous function on R'^* with support contained in G'. If $\hat{\varphi}(A) \cap G' = \emptyset$, then $\lim f' \circ \varphi = 0$ $d\omega^W$-almost everywhere on $\pi^{-1}(A) \cap \Gamma^W$, where $\hat{\varphi}(A) = \cup \{\hat{\varphi}(x);\ x \in A\}$.

It is obvious that φ is extended continuously on $R \cup \tilde{F}$. The extension will be denoted by $\overline{\varphi}$.

<u>Lemma B.</u> Let G' be an open subset of R'^* such that $G' \cap R' \notin O_G$, and let $G = \overline{\varphi}^{-1}(G')$. If the set $\tilde{C}$ is a compact subset of $G \cap \Gamma^W$, then $\hat{\varphi}(\pi(\tilde{x})) = \tilde{\varphi}(\tilde{x})$ for $d\omega^W$-almost all points $\tilde{x} \in \tilde{C}$.

From above lemmas we can derive easily:

<u>Theorem A.</u> $\omega^W(\pi^{-1}(\Delta^M \setminus \hat{P}) \cap \tilde{P}) = 0$.

<u>Theorem B.</u> If we set $\tilde{F}_1 = \{\tilde{x} \in \tilde{F} : \tilde{\varphi}(\tilde{x}) = \hat{\varphi}(\pi(\tilde{x}))\}$, then $\omega^W(\tilde{F} \setminus \tilde{F}_1) = 0$.

4. The main theorem

<u>Main Theorem.</u> There exist a $d\omega^M$-null set N and a $d\omega^W$-null set

---

4) Cf. [3], Cor. 6.1 and 6.2 to Th. 6.4.

$\widetilde{N}$ such that if $x \in \Delta_1^M \setminus N$ then $\hat{\varphi}(x) = \widetilde{\varphi}(\widetilde{x})$ is valid for every $\widetilde{x} \in (\pi^{-1}(x) \cap \Gamma^W) \setminus \widetilde{N}$.

Proof. Let $\widetilde{N} = [\pi^{-1}(\Delta_1^M \setminus \hat{P}) \cap \widetilde{P}] \cup [\widetilde{F} \setminus \widetilde{F}_1]$. By Theorem A and B, we have $\omega^W(\widetilde{N}) = 0$. Next, let $N = \Delta^M \setminus (\hat{P} \cup \hat{F})$. We shall show that $\omega^M(N) = 0$. For, otherwise we have $\omega^W(\pi^{-1}(N) \cap \Gamma^W) > 0$[5]. Then, we may find a point $\widetilde{x} \in (\pi^{-1}(N) \cap \Gamma^W) \setminus \widetilde{N}$. Since $\widetilde{x} \notin \widetilde{P}$ we have $\widetilde{x} \in \widetilde{F}$. From this and $\widetilde{x} \notin \widetilde{F} \setminus \widetilde{F}_1$, $\widetilde{x} \in \widetilde{F}_1$ is derived, that is, $\widetilde{\varphi}(\widetilde{x}) = \hat{\varphi}(\pi(\widetilde{x}))$. This implies $\pi(\widetilde{x}) \in \hat{F}$, which contradicts $\widetilde{x} \in \pi^{-1}(N)$. Thus the sets N and $\widetilde{N}$ fulfill the requirement of the theorem.

5. Applications

1. Theorem of Plessner type [2],[6]. The following alternative holds for $d\omega^M$-almost all points of Δ^M:

$\hat{\varphi}(x)$ is either R'^* or a set consisting of one point.

In fact, as in the proof of the main theorem, $\omega^M(\Delta^M \setminus (\hat{P} \cup \hat{F})) = 0$.

Theorem [2],[6]. An analytic mapping φ is a Fatou mapping if and only if $\omega^M(\Delta^M \setminus \hat{F}) = 0$.

Proof. φ is a Fatou mapping if and only if $\widetilde{P} = \emptyset$. If $\omega^M(\hat{P}) > 0$, then $\widetilde{P} \neq \emptyset$ by the main theorem. Conversely, if $\widetilde{P} \neq \emptyset$ then $\omega^W(\widetilde{P}) > 0$ since $\widetilde{P}$ is open in Γ^W, and $\omega^M(\hat{P}) > 0$.

2. Theorem of Riesz type [2],[5]. Let A be a subset of Δ_1^M. If $\cup\{\hat{\varphi}(x);\ x \in A\}$ is a polar subset of R'^* then $\omega^M(A) = 0$.

Proof. Let $\widetilde{A} = [\pi^{-1}(A \setminus N) \cap \Gamma^W] \setminus \widetilde{N}$, where N and $\widetilde{N}$ are sets described in the main theorem. Since $\widetilde{\varphi}(\widetilde{x}) \subset \cup\{\hat{\varphi}(x);\ x \in A\}$ for every $\widetilde{x} \in \widetilde{A}$, $\cup\{\widetilde{\varphi}(\widetilde{x});\ \widetilde{x} \in \widetilde{A}\}$ is a polar subset of R'^*, whence $\omega^W(\widetilde{A}) = 0$. Thus we conclude that $\omega^M(A) = 0$, since the set $\pi^{-1}(A) = \widetilde{A} \cup \pi^{-1}(N) \cup \widetilde{N} \cup (\Delta^W \setminus \Gamma^W)$ is of $d\omega^W$-measure zero.

3. Mappings of type Bl.

Theorem [7].

(i) if $R' \in O_G$, then φ is of type Bl if and only if $\hat{\varphi}(x) = R'^*$ for $d\omega^M$-almost all points $x \in \Delta^M$;

(ii) if $R' \notin O_G$, then φ is of type Bl if and only if $\hat{\varphi}(x) \cap R' = \emptyset$ for $d\omega^M$-almost all points $x \in \Delta^M$.

Proof. (i): φ is of type Bl if and only if $\Gamma^W = \widetilde{P}$, which is the same thing as $\omega^M(\Delta^M \setminus \hat{P}) = 0$ by the main theorem.

(ii): In this case φ is a Fatou mapping, therefore $\omega^M(\Delta^M \setminus \hat{F}) = 0$.

---

5) Cf. [4], Cor. to Th. 1.2, p. 41.

By the main theorem, for $d\omega^M$ - almost all points $x \in \Delta^M$ there exists $\tilde{x} \in \pi^{-1}(x) \cap \Gamma^W$ satisfying $\tilde{\varphi}(\tilde{x}) = \hat{\varphi}(x)$. If φ is of type Bl, then $\hat{\varphi}(x) \cap R' = \emptyset$ for $d\omega^M$ - almost all x since $\tilde{\varphi}(\tilde{x}) \cap R' = \emptyset$ for every $\tilde{x} \in \Gamma^W$. Next, suppose that $\hat{\varphi}(x) \cap R' = \emptyset$ for $d\omega^M$ - almost all $x \in \Delta^M$. If there exists $\tilde{x} \in \Gamma^W$ such that $\tilde{\varphi}(\tilde{x}) \in R'$, then from continuity of the extension $\overline{\varphi}$ of φ onto R^W we may find a neighborhood $\tilde{U}(\tilde{x})$ of $\tilde{x}$ carried into R' by $\overline{\varphi}$. Since $\tilde{U}(\tilde{x}) \cap \Gamma^W$ is of $d\omega^W$ - measure positive we are led to a contradiction by the main theorem.

6. Remarks

1. Let R^* be an arbitrary resolutive compactification of R. Let $\Delta^* = R^* \setminus R$ and $\Delta_1^* = \{\xi \in \Delta^*;\ \pi^{*-1}(\xi) \cap \Gamma^W \neq \emptyset\}$, where π^* denotes the canonical mapping of R^W onto R^* fixing each point of R. We define for $\xi \in \Delta_1^*$

$$\varphi^*(\xi) = \cup \{\overline{\varphi(\tilde{G} \cap R)};\ \tilde{G} \text{ is an open neighborhood of } \pi^{*-1}(\xi) \cap \Gamma^W\}.$$

Then we obtain the following theorem of Riesz type:

Let A^* be a subset of Δ^* and A' be a polar subset of R'^*. If $\varphi^*(\xi) \subset A'$ for every $\xi \in A^*$ then $\omega^*(A^*) = 0$, where ω^* is the harmonic measure on R^*.

2. We can also extend the theorem on mappings of Type Bl on the above compactification R^*:

when $R' \in O_G$, φ is of type Bl if and only if $\varphi^*(\xi) = R'^*$ $d\omega^*$ - a.e. on Δ^*;

when $R' \notin O_G$, φ is of type Bl if and only if $\varphi^*(\xi) \cap R' = \emptyset$ $d\omega^*$ - a.e. on Δ^*.

In the latter case, it is proved further that $\varphi^*(\xi)$ is contained in the harmonic boundary of R'^* for $d\omega^*$ - almost all points ξ.

3. It is possible to extend the results in this paper for a harmonic map of harmonic spaces under the suitable conditions.

References

[1] Constantinescu, C. and Cornea, A.: Über das Verhalten der analytischer Abbildungen Riemannscher Flächen auf dem idealen Rand von Martin. Nagoya math. J. 17 (1960), 1 - 87.

[2] Constantinescu, C. and Cornea, A.: Ideale Ränder Riemannscher Flächen. Springer-Verlag, Berlin - Göttingen - Heidelberg, 1963.

[3] Constantinescu, C. and Cornea, A.: Compactifications of harmonic spaces. Nagoya math. J. 25 (1965), 1 - 57.

[4] Ikegami, T.: Relations between Wiener and Martin boundaries. Osaka J. Math. 4 (1967), 37 - 67.

[5] Ikegami, T.: Theorems of Riesz type on the boundary behavior of harmonic maps. Osaka J. Math. 10 (1973), 247 - 264.

[6] Ikegami, T.: On the boundary behavior of harmonic maps. Osaka J. Math. 10 (1973), 641 - 653.

[7] Ikegami, T.: On the characterization of harmonic maps of type Bl at the Martin boundary points of a harmonic space. Osaka J. Math. 13 (1976), 67 - 82.

Osaka City University
Department of Mathematics
Osaka
Japan

THE SCHOTTKY SPACE IN DIMENSIONS GREATER THAN TWO

D. Ivaşcu

0. Introduction. In this paper, some results, concerning the Schottky space, are extended to the case of a dimension greater than two. The first section contains some general results concerning the discontinuous subgroups of the Moebius group $GM(n)$, $n \geq 3$. Our main result, in this section, is the construction, on the domain of discontinuity of any discontinuous subgroup G of $GM(n)$, $(n \geq 3)$, of a G-invariant, Riemannian metric, conformally equivalent with the Euclidean one.

In section 2 we introduce and study the Schottky space of genus g in dimensions greater than two. Just like in the two-dimensional case, it can be seen that, this is the natural parametrization space for the family of different conformal structures that can be introduced on a compact n-dimensional variety uniformized by a discontinuous Moebius group of Schottky type of genus g.

1. Discontinuous subgroups of $GM(n)$.

1.1. Definition. The one-point compactification of the Euclidean n-space R^n is denoted by $M(n)$ and called the Moebius space of dimension n.

1.2. Definition. The group of homeomorphisms of $M(n)$ generated by reflections in the $(n-1)$-spheres and planes of R^n is denoted by $GM(n)$ and called the Moebius group in dimension n.

The space $M(n)$ has a natural conformal structure that can be obtained by projecting stereographically the sphere $\xi_1^2 + \cdots + \xi_{n+1}^2 = 1$ onto $M(n)$ identified with the one-point compactification of the plane $\xi_{n+1} = 0$. The group $GM(n)$ can be characterized as follows.

1.3. Theorem. [6] $GM(n)$ is the group of all conformal transformations of the Moebius space $M(n)$.

If the stereographic projection of the sphere $S^n : \xi_1^2 + \cdots + \xi_{n+1}^2 = 1$ onto $M(n)$ is denoted by π the group $\pi^{-1} \circ GM(n) \circ \pi$ will be the group of all conformal transformations of S^n. By identifying the sphere S^n with the projective variety $y_0^2 - y_1^2 - \cdots - y_{n+1}^2 = 0$ the following theorem can be proved.

1.4. Theorem. [6] The group $\pi^{-1} \circ GM(n) \circ \pi$ as a group of transformations of the projectiv variety $y_0^2 - y_1^2 - \cdots - y_{n+1}^2 = 0$ is isomorphic with the orthogonal group $O(1,n+1)/(\pm I)$ of the quadratic form $y_o^2 - y_1^2 - \cdots - y_{n+1}^2$. Consequently the group $GM(n)$ can be identified with a group of matrix.

It is easy to see that $GM(n-1)$ can be identified with the subgroup of $GM(n)$ that leaves invariant the unit ball $B^n = \{x \mid x \in R^n, x_1^2 + \cdots + x_n^2 < 1\}$. If we consider the hyperbolic metric $ds^2 = (dx_1^2 + \cdots + dx_n^2)/(1 - |x|^2)^2$ of the ball B^n the following theorem can be proved.

<u>1.5. Theorem.</u> [6] The hyperbolic metric of the unit ball B^n is invariant in respect to the subgroup of $GM(n)$ which stabilizes B^n.

For any $T \in GM(n)$ the point $T^{-1}(\infty)$ is denoted by O_T. If $T \in GM(n)$ and $O_T \neq \infty$, it is easy to see that there exists a unic sphere S_T centered in O_T, such that the restriction of T to S_T is an isometric transformation, in respect with the euclidean metric of S_T. S_T is called the isometric sphere of T. Further B_T is the ball having S_T as its own boundary and r_T the radius of S_T. Obviously $TS_T = S_{T^{-1}}$, $TB_T = \operatorname{Ext} B_{T^{-1}}$.

The following lemma proves to be a very useful tool for our later considerations.

<u>1.6. Lemma.</u> If $T \in GM(n)$ and $O_T \neq \infty$ there exists an orthogonal transformation $A_T \in O(n)$ so that

$$Tx = O_{T^{-1}} + \frac{r_T^2 A_T (x - O_T)}{|x - O_T|^2}$$

As a direct consequence of the previous representation we obtain the formula

$$J_T(x) = r_T^{2n}/|x - O_T|^{2n}$$

where $J_T(x)$ is the modulus of the Jacobian of the transformation T in x.

Let us now remind the classification of the elements of $GM(n)$ as given in [7]. An element T of $GM(n) - I$ will be called parabolic if it has a unique fixed point. Any parabolic element of $GM(n)$ is conjugated to a transformation U of the form $U(x) = V(x) + a$ with $V \in O(n)$ and $a \in R^n$. T will be called loxodromic if it is conjugated to a transformation U of the form $U(x) = \lambda V(x)$ with $V \in O(n)$ and $\lambda > 1$. Finally every element of $GM(n) - I$ that is conjugated to a transformation $U \in O(n)$ will be called elliptic.

Let G be a subgroup of $GM(n)$. A point $x \in M(n)$ will be called a limit point of the group G if there exists an infinite sequence

$T_n \in G$ so that $T_n y \to x$ for some $y \in M(n)$. The set of limit points of G will be denoted by $\Lambda(G)$.

1.7. Definition. A subgroup G of $GM(n)$ will be called discontinuous if $\Lambda(G) \neq M(n)$. For such a group we denote by $\Omega(G)$ the complementary set of the set $\Lambda(G)$. The set $\Omega(G)$ is G-invariant and open. If K is a compact subset of $\Omega(G)$, then the set $G_K = \{T \mid T \in G, TK \cap K \neq \phi\}$ will be finite.

$\Lambda(G)$ is a perfect set if it has more than two points. In addition $\Lambda(G)$ can be proved to be the cluster set of the orbit Gx of every point $x \in \Omega(G)$.

1.8. Definition. Let G be a discontinuous subgroup of $GM(n)$. We say that the open subset D of $\Omega(G)$ is a fundamental domain for G in the case it satisfies the conditions

i) $TD \cap D = \phi$ if $T \in G - I$

ii) $\bigcup_{T \in G} T\overline{D} = \Omega(G)$, $\overline{D}$ being the closure of D in $\Omega(G)$.

If G is a discontinuous Moebius group acting on B^n and x_o a point of B^n so that $G_{x_o} = I$ the set

$$P = \{x \in B^n \mid d(x,x_o) < d(x,Tx_o) \quad \text{for} \quad T \in G - I\}$$

will be a fundamental domain for G in B^n see [5]. (d being the hyperbolic distance function on B^n). Clearly P is a hyperbolic convex polyhedron whose faces are pairwise G-equivalent by transformations $T_j \in G$ that generate the group.

In fact we can prove a more general result.

1.9. Proposition. Every discontinuous subgroup G of $GM(n)$ has a polyhedral fundamental domain (not necessarily connected).

Proof. We may suppose that $\infty \in \Omega(G)$ and $G_\infty = I$. In this case (just as in the case of a Kleinian group) it can be proved that the polyhedron $D(G) = \text{Int}\,(\bigcap_{T \in G-I} \text{Ext}\, B_T)$ is a fundamental domain for G.

If $T \in G - I$ we denote by $S_{T,G}$ the interior of the set $S_T \cap \text{Fr}\, D(G)$ as a subset of the sphere S_T. If $S_{T,G}$ is not void then it is a polyhedral $(n - 1)$-dimensional domain (not necessarily connected). The sets $\{S_{T,G}\}_{T \in G - I}$ are called the faces of $D(G)$.

Now, using the same arguments as in the case of a Kleinian group it can be proved the following proposition.

1.10. Proposition. The faces of the fundamental polyhedron $D(G)$ of a discontinuous group $G \subset GM(n)$ are pairwise G-equivalent. If G has an invariant domain Ω, then the transformations of G that conjugate the faces of $D(G) \cap \Omega$ generate the group.

Further we only consider those discontinuous subgroups of $GM(n)$ satisfying the conditions $\infty \in \Omega(G)$ and $G_\infty = I$.

1.11. Theorem. If G is a discontinuous subgroup of $GM(n)$ then the series $\sum_{T\in G-I} r_T^{2n}$ is convergent.

Proof. Let B be a ball so that $B \subset D(G)$ and $\overline{B} \cap G(\infty) = \phi$. Clearly we can find an $R > 0$ for which $TB \subset B_R = B_R(0)$ if $T \in G$ and $G(\infty) \subset B_R$. As $TB \cap UB = \phi$ if $T \neq U$ $(T,U \in G)$, then

$$\sum_{T\in G-I} \int_{TB} dv \leq V(B_R)$$

(dv being the Lebesque measure in R^n and $V(B_R)$ the volume of the ball B_R)

Since
$$\int_{TB} dv = \int_B (r_T^{2n}/|x - O_T|^{2n})dv$$

it follows that

$$\sum_{T\in G-I} \int_{TB} dv = \int_B \Big(\sum_{T\in G-I} r_T^{2n}/|x - O_T|^{2n}\Big)dv \leq V(B_R)$$

Because $|x - O_T| \leq |x| + |O_T| \leq 2R$ we can write

$$\Big[\Big(\sum_{T\in G-I} r_T^{2n}\Big)/(2R)^{2n}\Big] \int_B dv \leq \int_B \Big(\sum_{T\in G-I} r_T^{2n}/|x - O_T|^{2n}\Big)dv \leq V(B_R)$$

As a consequence $\sum_{T\in G-I} r_T^{2n} \leq (2R)^n V(B_R)/V(B)$ and thus the theorem is proved.

Further we construct a G-invariant metric on the domain of discontinuity of a group $G \subset GM(n)$. This metric can be used in order to give a natural Riemannian structure on the orbit space of G. Such a construction is necessary because, in dimensions greater than two, there is no analogue for the hyperbolic metric of a plane domain.

1.12. Theorem. If G is a discontinuous subgroup of $GM(n)$, that has no fixed points in $\Omega(G)$, there exists a G-invariant Riemannian metric on $\Omega(G)$, conformally equivalent to the Euclidean metric of $\Omega(G)$.

Proof. As usually it is supposed that $\infty \in \Omega(G)$ and $G_\infty = I$. First of all we define the function $\lambda_G : \Omega(G) \to R_+$ by the formula

$$\lambda_G(x) = \left(\sum_{T\in G} \Big[\frac{1}{1 + |Tx|^2}\Big]^n J_T(x)\right)^{2/n} \tag{1}$$

Since the series $\sum_{T\in G} J_T(x) = 1 + \sum_{T\in G-I} r_T^{2n}/|x - O_T|^{2n}$ is uniformly convergent on each compact subset $K \subset \Omega(G) - G(\infty)$ (1) is a good definition for λ_G on $\Omega(G) - G(\infty)$. If $p \in G(\infty)$, it is easy to see that

$$\lim_{x\to p} \left[\frac{1}{1 + |Tx|^2}\right]^n J_T(x) = 1/r_T^{2n}$$

As for any compact subset K of $\Omega(G)$ the set $K \cap G(\infty)$ is finite, the relation (1) defines the function λ_G for each point of $\Omega(G)$. Now we can consider the metric $ds_G^2(x) = \lambda_G(x) \sum_{i=1}^n dx_i^2$ defined on $\Omega(G)$. If $T \in G$ and $y = Tx$ then

$$T^*ds_G^2(y) = \lambda_G(Tx) J_T^{2/n}(x) \sum_{i=1}^n dx_i^2$$

$$= \left[\sum_{U\in G} \left(\frac{1}{1 + |U \circ T(x)|^2}\right)^n J_U(Tx)\right]^{2/n} \cdot J_T^{2/n}(x) \sum_{i=1}^n dx_i^2$$

$$= \left[\sum_{U\in G} \left(\frac{1}{1 + |U \circ T(x)|^2}\right)^n J_{U\circ T}(x)\right]^{2/n} \sum_{i=1}^n dx_i^2 = \lambda_G(x) \sum_{i=1}^n dx_i^2 = ds_G^2(x) .$$

Therefore the metric ds_G^2 is G-invariant.

In the case G has no fixed point in $\Omega(G)$ obviously the connected components of $\Omega(G)/G$ are n-dimensional varieties becoming Riemannian varieties by projecting the metric ds_G^2 onto $\Omega(G)/G$. Clearly the canonical projection will be a locally isometric mapping.

2. The Schottky space

As it is well known a finitely generated Fuchsian group of the first kind and also a Schottky group of genus g may be used for obtaining an uniformization of a compact Riemann surface of genus g. By employing the technique of Ahlfors and Bers one can utilize the space of quasi-conformal deformations of the group to obtain the moduli space of the surface.

Let us consider an analogous problem concerning a n-dimensional compact Riemannian variety uniformized by a discontinuous subgroup G of GM(n).

If the Moebius groups G, G' act discontinuously on the semispace $H = \{x \mid x \in R^n, x_n > 0\}$ $(n \geqq 3)$ and H/G, H/G' are compact quasi-conformally equivalent varieties, then, as a consequence of Mostow rigidity theorem (see [6]), they are isometrically equivalent. (The metrics of the varieties H/G, H/G' being the canonical projections of the hyperbolic metric of H). As a result, the moduli space of such a variety (the space of distinct conformal structures) is reduced to a point.

The situation is quite different for a compact n-dimensional variety ($n \geq 3$) uniformized by a Schottky type group. In this case the analogy with the two-dimensional case is much more significant. With the wiew of a more detailed discussion, the following definitions are necessary.

2.1. Definition. The group $G = \langle T_1,\ldots,T_g\rangle \subset GM^+(n)$* will be called a marked Schottky group of genus g if there are $2g$ domains B_j, topologically equivalent to the closed unit ball of the space R^n, such that $T_j(CB_j) = \mathrm{Int}\, B_{g+j}$ $(1 \leq j \leq g)$. ($\langle T_1,\ldots,T_g\rangle$ being the group generated by the set $\{T_1,\ldots,T_g\}$).

2.2. Definition. The marked Schottky group $G = \langle T_1,\ldots,T_g\rangle$ will be called classical if the domains B_j in the preceding definition can be chosen to be balls.

It is easy to see that any marked Schottky group $G = \langle T_1,\ldots,T_g\rangle$ is a free discontinuous subgroup of $GM^+(n)$, having the domain $D = C(\bigcup_{j=1}^{2g} B_j)$ as a fundamental domain. Since the group G acts freely on $\Omega(G)$ and D is relatively compact in $\Omega(G)$, the space $V_G = \Omega(G)/G$ will be a compact n-dimensional variety. Topologically the variety V_G is completely determined by the genus g of the group G. The homology of such a variety can be easily seen to be $H_o(V_G) \cong H_n(V_G) \cong Z$, $H_1(V_G) \cong H_{n-1}(V_G) \cong Z^g$, $H_i(V_G) = 0$ if $1 < i < n-1$. ($n \geq 3$).

From now on we shall consider V_G with the conformal structure defined by the projection of the metric ds_G^2 on V_G. Clearly, if $n \geq 3$, the domain $\Omega(G)$ is simply connected. As a consequence the projection $\pi : \Omega(G) \to V_G$ will be the universal covering of the variety V_G.

2.3. Proposition. Let G, G' be Schottky groups of the same genus g. The varieties V_G, $V_{G'}$ are conformally equivalent if and only if $G' = U \circ G \circ U^{-1}$ with $U \in GM^+(n)$.

Proof. The mappings $\pi : \Omega(G) \to V_G$, $\pi' : \Omega(G') \to V_{G'}$ being the universal coverings of the varieties V_G, $V_{G'}$ each conformal mapping $u : V_G \to V_{G'}$ can be lifted to a conformal mapping $U : \Omega(G) \to \Omega(G')$ such that $G' = U \circ G \circ U^{-1}$. As it is well known, if $n \geq 3$, $U \in GM^+(n)$ and the proposition is proved.

It is easy to find two classical Schottky groups that are not conjugated in $GM^+(n)$. Since, as we shall prove further, every two classical Schottky groups of the same genus are quasiconformally equivalent, the Mostow rigidity theorem is not true for this kind of groups. Accordingly

---

* $GM^+(n)$ will be the group of all orientation preserving Moebius transformations.

there are many different conformal structures on a given variety uniformized by a Schottky group. Like in the case of a Riemann surface, the family of all distinct conformal structures on such a variety can be parametrized by means of the Schottky space.

The marked Schottky groups $G = \langle T_1,\dots,T_g\rangle$, $H = \langle U_1,\dots,U_g\rangle$ will be called equivalent if there exists a transformation $U \in GM^+(n)$ such that $U \circ T_j \circ U^{-1} = U_j$ for all $1 \leq j \leq g$.

<u>2.4. Definition.</u> The set $S_g(n)$ of all equivalence classes of Schottky subgroups of $GM^+(n)$ of genus g will be called the Schottky space of genus g in dimension n.

In order to define the topological structure of the space $S_g(n)$, the set $S_g^*(n)$ of all marked Schottky groups of genus g is identified with an open subset of the space $GM^+(n)^g$ by means of the mapping $\varphi : S_g^*(n) \to GM^+(n)^g$, given by the formula $\varphi(\langle T_1,\dots,T_g\rangle) = (T_1,\dots,T_g)$. Then we project the topology of the set $S_g^*(n)$ onto $S_g(n)$.

The set $S_g^c(n)$ of all equivalence classes of classical Schottky groups, of genus g, will be called the classical Schottky space. Clearly $S_g^c(n)$ is an open subset of the space $S_g(n)$.

As it is well known (see [2]) the space $S_g(2)$ is linearly connected. The method used to prove this result cannot be extended to greater dimensions. For the case of an arbitrary dimension we have the following proposition.

<u>2.5. Proposition.</u> The classical Schottky space $S_g^c(n)$ $(n \geq 2)$ is linearly connected.

In order to prove this proposition we use the notations $\tilde{S}_g = \{(x,r) \mid (x,r) \in (R^n)^{2g} \times (R_+)^{2g},\ |x_j - x_i| > r_i + r_j,\ \text{if } i \neq j\}$ $S_g^{*c} = \{\langle T_1,\dots,T_g\rangle,\ \langle T_1,\dots,T_g\rangle$ is a classical Schottky group$\}$. If $(x,r) \in \tilde{S}_g$ we denote by $S_g(x,r)$ the set

$$S_g(x,r) = \{\langle T_1,\dots,T_g\rangle \mid T_j(\{y \mid |y - x_j| \geq r_j\}) = \{u \mid |u - x_{g+j}| \leq r_{g+j}\}\}$$

If $(x,r) \in \tilde{S}_g$ let $T_j(x,r)$ be the Moebius transformation $\zeta_j \circ \sigma_j \circ s_j$ where $\zeta_j(y) = r_{g+j}(y - x_j)/r_j + x_{g+j}$, σ_j is the reflexion in the sphere of radius r_j centered in x_j and s_j is the reflexion in the hyperplane $\{x \mid x_n = x_{j,n}\}$

<u>2.5! Lemma.</u> $S_g(x,r)$ is a linearly connected subset of the space S_g^*.

<u>Proof.</u> For any ball $B \subset R^n$, we denote by $GM^+(n,B)$ the group of all Moebius transformations $T \in GM^+(n)$ that leave B invariant. If

$\langle T_1,\ldots,T_g\rangle \in S_g(x,r)$ then clearly

$$\varphi(S_g(x,r)) = T_1 \circ GM^+(n,B_1) \times \cdots \times T_g \circ GM^+(n,B_g)$$

(B_j being the ball of radius r_j centered in x_j). As the group $GM^+(n,B_j)$ is linearly connected and φ is a homeomorphism the lemma is proved. Further we denote by π the mapping $\pi : \tilde{S}_g \to S_g^{*c}$ defined by the formula $\pi((x,r)) = \langle T_1(x,r),\ldots,T_g(x,r)\rangle$. Now we can prove the proposition 2.5.

Since $\tilde{S}_g$ is an open connected subset of $(R^n)^{2g} \times (R_+)^{2g}$ and π a continuous mapping $\pi(\tilde{S}_g)$ is a linearly connected subset of the space S_g^{*c}. It can be noted that $S_g^{*c} = \bigcup_{(x,r)\in\tilde{S}_g} S_g(x,r)$ and $\pi((x,r)) \in S_g(x,r) \cap \pi(\tilde{S}_g)$. As a result the space S_g^{*c} is linearly connected. The space $S_g^c(n)$ being a quotient space of S_g^{*c} the proposition is proved. As a matter of fact we can prove that $S_g^c(n)$ is a differentiable manifold of dimension $(n + 1)(n + 2)(g - 1)/2$.

2.6. Definition. The Schottky group $G = \langle T_1,\ldots,T_g\rangle$ is quasiconformal equivalent to the group $H = \langle U_1,\ldots,U_g\rangle$ if there is a quasiconformal mapping $w : \Omega(G) \to \Omega(H)$ such that $w \circ T_j \circ w^{-1} = U_j$. As it is well known, in the two-dimwnsional case, any two marked Schottky groups of the same genus are quasiconformally equivalent. Now we shall prove an analogus result for the classical Schottky groups in dimensions greater than two. It can be stated as follows.

2.7. Theorem. If $G = \langle T_1,\ldots,T_g\rangle$, $H = \langle U_1,\ldots,U_g\rangle$ are classical Schottky groups then there is a quasiconformal mapping $w : M(n) \to M(n)$ such that $w \circ T_j \circ w^{-1} = U_j$.

This theorem is obtained as a consequence of some auxiliary results that we shall prove further. From now on any domain $D \subset M(n)$ whose complementary set consists of $2g$ disjoints closed balls will be called a domain of type g. If $G = \langle T_1,\ldots,T_g\rangle$ is a classical Schottky group we denote by D a fundamental domain for G of type g. For such a domain D let $CD = \bigcup_{j=1}^{2g} D_j$ with $CD_{g+j} = T_jD_j$ when $1 \leq j \leq g$. If Λ is another domain of the type g so that $\Delta \subset D$, we denote by Δ_j the component of $C\Lambda$ that satisfies the condition $D_j \subset \Delta_j$. If a, $b \in R^n$, $r_1, r_2 \in R_+$ and $|a - b| < r_1 - r_2$, $r_1 - r_2 > 0$, we denote by $R(a,b,r_1,r_2)$ the domain $R(a,b,r_1,r_2) = \{x \mid |x - a| < r_1,\ r_2 < |x - b|\}$ and by S_1 (S_2) the sphere $\{x \mid |x - a| = r_1\}$ ($\{x \mid |x - b| = r_2\}$). Let a, b, r_1, r_2 and a', b', r_1', r_2' be like above. If T_1, T_2 are Moebius transformations such that $T_1(B_{r_1}(a)) = B_{r_1'}(a')$,

$T_2(B_{r_2}(b)) = B_{r_2'}(b')$ then it can be found a homeomorphism $w:\overline{R(a,b,r_1,r_2)} \to \overline{R(a',b',r_1',r_2')}$ satisfying the conditions $w|S_1 = T_1$, $w|S_2 = T_2$, $w|R(a,b,r_1,r_2)$ is a quasiconformal mapping. Now we can prove the following proposition.

2.8. Proposition. Let $G = \langle T_1,\ldots,T_g\rangle$, $H = \langle U_1,\ldots,U_g\rangle$ be classical Schottky groups and B, D the fundamental domains of type g for G and H respectively. If

i) the plane $\{x|x_n = 0\}$ is G-invariant

ii) there are two domains E, F of the type g, such that $E \subset B$, $F \subset D$ and a transformation $T \in GM^+(n)$ satisfying the condition $TE_j = F_j$

then there is a quasiconformal mapping $w: M(n) \to M(n)$ for which $w \circ T_j \circ w^{-1} = U_j$.

Proof. First of all we choose g elements, V_j $(1 \leq j \leq g)$ of the group $GM^+(n)$ satisfying the condition $V_jB_j = D_j$ for any j. If we denote by R_j (R_j') the domain $\text{Int}\, E_j \cap \text{Ext}\, B_j$ $(\text{Int}\, F_j \cap \text{Ext}\, D_j)$ then, a homeomorphism $w_j: \bar{R}_j \to \bar{R}_j'$ may be found, (as has been already mentioned) such that

i) $w_j|\text{Fr}B_j = V_j$, $w_{g+j}|\text{Fr}B_{g+j} = U_j \circ V_j \circ T_j^{-1}$ and $w_j|\text{Fr}E_j = T$ if $1 \leq j \leq g$

ii) $w_j|R_j$ is a quasiconformal mapping for each j.

Let $w_o: B \to D$ be the mapping defined as follows $w_o|\bar{E} = T$, $w_o|\bar{R}_j = w_j$. Since w_o is obviously quasiconformal in $B - \text{Fr}\, E$ it results that it is quasiconformal everywhere. If we denote by $\psi: G \to H$ the isomorphism defined by the relations $\psi(T_j) = U_j$ we can extend w_o to $\Omega(G)$ by the usual procedure, namely $w_o|T\bar{B} = \psi(T)|\bar{D} \circ w_o \circ T^{-1}|T\bar{B}$.

This extension, also denoted by w_o, is a homeomorphism of $\Omega(G)$ on $\Omega(H)$, that is quasiconformal on $\Omega(G) - G(\text{Fr}\, B)$. As $G(\text{Fr}\, B)$ is a countable union of $(n-1)$-dimensional spheres which do not accumulate in $\Omega(G)$, it follows that w_o is a quasiconformal mapping of $\Omega(G)$ onto $\Omega(H)$. The sets $\Lambda(G)$, $\Lambda(H)$ being totally disconnected, we can extend w_o to a homeomorphism $w: M(n) \to M(n)$. Since $\Lambda(G) \subset \{x|x_n = 0\}$ this last extension is a quasiconformal one. Finally we remark that $w \circ T \circ w^{-1} = \psi(T)$ for $T \in G$. Consequently the proposition is proved.

2.9. Remark. If G and H satisfy only the condition (ii) of the proposition 2.8. there is a homeomorphism $w: M(n) \to M(n)$ such that $w|\Omega(G)$ is a quasiconformal mapping and $w \circ T \circ w^{-1} = \psi(T)$.

2.10. Lemma. If G_o is a classical Schottky group that leaves invariant the plane $\{x|x_n = 0\}$ the set of classical Schottky groups, quasiconformally equivalent to G_o, is open.

Proof. Let B_o be a fundamental domain of type g for G_o and E_o be a relatively compact subdomain of B_o of the same type. The set $S(E_o) = \{G | G = \langle T_1,\ldots,T_g\rangle \in S_g^c,\ T_j(CB_{o,j}) \subset \text{Int } E_{o,g+j}\}$ is clearly an open subset of the space S_g^c. Let us now consider a classical Schottky group G, which is a quasiconformal deformation of G_o, hence $G = w_o \circ G_o \circ w_o^{-1}$ with $w_o : M(n) \to M(n)$ a quasiconformal mapping. If B is a fundamental domain of type g for G and E is a relatively compact subdomain of B of the genus g, as in 2.9., for any $\tilde{G} \in S(E)$ it can be found a homeomorphism $w : M(n) \to M(n)$, such that $w|\Omega(G)$ is a quasiconformal mapping and $w \circ T_j \circ w^{-1} = U_j$, $\langle T_1,\ldots,T_g\rangle$, $\langle U_1,\ldots,U_g\rangle$ being the markings of G and $\tilde{G}$ respectively. Since the mapping $w \circ w_o$ is quasiconformal everywhere and $\tilde{G} = (w \circ w_o) \circ G_o \circ (w \circ w_o)^{-1}$, the proposition is proved.

2.11. Lemma. If G is a classical Schottky group leaving invariant the plane $\{x | x_n = 0\}$ the set of all classical Schottky groups quasiconformally equivalent to G is closed.

Proof. Let $G_k = w_k \circ G \circ w_k^{-1}$ be a convergent sequence of the space $S_g^c(n)$ with $w_k : M(n) \to M(n)$ being quasiconformal mappings. If G_o is the limit group of the sequence G_k, then, as we have already observed there is a neighborhood V of G_o, such that for any $G_k \in V$ we can find a homeomorphism $w : M(n) \to M(n)$ that verifies the conditions $w|\Omega(G_k)$ is a quasiconformal mapping and $G_o = w \circ G_k \circ w^{-1}$. Since $G_o = (w \circ w_k) \circ G \circ (w \circ w_k)^{-1}$ and $w \circ w_k$ is a quasiconformal mapping, the lemma is proved. The set of classical Schottky groups quasiconformally equivalent to a fixed group G being simultaneously closed and open and the space S_g^c being connected the theorem 2.7. is proved. As a consequence of the theorem 2.7. the space $S_g^c(n)$ has a natural Teichmüller metric. As a matter of fact all the previous results seem to be true for the space $S_g(n)$.

Finally we remark that almost all the results in [2] are also true in dimensions greater than two. As a consequence the Chuckrow's characterization of the boundary points of the Schottky space is true in any dimension. It can be stated as follows.

2.12. Theorem. Every boundary point of the space $S_g(n)$ (in $GM^+(n)^g/\sim$) is a free group with g generators, without elliptic elements.

References

[1] Ahlfors, L. V.: Two lectures on Kleinian groups, in "Proceedings

of the Romanian - Finnish Seminar on Teichmüller spaces and quasiconformal mappings, Braşov 1969". Publishing House of the Academy of the Socialist Republic of Romania, Bucharest (1971), 49 - 64.

[2] Chuckrow, V.: On Schottky groups with applications to Kleinian groups. Ann. of Math., II. Ser. 88 (1968), 45 - 61.

[3] Gehring, F. W., Palka, B. P.: Quasiconformally homogenous domains. J. Analyse math. 30 (1976), 172 - 199.

[4] Ivascu, D.: On Klein-Maskit combination theorems. To appear in "Proceedings of the III Romanian - Finnish Seminar on Complex Analysis, Bucharest 1976".

[5] Martio, O., Srebro, U.: Automorphic quasimeromorphic mappings in R^n. Acta math. 135 (1975), 221 - 247.

[6] Mostow, G. D.: Quasiconformal mappings in n-space and the rigidity of hyperbolic space forms. Inst. haut. Etud. sci., Publ. math. 34 (1968), 53 - 104.

[7] Tukia, P.: Multiplier preserving isomorphisms between Möbius groups. Ann. Acad. Sci. Fenn., Ser. A I 1 (1975), 327 - 341.

[8] Väisälä, J.: Lectures on n-dimensional quasiconformal mappings. Lecture Notes in Mathematics 229, Springer-Verlag, Berlin - Heidelberg - New York (1971).

Institutul de Matematică
Universitatea din Bucureşti
Str. Academiei 14
70109 Bucureşti
Romania

SOME EXCEPTIONAL SETS IN POTENTIAL THEORY

Howard L. Jackson

1. Preliminary remarks.

This discussion shall be devoted mainly, though not exclusively, to four different kinds of exceptional sets in classical potential theory which also have applications to certain questions in complex analysis. In future we shall restrict ourselves to a region (open, connected set) $\Omega \subset \mathbb{R}^p$ $(p \geq 2)$ that tolerates a Green function G in the usual sense. In relation to this region the four kinds of exceptional sets that shall be the principal object of our attention are now listed as follows:

Type 1: Sets which are classically thin at points of Ω, or possibly at Euclidean boundary points of Ω. This kind of thinness is also referred as internal thinness, or ordinary thinness.

Type 2: Sets which are thin at minimal Martin boundary points of Ω. We shall refer to this type of thinness as minimal thinness.

Type 3: Sets which are thin at extremal Kuramochi boundary points of Ω, henceforth to be referred as full-thin sets.

Type 4: Sets which M. Essén and I (cf. [7]) have recently defined to be rarefied at boundary points of Ω in the special case where Ω is a half space.

Other exceptional sets that will be briefly mentioned include semithin sets defined initially by J. Lelong-Ferrand (cf. [13], p. 134) at boundary points of a half space as well as sets which she also defined to be "rarefied" there. Since it appears that a rarefied set in the sense of J. Lelong-Ferrand closely resembles a semithin set and since the rarefied set referred to in Type 4 closely resembles a minimally thin set, therefore Essén and I (cf. [7], p. 19) decided to call a set semirarefied at a boundary point of a half space iff it is rarefied there in the sense of J. Lelong-Ferrand. It is doubtful if this change of terminology will lead to any appreciable confusion in view of the fact that very little work has been published concerning this kind of exceptional set up to the present time. One other point that should be mentioned here is that it is possible to refer to an ordinary thin set (Type 1) at ∞ (cf. Brelot [4]) if Ω is an unbounded region.

2. The thin sets and their relationships.

Ordinary thin sets at either Euclidean points or at infinity were defined and their properties developed in the early 1940's by M. Brelot (cf. [3], [4]). A limit along a filter of neighbourhoods of a point each one of which excluded this type of thin set as an exceptional set was called by Brelot a pseudo-limit at this point. If $u \in S^+(\Omega)$, where $S^+(\Omega)$ is the cone of positive superharmonic functions on Ω and if $x_0 \in \Omega$ the pseudo-limit of u at x_0 is $u(x_0) = \liminf_{x\in\Omega, x\to x_0} u(x)$. At about the same time, H. Cartan noticed that a pseudo-limit at a point was in fact the limit along the neighbourhood filter of that point with respect to the least topology on Ω that made continuous every member of the cone $S^+(\Omega)$. Limits with respect to this topology, now called the classical fine topology, are usually referred to as ordinary (or classical) fine limits rather than as pseudo-limits. With respect to the point at infinity there is a considerable difference in the theory of ordinary thin sets in the planar case as opposed to the higher dimensional cases. Under an inversion map about the unit sphere (i.e. a circle in the plane) an ordinary thin set at 0 is always mapped onto an ordinary thin set at ∞ in the plane but this fact no longer holds in higher dimensions.

The second type of thin set was introduced in 1949 by J. Lelong-Ferrand (cf. [13], p. 130) at the boundary points of a half space $\Omega = \{x = (x_1,\ldots,x_p) \in \mathbb{R}^p \ (p \geqq 2) : x_1 > 0\}$ and this kind of thinness together with its associated fine topology was extended onto the Martin compactification of a Green space by Naim in her thesis (cf. [15]). Since the entire Green space is thin at any Martin boundary point that is non-minimal (cf. [15]), we shall therefore concentrate on those sets in the space that are thin at the minimal ones. This explains why we have employed the term "minimal thinness" to describe thin sets of this type. If Ω is a half space and $\hat{x} \in \partial\Omega \cup \{\infty\}$ then the minimal harmonic function that naturally corresponds to $\hat{x}$ is

$$h_{\hat{x}} = \begin{cases} x_1 |x - \hat{x}|^{-p} & \text{if } |\hat{x}| < +\infty \\ x_1 & \text{if } |\hat{x}| = +\infty . \end{cases}$$

We could also say that $\hat{x}$ is the pole for $h_{\hat{x}}$. If $\Delta_1(\Omega)$ is the minimal Martin boundary of a Green space Ω, each point of which can be represented by a minimal harmonic function h, then $E \subset \Omega$ is minimally thin with respect to (or at) h iff the regularized réduite

$$\hat{R}_h^E \equiv \inf\{u \in S^+(\Omega) : u \geqq h \text{ quasi-everywhere on } E\}$$

is a Green potential on Ω (which may vanish identically on Ω). If Ω

is a half space, $u \in S^+(\Omega)$, $h_{\hat{x}} \in \Delta_1(\Omega)$, then one of the key results of J. Lelong-Ferrand implies that minimally thin sets at $h_{\hat{x}}$ are the good exceptional sets when one attempts to form the limit of the quotient $u/h_{\hat{x}}$ as $x \to \hat{x} \in \partial D \cup \{\infty\}$ where $\hat{x}$ is the (unique) pole of $h_{\hat{x}}$. Her result can now be interpreted as meaning that the minimal fine limit of $u/h_{\hat{x}}$ at $\hat{x}$ equals $\liminf_{x \to \hat{x}} (u(x)/h_{\hat{x}}(x))$ in the ordinary topology. In parallel with a theorem of H. Cartan who demonstrated that a classical fine limit is just the ordinary limit taken over the complement of a suitably chosen internally thin set, Brelot has observed that a minimal fine limit is the ordinary limit taken over the complement of a suitably chosen minimally thin set. Such theorems have of course been generalized by L. Naim, J.L. Doob and others. Another feature of minimally thin sets worth mentioning concerns invariance under conformal mappings when Ω is a planar region. This property allowed C. Constantinescu and A. Cornea as well as J.L. Doob to obtain fine boundary limit theorems for analytic mappings on Riemann surfaces at Martin boundary points. Strictly speaking Constantinescu and Cornea (cf. [6], p. 146) defined their cluster set along the subfilter of the filter of minimally fine neighbourhoods of a Martin boundary point $\hat{p} \in \Delta_1(\Omega)$ whose filterbase consisted of open sets in the usual topology of Ω. If f is a continuous function then their cluster set, denoted by $\hat{f}(\hat{p})$, is the same as the minimally fine cluster set of f at $\hat{p}$, but this is not true in general (cf. [6], p. 152).

It was natural to compare the minimal fine topology with its ordinary counterpart. The situation is fairly easy at isolated points (cf. [5], p. 148). At any finite (polar in our case) point $a \in \Omega$, or if $a = \infty$ in the planar case, then $E \subset \Omega - \{a\}$ is ordinary thin at a iff E is minimally thin at the associated Martin boundary point of $\Omega - \{a\}$ whose minimal harmonic function has its pole at a. At $a = \infty$, and if $p \geq 3$, minimally thin sets are, in general, much larger than ordinary ones. Under an inversion map about a sphere a minimally thin set is preserved, but this is not true of an ordinary thin set as we have already pointed out.

At the boundary points of a half space Ω the situation is not so easy. If $E \subset \Omega$, J. Lelong-Ferrand ([13], p. 132) proved that ordinary thin sets are the same as the minimally thin ones at the origin with respect to Ω as long as E is restricted to a Stolz domain with vertex there, and if the dimension $p \geq 3$. She also claimed, but did not prove that ordinary thinness always implies minimal thinness at the origin as long as $p \geq 3$. We shall return to this case later. If Ω is the half plane it was believed for a long time that ordinary thin sets could not be compared with minimally thin sets at the boundary points except in a statistical sense (cf. [5], p. 153). In 1970, I was able to correct this error (cf. [10]) by demonstrating that ordinary thinness always implies the minimal one at boundary points of a half plane and that, in contrast with the higher dimensional case, this implication might well be strict if one is restricted to a Stolz domain. Brelot pointed out that such implications also hold for semithin sets. In a later paper (cf. [12]), I was able to show that for a wedge-shaped region this implication still holds, but that it will not necessarily hold at the boundary point of a Jordan region if the region is sufficiently narrow there. Returning to the higher dimensional cases Choquet proved, but did not publish the claim of J. Lelong-Ferrand that ordinary thinness always implies minimal thinness at the boundary points of a half space, and in the Brelot-Choquet-Deny seminar of 1972, I gave another proof of this fact (cf.[11]) along with a different proof of the planar case. Later, I discovered a very simple proof of the higher dimensional case which will now be presented here.

Theorem. If E is ordinary thin set at 0, then E is minimally thin at 0 with respect to the half space

$\Omega = \{x = (x_1,\dots,x_p) \in \mathbb{R}^p \ (p \geq 3) : x_1 > 0\}$.

Proof. We first fix s such that $0 < s < 1$, let $I_n = \{x : s^{n+1} < |x| \leq s^n\}$ and let $E_n = E \cap I_n$. If c_n is the ordinary outer capacity of E_n then E is ordinary thin at 0 iff $\sum_n c_n/s^{n(p-2)} < \infty$. Now let $\hat{R}_{x_1}^{E_n}$ (in Ω) be the Green potential of the mass distribution λ_n whose energy (λ_n,λ_n) is denoted by γ_n. We recall (cf. [13], p. 129) that J. Lelong-Ferrand used the term "outer power" of E_n for γ_n. Then E is minimally thin at 0 in Ω iff $\sum_n \gamma_n/s^{np} < +\infty$. We shall demonstrate that $\gamma_n \leq c_n s^{2n}$ except for a constant factor that depends only on the dimension p. If G is the Green kernel on $\Omega \times \Omega$, and y^* is the reflection of y about $\partial\Omega$, then elementary calculations (cf. [11]) indicate that

$$G(x,y) \simeq x_1 y_1 |x-y|^{2-p} |x-y^*|^{-2} \tag{2.1}$$

where the constants of comparison are global, and depend only on the dimension p. Since $|x-y^*| \leq \operatorname{diam} I_n$ if $(x,y) \in \overline{E_n} \times \overline{E_n}$, therefore

$$\int_{\overline{E_n}} G(x,y)\,d\lambda_n(y) \geq s^{-2n} x_1 \int_{\overline{E_n}} \frac{y_1\,d\lambda_n(y)}{|x-y|^{p-2}} \quad \text{(constant)} . \tag{2.2}$$

Hence

$$s^{2n} \geq \int_{\overline{E_n}} \frac{y_1\,d\lambda(y)}{|x-y|^{p-2}}$$

on the fine closure of E_n up to a constant factor. If ω_n is the ordinary capacitary distribution for E_n and we integrate with respect to ω_n, then

$$s^{2n}c_n \geq \int_{\overline{E}_n}\int_{\overline{E}_n} \frac{y_1 d\lambda(y) d\omega_n(x)}{|x-y|^{p-2}} \quad \text{(up to a constant factor).} \tag{2.3}$$

A reversal of the order of integration on the right gives the inequality $\gamma_n \leq s^{2n}c_n$ and hence the theorem.

We might also mention here that if $E \subset \Omega$ is minimally thin at ∞ in $\mathbb{R}^p$ $(p \geq 2)$ then it is also minimally thin there with respect to Ω itself, and this implication is strict in general. Even for Jordan regions, however, it is not clear just how large the boundary set can be at which ordinary thin sets cannot be compared with minimally thin sets. The most general results in this direction have been obtained by Brelot (cf. [5], p. 153).

The third type of thin set to be discussed is called a full-thin (or N-thin) set at a Kuramochi boundary point of Ω. In order to motivate this particular theory we shall first assume that Ω is a bounded region in $\mathbb{R}^p$ $(p \geq 2)$ with a finite number of smooth boundary components. Let K_o be any compact ball in Ω, and then let us concentrate on $\Omega_o = \Omega - K_o$. The Kuramochi function for Ω_o with pole at $p \in \Omega_o$, and denoted by N_p, is the mixed Green function for Ω_o with pole at p which vanishes on ∂K_o and whose normal derivative vanishes on $\partial\Omega$. As a symmetric kernel N is naturally extendable onto $(\Omega_o \cup \partial\Omega) \times (\Omega_o \cup \partial\Omega)$ and there is a one-one correspondence between any point $\hat{p} \in \partial\Omega$ and the Kuramochi function $N_{\hat{p}}$ with pole at $\hat{p}$. In fact $\{N_{\hat{p}} : \hat{p} \in \partial\Omega\}$ could be viewed as a realization of the Kuramochi boundary $\Delta_N(\Omega)$ of Ω (modulo K_o). The cone of (positive) full-harmonic functions on Ω_o, denoted by $\widetilde{H}^+(\Omega_o)$, is the subcone of the cone of positive harmonic functions $H^+(\Omega_o)$ with vanishing normal derivative on $\partial\Omega$. The cone of N-potentials of mass distributions on $\Omega_o \cup \partial\Omega$ shall be called full-potentials, denoted by $\widetilde{P}(\Omega_o)$, and the direct sum of $\widetilde{H}^+(\Omega_o)$ and $\widetilde{P}(\Omega_o)$ shall be called the cone of (positive) full-superharmonic functions on Ω_o, denoted by $\widetilde{S}^+(\Omega_o)$. Each member of $\widetilde{S}^+(\Omega_o)$ is defined on $\Omega_o \cup \partial\Omega$ and is lower semicontinuous there. A set $E \subset \Omega \cup \partial\Omega$ is defined to be full-thin at $\hat{a} \in \partial\Omega$ iff for every Ω_o, there exists $u \in \widetilde{S}^+(\Omega_o)$ (or even $u \in \widetilde{P}(\Omega_o)$) such that $u(\hat{a}) < \liminf_{x \in E, x \to \hat{a}} u(x)$. In order to define E to be full-thin at $\hat{a} \in \partial\Omega$ we can even require that $u \in \widetilde{P}(\Omega_o)$ can always be found such that $u(\hat{a}) < +\infty$ and $\lim_{x \in E, x \to \hat{a}} u(x) = +\infty$. Notice that full-thinness is independent of the deleted K_o. We also mention that for eac

$\hat{p} \in \partial\Omega$, $N_{\hat{p}}$ is extremal in the sense that if

$$N_{\hat{p}}(x) \equiv \int_{\partial\Omega} N(x,y)\,d\mu(y),$$

then $\mu(\partial\Omega) = \mu(\{\hat{p}\})$ where μ is any mass distribution on $\partial\Omega$. For general regions, the standard construction of the Kuramochi function for Ω_0 with pole at $a \in \Omega_0$ involves the use of the Dirichlet principle (cf. [6], or [14]). We do not go into the details here except to give the actual construction. Let $HD_o(\Omega_0)$ be the Hilbert space of Dirichlet finite (or BLD) harmonic functions on Ω_0 which vanish continuously on ∂K_o. For each $a \in \Omega_0$ the evaluation functional $u \to u(a)$ is a bounded linear functional on $HD_o(\Omega_0)$ and therefore can be represented by a member $U_a \in HD_o(\Omega_0)$. We shall choose U_a such that

$$(U_a, u) = \varphi_p u(a) \tag{2.4}$$

where φ_p is the total flux of the fundamental function over large spheres. For example $\varphi_p = 2\pi$ if $p = 2$, and equals $(p - 2)$(area of the unit sphere if $p \geqq 3$). If G_a is the ordinary Green function for Ω_0 with pole at $a \in \Omega_0$ then we define

$$N_a = G_a + U_a. \tag{2.5}$$

There are various ways of attaching the Kuramochi boundary $\Delta_N(\Omega)$ to the region Ω (cf. [6], p. 167). For our purposes, it is sufficient to observe that $a \in \Delta_N(\Omega)$ iff for any K_o, there exists a unique harmonic function N_a on Ω_0 such that $N_a = \lim_{n\to\infty} N_{a_n}$ for some sequence $\{a_n\}$ in Ω (and hence in Ω_0 in this case) that eventually leaves all compact subsets of Ω. We must now think of full-potentials, again denoted by $\tilde{P}(\Omega_0)$, as being defined on $\Omega_0 \cup \Delta_N(\Omega)$. We now proceed to some specific examples and compare full-thin sets with the other types considered thus far.

The case where Ω is the unit disk is especially easy. If we let $x^* = |x|^{-2}x$ be the geometric inverse of x with respect to the unit circle $\partial\Omega$ and delete $K_o = \{z : |z| \leq 1/2\}$ then we consider the double region $\tilde{\Omega}_0 = \{z : 1/2 < |z| < 2\} = \Omega_0 \cup \partial\Omega \cup \Omega_0^*$ where $\Omega_0^* = \{x : x^* \in \Omega_0\}$. For any $a \in \Omega_0$, the Kuramochi function N_a for Ω_0 with pole at a is the restriction to $\Omega_0 \cup \partial\Omega$ of $\tilde{G}_a + \tilde{G}_{a^*}$ where $\tilde{G}_a$ (resp. $\tilde{G}_{a^*}$) is the ordinary Green function for $\tilde{\Omega}_0$ with pole at a (resp. a^*), and the function U_a mentioned in (2.5) is the restriction to $\Omega_0 \cup \partial\Omega$ of $2\tilde{G}_{a^*}$. Full-potentials on Ω_0 are therefore the restriction to $\Omega_0 \cup \partial\Omega$ of Green potentials on $\tilde{\Omega}_0$ which are symmetric about the unit circle via the inversion map (cf. [6], p. 233, or [8]). One can therefore conclude that $E \subset \Omega_0 \cup \partial\Omega$ is full-thin at $e^{i\alpha} \in \partial\Omega$ in Ω iff it is thin there in the ordinary sense. There is a significant difference, however, between ordinary thinness and full-thinness due to the fact that full-thinness like minimal thinness is a conformal invariant. Hwang and I (cf. [8]) have recently demonstrated that one cannot compare ordinary thin sets and full-thin sets in general, even for a Jordan domain. The reason generally depends on the fact that ordinary thinness fails to be a conformal invariant. On the other hand full-thinness always implies minimal thinness in any simply connected region strictly contained in the plane where we can set up a natural correspondence between prime ends, Martin boundary points, and Kuramochi boundary points via a conformal map from the unit disk. If f is a function from the unit disk (open or closed) into a topological space then we let (cf. [6], p.221) $\check{f}(e^{i\alpha})$ be the cluster set of f at $e^{i\alpha}$ along the filter of complements of full-thin sets at $e^{i\alpha}$ with respect to $e^{i\alpha}$. We can now say (cf. [8]) that $\check{f}(e^{i\alpha}) \subset \hat{f}(e^{i\alpha})$ always. Constantinescu and Cornea (cf. [6], p. 231) apparently did not notice this fact. Even more recently, Hwang and I (cf. [9]) have noticed that in higher dimensions there is again equivalence between ordinary thin sets and full-thin sets at all boundary points of a ball or half space including the point at ∞. Each of these cases can be handled by a doubling technique, but special difficulties must be treated that were not present in the two dimensional case. In the half space, for example, one must carefully take into account the harmonic measure of $\{\infty\}$ with respect to the doubled region $\tilde{\Omega}_0$.

3. Rarefied sets and their covering properties.

In this section I wish to discuss some recent results obtained in a joint work with M. Essén (cf. [7]) where $\Omega \subset \mathbb{R}^p$ $(p \geq 2)$ is always a half space. If $h(x) = x_1$ is the minimal harmonic function for Ω with pole at ∞, and $K(\hat{y},x)$ is the Poisson kernel defined on $\partial\Omega \times \Omega$ then any $u \in S^+(\Omega)$ has a canonical decomposition of the form

$$u(x) = G\lambda(x) + \int_{\partial D} K(y,x)d\mu(y) + \alpha h(x) \tag{3.1}$$

where $G\lambda$ is the Green potential of a measure λ on Ω and $\alpha \geq 0$. Another way of phrasing (3.1) is to say that any $u \in S^+(\Omega)$ has a unique canonical measure ν on $\Omega \cup \partial\Omega \cup \{\infty\}$ such that $\nu|_\Omega = \lambda$, $\nu|_{\partial\Omega} = \mu$ and $\nu\{\infty\} = \alpha$. We now let $S_1^+(\Omega) \subset S^+(\Omega)$ be the subcone of positive superharmonic functions on Ω each of whose canonical measure ν does not charge the point at ∞ (i.e. $\nu\{\infty\} = \alpha = 0$). In response to the general observation that the growth order of any $u \in S^+(\Omega)$ at ∞ is in many ways no larger than the growth order of the minimal harmonic function h, there have been two basic approaches in studying the detailed growth behaviour of any $u \in S^+(\Omega)$ as $x \to \infty$.

In the first approach one forms the quotient u/h where $u \in S_1^+(\Omega)$ and then attempts to precisely describe the exceptional set $E \subset \Omega$ on which $u(x)/h(x) \to 0$ as $x \to \infty$, $x \in \Omega - E$. This was the approach of J. Lelong-Ferrand whose results imply that the right exceptional set E for this quotient is in fact a suitably chosen set that is minimally thin at ∞ in Ω. In the second approach one asks the analogous question for the quotient u/s where $u \in S_1^+(\Omega)$ as before, and $s(x) = |x| = r$ which is a subharmonic function on Ω whose growth order at ∞ coincides with h on any ray or even in any Stolz domain in Ω. Since $u/s \leq u/h$ everywhere on Ω it is clear that any set $E \subset \Omega$ that is exceptional for u/s at ∞ in the sense that $\liminf u(x)/s(x) > 0$ as $x \to \infty$, $x \in E$, will also be exceptional there for u/h, and therefore will be minimally thin at ∞ with respect to Ω. It is also evident that if ω is any Stolz domain in Ω with vertex at the origin then $E \subset \omega$ will be exceptional for u/s at ∞ iff it is exceptional there for u/h. The question of characterizing the exceptional set E for u/s was first considered by Ahlfors and Heins (cf. [1]) whose work was restricted to the planar case. They demostrated that if ω is any Stolz domain in Ω then the image set which is obtained by the circular projection of $E \cap \omega$ into the positive real axis has finite logarithmic length. This property which is also possessed by minimally thin sets in a Stolz domain does not characterize the exceptional set E even when E itself lies on the positive real axis (cf. [10]). In 1956 W. K. Hayman, whose work was also restricted to the two dimensional case, demonstrated that if the exceptional set E itself for u/s is projected onto the positive real axis via circular projection, then the projected set will again have finite logarithmic length. This proved that E is much smaller than a minimally thin set at ∞ in Ω even though the two kinds of exceptional sets always coincide on every Stolz domain. During the middle 1960's, V. Azarin (cf. [2]) whose work was unrestricted by the dimension p, generalized Hayman's result by demonstrating that the exceptional set E for u/s can be covered by a sequence of balls $\{B_n\}$ such that if $B_n = (r_n, R_n)$ where r_n is the radius of B_n and R_n is the modulus of its centre then

$$\sum_n (r_n/R_n)^{p-1} < +\infty . \tag{3.2}$$

It turns out that none of these conditions will characterize the exceptional set E for u/s except that in a certain sense we shall see that the conditions of Hayman and Azarin do characterize that part of E that is sufficiently close to the boundary $\partial\Omega$. Recently Essén and I (cf. [7]) have completely characterized the exceptional set for u/s. We have called such an exceptional set a rarefied set at ∞ in Ω, and our results indicate (cf. [7], Theorem 3.3) that $E \subset \Omega$ is rarefied at ∞ in Ω iff $\hat{R}_r^E \in S_1^+(\Omega)$. At the origin a set $E \subset \Omega$ is rarefied in Ω iff $\hat{R}^E_{r^{1-p}} \in S^+(\Omega)$ whose canonical measure fails to charge $\{0\}$.

We now fix $\sigma \geq 1$, let $I_n = \{x \in \Omega : \sigma^n < |x| \leq \sigma^{n+1}\}$ and let $E_n = E \cap I_n$. We have also proved (cf. [7], Lemma 3.1) that E is rarefied at ∞ in Ω iff for any $a \in \Omega$

$$\sum_n \hat{R}_r^{E_n}(a) < +\infty . \tag{3.3}$$

If $\lim_{n\to\infty} \hat{R}_r^{E_n}(a) = 0$ then we have defined E to be semirarefied at ∞ in Ω. As we have already pointed out, our semirarefied sets are the same as the rarefied sets of J. Lelong-Ferrand. For comparison purposes we recall that a set $E \subset \Omega$ is minimally thin at ∞ in Ω iff $\sum_n \hat{R}_{x_1}^{E_n}(a) < \infty$, and semithin there iff $\lim_{n\to\infty} \hat{R}_{x_1}^{E_n} = 0$. We now consider some specific examples.

Let us now describe a sequence of balls $B_n = (t_n, r_n, R_n)$ where r_n is the radius of B_n, R_n is the modulus of its centre and $t_n > 0$ is the first coordinate of its centre. For the discussion here we shall require $t_n \geq r_n$ so that all of each B_n is contained in Ω, but in general this condition is not necessary. Let us now suppose that $E = \cup_{n=1}^{\infty} E_n$ is structured so that each E_n consists of exactly one ball $B_n = (t_n, r_n, R_n)$. In the higher dimensional cases, our results indicate that E is rarefied at ∞ in Ω iff

$$\sum_n \left(\frac{t_n}{R_n}\right)\left(\frac{r_n}{R_n}\right)^{p-2} < +\infty, \quad (p \geq 3) \tag{3.4}$$

and that E is minimally thin there iff

$$\sum_n \left(\frac{t_n}{R_n}\right)^2 \left(\frac{r_n}{R_n}\right)^{p-2} < +\infty, \quad (p \geq 3) . \tag{3.5}$$

Since $t_n \geq r_n$ it is clear that (3.4) implies Azarin's condition (3.2), and coincides with it in the very special case where r_n and t_n are all of comparable size which means that E is close to the boundary $\partial\Omega$ in a certain sense. Our results also indicate that if $E \subset \Omega \subset \mathbb{R}^p$ $(p \geq 3)$ is arbitrary and can be covered by a sequence of balls $\{B_n\}$, $B_n = (t_n, r_n, R_n)$ such that condition (3.4) (resp. (3.5)) is satisfied then E is rarefied (resp. minimally thin) at ∞ in Ω. The converse does not hold but our results do indicate that if $p \geq 3$, and E is rarefied at ∞ in Ω then E can be covered by a sequence $\{B_n\}$ where $B_n = (t_n, r_n, R_n)$, $t_n \geq (\text{constant}) r_n$ such that

$$\sum_n \left(\frac{t_n}{R_n}\right)\left(\frac{r_n}{R_n}\right)^{\beta} < +\infty \quad \text{where} \quad \beta > p-2. \tag{3.6}$$

If E is minimally thin at ∞ the covering result states that

$$\sum_n \left(\frac{t_n}{R_n}\right)^2 \left(\frac{r_n}{R_n}\right)^{\beta} < +\infty \quad \text{where} \quad \beta > p-2. \tag{3.7}$$

In the two dimensional case our results are surprisingly different. For the plane we must replace condition (3.4) by

$$\sum_n \left(\frac{t_n}{R_n}\right)\left(\log\left(\frac{2t_n}{r_n}\right)\right)^{-1} < +\infty \tag{3.8}$$

and condition (3.5) by

$$\sum_n \left(\frac{t_n}{R_n}\right)^2 \left(\log\left(\frac{2t_n}{r_n}\right)\right)^{-1} < +\infty . \tag{3.9}$$

For the converse, condition (3.6) should be replaced by

$$\sum_n \left(\frac{t_n}{R_n}\right)\left(\log\left(\frac{2t_n}{r_n}\right)\right)^{-\beta} < +\infty \quad \text{where} \quad \beta > 1. \tag{3.10}$$

One could ask if the two conditions of minimal thinness and Azarin's condition taken together are sufficient to characterize a rarefied set at ∞ in Ω. Our results indicate that this is not the case.

References

[1] Ahlfors, L. V., Heins, M.: Questions of regularity connected with the Phragmén-Lindelöf principle. Ann. of math., II. Ser. 50 (1949), 341 - 346.

[2] Azarin, V.: Generalization of a theorem of Hayman on subharmonic functions in an m-dimensional cone. Amer. math. Soc., Translat., II. Ser. 80 (1969), 119 - 138.

[3] Brelot, M.: Sur les ensembles effilés. Bull. Sci. math., II. Sér. 68 (1944), 12 - 36.

[4] Brelot, M.: Sur le rôle du point à l'infini dans la theórie des fonctions harmoniques. Ann. sci. École norm. sup., III. Sér. (1944), 301 - 332.

[5] Brelot, M.: On topologies and boundaries in potential theory. Lecture Notes in Mathematics 175, Springer-Verlag, Berlin - Heidelberg - New York (1971).

[6] Constantinescu, C., Cornea, A.: Ideale Ränder Riemannscher Flächen. Springer-Verlag, Berlin - Göttingen - Heidelberg (1963).

[7] Essén, M., Jackson, H. L.: On the covering properties of certain exceptional sets in a half space. Royal Institute of Technology, Stockholm, Sweden. Preprint.

[8] Hwang, J. S., Jackson, H. L.: Some results on Kuramochi thin sets. An. Acad. Bras. Ciênc. 50, 4 (1978).

[9] Hwang, J. S., Jackson, H. L.: On the relationship between ordinary thin sets and full-thin sets. Centre de recherches mathématiques, Université de Montréal, report 845. Preprint.

[10] Jackson, H. L.: Some results on thin sets in a half plane. Ann. Inst. Fourier 20, 2 (1970), 201 - 218.

[11] Jackson, H. L.: Sur la comparaison entre deux types d'effilement. Sémin. Théorie Potent., 15e année 1972, Exposé 23 (1973).

[12] Jackson, H. L.: On the mapping properties of certain exceptional sets in $\mathbb{R}^2$. Canadian J. Math. 27 (1975), 44 - 49.

[13] Lelong-Ferrand, J.: Etude au voisinage de la frontière des fonctions surharmoniques positives dans un demi-espace. Ann. sci. École norm. sup., III. Sér. 66 (1949), 125 - 159.

[14] Maeda, F. Y., Ohtsuka, M.: Kuramochi boundaries of Riemann surfaces. Lecture Notes in Mathematics 58, Springer-Verlag, Berlin - Heidelberg - New York (1968).

[15] Naïm, L.: Sur le rôle de la frontière de R. S. Martin dans la théorie du potentiel. Ann. Inst. Fourier 7 (1957), 183 - 281.

Department of Mathematics
McMaster University
Hamilton, Ontario L8S 4K1
Canada

ZUR WERTEVERTEILUNG DER LÖSUNGEN LINEARER DIFFERENTIALGLEICHUNGEN

Otto Knab

1. Übersicht. Gegeben ist die Differentialgleichung

$$w^{(n)} + a_{n-1}(z)w^{(n-1)} + \cdots + a_o(z)w = 0, \tag{D}$$

in welcher die a_{n-j} für $j = 1,\cdots,n$ Polynome der Form

$$a_{n-j}(z) = A_{n-j}z^{\alpha_{n-j}}(1 + o(1)) \tag{1}$$

für $z \to \infty$ sind.

$w(z)$ bezeichne stets eine transzendente Lösung von (D) mit der Wachstumsordnung $\lambda = \lambda(w)$ und dem Typus $\sigma = \sigma(w)$, wobei Wachstumsordnung und Typus in üblicher Weise über den Maximalbetrag von w definiert werden. Bekannterweise gilt $0 < \lambda, \sigma < \infty$, und die möglichen Werte von λ und σ sind durch das zu (D) gehörende Puiseux-Diagramm Q, das in 2. angegeben wird, festgelegt (hierzu vergleiche man Nikolaus [8], Pöschl [9] und Wittich [14]).

Im Anschluß an Lepson [7] und Frank [2] wurde in [4] der (α,x)-Index von w eingeführt und dessen asymptotische Darstellung für den Fall $1 - \lambda < \alpha \leq 1$ in [5] und [6] abschließend behandelt. In der vorliegenden Arbeit wird eine entsprechende Darstellung des Index für $\alpha > 1$ hergeleitet.

Möchte man nun die Anzahl der c-Stellen $(c \in \mathbb{C})$ von w innerhalb von Kreisscheiben nach oben abschätzen, dann bietet sich als elementares Hilfsmittel die Jensensche Formel an. Mit ihrer Hilfe gelingt über den lokalen Maximalbetrag und das Maximalglied der Potenzreihenentwicklung von w eine Abschätzung der Anzahlfunktion der c-Stellen durch den Zentralindex eben dieser Potenzreihenentwicklung.

Schreiben wir für $|\zeta| = r$, $\alpha \in \mathbb{R}$, $x,\tau > 0$

$$S_\zeta(\tau r^\alpha) = \{z : |z - \zeta| \leq \tau r^\alpha\},\ K_\zeta(xr^\alpha) = \{z : |z - \zeta| = xr^\alpha\} \tag{2}$$

und führen diese Betrachtungen in den Kreisscheiben $S_\zeta(\tau r^\alpha)$ für die Funktion

$$g_\zeta(z;c) = (w(z) - c)/(z - \zeta)^m \tag{3}$$

durch, wobei m die c-Stellenmultiplizität von w im Punkte $z = \zeta$ angibt, dann kann der Zentralindex $\nu_\zeta(xr^\alpha, g_\zeta)$ der Potenzreihenentwicklung um ζ auf dem Kreis $K_\zeta(xr^\alpha)$ durch den (α,x)-Index von w nach oben abgeschätzt werden.

Dies ergibt in 4. eine allgemeine Abschätzung der Anzahl $n_\zeta(\tau r^\alpha, w - c)$ der c-Stellen von w in $S_\zeta(\tau r^\alpha)$ nach oben, wobei jede c-Stelle entsprechend ihrer Vielfachheit gezählt wird. Hieraus werden einige spezielle Ergebnisse über die lokale Werteverteilung der transzendenten Lösungen von (D) hergeleitet, wobei insbesondere auf das Diskonjugiertheitsproblem eingegangen wird. Außerdem geben wir in 5. eine Abschätzung für die globale Werteverteilung von w an.

2. Das Puiseux-Diagramm Q von (D) und der (α,x)-Index der transzendenten Lösungen. Für Wachstumsordnung und Typus einer transzendenten Lösung w von (D) liegt die folgende geometrische Kennzeichnung durch das Puiseux-Diagramm vor:

Mit (1) und

$$g = \max_{j=1}^{n} \{\alpha_{n-j} + j\} \tag{4}$$

tragen wir die Punkte

$$Q_o = (0,g),\; Q_j = (j, g - (\alpha_{n-j} + j))$$

für $j = 1,\cdots,n$ in einem kartesischen Koordinatensystem auf und konstruieren den kleinsten konvexen Polygonzug, welcher die Punkte Q_j für $j = 0,1,\cdots,n$ von unten umspannt. Dieser Polygonzug besitzt einen in j streng monoton fallenden Teilpolygonzug, der das Puiseux-Diagramm Q von (D) genannt wird.

Ist nun λ die Wachstumsordnung einer transzendenten Lösung von (D), dann ist $-\lambda = -\lambda_i$ die Steigung eines Streckenzuges von Q, und mit dem Typus σ von w ist $\sigma \cdot \lambda$ nach Nikolaus [8], Pöschl [9] und [6] der Betrag einer Lösung der Gleichung

$$\sum_{j\in N_i} A_{n-j} t^{j_i - j} = 0 , \tag{5}$$

wobei

$$N_i = \{j_{i-1},\cdots,j_i : j_{i-1} < \cdots < j_i\}$$

die Menge derjenigen ganzen Zahlen angibt, für welche die Punkte Q_j auf der Strecke von Q mit der Steigung $-\lambda_i$ liegen. Aus dem Betragssummensatz ergibt sich mit (1) aus (5)

$$\sigma \cdot \lambda \leqq A = \frac{|A_{n-j_{i-1}}| + \max\{|A_{n-j}| : j \in N_i,\; j \neq j_{i-1}\}}{|A_{n-j_{i-1}}|} \tag{6}$$

wobei in Falle $j_{i-1} = 0$ stets $A_n = 1$ zu setzen ist.

Entwickeln wir nun w um den Punkt $z = \zeta$ in die Taylor-Reihe

$$w(z) = \sum_{j=0}^{\infty} (w^{(j)}(\zeta)/j!) \cdot (z - \zeta)^j ,$$

dann existiert wegen der Konvergenz dieser Reihe nach (2) für den Kreis $K_\zeta(xr^\alpha)$ eine größte Zahl $\nu = \nu_\zeta(xr^\alpha,w)$, so daß

$$\mu = \frac{|w^{(\nu)}(\zeta)|}{\nu!} \cdot (xr^\alpha)^\nu \geqq \frac{|w^{(j)}(\zeta)|}{j!} \cdot (xr^\alpha)^j$$

für $j = 0,1,\cdots$ gilt. $\mu = \mu_\zeta(xr^\alpha,w)$ wird das Maximalglied und ν der Zentralindex der Potenzreihenentwicklung von w bezüglich des Punktes $z = \zeta$ auf dem Kreis $K_\zeta(xr^\alpha)$ genannt.

Nach Lepson [7] und Frank [2] wird der (α,x)-Index $J(xr^\alpha,w)$ von w auf $|\zeta| = r$ nach [4] und [5] durch

$$J(xr^\alpha,w) = \max_{|\zeta|=r} \{\nu_\zeta(xr^\alpha,w)\}$$

definiert. Mit

$$t(x) = xr^{-(\lambda-1+\alpha)}(1 + o(1)) \tag{7}$$

für $r \to \infty$ gilt nach [6] für $1 - \lambda < \alpha < 1$

$$J(xr^\alpha,w) \begin{cases} \leqq n & \text{für } x \in (0,t(n/\sigma\lambda)) \\ = \sigma\lambda xr^{\lambda-1+\alpha}(1 + o(1)) & \text{für } x \geqq t(n/\sigma\lambda) . \end{cases} \tag{8}$$

Diese Darstellung trägt der Tatsache Rechnung, daß w Nullstellen der Ordnung $n - 1$ besitzen kann. Weiter ist die Voraussetzung $\alpha > 1 - \lambda$ notwendig, da (8) für $x \geqq t(n/\sigma\lambda)$ nur gilt, wenn der Index für $r \to \infty$ unbeschränkt ist. Die einschränkende Bedingung $x \leqq 1$ aus [4] und [5] entfällt hier, da $\alpha < 1$ vorausgesetzt und w eine ganze Funktion ist.

Für $\alpha = 1$ sind die Dinge nicht ganz so einfach. Ist hier $b \neq 0$ die betragskleinste Nullstelle der Diskriminante der algebraischen Gleichung

$$\sum_{j \in N_1} y^{j_i - j} A_{n-j} P_{n-j}(x) = 0$$

(die Polynome $P_{n-j}(x)$ sind in [5], Beweis von Satz 2, angegeben, wobei $B_{J+1-m} = 1$ auf Seite 308 zu setzen ist), dann gilt mit $B = |b|$

$$J(xr,w) \begin{cases} \leqq n & \text{für } x \in (0,t(n/\sigma\lambda)) \\ = \sigma\lambda xr^\lambda(1 + o(1)) & \text{für } x \in [t(n/\sigma\lambda), B) \end{cases} \tag{9}$$

für $r \to \infty$, wobei zu beachten ist, daß $t(n/\sigma\lambda) < B$ nach (7) wegen $\lambda > 0$ für hinreichend große r gilt. Die Sonderstellung von $\alpha = 1$ liegt darin begründet, daß hier die in 2. angesprochene Konvexität des Puiseux-Diagramms Q nicht so stark wie im Falle $\alpha < 1$ zum Tragen kommt.

Im Anschluß an Frank [2], der den $(\alpha,1)$-Index von w untersuchte, betrachten wir nun den Fall $\alpha > 1$. Mit

$$t^*(x) = x^{1/\lambda} r^{-\alpha}(1 + o(1)) \tag{10}$$

für $r \to \infty$ erhalten wir entsprechend den Darstellungen (8) und (9) des Index

Satz 1. Es gilt für beliebiges, aber festes $x > 0$

$$J(xr^\alpha, w) \begin{cases} \leq n & \text{für} \quad x \in (0, t^*(n/\sigma\lambda)) \\ = \sigma\lambda(xr^\alpha)^\lambda(1 + o(1)) & \text{für} \quad x \geq t^*(n/\sigma\lambda) \end{cases}$$

für $r \to \infty$ und $\alpha > 1$.

Zum Beweis dieses Satzes wird wiederum wesentlich die Konvexität des Puiseux-Diagramms von (D) ausgenützt.

3. Der Zentralindex der Funktion $g_\zeta(z;c)$. Für die in (3) erklärte Funktion g_ζ untersuchen wir nun den Zentralindex $\nu_\zeta(xr^\alpha, g_\zeta)$ auf Kreisen $K_\zeta(xr^\alpha)$.

Hierzu setzen wir im Vergleich zu (8), (9) und Satz 1

$$\phi(r,x;\alpha) = \begin{cases} xr^{\lambda-1+\alpha} & \text{für} \quad 1 - \lambda < \alpha \leq 1 \\ (xr^\alpha)^\lambda & \text{für} \quad \alpha > 1 \end{cases} \tag{11}$$

und beachten, daß mit (1) und

$$N_c = \begin{cases} n + 1 & \text{für} \quad c = 0 \\ n + \alpha_o & \text{für} \quad c \neq 0 \end{cases}$$

nach (6) und (7) stets

$$t(1/A) \leq t((N_c + 1)/\sigma\lambda)$$

für hinreichend große r gilt. Wir erhalten dann unter Beachtung der möglichen Nullstellenmultiplizität der Funktion $w(z) - c$ aus der Differentialgleichung (D) und durch Vergleich mit dem Index von w

<u>Satz 2.</u> Für $\nu_\zeta(xr^\alpha, g_\zeta)$ gelten in Abhängigkeit von $\lambda = \lambda(w)$ und $\alpha > 1 - \lambda$ für $r \to \infty$ die folgenden Abschätzungen:

a) $\alpha \leq 1$, λ beliebig:

$$\nu_\zeta(xr^\alpha, g_\zeta) \begin{cases} = 0 & \text{für } x \in (0, t(1/A)) \\ \leq N_c + 1 & \text{für } x \in [t(1/A), t((N_c+1)/\sigma\lambda)) \\ \leq \sigma\lambda\phi(r,x;\alpha)(1 + o(1)) & \text{für } x \geq t((N_c+1)/\sigma\lambda) \end{cases}$$

wobei im Falle $\alpha = 1$ stets $x < B$ vorauszusetzen ist.

b) $\alpha > 1$

b.1) $\lambda \geq 1$

$$\nu_\zeta(xr^\alpha, g_\zeta) \begin{cases} = 0 & \text{für } x \in (0, t(1/A)) \\ \leq N_c + 1 & \text{für } x \in [t(1/A), t^*((N_c+1)/\sigma\lambda)) \\ \leq \sigma\lambda\phi(r,x;\alpha)(1 + o(1)) & \text{für } x \geq t^*((N_c+1)/\sigma\lambda) \end{cases}$$

b.2) $0 < \lambda < 1$

$$\nu_\zeta(xr^\alpha, g_\zeta) \begin{cases} = 0 & \text{für } x \in (0, t(1/A)) \\ \leq \sigma\lambda\phi(r,x;\alpha)(1 + o(1)) & \text{für } x \geq t(1/A) \end{cases},$$

wobei t nach (7) und t^* nach (10) zu wählen sind.

Nach Definition des Index in 2. ist diese Abschätzung für solche x nicht zu verbessern, für welche die Abschätzung $\leq \sigma\lambda\phi(r,x;\alpha)(1 + o(1))$ eintritt. Hierzu beachten wir (8), (9) und Satz 1. Denkbar ist eine Verbesserung der Intervallgrenze $t(1/A)$, also unter Berücksichtigung von $g_\zeta(\zeta;c) \neq 0$ eine bessere Abschätzung der ersten Sprungstelle des Zentralindex der betrachteten speziellen Lösung w nach unten. In den Intervallen, für die $x > t(1/A)$ gilt und $\nu_\zeta(xr^\alpha, g_\zeta)$ beschränkt ist, spielen Glieder höherer Ordnung der Potenzreihenentwicklung von g_ζ eine Rolle, die wir mit der hier verwendeten Methode nicht in der Hand haben. Deswegen wird ν_ζ durch $N_c + 1$ abgeschätzt.

<u>4. Abschätzung der c-Stellenanzahl von w in $S_\zeta(\tau r^\alpha)$.</u> Mit Hilfe von 3. wollen wir nun die Abschätzungen aus [4] vereinfachen und verbessern sowie auf den Fall $\alpha > 1$ ausdehnen.

Gibt $n_\zeta(\tau r^\alpha, w - c)$ für $c \in \mathbb{C}$ die Anzahl der c-Stellen von w nach (2) in den Kreisscheiben $S_\zeta(\tau r^\alpha)$ an, wobei jede c-Stelle entsprechend ihrer Vielfachheit gezählt wird, dann kann n_ζ mit Hilfe der Jensenschen Formel durch Quotienten von Maximalbeträgen von g_ζ nach oben abgeschätzt werden. Eine weitere Abschätzung wird in 7. durchgeführt und ergibt nach der Definition von g_ζ und Pólya, Szegö [11], S. 5,

Hilfssatz 1. Für $0 < \tau < k < K$ gilt

$$n_\zeta(\tau r^\alpha, w - c) \leqq N_c + \frac{1}{\log(k/\tau)} \left\{ \log \frac{K}{K-k} + \int_{t(1/A)}^{K} \frac{\nu_\zeta(xr^\alpha, g_\zeta)}{x} dx \right\},$$

wobei das auftretende Integral für $K \leqq t(1/A)$ gleich 0 zu setzen ist und für $\alpha = 1$ stets $K < B$ vorausgesetzt werden muß.

Zusammen mit Satz 2 erhalten wir mit Hilfssatz 1 eine recht allgemeine Abschätzung von n_ζ, wobei wir sehen, daß hierbei die in Satz 2 angegebene Abschätzung der ersten Sprungstelle des Zentralindex von g_ζ wesentlich ist.

Wir wollen nun aus Satz 2 und Hilfssatz 1 einige spezielle Ergebnisse über die lokale Werteverteilung von w herleiten.

Setzen wir nach (7) und (10) in Zusammenhang mit Satz 2

$$C = \begin{cases} t((N_c + 1)/\sigma\lambda) & \text{für } 1 - \lambda < \alpha \leqq 1 \\ \max\{t(1/A), t^*((N_c + 1)/\sigma\lambda) & \text{für } \alpha > 1, \end{cases} \tag{12}$$

dann ist $\nu_\zeta(xr^\alpha, g_\zeta)$ für $x > C$ und $|\zeta| = r \to \infty$ unbeschränkt, und eine einfache Extremwertbetrachtung über die integrierte Abschätzung in Hilfssatz 1 ergibt mit (11)

Satz 3. Mit $\tau > C$ nach (12) und $\alpha > 1 - \lambda$, $\alpha \neq 1$, gilt

$$n_\zeta(\tau r^\alpha, w - c) \leqq e\sigma\lambda\phi(r, \tau; \alpha)(1 + o(1))$$

für $r \to \infty$. Für $\alpha = 1$ ist $C < \tau < B$ vorauszusetzen, und dies ergibt

$$n_\zeta(\tau r^\alpha, w - c) \leqq \begin{cases} e\sigma\lambda\phi(r, \tau; 1)(1 + o(1)) & \text{für } \tau \in (C, B/e) \\ \sigma\lambda\phi(r, B; 1)(1 + o(1))/\log(B/\tau) & \text{für } \tau \in [B/e, B) \end{cases}$$

unter Beachtung von (9) und (12).

Für $\alpha = 1 - \lambda$, also $C = t((N_c + 1)/\sigma\lambda)$ wegen $\lambda > 0$ in (12), erhalten wir aus der Monotonie der Anzahlfunktion n_ζ und des Index von w in α für $\tau \to C$ aus Satz 3 insbesondere

Korollar 3.1. Es gilt

$$n_\zeta((N_c + 1)r^{1-\lambda}(1 + o(1))/\sigma\lambda, w - c) \leqq N_c + e(N_c + 1)(1 + o(1))$$

für $r \to \infty$.

Hier gelingt also eine Charakterisierung der Kreisscheiben $S_\zeta(\tau r^\alpha)$, in denen alle Lösungen von (D) asymptotisch beschränkte Werteverteilung besitzen, wobei bemerkenswert ist, daß im Falle $0 < \lambda(w) < 1$ die Radien dieser Kreisscheiben für $r \to \infty$ unbeschränkt sind.

Als Beispiel betrachten wir die Funktion $w(z) = \exp(2\pi i z^p)$, $p \in \mathbb{N}$, die der Differentialgleichung

$$w' - 2\pi i p z^{p-1} w = 0$$

genügt. Für positive reelle $\zeta = r$ ergibt die Abzählung der positiven reellen 1-Stellen $k^{1/p}$, $k \in \mathbb{N}$,

$$n_\zeta(\tau r^\alpha, w - 1) \geqq \begin{cases} 2p\tau r^{p-1+\alpha}(1 + o(1)) & \text{für } \alpha \leqq 1,\ \tau < 1 \\ (2r)^p(1 + o(1)) & \text{für } \alpha = 1,\ \tau = r/2\ , \end{cases}$$

und dies zeigt, daß für $\alpha \leqq 1$ der Exponent $\lambda - 1 + \alpha$ in Satz 3 zusammen mit (8) und (11) durch keinen kleineren ersetzt werden kann und allgemein für $\alpha > 1 - \lambda$ für $r \to \infty$ keine beschränkte Werteverteilung vorliegen kann. Im Falle $\alpha = 1$ zeigt dieses Beispiel weiter, daß die in (9) angegebene Schranke $x < B$ nicht nur beweistheoretischer Natur ist. Aus der Monotonie des (α,x)-Index in α (vergl. Lepson [7]) folgt damit aber auch, daß der Exponent $\alpha\lambda$ für $\phi(r,x;\alpha)$ in Satz 3 für $\alpha > 1$ durch keinen kleineren zu ersetzen ist.

Im Anschluß an Turán wollen wir nun die Frage behandeln, wie τ nach Korollar 3.1 zu wählen ist, damit $n_\zeta(\tau r^{1-\lambda}, w - c)$ durch N_c nach oben beschränkt ist. Man sagt dann, daß w in solchen Kreisscheiben $S_\zeta(\tau r^{1-\lambda})$ diskonjugiert sei. Es ergibt sich

<u>Satz 4.</u> Für $\tau \in (0, t(1/A)/4)$ nach (6) gilt

$$n_\zeta(\tau r^{1-\lambda}, w - c) \leqq N_c$$

für hinreichend große r.

Im Falle $c = 0$ wurde das Diskonjugiertheitsproblem von Hayman [3] behandelt und sein Ergebnis von Rahman, Stankiewicz [12] verbessert. Es lautet

<u>Satz A.</u> Die Koeffizienten $a_{n-j}(z)$ der Differentialgleichung

$$w^{(n)} + a_{n-1}(z)w^{(n-1)} + \cdots + a_o(z)w = 0$$

seien in der Kreisscheibe $|z - \zeta| < R$ holomorph und beschränkt, und t_o sei die positive Wurzel der Gleichung

$$\sum_{j=1}^{n} \beta_{n-j}t^j = 1, \ \beta_{n-j} = \sup \{|a_{n-j}(z)| : |z - \zeta| < R\} .$$

Dann besitzt jede Lösung w in

$$|z - \zeta| \leqq R_1 = \min \{t_o\sqrt{n}/e\sqrt{10}, R/2e\sqrt{10n}\}$$

höchstens $n - 1$ Nullstellen.

Im Gegensatz zu Satz 4 liefert dieser Satz eine Diskonjugiertheitsaussage in der ganzen Ebene, besitzt jedoch den Nachteil, daß R_1 wesentlich von der Ordnung n der Differentialgleichung abhängt. Diese Abhängigkeit ist für Lösungen von Differentialgleichungen (D) mit Polynomkoeffizienten nicht gegeben, da die Ordnung n beliebig groß gewählt werden kann und das asymptotische Verhalten der Anzahlfunktion n_ζ von w weitgehend von dem Anwachsen von $|w(z)|$ für $|z| \to \infty$ bestimmt wird.

Daß tatsächlich λ und σ und damit die a_{n-j}, $j \in N_i$, nach 2. das Anwachsen von n_ζ bestimmen, zeigt auch die Übertragung von Satz A auf die Differentialgleichung (D). Wir erhalten

Korollar A.1. Für festes $\tau > 0$ besitzt jede Lösung von (D) für hinreichend große $|\zeta| = r$ in Kreisscheiben

$$|z - \zeta| < r^{1-\lambda} \cdot \min \{D\sqrt{n}/e\sqrt{10}, \tau/2e\sqrt{10n}\}$$

höchstens $n - 1$ Nullstellen, wobei

$$1/A \leqq D \leqq (|A_{n-j_i}| + \max \{|A_{n-j}| : j \in N_i, \ j \neq j_i\})/|A_{n-j_i}|$$

unter Beachtung von (6) gilt.

Zum Beweis von Korollar A.1 wird wieder die Konvexität des Puiseux-Diagramms herangezogen, und dies hat natürlich eine asymptotische Abschätzung zur Folge.

Vergleichen wir Satz 4 mit Korollar A.1, dann zeigt sich, daß für

$$n > \max \{\tau/2D, (2A\tau/e)^2/10\}$$

Satz 4 die bessere Abschätzung liefert. Auf der anderen Seite besitzt das Haymansche Ergebnis natürlich den Vorteil, eine globale Diskonjugiert heitsaussage für lineare Differentialgleichungen mit sehr viel allgemeineren Koeffizienten als diejenigen in (D) zu geben.

Zur Verschärfung von Satz 4 ist eine gegenüber Hilfsatz 1 verbesserte Abschätzung aus der Jensenschen Formel zu diskutieren. Hierzu sei bemerkt, daß die verfeinerte Interpretation von Tijdeman [13] kein besseres Ergebnis ergibt.

Abschließend sei erwähnt, daß sich die Sätze 3 und 4 auf sich unbestimmt verhaltende Lösungen von Differentialgleichungen

$$a_n(z)w^{(n)} + a_{n-1}(z)w^{(n-1)} + \cdots + a_o(z)w = 0$$

übertragen lassen, in denen die a_{n-j} für $j = 0,1,\ldots,n$ Polynome mit $a_n(z) \not\equiv$ const. sind und die genau eine Stelle der Unbestimmtheit besitzen. Man kann diese Unbestimmtheitsstelle als $z = \infty$ annehmen und die sich dort unbestimmt verhaltenden Lösungen für hinreichend große $|z| = r$ auf der zu $\log z$ gehörenden Riemannschen Fläche in eindeutiger Weise erklären. Nach [4], [5] und [6] gelten die hier bewiesenen Ergebnisse für jeden Zweig von w, wenn man dafür sorgt, daß die Kreisscheiben $S_\zeta(\tau r^\alpha)$ keine Nullstellen von $a_n(z)$ enthalten.

5. Zur globalen Werteverteilung der Lösungen von (D). In dem anläßlich des Colloquium on Complex Analysis gehaltenen Vortrag wurde auch die globale Werteverteilung von w untersucht.

Gibt $n_o(r,f - c)$ die Anzahl der c-Stellen einer ganz transzendenten Funktion f in $|z| \leq r$ an, wobei jede c-Stelle entsprechend ihrer Vielfachheit gezählt wird und gilt für die Wachstumsordnung $\lambda = \lambda(f) > 0$ sowie für den Typus $\sigma = \sigma(f) < \infty$, dann erhält man nach Boas [1], S. 16, die scharfe Abschätzung

$$L = \limsup_{r\to\infty} (n_o(r,f-c)/r^\lambda) \leqq e\sigma\lambda \ .$$

Für die Lösungen w von (D) erhält man genauer

Satz 5. Für jede transzendente Lösung von w gilt $L \leqq \sigma\lambda$.

Der Beweis dieses Satzes erfordert die Anwendung der in [5] und [6] durchgeführten Substitutionen. Diese Betrachtungen erfordern weitere Hilfsmittel, auf deren Bereitstellung hier verzichtet werden soll.

Die Konstante $\sigma\lambda$ in Satz 5 ist durch keine kleinere zu ersetzen. Dies zeigt die Funktion

$$w(z) = \sum_{j=1}^{m} \exp(\eta_j z), \ \eta_j = \exp(2\pi i j/m) \ ,$$

für die $\sigma = \lambda = 1$ gilt. Nach Pólya [10] und Schwengeler zeigt man mit Hilfe des Indikatordiagramms von w

$$n_o(r,w-c) = \frac{m}{\pi}\cdot\sin\left(\frac{\pi}{m}\right) + O(\log r)$$

für $r \to \infty$, und man kann durch hinreichend große Wahl von m der durch Satz 5 gelieferten Schranke 1 beliebig nahe kommen.

6. Der Zentralindex von g_ζ in $z = \zeta$. Wir betrachten hier den Fall $c \neq 0$, also $N_c = n + \alpha_o$, und führen die Funktion

$$h(z;c) = w(z) - c$$

ein.

Nach der Definition von g_ζ in (3) gilt

$$\nu_\zeta(xr^\alpha,g_\zeta) \leqq \nu_\zeta(xr^\alpha,h) \leqq J(xr^\alpha,h) \tag{13}$$

unter Beachtung der Maximalitätseigenschaft des Index. Da h einer homogenen linearen Differentialgleichung mit Polynomkoeffizienten der Ordnung $n + \alpha_o + 1$ genügt, kann man entsprechend (8), (9) und Satz 1 den (α,x)-Index von $h(z;c)$ berechnen. Dies ergibt mit (7) und (10)

Hilfssatz 2. Es gilt $J(xr^\alpha,h) \leqq N_c + 1$ für $x < t((N_c + 1)/\sigma\lambda)$ bzw. $x < t^*((N_c + 1)/\sigma\lambda)$, während $J(xr^\alpha,h) = J(xr^\alpha,w)$ für alle größeren x eintritt, wobei wir im Falle $\alpha = 1$ wieder $x < B$ voraussetzen müssen.

Für $\alpha \leq 1$ wurde dies in [4] gezeigt, und der Beweis wird im Falle $\alpha > 1$ in 8. mit den bei der Herleitung von Satz 1 verwendeten Methoden analog durchgeführt.

Zum Beweis von Satz 2 in 3. müssen wir nun noch die erste Sprungstelle des Zentralindex der Potenzreihenentwicklung von g_ζ um den Punkt $z = \zeta$ abschätzen. Dies geschieht wieder mit Hilfe des Zentralindex von $h(z;c)$, und es ergibt sich bei Beachtung von $g_\zeta(\zeta;c) \neq 0$

Hilfssatz 3. Für $x \in (0,t(1/A))$ gilt $\nu_\zeta(xr^\alpha,g_\zeta) = 0$.

Vergleicht man nun noch $t(1/A)$ mit $t^*((N_c + 1)/\sigma\lambda)$ im Falle $\alpha > 1$ in Abhängigkeit von λ, dann folgt aus diesen beiden Hilfssätzen zusammen mit (8), (9), (13) und Satz 1 auch Satz 2.

7. Auswertung der Jensenschen Formel. Wir führen nun zur Herleitung von Hilfssatz 1 ähnliche Betrachtungen wie in [4] durch:

Für $k > 0$ wenden wir für g_ζ die Jensensche Formel auf die Kreisscheibe $S_\zeta(kr^\alpha)$ an und erhalten wegen $g_\zeta(\zeta;c) \neq 0$

$$\int_0^k \frac{n_\zeta(xr^\alpha, g_\zeta)}{x}\,dx = \frac{1}{2\pi}\int_0^{2\pi} \log |g_\zeta(\zeta + kr^\alpha \cdot e^{i\phi};c)|\,d\phi - \log |g_\zeta(\zeta;c)| .$$

Mit $0 < \tau < k$ gilt

$$\int_0^k \frac{n_\zeta(xr^\alpha, g_\zeta)}{x}\,dx \geqq n_\zeta(\tau r^\alpha, g_\zeta) \cdot \log (k/\tau) ,$$

und mit (2) und

$$m_\zeta(kr^\alpha, g_\zeta) = \max \{|g_\zeta(z;c)| : z \in K_\zeta(kr^\alpha)\}$$

ergibt sich

$$n_\zeta(\tau r^\alpha, g_\zeta) \leqq (\log (m_\zeta(kr^\alpha)/|g_\zeta(\zeta)|))/\log (k/\tau) . \tag{14}$$

Nach [4] folgt aus der Potenzreihenentwicklung von g_ζ um $z = \zeta$ für $k < K$

$$m_\zeta(kr^\alpha, g_\zeta) \leqq K\mu_\zeta(Kr^\alpha, g_\zeta)/(K - k) , \tag{15}$$

wenn μ_ζ nach 2. das Maximalglied dieser Potenzreihenentwicklung angibt. Beachtet man noch, daß nach der Definition von g_ζ und m in (3)

$$n_\zeta(\tau r^\alpha, w - c) = m + n_\zeta(\tau r^\alpha, g_\zeta) \leqq N_c + n_\zeta(\tau r^\alpha, g_\zeta)$$

gilt, weil für $h = w - c$ nach 6. die Nullstellenmultiplizität durch N_c beschränkt ist, dann erhält man aus (14) und (15)

$$n_\zeta(\tau r^\alpha, w - c) \leqq N_c + \frac{1}{\log (k/\tau)}\left\{ \log \left|\frac{K}{K - k}\right| + \log\left|\frac{\mu_\zeta(Kr^\alpha, g_\zeta)}{|g_\zeta(\zeta;c)|}\right|\right\} . \tag{16}$$

Nun gilt nach Pólya, Szegö [11], S. 5,

$$\log (\mu_\zeta(Kr^\alpha)/|g_\zeta(\zeta)|) = \int_0^K (\nu_\zeta(xr^\alpha, g_\zeta)/x)\,dx ,$$

und dies ergibt nach (16) unter Beachtung von Hilfssatz 3 gerade Hilfssatz 1.

8. Beweis der Sätze

8.1 Beweis von Satz 1. Differenziert man (D) nach (4) mindestens $(n + g)$-mal und beachtet man, daß $a_{n-k}^{(m-k)} \equiv 0$ für $m > \alpha_{n-k} + k$ gilt, dann erhält man für hinreichend große $|\xi| = r$ in den Punkten ξ, in denen $\nu_\xi(xr^\alpha, w) = J = J(xr^\alpha, w)$ nach 2. eintritt, nach [5] für

$J > g + n$ die Gleichung

$$\sum_{j=0}^{n} J^{n-j} \left\{ \sum_{k=0}^{j} \left\{ \sum_{m=k}^{\alpha_{n-k}+k} B_{J+\kappa-m} \cdot (x\xi^{\alpha})^{m} \cdot \beta_{n-j}^{(m,k)} \cdot a_{n-k}^{(m-k)}(\xi) \right\} \right\} = 0 \quad (17)$$

mit rationalen Zahlen $\beta_{n-j}^{(m,k)}$, $\beta_{n-j}^{(m,j)} = 1$, und Funktionen $B_{J+\kappa-m}(\xi,x)$, für die $B_J = 1$ und $|B_{J+\kappa-m}| \leqq 1$ gilt, wobei $\kappa \in \mathbb{N} \cup \{0\}$ frei gewählt werden kann.

Wir setzen nun $\alpha > 1$ voraus und beachten, daß aus der Konvexität des Puiseux-Diagramms Q von (D) in (1)

$$\alpha_{n-k} = \alpha_{n-j_{i-1}} + (k - j_{i-1})(\lambda - 1) - \delta_k \quad (18)$$

mit $\delta_k \geqq 0$ ($\delta_k = 0$ genau für $k \in N_i$) folgt. (17) reduziert sich somit für $\xi \to \infty$ zu

$$\sum_{j\in N_i} J^{n-j} A_{n-j} B_{J+\kappa-\alpha_{n-j}-j} (x\xi^{\alpha})^{j\lambda} (1 + o(1)) = 0 . \quad (19)$$

Setzt man nacheinander $\kappa = \alpha_{n-j} + j$, $j \in N_i$, dann sieht man wie in [4], daß die Lösungen von (19) von der Form

$$J(\xi) = d_i(x)\xi^{\alpha\lambda}(1 + o(1))$$

sind, wobei $d_i(x)$ Lösung der algebraischen Gleichung

$$\sum_{j\in N_i} y^{j_i - j} A_{n-j} x^{j\lambda} = 0 \quad (20)$$

ist. Beachtet man die Stetigkeit der Lösungen von (20), dann sieht man, daß $x = 0$ die einzige Singularität in $\mathbb{C}$ ist. Damit muß $d_i(x) = d_i x^{\lambda}$ gelten, und d_i ist der Betrag einer Wurzel der Gleichung (5). Da der Index positive reelle Lösung von (17) ist, folgt

$$J(xr^{\alpha},w) = |d_i|(xr^{\alpha})^{\lambda}(1 + o(1))$$

für $r \to \infty$.

Die gleichen Betrachtungen wurden im Falle $1 - \lambda < \alpha \leqq 1$ in [4], [5] und [6] durchgeführt. Für die Lösungen von (17) erhält man dann

$$J(\xi) = d_i(x)\xi^{\lambda-1+\alpha}(1 + o(1))$$

für $\xi \to \infty$, und im Falle $\alpha < 1$ ist $d_i(x)$ die Lösung der algebraischen Gleichung

$$\sum_{j\in N_i} y^{j_i - j} A_{n-j} x^{j} = 0 . \quad (21)$$

In [6] wird dann $|d_i(x)| = |d_i|x = \sigma\lambda x$ gezeigt, und damit ist $\sigma\lambda$ der Betrag einer Wurzel von (5). Im Falle $\alpha = 1$ sind in (21) die Monome in x durch Polynome zu ersetzen. Damit kann (21) außer $x = 0$ noch weitere endliche singuläre Punkte besitzen, und dies ergibt die Einschränkung $x < B$ in (9).

Vergleicht man nun noch das Anwachsen des Maximalbetrages von w für $r \to \infty$ mit dem Anwachsen des Index, dann zeigt eine eingehende Betrachtung mit den in [6] verwendeten Methoden, daß

$$\lim_{r\to\infty} J(xr^{\alpha},w)/r^{\alpha\lambda} = \sigma\lambda x^{\lambda}$$

auch im Falle $\alpha > 1$ gelten muß. Also gilt $d_i(x) = \sigma\lambda x^{\lambda}$ in (20) ohne einschränkende Bedingungen an x.

Wie in [5] zeigt man abschließend noch, daß die in Satz 1 für $n \leqq J(xr^{\alpha},w) \leqq n + g$ angegebene Darstellung des Index gilt.

<u>8.2 Beweis von Hilfssatz 2.</u> Setzt man $w(z) = h(z;c) + c$ in (D) ein und differenziert $(n + \alpha_o + 1)$-mal, dann erhält man im Falle $c \neq 0$ für h die Differentialgleichung

$$\sum_{j=0}^{n+\alpha_o+1} h^{(n+\alpha_o+1-j)} \left\{ \sum_{k=\max\{0,j-(\alpha_o+1)\}}^{\min\{j,n\}} \binom{\alpha_o+1}{\alpha_o+1-j+k} a_{n-k}^{(j-k)}(z) \right\} = 0 \,. \quad (22)$$

Berechnet man unter Berücksichtigung von (18) das Puiseux-Diagramm der Differentialgleichung (22), dann erkennt man, daß ein Streckenzug mit der Steigung $-\lambda$ auftritt. Da Wachstumsordnung und Typus von w und h übereinstimmen, folgt Hilfssatz 2 aus (8), (9) und Satz 1, wenn man noch beachtet, daß h Nullstellen der Ordnung N_c besitzen kann.

<u>8.3 Beweis von Hilfssatz 3.</u> Nach der Definition von h in 6. und g_ζ in (3) genügt es, nach (13) die erste Sprungstelle des Index von h im Punkte $z = \zeta$ abzuschätzen, da sich die Summanden der entsprechenden Potenzreihenentwicklungen um ζ nur durch einen konstanten Faktor unterscheiden.

Wegen $w(\zeta) = c$ und $w^{(k)}(\zeta) = h^{(k)}(\zeta;c)$ für $k \in \mathbb{N}$ besitzt h nach 6. die Potenzreihenentwicklung

$$h(z;c) = \frac{w^{(m)}(\zeta)}{m!}(z-\zeta)^m + \sum_{j=q}^{\infty} \frac{w^{(j)}(\zeta)}{j!}(z-\zeta)^j$$

mit einem gewissen $q \in \mathbb{N}$, $q > m$. Nach der Definition des Zentralindex in 2. existiert ein $a > 0$ so, daß nach (7)

$$\nu_\zeta(xr^{\alpha},h) \begin{cases} = m & \text{für } x \in (0,t(a)) \\ \geqq q & \text{für } x \geqq t(a) \end{cases}$$

gilt.

Außerdem genügt h der Differentialgleichung (22), die wir nun auf $K_\zeta(t(a)r^\alpha) = K_\zeta(ar^{1-\lambda}(1 + o(1))$ betrachten. Dividieren wir durch das Maximalglied $w^{(m)}(\zeta)(z-\zeta)^m/m!$ der Potenzreihenentwicklung von h um $z = \zeta$ auf $K_\zeta(t(a)r^\alpha)$, dann erhalten wir nach einigen Umformungen

$$\sum_{j=0}^{n+\alpha_o+1} a^j \cdot \left\{ B_{n+\alpha_o+1-j} \cdot \zeta^{j(1-\lambda)}(1 + o(1)) \cdot \left\{ \sum_{k=\max\{0,j-(\alpha_o+1)\}}^{\min\{j,n\}} \gamma_{j,k} \cdot \right.\right.$$

$$\left.\left. \cdot \prod_{\mu=0}^{n-k-1} (n + \alpha_o + 1 - \mu) \; a_{n-k}^{(j-k)}(\zeta) \right\}\right\} = 0 \qquad (23)$$

wieder mit rationalen Zahlen $\gamma_{j,k}$, $\gamma_{j,j} = 1$, und Funktionen $B_{n+\alpha_o+1-j}(\zeta,a)$ mit $|B_{n+\alpha_o+1-j}| \leqq 1$. Beachten wir (18), dann folgt für a die Gleichung

$$\sum_{j\in N_i} a^j \cdot B_{n+\alpha_o+1-j} \cdot A_{n-j} \cdot \prod_{\mu=0}^{n-j-1} (n + \alpha_o + 1 - \mu)(1 + o(1)) = 0 \qquad (24)$$

für $\zeta \to \infty$.

Nun ist a positive reelle Lösung von (24) und hat die Aufgabe, in (23) die für $\zeta \to \infty$ am stärkesten anwachsenden Summanden zu neutralisieren, und damit wird a umso kleiner ausfallen, je schneller die Koeffizienten in (24) anwachsen. Vergleichen wir also a mit den von Null verschiedenen Lösungen y von

$$\sum_{j\in N_i} y^{j_i-j} A_{n-j} \cdot \prod_{\mu=0}^{n-j-1} (n + \alpha_o + 1 - \mu) = 0 ,$$

dann ergibt sich $a \geqq |y| \geqq 1/A$ mit (6) und damit auch Hilfssatz 3.

Abschließend sei bemerkt, daß diese Betrachtungen unabhängig von der Wahl von α sind. Hieraus ergeben sich die Fallunterscheidungen in Satz 2.

<u>8.4 Beweis von Satz 4.</u> Wir können $1 - \lambda < \alpha < 1$ voraussetzen und wählen $K \leqq t(1/A)$ in Hilfssatz 1, damit das Integral nach Satz 2 keinen Beitrag liefert. Wir erhalten dann

$$n_\zeta(\tau r^\alpha, w - c) \leqq N_c + \log\left(\frac{K}{K-k}\right) / \log(k/\tau)$$

für $0 < \tau < k < K$. Da die rechte Seite in K monoton fallend ist, können wir

$$n_\zeta(\tau r^\alpha, w - c) \leqq N_c + \log\left(\frac{t(1/A)}{t(1/A) - k}\right) / \log(k/\tau) \qquad (25)$$

setzen. Diskonjugiertheit tritt für

$$k^2 - kt(1/A) + \tau t(1/A) < 0$$

ein. Diese Ungleichung besitzt nur für $\tau \leqq t(1/A)/4$ reelle Lösungen, und für $\tau = t(1/A)/4$, $k = t(1/A)/2$, erhalten wir mit (7) aus (25)

$$n_\zeta\left(\frac{1}{4A}r^{1-\lambda}(1 + o(1)\right) \leqq N_c + 1 .$$

Also liegt für $\tau < t(1/A)/4$ Diskonjugiertheit vor.

8.5 Beweis von Korollar A.1. Für die Polynome a_{n-j} in (D) ergibt sich in Satz A für die Kreisscheiben $S_\zeta(\tau r^{1-\lambda})$ wegen $\lambda > 0$

$$\beta_{n-j} = |A_{n-j}|r^{\alpha_{n-j}}(1 + o(1))$$

für $|\zeta| = r \to \infty$. t_o ist hier also Lösung der Gleichung

$$\sum_{j=1}^{n} |A_{n-j}|r^{\alpha_{n-j}}t^j(1 + o(1)) = 1 .$$

Für $j_{i-1} > 0$ in (5) erhalten wir mit (18) wieder

$$\sum_{j\in N_i} |A_{n-j}|r^{j(\lambda-1)}t^j(1 + o(1)) = 0 .$$

Für t_o folgt damit die Darstellung $t_o = D \cdot r^{1-\lambda}(1 + o(1))$, und D ist der Betrag einer Wurzel von

$$\sum_{j\in N_i} |A_{n-j}|x^{j-j_{i-1}} = 0 .$$

Wie in (6) folgt damit die Abschätzung von D in Korollar A.1. Im Falle $j_{i-1} = 0$ setzen wir $\alpha_n = 0$, $A_n = 1$. Dann ist D der Betrag einer Wurzel von

$$\sum_{j\in N_i} |A_{n-j}|x^j - |A_n| = 0 \quad (j \neq 0) ,$$

und es ergibt sich die gleiche Abschätzung.

Literaturverzeichnis.

[1] Boas, R.P.: Entire functions. Academic Press, New York, N.Y. (1954).

[2] Frank, G.: Zur lokalen Werteverteilung der Lösungen linearer Differentialgleichungen. Manuscripta math. 6 (1972), 381 - 404.

[3] Hayman, W.K.: Differential inequalities and local valency. Pacific J. Math. 44 (1975), 117 - 137.

[4] Knab, O.: Über lineare Differentialgleichungen mit rationalen Koeffizienten. Dissertation, Karlsruhe (1974).

[5] Knab, O.: Wachstumsordnung und Index der Lösungen linearer Differentialgleichungen mit rationalen Koeffizienten. Manuscripta math. 18 (1976), 299 - 316.

[6] Knab, O.: Über Wachstumsordnung und Typus der Lösungen linearer Differentialgleichungen mit rationalen Koeffizienten. Arch. der Math. 31 (1978), 61 - 69.

[7] Lepson, B.: Differential equations of infinite order, hyperdirichlet series and entire functions of bounded index, in "Entire functions and related parts of analysis". Amer. Math. Soc., Providence, R. I. (1968), 298 - 307.

[8] Nikolaus, J.: Über ganze Lösungen linearer Differentialgleichungen. Arch. der Math. 18 (1967), 618 - 626.

[9] Pöschl, K.: Über Anwachsen und Nullstellenverteilung der ganzen transzendenten Lösungen linearer Differentialgleichungen I. J. reine angew. Math. 199 (1958), 121 - 138.

[10] Pólya, G.: Untersuchungen über Lücken und Singularitäten von Potenzreihen. Math. Z. 29 (1929), 549 - 640.

[11] Pólya, G., Szegö, G.: Aufgaben und Lehrsätze aus der Analysis, Bd. 2. 3. Auflage. Springer-Verlag, Berlin - Heidelberg - New York (1964).

[12] Rahman, Q.J., Stankiewicz, J.: Differential inequalities and local valency. Pacific J. Math. 54 (1974), 165 - 181.

[13] Tijdeman, R.: On the number of zeros of general exponential polynomials. Nederl. Akad. Wet., Proc., Ser A 74 (1971), 1 - 7.

[14] Wittich, H.: Neuere Untersuchungen über eindeutige analytische Funktionen. Springer-Verlag, Berlin - Göttingen - Heidelberg (1955).

Mathematisches Institut I
der Universität Karlsruhe (TH)
Englerstraße 2
D-7500 Karlsruhe
BR Deutschland

MÜNTZ APPROXIMATION OF ARCS AND MACINTYRE EXPONENTS

Jacob Korevaar

1. Introduction

This is a survey of work, dealing with uniform Müntz-type approximation (or the impossibility thereof) on arcs, and the related Macintyre problem for entire functions with lacunary power series. Some suggestions are included.

For the classes of increasing sequences of positive integers

$$\Lambda : \lambda_1 < \lambda_2 < \cdots < \lambda_n < \cdots \tag{1.1}$$

described below, the following inclusion relations have been established:

$$I \subset PSF \subset M \subset CC \ , \tag{1.2}$$

$$I \subset PSN \subset PNJ \subset PNLLR \subset PNA = CC \ . \tag{1.3}$$

Here M denotes the class of Macintyre sequences (Macintyre exponents) Λ, that is, sequences $\{\lambda_n\}$ such that every nonconstant entire function of the form

$$f(z) = \sum a_n z^{\lambda_n}$$

is necessarily unbounded on every curve to infinity. CC is simply the convergence class:

$$\sum 1/\lambda_n < \infty \ . \tag{1.4}$$

Macintyre [16] proved that $M \subset CC$ and conjectured equality. Since then, Kövari [11, 12] has shown that the condition

$$\lambda_n > n \log n \, (\log \log n)^{2+\delta}, \ n > n_0 \tag{1.5}$$

is sufficient for Λ to be in M. (Cf. also the value distribution results of Hayman [7] under this condition.) Later contributions by Pavlov [20] and Korevaar-Dixon [9] are reviewed below.

I stands for a class of interpolation sequences Λ. The requirement is that every bounded sequence of complex numbers can be interpolated on Λ by entire functions of restricted growth, corresponding to order one, convergence type (sec. 6). Such interpolation was introduced by Pavlov [20] in his work on the Macintyre problem: he proved essentially that $I \subset M$. The interpolation enables one to relate the maximum modulus of a lacunary power series to its modulus on a small eccentric set. Kövari had used Turán's lemma for that purpose.

Pavlov has shown that regular sequences Λ:

$$\lambda_n = n\, L(n),\ 0 < L(n) \uparrow \tag{1.6}$$

of convergence class are interpolation sequences, and hence Macintyre sequences. Suppose now that

$$\lambda_n \geq n\, L(n),\ 0 < L(n) \uparrow \infty\ . \tag{1.7}$$

By the work of Korevaar-Dixon, the condition

$$\sum \frac{\log L(n)}{n\, L(n)} < \infty \tag{1.8}$$

is sufficient for Λ to be in I and hence in M. Observe that (1.8) represents a slight improvement of (1.5). Assuming (1.7), Berndtsson [1] has shown that condition (1.8) for Λ in I is sharp.

PSF in (1.2) denotes the class of sequences Λ for which the associated set of powers

$$P_\Lambda = \{z^{\lambda_n}\} \tag{1.9}$$

is strongly free: for each power z^{λ_k}, the approximation in $C(\gamma)$ by combinations of the other powers is uniformly bad for all curves γ, extending from one circle about the origin to another (sec. 5). The inclusions $I \subset PSF \subset M$ are in Korevaar-Dixon [9]; it can be shown that $I \neq PSF$, hence $I \neq M$.

A sequence Λ is said to be in PNA if the set of powers P_Λ is nonspanning in $C(\gamma)$ for every analytic arc γ. Similarly, Λ in PNJ and Λ in PNLLR mean, respectively, that the set of powers P_Λ is nonspanning for all Jordan arcs, and for all arcs of locally limited rotation. We say that γ is of locally limited rotation if for some (rectifiable) subarc, all oriented chordal directions fall within an angle less than π. In particular, every C^1 arc is of locally limited rotation. Finally, Λ is said to be in PSN if the set of powers P_Λ is strongly nonspanning: for each positive integer ν not in Λ, the approximation to z^ν by combinations of the powers z^{λ_n} is uniformly bad for all curves γ, extending from one circle about the origin to another.

Malliavin-Siddiqi [17] and the author [8] have proved that PNA = CC; the inclusions $I \subset PSN \subset PNJ$ are in Korevaar-Dixon [9]; $I \neq PSN$. Let us now suppose again that Λ satisfies condition (1.7). Then by recent work of Korevaar-Dixon [10], the condition

$$\sum \frac{1}{n\,L(n)} < \infty \tag{1.10}$$

is sufficient for Λ to be in PNLLR. Observe that the stronger condition (1.8) is sufficient for Λ to be in (I and hence in) PNJ.

The big questions are of course if actually $M \neq CC$ and $PNJ \neq CC$. Restricted to regular sequences Λ (1.6), the classes in (1.2) and (1.3) all coincide (a regular sequence of class CC is in I).

2. Müntz approximation on intervals.

By Weierstrass's theorem of 1885, all continuous functions on a bounded closed interval $[a,b]$ can be uniformly approximated by polynomials. In other words, the linear combinations of the powers $1, x, x^2$, ... lie dense in $C[a,b]$; the powers span $C[a,b]$. Stating the result in that way, the following are natural questions: Are all these powers required? Could one use more general exponents $\rho_n > 0$ instead of the positive integers?

Answers were provided by Müntz's theorem of 1914 [19] of which we quote the principal assertion:

Theorem: Distinct powers $x^0 = 1$ and

$$x^{\rho_n},\ n = 1, 2, \dots, \tag{2.1}$$

where $\rho_n \geqq \delta > 0$, span $C[0,1]$ if and only if

$$\sum \frac{1}{\rho_n} = \infty\ . \tag{2.2}$$

The key ingredient in Müntz's proof was a beautiful explicit formula for approximation in $L^2(0,1)$:

$$\inf_c \int_0^1 |x^\rho - \sum c_n x^{\rho_n}|^2 \frac{dx}{x} = \frac{1}{2\rho} \prod \left(\frac{\rho - \rho_n}{\rho + \rho_n}\right)^2 .$$

The contemporary approach to the question is to look for continuous linear functionals which vanish on the powers x^{ρ_n}. The problem thus becomes one about Laplace transforms with certain zeros, and methods of complex analysis can be applied.

Müntz's proof for $C[0,1]$ was simplified by Szász [25], who also considered complex exponents. For intervals $[a,b]$ with $a > 0$ there is a cleaner result: Distinct powers (2.1), with $\rho_n > 0$, span $C[a,b]$ if and only if condition (2.2) is satisfied. However, this was only proved in 1943, by Clarkson-Erdös [2] and Schwartz [22]. Their principal interest was to characterize the approximable functions in case (2.2) is violated. We describe the main result for the case of positive integral exponents (1.1):

Suppose the convergence condition (1.4) is satisfied. Let f be a uniform limit, on $[a,b]$ $(a > 0)$, of linear combinations of powers x^{λ_n}. Then f has a holomorphic extension F to the disc $D(0,b)$, and the power series for F involves only powers z^{λ_n}.

This analyticity result is a consequence of the strong linear independence of the set $\{x^{\lambda_n}\}$ in $C[a,b]$ under condition (1.4):

$$d_k = \inf_c \|x^{\lambda_k} - \sum{}' c_n x^{\lambda_n}\|_{[a,b]} \gtrsim (b - \varepsilon)^{\lambda_k}, \quad k \to \infty . \tag{2.3}$$

One can prove (2.3) by constructing suitable continuous linear functionals on $C[a,b]$ which vanish on the powers x^{λ_n}, $n \neq k$. This can be done with the aid of appropriate entire functions of exponential type, cf. Luxemburg-Korevaar [15].

A number of authors have studied the degree of approximation by Müntz polynomials, cf. Ganelius-Newman [6].

3. Müntz-type approximation on arcs.

Walsh [26] has extended Weierstrass's theorem to arcs: The powers z^n, $n = 0, 1, 2, \ldots$ span $C(\gamma)$ for every Jordan arc γ in $\mathbb{C}$.

Question: Is there a Müntz theorem for arcs?

It is convenient to consider sets of (distinct) exponentials

$$\{e^{\rho_n z}\}, \quad \rho_n > 0 \tag{3.1}$$

instead of powers, in order to avoid multivalued functions. Also, we should be careful about verticality in our arcs Γ. Indeed, suppose Γ contains two points z_1, z_2 with the same real part. Then for bad choice of $c > 0$, the exponentials e^{cnz}, $n = 1, 2, \ldots$ will all have the same value at z_2 as they have at z_1. In that case they can not span $C(\Gamma)$, even though $\sum 1/cn = \infty$.

A true two-sided Müntz-type result:

$$\{e^{\rho_n z}\} \text{ spans } C(\Gamma) \rightleftarrows \sum 1/\rho_n = \infty \tag{3.2}$$

is known only for some special classes of arcs Γ. One such class is that of the polygonal lines Γ (arcs consisting of a finite number of straight line segments) without vertical chords. In the proof, one considers the Laplace transforms, of complex Borel measures μ on Γ, which vanish on the sequence $\{\rho_n\}$. Rouché's theorem and the Ahlfors-Heins theorem for bounded functions in a half-plane enable one to reduce the case of a polygonal line to that of a single line segment.

There is a positive result (more precisely, $\leftarrow$ holds in (3.2)) for those (rectifiable) arcs Γ whose oriented chordal directions fall within the angle $|\arg z| \leq \pi/4$ (Korevaar [8]; related results have been obtained by Leont'ev [14] and Malliavin-Siddiqi [18]). For sufficiently regular exponents, one can extend this positive result to all arcs Γ whose oriented chordal directions fall within an angle $|\arg z| \leq \frac{1}{2}\pi - \varepsilon$, $\varepsilon > 0$. The key step in the proof is division of Laplace transforms, which vanish on the sequence $\{\rho_n\}$, by $\Pi(1 - s^2/\rho_n^2)$. When the quotient has to be considered on rays close to the real axis, as in the extension, regularity of the sequence $\{\rho_n\}$ is helpful.

Various negative results are known (cf. sec. 1). We elaborate here on the oldest result, concerning nonspanning sets on analytic arcs. The condition $\sum 1/|s_n| < \infty$ for complex numbers s_n implies that the set of exponentials

$$\{e^{s_n z}\} \tag{3.3}$$

fails to span $C(\Gamma)$ for every analytic arc Γ (Malliavin-Siddiqi [17], Korevaar [8]). On such arcs there exist many C^∞ functions φ of compact support. Operating on suitable φ by $\Pi(1 - D^2/s_n^2)$, one constructs a nonzero measure μ on Γ orthogonal to the set (3.3).

Conjecture. Assertion (3.2) is valid for all piecewise analytic arcs Γ whose oriented chordal directions fall within an angle $|\arg z| \leq \frac{1}{2}\pi - \varepsilon$, $\varepsilon > 0$.

For measures on such arcs, the Laplace transforms are of regular growth in the angle $|\arg z| < \varepsilon$ in the sense of Pfluger and Levin [3]. The growth may well be as regular as that given by the Ahlfors-Heins theorem for a half-plane.

We finally mention an analyticity result of Dixon-Korevaar [3] for the approximable functions in the case of nonspanning sets of powers on arcs. Let γ be a piecewise smooth arc on the closed disc $\overline{D}(0,b)$ which meets the circle $C(0,b)$ but is not part of it. Suppose that the set of powers (1.9) is nonspanning for every subarc of γ. Then every function f in $C(\gamma)$ which belongs to the closed span of the set (1.9) has an analytic extension F to the disc $D(0,b)$; the power series for F contains only powers z^{λ_n}.

4. The Macintyre problem.

Suppose f is a nonconstant entire function with a "lacunary" power series,

$$f(z) = \sum a_n z^{\lambda_n} . \tag{4.1}$$

Roughly speaking, lacunarity should imply that f has the same behavior in all directions, the same growth on all curves to infinity, minimum modulus not much smaller than the maximum modulus most of the way out to infinity. In particular, there should be no asymptotic paths.

Pólya [21] showed that for entire functions (4.1) of finite order, the Fabry condition

$$\lambda_n/n \to \infty$$

is a good lacunarity condition. His results have been augmented by a number of authors, among them Fuchs [5], Sons [24] and Hayman [7].

For entire functions (4.1) of unlimited growth, the situation is more complicated. Macintyre [16] proved that whenever

$$\sum 1/\lambda_n = \infty ,$$

there exist nonconstant entire functions (4.1) which tend to zero along the positive real axis. A very nice elementary proof is in Kövari [13]. On the other hand, as observed by Erdös and Macintyre (cf. [16]), the Müntz theorem immediately shows that the condition

$$\sum 1/\lambda_n < \infty \tag{4.2}$$

precludes the possibility of radial asymptotic paths. (Cf. the proof of the lemma in sec. 5.) Thus what is often called the

Macintyre conjecture. Condition (4.2) is necessary and sufficient in order that every nonconstant entire function of the form (4.1) be unbounded on every curve extending to infinity.

In sec. 1, we already referred to work of Kövari, Hayman, Pavlov and Korevaar-Dixon concerning lacunarity conditions for entire functions (4.1) of unlimited growth.

5. Strongly free and strongly nonspanning sets of powers.

For the Macintyre problem, a strong kind of linear independence is of interest. Let Λ be a sequence of positive integers as in (1.1), P_Λ the corresponding set of powers (1.9). Making precise what was said in sec. 1, P_Λ will be called strongly free (and we write $\Lambda \in PSF$) if for every $a > 1$ and every k,

$$\inf_{\gamma_a} \inf_{c} \|z^{\lambda_k} - \sum{}' c_n z^{\lambda_n}\|_{\gamma_a} = \delta_k(a) > 0 . \qquad (5.1)$$

Here the norm is the supremum norm, the inner infimum is taken over all finite sums $\sum' c_n z^{\lambda_n}$ with $n \neq k$, and the outer infimum is taken over all curves γ_a extending from a point on the unit circle $C(0,1)$ to a point on $C(0,a)$.

Lemma. Suppose P_Λ is strongly free. Then Λ is a Macintyre sequence (sec. 1).

Proof. Suppose, on the contrary, that there is a nonconstant entire function (4.1) which is bounded on a certain curve Γ from 0 to infinity:

$$\|\sum a_n z^{\lambda_n}\|_{\Gamma} = B < \infty .$$

Then for given $a > 1$ and every $R > 0$, there is an index N_R such that on a subarc $\Gamma(R,aR)$ of Γ which extends from $C(0,R)$ to $C(0,aR)$,

$$0 < \|\sum_1^N a_n z^{\lambda_n}\|_{\Gamma(R,aR)} \leqq 2B$$

whenever $N \geqq N_R$. Let a_k be a nonzero coefficient in the power series, N also $\geq k$. Then we find, setting $z = R\zeta$,

$$2B \geqq |a_k| \, \|z^{\lambda_k} - \sum{}' \cdot\cdot \, z^{\lambda_n}\|_{\Gamma(R,aR)} \qquad (5.2)$$

$$= |a_k| \, R^{\lambda_k} \, \|\zeta^{\lambda_k} - \sum{}' \cdot\cdot \, \zeta^{\lambda_n}\|_{\gamma_a} ,$$

where $\gamma_a = \gamma_a(R)$ extends form $C(0,1)$ to $C(0,a)$. For large R, (5.2) will contradict (5.1).

Remark. Suppose that in (5.1),

$$\delta_k(a) \geqq \delta = \delta(a) > 0 \quad \text{for all } k, \qquad (5.3)$$

as one has when Λ is an interpolation sequence (see sec. 6). Then by an argument as above,

$$\|f\|_{\Gamma(R,aR)} \geqq \tfrac{1}{2}\delta \, |a_k| \, R^{\lambda_k} \quad \text{for all } k. \qquad (5.4)$$

This will in particular hold for the index of the maximum term. Conclusion: if (5.3) holds, the left-hand side of (5.4) can not be very much smaller than the maximum modulus $M(R)$.

A set P_Λ will be called <u>strongly nonspanning</u> (and we write $\Lambda \in$ PSN) if for every $a > 1$ and every positive integer ν not in Λ,

$$\inf_{\gamma_a} \inf_{c} \| z^\nu - \sum c_n z^{\lambda_n} \|_{\gamma_a} = \varepsilon_\nu(a) > 0 . \tag{5.5}$$

Suppose $\Lambda \in$ PSN. Then P_Λ fails to span $C(\gamma)$ for every arc γ of the form γ_a, $a > 1$, and hence (by change of scale) for every arc γ that does not lie on a circle $C(0,r)$. Taking $\gamma_a = [1,a]$, Müntz's theorem shows that one must have $\sum 1/\lambda_n < \infty$. Hence by sec. 3, P_Λ also fails to span $C(\gamma)$ for $\gamma \subset C(0,r)$ (such arcs are analytic). Thus P_Λ is nonspanning for every Jordan arc; PSN $\subset$ PNJ (sec. 1).

6. Interpolation sequences.

We will at first allow more general sequences than in sec. 1. A sequence S of distinct complex numbers s_n is called an <u>interpolation sequence</u> if there exists a positive increasing function $\omega = \omega(r,S)$ on $[0,\infty)$ with the properties

$$\int_1^\infty r^{-2}\omega(r)\, dr < \infty, \quad \omega(r)/r \downarrow , \tag{6.1}$$

such that the following is true. For every sequence of complex numbers $\{b_n\}$ with $|b_n| \leq 1$, there is an entire function g for which

$$g(s_n) = b_n \quad \text{for all } n, \; M(r,g) \leq e^{\omega(r)}, \; r \geq 0 . \tag{6.2}$$

(Observe that g must be of exponential type 0).

Suppose S is an interpolation sequence. Then by Jensen's formula, applied to a function g corresponding to $b_1 = 1$, $b_n = 0$ for $n \neq 1$, one must have $\sum' 1/|s_n| < \infty$. A subsequence of S is also an interpolation sequence, as is any sequence obtained by adjoining an element different from all s_n.

<u>Lemma</u> (cf. [20], [10]). Suppose $0 < \rho_n/n \uparrow$ $(n = 1,2,\ldots)$ and $\sum 1/\rho_n < \infty$. Then

$$\ldots, -\rho_2, -\rho_1, 0, \rho_1, \rho_2, \ldots$$

is an interpolation sequence.

<u>Proof.</u> Introduce

$$F(z) = z \prod_1^\infty (1 - z^2/\rho_k^2) .$$

Writing $\rho_{-n} = -\rho_n$, one can show that

$$|F'(\rho_n)| \geqq e^{-3|n|}, \quad M(r,F) \leqq e^{\omega(r)}$$

with an ω as in (6.1). Estimating with care, it turns out that one may define

$$g(z) = b_0 + \sum_{-\infty}^{\infty}{}' b_n \frac{F(z)}{F'(\rho_n)(z-\rho_n)} \left(\frac{z}{\rho_n}\right)^{4|n|} .$$

Berndtsson [1] has characterized interpolation sequences Λ of positive integers (1.1) by a uniform separation condition: Λ belongs to I if and only if for an ω as in (6.1),

$$n(r) = \sum_{\lambda_n \leq r} 1 \leq \omega(r), \quad r \geq 0$$

and

$$\prod_{\frac{1}{2}\lambda_k<\lambda_n<2\lambda_k}{}' \left|1 - \frac{\lambda_k}{\lambda_n}\right| \geqq \exp\{-\omega(\lambda_k)\}, \quad k = 1,2,\ldots$$

(in the product, $\lambda_n \neq \lambda_k$). Berndtsson's elegant proof is based on Hörmander's estimates for solutions of the $\bar{\partial}$-equation. (The result may also be established classically.) One readily verifies the above conditions for sequences Λ satisfying (1.7) and (1.8).

Theorem [9]. Let Λ be an interpolation sequence (1.1) of positive integers; we let ω be the associated function as in (6.1), (6.2). Then the corresponding set of powers P_Λ (1.9) is strongly free (sec. 5); it is even true that for every $a > 1$,

$$\inf_k \inf_{\gamma_a} \inf_c \|z^{\lambda_k} - \sum{}' c_n z^{\lambda_n}\|_{\gamma_a} = \delta = \delta(a,\omega) > 0 . \qquad (6.3)$$

Outline of the proof. One considers polynomials of the form

$$p(z) = \sum a_n z^{\lambda_n} \quad \text{with} \quad a_k = 1/\varepsilon, \ \varepsilon > 0 ,$$

and assumes that $|p|$ is bounded by 1 on γ_a. This implies a bound for $|p|$ in the vicinity of a point $z \in \gamma_a \cap C(0,r)$ which involves the maximum modulus $M(re^\eta)$ to a power less than 1. The crucial step is to go from here to a bound for $M(r)$ in terms of $M(re^\eta)$. In such a situation Kövari used Turán's lemma, but interpolation appears to work even a little better. For suitable choice of an interpolating function g,

$$M(r) = p^*(z) = \sum a_n g(\lambda_n) z^{\lambda_n} = -\frac{1}{2\pi i} \int_C p\left(\frac{z}{\zeta}\right) G(\zeta) \frac{d\zeta}{\zeta},$$

where G is the Leau-Wigert transform of g:

$$G(\zeta) = \sum_0^\infty g(\nu)\zeta^\nu \quad \text{for} \quad |\zeta| < 1, \; -\sum_{-\infty}^{-1} g(\nu)\zeta^\nu \quad \text{for} \quad |\zeta| > 1;$$

G is holomorphic on $C^* \setminus \{1\}$. For C one may take a small circle about the point 1; an integral representation for G in terms of the Borel transform of g yields an estimate for $|G|$ in terms of ω.

One thus obtains an estimate for $M(r)$ in terms of $M(re^\eta)$ and ω. A very small value for ε would force $M(1)$ to be very large; repeated application of the estimate would then show that $M(a)$ must be infinite. Thus ε can not be too small and (6.3) follows.

7. Nonspanning sets of powers and exponentials.

Let Λ be an interpolation sequence of positive integers (1.1), ν any positive integer not in Λ. Then the sequence Λ^* obtained by adjoining ν is also an interpolation sequence. Thus by sec. 6 the set of powers P_{Λ^*} is strongly free; in particular (5.5) will be satisfied. It follows that the set P_Λ is strongly nonspanning, and hence nonspanning on every Jordan arc γ.

Before Korevaar-Dixon [9] obtained this result for arbitrary Jordan arcs, Erkama [4] had observed that for arcs γ which satisfy a mild smoothness condition at two points, the condition

$$\lambda_n > cn(\log n)^2, \; c > 0$$

assures a nonspanning set P_Λ in $C(\gamma)$. In his proof, a suitable C^∞ function φ on γ (extendable to one of compact support) was obtained by restricting an analytic function. Such an approach had also been advocated by Beurling; it is now known that it can be used to give more refined conditions for nonspanning sets.

The result of Korevaar-Dixon [10] for arcs of locally limited rotation, stated in sec. 1, was obtained in a different way. Starting point was the method of proof sketched in sec. 6. That method can also be applied to interpolation sequences of positive and negative integers. In particular, let Λ be a sequence of positive integers (1.1), subject to the regularity condition (1.6) and the convergence condition (1.4), and now define $\lambda_{-n} = -\lambda_n$, $\lambda_0 = 0$. Then the extended sequence $\{\lambda_n\}$ is an interpolation sequence (sec. 6). One may deduce that for every $a > 1$,

$$\inf_{\gamma_a} \inf_{c} \|1 - \sum{}' c_n z^{\lambda_n}\|_{\gamma_a} = \varepsilon_0(a) > 0 ,$$

where the sum involves both positive and negative powers. Replacing z by $e^{\eta z}$ one obtains the following

Lemma. Let Λ be a regular sequence of positive integers (1.1), (1.6) subject to the convergence condition (1.4). Then the exponentials

$$\exp(\pm \eta\lambda_n z),\ n = 1,2,\ldots$$

fail to span $C(\Gamma)$ for every arc Γ and every constant $\eta > 0$.

One more observation is needed. For arcs Γ that are close to horizontal, it is roughly correct that the approximation to 1 by combinations of real exponentials $\exp(\pm \rho_n z)$ becomes worse when one moves the exponents further out. What is involved is the following. For measures μ on Γ orthogonal to the functions $\exp(\pm \rho_n z)$, one divides the Laplace transform by $\Pi(1 - s^2/\rho_n^{\ 2})$, while multiplying by $\Pi(1 - s^2/s_n^{\ 2})$, where $|s_n| > \rho_n$. This works even for arcs Γ that are not so close to horizontal, but then the ρ_n should be of regular behavior, and the $|s_n|$ much larger than the ρ_n. Thus from the lemma one obtains the following

Theorem (cf. [10]). Let Γ be an arc whose oriented chordal directions fall within an angle $|\arg z| \leqq \frac{1}{2}\pi - \varepsilon,\ \varepsilon > 0$. Let the $s_n,\ n = 1,2,\ldots$ be complex numbers such that

$$|s_n| \geqq n\,L(n),\ 0 < L(n) \uparrow,\ \sum 1/n\,L(n) < \infty .$$

Then the exponentials $\exp(\pm s_n z)$ fail to span $C(\Gamma)$.

8. Problems and suggestions.

The big open question, of course, is the Macintyre conjecture itself. The approach pursued here was to try and solve

Problem 1. Characterize the class M of Macintyre sequences Λ (1.1) in terms of a strong linear independence property of the associated sets of powers P_Λ (1.9).

The author does not know if the strongly free sets introduced in sec. 5 provide the correct answer.

The class I of interpolation sequences is smaller than M. Perhaps there is a weaker interpolation property which suffices for Macintyre sequences. For sequences subject to condition (1.7), the present interpolation method gives a slightly better result than the earlier method, based on Turán's lemma. Nevertheless, that lemma works better than interpolation in some cases of "mild clustering". (The author's example which shows that $I \neq PSF$ is based on Turán's lemma.)

Problem 2. Develop a more suitable method to estimate the maximum modulus of a lacunary power series in terms of the modulus on a small eccentric set (for example, a disc).

Such a method should at least combine the advantages of interpolation with those of Turán's lemma.

In the theorem of sec. 6, the badness of approximation on $\gamma = \gamma_a$ is measured against the "radial extent" of the curve. For a proof of the Macintyre conjecture, an estimate against the "angular extent" might be very useful.

Problem 3. Is there, for a suitably large class of sequences Λ, and for γ outside $D(0,1)$, say, an estimate from below for

$$\inf_{c} \|z^{\lambda_k} - \sum{}' c_n z^{\lambda_n}\|_{\gamma}$$

in terms of the angular extent of γ? In terms of the diameter of γ?

What could one say if the Macintyre conjecture would be false? In that case, there would be sequences Λ of convergence class (1.4) for which P_Λ fails to be strongly free. There would then be a constant $a > 1$, a power z^{λ_k} and a sequence of curves γ_a from $C(0,1)$ to $C(0,a)$ on which the approximation to z^{λ_k}, by combinations of the other powers, becomes arbitrarily good. Perhaps there is then some set E_a, meeting every circle $C(0,r)$ with $1 \leq r \leq a$, on which z^{λ_k} is in the closed span of the other powers. Perhaps there would even be a Jordan arc γ_a' with that property. If such an arc would be rectifiable, the set of powers P_Λ would have to span $C(\gamma_a')$. (If there would be a Laplace transform that vanishes on Λ without vanishing identically, one could divide out the zero at λ_k to get a contradiction.)

Problem 4. Is there a sequence Λ of convergence class such that the set of powers P_Λ spans $C(\gamma)$ for some Jordan arc γ?

For arcs Γ whose oriented chordal directions fall within an angle $|\arg z| \leq \frac{1}{2}\pi - \varepsilon$, $\varepsilon > 0$, there is the following counterpart.

Problem 5. Is there a sequence Λ of divergence class such that the set of exponentials $\{\exp(\lambda_n z)\}$ fails to span $C(\Gamma)$ for some arc Γ "of bounded slope"?

References

[1] Berndtsson, B.: A note on Pavlov-Korevaar-Dixon interpolation. Nederl. Akad. Wet., Proc., Ser. A 81 (1978), 409 - 414.

[2] Clarkson, J. A., Erdös, P.: Approximation by polynomials. Duke math. J. 10 (1943), 5 - 11.

[3] Dixon, M., Korevaar, J.: Nonspanning sets of powers on curves: analyticity theorem. Duke math. J. 45 (1978), 543 - 559.

[4] Erkama, T.: Classes non quasi-analytiques et le théorème d'approximation de Müntz. C. r. Acad. Sci., Paris, Sér. A 283 (1976), 595 - 597.

[5] Fuchs, W. H. J.: Proof of a conjecture of G. Pólya concerning gap series. Illinois J. Math. 7 (1963), 661 - 667.

[6] Ganelius, T., Newman, D.: Müntz-Jackson theorems in all L^p spaces with unrestricted exponents. Amer. J. Math. 98 (1976), 295 - 309.

[7] Hayman, W. K.: Angular value distribution of power series with gaps. Proc. London math. Soc., III. Ser. 24 (1972), 590 - 624.

[8] Korevaar, J.: Approximation on curves by linear combinations of exponentials, in "Approximation theory". Academic Press, New York and London (1973), 387 - 393.

[9] Korevaar, J., Dixon, M.: Interpolation, strongly nonspanning powers and Macintyre exponents. Nederl. Akad. Wet., Proc., Ser. A 81 (1978), 243 - 258.

[10] Korevaar, J., Dixon, M.: Nonspanning sets of exponentials on curves. Acta math. Acad. Sci. Hungar. 33 (1979), 89 - 100.

[11] Kövari, T.: On the asymptotic paths of entire functions with gap power series. J. Analyse math. 15 (1965), 281 - 286.

[12] Kövari, T.: A gap theorem for entire functions of infinite order. Michigan math. J. 12 (1965), 133 - 140.

[13] Kövari, T.: On a result of A. J. Macintyre, in "Mathematical essays dedicated to A. J. Macintyre". Ohio University Press, Athens, Ohio (1970), 217 - 222.

[14] Leont'ev, A. F.: On the completeness of a system of exponentials on a curve (Russian). Sibir. mat. Žurn. 15 (1974), 1103 - 1114.

[15] Luxemburg, W. A. J., Korevaar, J.: Entire functions and Müntz-Szász type approximation. Trans. Amer. math. Soc. 157 (1971), 23 - 37.

[16] Macintyre, A. J.: Asymptotic paths of integral functions with gap power series. Proc. London math. Soc., III. Ser. 2 (1952), 286 - 296.

[17] Malliavin, P., Siddiqi, J. A.: Approximation polynomiale sur un arc analytique dans le plan complexe. C. r. Acad. Sci., Paris, Sér. A 273 (1971), 105 - 108.

[18] Malliavin, P., Siddiqi, J. A.: Classes de fonctions monogènes et

approximation par des sommes d'exponentielles sur un arc rectifiable de ℂ. C. r. Acad. Sci., Paris, Sér. A 282 (1976), 1091 - 1094.

[19] Müntz, C. H.: Über den Approximationssatz von Weierstrass. H. A. Schwarz Festschrift, Berlin (1914), 303 - 312.

[20] Pavlov, A. I.: The growth along curves of entire functions that are defined by lacunary power series (Russian). Sibir. mat. Žurn. 13 (1972), 1169 - 1181.

[21] Pólya, G.: Untersuchungen über Lücken und Singularitäten von Potenzreihen. Math. Z. 29 (1929), 549 - 640.

[22] Schwartz, L.: Étude des sommes d'exponentielles réelles. Hermann, Paris (1943).

[23] Sons, L. R.: On the Macintyre conjecture. Illinois J. Math. 14 (1970), 613 - 620.

[24] Sons, L. R.: An analogue of a theorem of W. H. J. Fuchs on gap series. Proc. London math. Soc., III. Ser. 21 (1970), 525 - 539.

[25] Szász, O.: Über die Approximation stetiger Funktionen durch lineare Aggregate von Potenzen. Math. Ann. 77 (1915), 482 - 496.

[26] Walsh, J. L.: Über die Entwicklung einer Funktion einer komplexen Veränderlichen nach Polynomen. Math. Ann. 96 (1927), 437 - 450.

Math. Institute
Roetersstraat 15
Amsterdam 1004,
Netherlands

AN ENTIRE FUNCTION WITH IRREGULAR GROWTH AND MORE THAN ONE DEFICIENT VALUE

Larry J. Kotman*

0. Introduction. W. K. Hayman in Research Problems in Function Theory [7] poses the question of existence of meromorphic functions of finite order with at least two deficient values and characteristic satisfying

$$\overline{\lim_{r\to\infty}} \frac{T(\sigma r)}{T(r)} = \infty \quad (\sigma > 1) \ .$$

A. A. Goldberg in "The Possible Magnitude of the Lower Order of an Entire Function with a Finite Deficient Value" [4] poses the question of existence of entire functions of infinite order, finite lower order, and having a finite deficient value. The answer to both questions is affirmative. We prove existence by constructing an explicit infinite product representation of an entire function with zero having positive deficiency to meet the requirements. Our methods include generalizing a result of B. Ja. Levin concerning particular entire functions with zeros evenly distributed on two rays. Next we exhibit a polynomial substitute for the exponential convergence factor which appears in the standard Weierstrass primary factor. Then we partition the complex plane into annular regions which are appropriate for our purposes of interpolating through a family of entire functions. Finally we take a comparison function by D. Drasin and generalize it to obtain a counting function for zeros which we use to construct our entire function.

1. Statement of result. In this article we take the standard notation of the Nevanlinna theory for granted (cf. [6]). Our first result resolves a problem posed by Hayman in [7; 1.10, p. 5].

Theorem 1. Let

$$1 < \rho < \infty \ . \tag{1}$$

There exists an entire function $f(z)$ of finite order ρ with

---

* This paper is a synopsis of the author's doctoral dissertation written under the direction of Professor David Drasin at Purdue University.

$\delta(0,f) > 0$

and whose Nevanlinna characteristic satisfies

$$\limsup_{r\to\infty} \frac{T(\sigma r,f)}{T(r,f)} = \infty$$

for each $\sigma > 1$.

The methods used also solve a problem proposed at a conference at Cornell University [11] and later by Goldberg [4].

Theorem 2. Let

$1 < \mu < \infty$.

There exists an entire function $f(z)$ of infinite order and finite lower order μ with

$\delta(0,f) > 0$.

Our method is to explicitly construct an infinite product representation of the required function $f(z)$. In particular, near each circle $\{|z| = r\}$, $f(z)$ is so constructed that $\log |f(z)|$ displays the sinusoidal behavior of $\log |f_{\lambda(r)}(z)|$ for some Lindelöf function $f_{\lambda(r)}(z)$ of order $\lambda(r)$, if $\lambda(r)$ is nonintegral, with its zeros distributed on some appropriate ray; or if $\lambda(r)$ is equal or nearly equal to an integer p, then $\log |f(z)|$ is to display the sinusoidal behavior of $\log |g_{\lambda(r)}(z)|$ for some entire function $g_{\lambda(r)}(z)$ with its zeros evenly distributed on two rays through the origin separated by the angular measure π/p. (This $g_{\lambda(r)}(z)$, $\lambda(r)$ not necessarily integral, arises from a generalization ((6.6), (6.7)) of a result of B. Ja. Levin ([9], p. 68) which we do not state in its complete generality.) Since $f(z)$ is to emulate functions of any order, we introduce in Section 2 a polynomial substitute for the standard Weierstrass primary factor and in Section 3 require that $\lambda(r)$ be continuous with $\limsup \lambda(r) = \infty$ and $\liminf \lambda(r) < \infty$ as $r \to \infty$. In section 4, the $n_j(r)$ $(j = 1,2)$ are given to count in Section 5 the substitute polynomial primary factors that make up $f(z)$. Very briefly in Sections 6 and 7 we indicate how $f(z)$ establishes the existence in Theorem 1 and Theorem 2. (A more detailed version of these results is to appear elsewhere.)

2. An alternative to the Weierstrass primary factor expansion. For

a fixed $\alpha > 4$, $q \geqq 0$, define polynomials

$$P(z,q,\alpha) = \begin{cases} \prod_{k=1}^{q} (1 + k^{-1}(z/\alpha)^k C_k(\alpha))^{\alpha^k} & (q \geqq 1) \\ 1 & (q = 0) \end{cases} \tag{2.1}$$

where the $C_k = C_k(\alpha)$ are chosen inductively so that for all z

$$\sum_{k=1}^{q} \sum_{\substack{j \geq 1 \\ jk \leq q}} (-1)^{j+1} j^{-1} \alpha^k \left[k^{-1}(z/\alpha)^k C_k(\alpha)\right]^j = z + \frac{1}{2} z^2 + \cdots + \frac{1}{q} z^q . \tag{2.2}$$

(The left side comes from $\log P(z,q,\alpha)$, $|k^{-1}(z/\alpha)^k C_k(\alpha)| < 1$). We can show that

$$C_1(\alpha) = 1$$

and in general, that

$$1 - \alpha^{-1} \leqq C_k(\alpha) \leq 1 + \alpha^{-1} \quad (k \geqq 1,\ \alpha > 4) .$$

Let

$$V(z,q,\alpha) = (1 - z)P(z,q,\alpha) . \tag{2.3}$$

A comparison with geometric series and an application of the triangle inequality relates V to the standard Weierstrass primary factor of genus q, $E(z/a,q)$, by

$$|\log V(z/a,q,\alpha^2) - \log E(z/a,q)| \leqq (q + 1)|z/\alpha a|^{q+1}$$
$$(|z| < |a|\alpha,\ \alpha > 4) \tag{2.4}$$

for any nonzero complex number a where

$$-\pi < \arg (1 - z/a) < \pi \quad (0 < \arg z - \arg a < 2\pi) .$$

The cancellation

$$(z/a)^{q+1} + (z/a\, e^{i\pi/(q+1)})^{q+1} = 0$$

yields the sharper estimate

$$|\log V(z/a,q,\alpha^2) + \log V(z/a\, e^{i\pi(q+1)},q,\alpha^2) - \log E(z/a,q)$$
$$- \log E(z/ae^{i\pi(q+1)},q)| \leqq 2(q + 1)|z/\alpha a|^{q+2} \quad (|z| < |a|\alpha,\ \alpha > 4) . \tag{2.5}$$

We associate with each $V(z/a,q,\alpha)$ an exceptional set that contains a neighborhood of the zeros of $(1 - z/a)\ P(z/a,q,\alpha)$. Thus we define

$$F(a,\alpha) = \{z : |\arg z/a| < (2\alpha)^{-1}\} \cap \{z : \alpha^{-1} < |z/a| < \alpha\} \cup$$

$$\left(\bigcup_{k=1}^{q} \bigcup_{j=1}^{k} \left\{ \{z : |\arg z/a - (2j+1)\pi/k| < (2kq\alpha)^{-1}\} \cap \right.\right. \tag{2.6}$$

$$\left.\left. \{z : (k/2|C_k|)^{1/k} < |z/a| < (2k/|C_k|^{1/k}\alpha^2\} \right\} \right)$$

and deduce, for Lebesque measure on $|z| = r$, that

$$\text{meas } \{z : z = re^{i\theta},\ z \in F(a,\alpha)\} \leq 2\alpha^{-1} r\ . \tag{2.7}$$

3. A partition of the real line and the definition of $\lambda(r)$. Let

$$\theta_m \geq 1;\ \varepsilon_m \equiv \log \hat{\varepsilon}_m \geq 0 \qquad (m \geq 1) \tag{3.1}$$

which we specify later. Define sequences $\{\omega_{m,j}\}_{j=1}^{m}$ and $\{\gamma_m\}$ $(m \geq 1)$ by

$$\gamma_1 = \exp\{\exp 16\} \tag{3.2}$$

$$\omega_{m,1} = \gamma_m\ (\geq 2\exp\{\exp(16m^2)\}) \tag{3.3}$$

$$\omega_{m,k+1} = 2m^5 (\gamma_m + \sum_{j=1}^{k} \omega_{m,j}) \tag{3.4}$$

$$\gamma_{m+1} = \max\{2\exp(\exp\{16(m+2)^2\}),\ \sum_{k=1}^{m} (\varepsilon_k + \gamma_k + \sum_{j=1}^{k} \omega_{k,j})\} \tag{3.5}$$

where $\varepsilon_k \geq 0$ as in (3.1). Thus γ_1 and $\omega_{1,1}$ are given at once by (3.2) and (3.3). In general, once γ_j and $\omega_{j,1}$ are known, the $\omega_{j,k} (2 \leq k \leq j)$ are given recursively by (3.4); (3.5) gives in turn γ_{j+1} which then yields $\omega_{j+1,1}$ in (3.3), etc. Now set

$$\Omega_{m,k} = \exp\{\sum_{j=1}^{k} \omega_{m,j}\} \qquad (k = 1,2,\ldots,m)\ , \tag{3.6}$$

$$\Gamma_k = \exp(\gamma_k) \qquad (k = 1,2,\ldots) \tag{3.7}$$

and define

$$r_1 = \Gamma_1 \quad \theta_1\ , \tag{3.8}$$

$$r_{m+1} = \Gamma_{m+1} \quad \theta_{m+1} \hat{\varepsilon}_m\ \Omega_{m,m}\ r_m \qquad (m \geq 1)\ . \tag{3.9}$$

The intervals $(\Gamma_m^{-1} r_m, \Omega_{m,m} r_m)$ $(m \geq 1)$ are pairwise disjoint. The θ_m (as well as $\hat{\varepsilon}_m$) which make the separation between successive intervals are selected to determine the order of our function $f(z)$.

Let $\Lambda(u)$ $(u \geq 0)$ be a nonnegative continuous function. In particular, for $k = 1,\dots,m$, $m = 1,2,\dots$ and $\Omega_{m,0} = 1$, define

$$\Lambda(\log r_m + \log \Omega_{m,k}) = m - k,\ \Lambda(\log r_m) = m\ , \tag{3.10}$$

$$\Lambda'(u) = -\omega_{m,k}^{-1},\ \log \Omega_{m,k-1} < u - \log r_m < \log \Omega_{m,k}\ , \tag{3.11}$$

$$\Lambda'(u) = m\gamma_m^{-1},\ \log r_m - \log \Gamma_m < u < \log r_m\ , \tag{3.12}$$

$$\Lambda(u) = 0,\ u \notin \bigcup_{m=1}^{\infty} (\log r_m - \log \Gamma_m, \log r_m + \log \Omega_{m,m})\ . \tag{3.13}$$

A result of the way in which Λ was constructed is that the sum of trapezoidal areas under the graph of $\Lambda(u)$ gives, for $1 \leq k \leq m$,

$$I_{m,k} \equiv \int_{\alpha}^{\beta} (\Lambda(u) - (m-k))du = (\gamma_m k^2/m + \sum_{j=1}^{k} \{2(k-j)+1\}\omega_{m,j})2^{-1} \tag{3.14}$$

where $\alpha = \log r_m - (k/m)\log\Gamma_m$, $\beta = \log r_m + \log \Omega_{m,k}$, and for $k \leq m - 1$.

$$\int_{\beta}^{\beta'} (\Lambda(u) - (m-k))du \leq -I_{m,k} - \exp(\exp 16m^2)\ , \tag{3.15}$$

where $\beta = \log r_m + \log \Omega_{m,k}$, $\beta' = \log r_m + \log \Omega_{m,k} + \omega_{m,k+1}/m^2$. With Λ given by (3.10)-(3.13) and $\rho > 1$ as in (1), define

$$\lambda(t)\ (=\lambda_\rho(t)) = \max(\rho, \Lambda(\log t))\ ; \tag{3.16}$$

hence

$$\lambda(t) \leq \lambda(r_m) = m \qquad (t \leq \Omega_{m,m} r_m) \tag{3.17}$$

and as is evident from (3.3)-(3.5) and (3.11)-(3.13)

$$|\lambda'(t)| < (m+1)(t\gamma_{m+1})^{-1} \qquad (t > \Omega_{m,m} r_m)$$

off a discrete set having no accumulation point in the plane.

Lemma 1. Let $\lambda(t)\ (= \lambda_\rho(t))$ be given in (3.16). If in (3.1)

$$\varepsilon_m = m \sum_{k=1}^{m} \omega_{k,k}$$

then

$$\liminf_{r\to\infty} (\log r)^{-1} \int_1^r \lambda_\rho(s) \frac{ds}{s} = \rho\ . \tag{3.18}$$

If in (3.1)

$$\theta_m \equiv 1$$

for all m, then

$$\limsup_{r\to\infty} (\log r)^{-1} \int_1^r \lambda_\rho(s) \frac{ds}{s} = \infty ; \tag{3.19}$$

if

$$\theta_m = \prod_{k=1}^{m} (\Gamma_k \Omega_{k,k})^{mk} \qquad (m = 1,2,\ldots) ,$$

then

$$\limsup_{r\to\infty} (\log r)^{-1} \int_1^r \lambda_\rho(s) \frac{ds}{s} = \rho . \tag{3.20}$$

In terms of r_m in (3.9) and γ_m in (3.2)-(3.5), let $A(t)$ $(t \geqq 0)$ be the positive nondecreasing continuous function given by

$$A(t) = \begin{cases} \log \gamma_1 & (0 \leqq t \leqq \Omega_{1,1} r_1) \\ \log \gamma_m & (t = \Omega_{m,m} r_m, m = 1,2,\ldots) \end{cases} \tag{3.21}$$

$$\frac{d\,A\,(t)}{d \log t} = \frac{\log \gamma_m - \log \gamma_{m-1}}{\log \Omega_{m,m} r_m - \log \Omega_{m-1,m-1} r_{m-1}} \tag{3.22}$$

where $\Omega_{m-1,m-1} r_{m-1} < t < \Omega_{m,m} r_m$, $m > 1$. Let $\alpha(t)$ be the integer-valued function given by

$$\alpha(t) = \begin{cases} [\sqrt{A(t)}] & ([\sqrt{A(t)}] \neq \sqrt{A(t)},\ t \geqq 0) \\ \lim\limits_{s \to t^-} \sqrt{A(s)} & ([\sqrt{A(t)}] = \sqrt{A(t)},\ t \geqq 0) . \end{cases} \tag{3.23}$$

The properties of $\alpha(r)$ and $\lambda(r)$ that are prominent in our construction of $f(z)$ as in Theorems 1 and 2 follow:

$$\alpha(r) \to \infty \qquad (r \to \infty) \tag{3.24}$$

$$\alpha^5(r)\, r |\lambda'(r)| = o(1) \qquad (r \to \infty) \tag{3.25}$$

$$\log \log \alpha(r) = o(\log r) \qquad (r \to \infty) \tag{3.26}$$

$$\lambda(r) \leqq \log \alpha(r) \qquad (r \geqq 0) \tag{3.27}$$

$$|\lambda(t) - \lambda(r)| = o(\alpha^{-3}(r)) \quad (\alpha^{-2}(r) r \leqq t \leqq \alpha^2(r) r, r \to \infty) . \tag{3.28}$$

4. The comparison and counting functions. With $\lambda(r)$ as in (3.16) we define a comparison function

$$n^*(r) = \exp \int_1^r \lambda(s) \frac{ds}{s} \qquad (r > 1) \tag{4.1}$$

(cf. [1], p. 130) so that

$$dn^*(t) = \lambda(t)\, n^*(t) \frac{dt}{t} \qquad (t > 1) \tag{4.2}$$

and

$$n^*(t) = \left(\exp \int_r^t \lambda(s) \frac{ds}{s}\right) n^*(r) . \tag{4.3}$$

Analysis of

$$(r/t)^{\lambda(r)} n^*(t)/n^*(r) - 1 = \exp\left(\int_r^t \{\lambda(s) - \lambda(r)\}\frac{ds}{s}\right) - 1$$

in light of (3.28) yields

$$n^*(t) = \{1 + o(\alpha^{-2}(r))\}(t/r)^{\lambda(r)} n^*(r) \qquad (t \in [\alpha^{-2}(r)r, \alpha^2(r)r]) \tag{4.4}$$

uniformly in t as $r \to \infty$. If $\Delta(t)$ is a nonnegative function of t then

$$n(r) = \int_1^r \Delta(t)\, dn^*(t) \qquad (r > 1)$$

is nondecreasing. If $t \in (r_m, r_{m+1}]$ $(m \geq 4)$, then let

$$\Delta_1(t) = \begin{cases} |\sin \pi\lambda(t)|\, \delta_m(t), & \lambda(t) - [\lambda(t)] \in (\frac{1}{m^2}, \frac{m^2-1}{m^2}) \\ 0, & \lambda(t) - [\lambda(t)] \notin (\frac{1}{m^2}, \frac{m^2-1}{m^2}) \end{cases} \tag{4.5}$$

where $\delta_m(t)$ is some appropriate continuous function. If $t \leq r_4$ then take $m = 4$ in (4.5). Define for $t \in (r_m, r_{m+1})$ $(m \geq 4)$

$$\Delta_2(t) = \begin{cases} 0, & \lambda(t) - [\lambda(t)] \in (\frac{2}{m^2}, \frac{m^2-2}{m^2}) \\ 1 - \delta_m(t), & \lambda(t) - [\lambda(t)] \in (\frac{1}{m^2}, \frac{2}{m^2}) \cup (\frac{m^2-2}{m^2}, \frac{m^2-1}{m^2}) \\ 1, & \lambda(t) - [\lambda(t)] \notin (\frac{1}{m^2}, \frac{m^2-1}{m^2}) \end{cases} \tag{4.6}$$

where $\delta_m(t)$ is the same as above. Again, if $t \leq r_4$, take $m = 4$ in (4.6). We now define

$$n_j^*(r) = \int_1^n \Delta_j(t)dn^*(t) \quad (r \geq 1,\ j = 1,2) . \tag{4.7}$$

5. Definition of the function. With $n^*_j(r)$ $(j = 1,2)$ defined in (4.7) set

$$n_j(r) = \begin{cases} [n_j^*(r)], & r \geq 1 \\ 0 & r < 1 \end{cases} \quad (j = 1,2) \tag{5.1}$$

and define the sequences of positive numbers

$$S_1 = \{a_n : \text{card}\ \{a_n \leq r\} = n_1(r)\} , \tag{5.2}$$

$$S_2 = \{a_n : \text{card}\ \{a_n \leq r\} = n_2(r)\} . \tag{5.3}$$

Observe from (4.5), (4.6), and (5.1)-(5.3) that if r is large and $\lambda(r)$ is nearly equal to an integer then $r \notin S_1$; if $r \in S_2$ for large r then $\lambda(r)$ is nearly integral.

With $V(z,q,\alpha)$, $\lambda(r)$, and $\alpha(r)$ given in (2.3), (3.16), and (3.23) define

$$U_1(z) = \prod_{t \in S_1} V(z/t \exp\{i\,\Psi(t)\},\ [\lambda(t)],\ \alpha^2(t)) \tag{5.4}$$

where

$$\Psi(r) = (1 + \lambda(r) - [\lambda(r)])\pi\lambda^{-1}(r) \quad (r > 0) \tag{5.5}$$

and

$$U_2(z) = \prod_{t \in S_2} V(z/t \exp\{-i\pi/2p(t)\},\ p(t) - 1,\ \alpha^2(t)) \cdot V(z/t \exp\{i\pi/2p(t)\},\ p(t) - 1,\ \alpha^2(t)) \tag{5.6}$$

where $p(t)$ is the integer-valued function ([ ], the usual greatest integer function)

$$p(t) = [\lambda(t) + \tfrac{1}{2}] . \tag{5.7}$$

Now with c_m and θ_m in (3.1) appropriately chosen by Lemma 1,

$$f(z) = U_1(z)U_2(z) \tag{5.8}$$

will be a candidate for either Theorem 1 or 2. One can show via (3.14),

(3.15), and (4.4)-(4.6) that the infinite products in (5.4) and (5.6) converge and define the entire functions $U_1(z)$ and $U_2(z)$; thereby $f(z)$ in (5.8) is also entire.

6. Asymptotic behavior. Let $F(a,\alpha^2(|a|))$, Ψ, λ, p be given in (2.6), (5.5), (3.16), (5.7). Let G be the open set given by

$$G = \bigcup_{t>0} \{z : z \in F(t\exp\{i\beta(t)\}, \alpha^2(t)); \ \beta(t) = \Psi(t), -\pi/2p(t), \pi/2p(t)\}. \tag{6.1}$$

Thus G contains the zero set of $f(z)$. With the discontinuities of Ψ taken into account, it follows from (2.6) and (2.7) that, for Lebesque measure on $|z| = r$,

$$\text{meas}\ \{z : z = re^{i\theta},\ z \in G\} \leq 12\ \{1 + o(1)\}\alpha^{-2}(r)r \quad (r \to \infty)\ . \tag{6.2}$$

The relationship between $V(z,q,\alpha)$ and the Weierstrass primary factor $E(z/q)$ in (2.4) and (2.5) and a lengthy analysis of $\log|U_1(z)|$ and $\log|U_2(z)|$, incorporating properties (3.24)-(3.28) with (4.1)-(4.7) and integration by parts, (cf [9], pp. 62 - 68, 98 - 107) yields

$$\log|U_1(z)| = -\Delta_1(r)n^*(r)\,\text{Re} \int_{r\alpha^{-1}(r)}^{r\alpha(r)} (t/r)^{\lambda(r)} \left\{ \frac{(z/t\exp\{i\Psi(r)\})^{[\lambda(r)+1]}}{1 - z/t\exp\{i\Psi(r)\}} \right\} \frac{dt}{t}$$
$$+\ o(n^*(r)) \quad (z = re^{i\theta} \notin G,\ r \to \infty) \tag{6.3}$$

and

$$\log|U_2(z)| = \Delta_2(r)n^*(r)\,\text{Re} \int_{r\alpha^{-1}(r)}^{r\alpha(r)} i(t/r)^{\lambda(r)}(z/t)^{p(r)}$$
$$\cdot\ \{(1 - z/t\exp\{i\pi/2p(r)\})^{-1} - (1 - z/t\exp\{-i\pi/2p(r)\})^{-1}\}\frac{dt}{t} + o(n^*(r)) \tag{6.4}$$
$$(z = re^{i\theta} \notin G,\ r \to \infty)$$

where Δ_1 and Δ_2 are given in (4.5) and (4.6). Further analysis of the integrals in (6.3) and (6.4) results in

Lemma 2. Let $z = re^{i\theta}$, $z \notin G$, where G is defined in (6.1). Let $U_1(z)$ and $U_2(z)$ be given in (5.4) and (5.6). Then

$$\log|U_1(re^{i\theta})| = n^*(r)\ \pi\{\sin(\pi\lambda(r))\}^{-1}\Delta_1(r)\ \cos\{\lambda(r)(\theta - \Psi(r) - \pi)\}$$
$$+\ o(n^*(r)) \quad (\Psi(r) < \theta < 2\pi + \Psi(r))\ , \tag{6.5}$$

$$\log|U_2(re^{i\theta})| = n^*(r)(2\pi - \pi/p(r))\Delta_2(r)\cos\{\lambda(r)\theta\} + o(n^*(r))$$

$$(-\pi/2p(r) < \theta < \pi/2p(r)) , \tag{6.6}$$

$$\log|U_2(re^{i\theta})| = -n^*(r)(\pi/p(r))\Delta_2(r)\cos\{\lambda(r)\theta\} + o(n^*(r))$$

$$(\pi/2p(r) < \theta < 2\pi - \pi/2p(r)) \tag{6.7}$$

where the $o(n^*(r))$ is independent of θ for $z \notin G$ as $r \to \infty$.

7. Proofs of Theorem 1 and 2. By first obtaining an estimate for the maximum modulus of $f(z)$ for $|z| = r$ we can show using a result of Nevanlinna ([10], p. 24) that

$$\log n^*(r) - \log\log\alpha(r) \leq \log T(r,f) \leq \log n^*(r) + O(\log\log\alpha(r)) . \tag{7.1}$$

Dividing (7.1) by $\log r$ and comparing this with (3.26) and (4.1), one observes that the order λ and lower order μ, of $f(z)$, are determined by

$$\lambda = \limsup_{r\to\infty}(\log r)^{-1}\int_1^r \lambda(s)\frac{ds}{s}, \quad \mu = \liminf_{r\to\infty}(\log r)^{-1}\int_1^r \lambda(s)\frac{ds}{s} .$$

Applying (3.16) and (3.18)-(3.20) to this, one can deduce, for an appropriate choice of ε_m and θ_m in the construction of $\lambda(r)$, that

$$\begin{cases} 1 < \rho = \mu = \lambda & (\theta_m \text{ as in } (3.20)) , \\ 1 < \rho = \mu,\ \lambda = \infty & (\theta_m = 1) . \end{cases} \tag{7.2}$$

It can be deduced with the help of the so-called Small Arcs Lemma ([3], p. 322) by Edrei-Fuchs, (6.1)-(6.2), (5.8), (6.5)-(6.7), and (4.4) that

$$\int |\log|f(\sigma re^{i\theta})||d\theta = \sigma^{\lambda(r)}o(n^*(r)) \qquad (\sigma \geq 1,\ r \to \infty)$$

where the integration is over the exceptional set G in (6.1). It can then be shown, since $f(z)$ is entire, that as $r \to \infty$

$$\frac{T(\sigma r,f)}{T(r,f)} = \frac{\frac{1}{2\pi}\int_0^{2\pi}\log^+|f(\sigma re^{i\theta})|d\theta}{\frac{1}{2\pi}\int_0^{2\pi}\log^+|f(re^{i\theta})|d\theta} = (1 + o(1))\sigma^{\lambda(r)} \qquad (\sigma > 1) \tag{7.3}$$

and

$$\delta(0,f) = \liminf_{r\to\infty} \frac{m(r,0,f)}{m(r,\infty,f)} \geq \frac{1}{3} . \tag{7.4}$$

Theorem 1 follows from (7.4), (7.3), (3.17), and an appropriate choice made in (7.2); Theorem 2 follows from (7.4) and (7.2).

References

[1] Drasin, D.: A flexible proximate order. Bull. London math. Soc. 6 (1974), 129 - 135.

[2] Edrei, A., Fuchs, W. H. J.: The deficiencies of meromorphic functions of order less than one. Duke math. J. 27 (1960), 233 - 49.

[3] Edrei, A., Fuchs, W. H. J.: Deficient values of meromorphic functions. Proc. London math. Soc., III. Ser. 12 (1962), 315 - 44.

[4] Goldberg, A. A.: The possible magnitude of the lower order of an entire function with a finite deficient value. Soviet Math., Doklady 5 (1964), 1631 - 1633.

[5] Hayman, W. K.: Some examples related to the $\cos \pi\rho$ theorem, in "Mathematical essays dedicated to A. J. Macintyre". Ohio University Press, Athens, Ohio (1970), 149 - 170.

[6] Hayman, W. K.: Meromorphic functions. Oxford University Press, Oxford (1964).

[7] Hayman, W. K.: Research problems in function theory. The Athlone Press, London (1967).

[8] Kotman, L. J.: An entire function which interpolates a family of functions with orders ranging between one and infinity. Ph. D. thesis, Purdue University (1977).

[9] Levin, B. Ja.: Distribution of zeros of entire functions. Amer. math. Soc., Providence, R.I. (1964).

[10] Nevanlinna, R.: Le théorème de Picard-Borel et la théorie des fonctions méromorphes. Gauthier-Villars, Paris (1929).

[11] Research problems. Bull. Amer. math. Soc. 68 (1962), 21.

University of Wisconsin - La Crosse
La Crosse, WI 54601
U.S.A.

ON THE CARATHÉODORY METRIC
ON THE UNIVERSAL TEICHMÜLLER SPACE*

Irwin Kra

Bers-Ehrenpreis [4] showed that every finite dimensional Teichmüller space is a domain of holomorphy by proving that the Universal Teichmüller space is convex with respect to a certain family of holomorphic functions. By considering essentially the same family of holomorphic functions we show that the Teichmüller and Carathéodory metrics are almost uniformly equivalent. From this fact follow all the known convexity properties of Teichmüller spaces.

§1. Introduction and summary of results. Let Γ be a Fuchsian group operating on the upper half plane U (thus also on the lower half plane U^*), and let $M(\Gamma)$ be the set of Beltrami coefficients for Γ; that is, the set of measurable functions μ on U with essential supremum $\|\mu\| < 1$ satisfying

$$(\mu \circ \gamma) \frac{\overline{\gamma'}}{\gamma'} = \mu, \quad \text{all } \gamma \in \Gamma.$$

Let $T(\Gamma)$ denote the Teichmüller space of the group Γ. We recall, for the convenience of the reader, the definition. For each $\mu \in M(\Gamma)$, let w^μ be the unique normalized (fixing $0,1,\infty$) automorphism of the extended complex plane $\mathbb{C} \cup \{\infty\}$ that is μ-conformal in U and conformal in U^*. Two Beltrami coefficients μ and ν are equivalent (modulo Γ) provided $w^\mu \mid U^* = w^\nu \mid U^*$. The set of equivalence classes of Beltrami coefficients forms the Teichmüller space.

Let φ^μ be the Schwarzian derivative of $w^\mu \mid U^*$. The mapping $\Phi : \mu \mapsto \varphi^\mu$ is a well defined, holomorphic, injective mapping from $T(\Gamma)$ onto a bounded open set in the Banach space $B_2(U^*,\Gamma)$ of holomorphic functions φ on U^* with

$$(\varphi \circ \gamma)(\gamma')^2 = \varphi, \quad \text{all } \gamma \in \Gamma,$$

$$\|\varphi\|_\infty = \sup_{z \in U^*} |(z - \bar{z})^2 \varphi(z)| < \infty.$$

The dilatation of a μ-conformal automorphism w of $\mathbb{C} \cup \{\infty\}$ is given by

---

*Research partially supported by NSF grant MCS 7604969A01.

$$K(w) = \frac{1 + \|\mu\|}{1 - \|\mu\|}.$$

This allows one to define, in addition to the metric $\|\cdot\|$ that $M(\Gamma)$ inherits as an open subset of a complex Banach space, the Teichmüller metric:

$$t_{M(\Gamma)}(\mu,\nu) = \frac{1}{2}\log K(w^{\mu} \circ (w^{\nu})^{-1}), \quad \mu,\nu \in M(\Gamma),$$

and the corresponding metric on the Teichmüller space

$$t_{T(\Gamma)}(\varphi,\psi) = \inf_{\substack{\mu\in\Phi^{-1}(\varphi)\\ \nu\in\Phi^{-1}(\psi)}} \{t_{M(\Gamma)}(\mu,\nu)\}, \quad \varphi,\psi \in T(\Gamma).$$

On any complex manifold X, we have two intrinsic metrics: c_X, the Carathéodory metric and k_X, the Kobayashi metric. The Schwarz-Pick lemma (see Kobayashi [9, pp. 50]) shows that $c_X \leq k_X$.

For the manifolds $M(\Gamma)$ we have

$$c_{M(\Gamma)} = k_{M(\Gamma)} = t_{M(\Gamma)}.$$

This equality follows from three observations:

1. There exist (many) right translations which are isometries in each of the metrics and which reduce all calculations to computing distances from the origin.

2. Fix $0 \neq \mu_0 \in M(\Gamma)$, then Teichmüller metric on the disk

$$\{z \frac{\mu_0}{\|\mu_0\|};\ z \in \mathbb{C},\ |z| < 1\}$$

agrees with the Poincaré metric on this disk (which is, of course, also the Carathéodory and Kobayashi metric on the disk).

3. If X is a (complex) Banach space and $x_0 \in X$, then there exists (by Hahn-Banach) a linear function $L : X \to \mathbb{C}$ of norm one with $L(x_0) = \|x_0\|$.

It is also quite elementary to show that

$$c_{T(\Gamma)} \leq k_{T(\Gamma)} \leq t_{T(\Gamma)}. \tag{1.1}$$

Only the last inequality needs verification. To establish it, we note that

$$\Phi : M(\Gamma) \to T(\Gamma)$$

is distance non-increasing, and therefore

$$k_{T(\Gamma)}(0,\varphi) \leqq \inf_{\mu\in\Phi^{-1}(\varphi)} \{k_{M(\Gamma)}(0,\mu)\} = \inf_{\mu\in\Phi^{-1}(\varphi)} \{t_{M(\Gamma)}(0,\mu)\}$$
$$= t_{T(\Gamma)}(0,\varphi).$$

It follows from (1.1) that for every $r > 0$ every Teichmüller ball of radius r is contained in a Carathéodory ball of the same radius. The converse is contained in

Theorem. There exists a continuous increasing function $\eta : \mathbb{R}_+ \to \mathbb{R}_+$ such that every Carathéodory ball of radius r in T is contained in a Teichmüller ball of radius $\eta(r)$. (Here $\mathbb{R}_+$ represents the positive reals.) Further, the function η satisfies $\lim_{r\to 0} \eta(r) = 0$.

In the above, T stands for the Universal Teichmüller Space, $T(\{1\})$. As usual, we supress the trivial group from our notation.

We shall also need another description of $T(\Gamma)$. For each $\mu \in M(\Gamma)$, w_μ denotes the unique normalized μ-conformal automorphism of U. The space $T(\Gamma)$ may be identified with the restrictions to the real axis $\mathbb{R}$ of these w_μ's (see, for example, Ahlfors [2, Chapter VI]).

§2. Consequences of the Theorem. We shall show in proving the theorem that there exists a continuous increasing function

$$h : \mathbb{R}_+ \to \mathbb{R}_+$$

with $\lim_{r\to 0} h(r) = 0$, $\lim_{r\to\infty} h(r) = \infty$ such that for all $x,y \in T$,

$$t_T(x,y) = r > 0 \Rightarrow c_T(x,y) \geq h(r). \tag{2.1}$$

The function η will, of course, be the inverse function to h.

Combining (1.1) with (2.1), we see that $x,y \in T$ with $t_T(x,y) = r$ implies that

$$h(r) \leqq c_T(x,y) \leqq r. \tag{2.2}$$

And conversely, $x,y \in T$ with $c_T(x,y) = r$ implies that

$$r \leqq t_T(x,y) \leqq \eta(r). \tag{2.3}$$

Corollary 1. (Earle [6]). The Carathéodory metric on $T(\Gamma)$ is complete.

Proof. The canonical inclusion

$$T(\Gamma) \hookrightarrow T$$

is distance decreasing with respect to the Carathéodory metrics (as well as the other two metrics under consideration). If $\{x_n\} \subset T(\Gamma)$ is a Cauchy sequence with respect to $c_{T(\Gamma)}$, it is also Cauchy with respect to c_T. Our theorem (see (2.3)) shows that the sequence is Cauchy with respect to t_T. Since the Teichmüller metric is complete, $\lim_{n\to\infty} x_n = x \in T$ with convergence in the Teichmüller metric t_T. Now $T(\Gamma)$ is closed in T (in every metric). Thus $x \in T(\Gamma)$. The $t_{T(\Gamma)}$-topology on $T(\Gamma)$ agrees with the t_T-topology on $T(\Gamma)$ because $B_2(U^*,\Gamma)$ is a closed Banach subspace of $B_2(U^*)$. Thus $\lim_{n\to\infty} x_n = x$ in the $t_{T(\Gamma)}$-metric. Now (2.2) shows that the same conclusion holds in the $c_{T(\Gamma)}$-metric.

Corollary 2 (Bers-Ehrenpreis [4]). Every finite dimensional Teichmüller space is a domain of holomorphy.

Proof. See Kobayashi [9, pp. 50 and 77] and Earle [6].

We have already remarked that

$$t_T \leq t_{T(\Gamma)} \quad \text{on} \quad T(\Gamma).$$

Conversely, if $X \subset T(\Gamma)$ is bounded in the t_T-metric, it must also be bounded in the $t_{T(\Gamma)}$-metric whenever $T(\Gamma)$ is finite dimensional. To show this, we may assume that X is closed. Thus X is a compact subset of T and hence also of $T(\Gamma)$. Since compact subsets of $T(\Gamma)$ are precisely the closed and bounded subsets, our assertion follows. (See §4 for a more detailed proof of this fact.) Thus we obtain an affirmative answer to a question of Earle [6]:

Corollary 3 (Krushkal [12]). If $T(\Gamma)$ is finite dimensional, then every closed $c_{T(\Gamma)}$-bounded set is compact.

Corollary 4 (Krushkal [12]). Every finite dimensional Teichmüller space is convex with respect to the family of bounded holomorphic functions.

Proof. This follows from Corollary 3 and the properties of the Carathéodory metric (see Kobayashi [9, pp. 55]).

§3. Proof of the Theorem. We recall some results of Ahlfors [1] and Beurling-Ahlfors [5]. Let σ be the Poincaré distance on $\mathbb{C} \setminus \{0,1\}$ and let ρ be the Poincaré distance on U. Let us introduce two functions f and g, by setting

$$f(x) = \sigma(-x,-1), \quad 1 \leq x < +\infty,$$

$$g(x) = f^{-1}(x), \quad 0 \leq x < +\infty.$$

Note that since the negative real axis is fixed under the automorphism

$z \mapsto \bar{z}$ of $\mathbb{C} \setminus \{0,1\}$, it is a geodesic for σ. Thus f is a strictly increasing, non-negative, continuous function on $[1,\infty)$. Furthermore

$$f(1) = 0,$$

and

$$\lim_{x\to\infty} f(x) = +\infty.$$

Then also g, the inverse function of f, is well defined and is a continuous strictly increasing function on $[0,\infty)$.

Let $\varphi = w_\mu \mid \mathbb{R}$ be the restriction to $\mathbb{R}$ of a μ-conformal normalized automorphism of U (here $\mu \in M = M(\{1\})$). Then (see, for example, Kra [11]) for all $x \in \mathbb{R}$, all $t > 0$,

$$\sigma\left(\frac{\varphi(x+t) - \varphi(x)}{\varphi(x-t) - \varphi(x)}, -1\right) \leqq \frac{1}{2}\log K, \tag{3.1}$$

provided $\|\mu\| \leqq k = \frac{K-1}{k+1}$. Note that $\frac{\varphi(x+t) - \varphi(x)}{\varphi(x-t) - \varphi(x)} < 0$. Further $\sigma(-y,-1) = \sigma(-\frac{1}{y},-1)$ for $0 < y < 1$ (since $z \mapsto \frac{1}{z}$ is an automorphism of $\mathbb{C} \setminus \{0,1\}$). Let $-y = \frac{\varphi(x+t) - \varphi(x)}{\varphi(x-t) - \varphi(x)}$. Then (3.1) reads

$$f(y) \leqq \frac{1}{2}\log K \quad \text{for } y \geqq 1,$$

and

$$f(\tfrac{1}{y}) \leqq \frac{1}{2}\log K \quad \text{for } 0 < y \leqq 1.$$

Applying g to both sides, we see that

$$y \leqq g(\tfrac{1}{2}\log K) \quad \text{for } y \geqq 1,$$

and

$$\frac{1}{y} \leqq g(\tfrac{1}{2}\log K) \quad \text{for } 0 < y \leqq 1.$$

We have re-proven that $\varphi = w_\mu \mid \mathbb{R}$ satisfies the M-condition (Beurling-Ahlfors [5])

$$M^{-1} \leqq \frac{\varphi(x+t) - \varphi(x)}{\varphi(x) - \varphi(x-t)} \leqq M$$

with

$$M = g(\tfrac{1}{2}\log K).$$

Conversely, Beurling-Ahlfors [5] have shown that every normalized homeomorphism φ of $\mathbb{R}$ that satisfies an M-condition is the restriction to $\mathbb{R}$ of a K-quasiconformal automorphism of U. According to Beurling-

Ahlfors [5] and Reed [13], w may be chosen so that

$$K(w) \leq \min(M^2, 8M).$$

Lemma 1. Let $\varphi \in T$ with $t_T(0,\varphi) = r > 0$. View φ as a normalized automorphism of $\mathbb{R}$. Then

$$\sup_{\substack{x\in\mathbb{R}\\ t>0}} \sigma\left(\frac{\varphi(x+t) - \varphi(x)}{\varphi(x-t) - \varphi(x)}, -1\right) \geq \begin{cases} f(\frac{1}{8}\exp 2r) & \text{for } r > \frac{1}{2}\log 8, \\ \\ f(\exp r) & \text{for any } r > 0. \end{cases} \tag{3.2}$$

Proof. Assume that

$$\sup_{\substack{x\in\mathbb{R}\\ t>0}} \sigma\left(\frac{\varphi(x+t) - \varphi(x)}{\varphi(x-t) - \varphi(x)}, -1\right) < f(\exp r).$$

Then, as we saw previously, this means that φ satisfies an M-condition with $M < g(f(\exp r)) = \exp r$. Thus we can choose a quasiconformal extension w of φ with $K(w) < \exp 2r$. In particular,

$$t_T(0,\varphi) \leq \frac{1}{2}\log K(w) < r.$$

To obtain the second estimate, we use the fact that w can be chosen with $K(w) \leq 8M$.

Lemma 2. Let $\varphi = \Phi(\mu_o)$, $\mu_o \in M$. If $t_T(0,\varphi) = r > 0$, then $c_T(0,\varphi) \geq \frac{1}{2} f(\exp r)$.

Proof. Let $\varphi = w_{\mu_o} \mid \mathbb{R}$ and define f^{μ_o} by $w^{\mu_o} = f^{\mu_o} \circ w_{\mu_o}$. Then $f^{\mu_o} \mid U$ is conformal. Since

$$\sup_{\substack{\zeta_1,\zeta_2,\zeta_3 \in f^{\mu_o}(\mathbb{R})\\ \zeta_2 \text{ between } \zeta_1 \text{ and } \zeta_3}} \sigma\left(\frac{(f^{\mu_o})^{-1}(\zeta_2) - (f^{\mu_o})^{-1}(\zeta_1)}{(f^{\mu_o})^{-1}(\zeta_3) - (f^{\mu_o})^{-1}(\zeta_1)}, \frac{\zeta_2 - \zeta_1}{\zeta_3 - \zeta_1}\right)$$

$$= \sup_{\substack{a,b,c\in\mathbb{R}\\ a<b<c}} \sigma\left(\frac{f^{\mu_o}(b) - f^{\mu_o}(a)}{f^{\mu_o}(c) - f^{\mu_o}(a)}, \frac{b-a}{c-a}\right),$$

we conclude from Lemma 1 and the triangle inequality for $\sigma(\cdot,\cdot)$ that either

$$\sup_{\substack{a,b,c\in\mathbb{R}\\ a<b<c}} \sigma\left(\frac{w^{\mu_o}(b) - w^{\mu_o}(a)}{w^{\mu_o}(c) - w^{\mu_o}(a)}, \frac{b-a}{c-a}\right) \geq \frac{1}{2} f(\exp r), \tag{3.3}$$

or

$$\sup_{\substack{a,b,c\in\mathbb{R}\\ a<b<c}} \sigma\left(\frac{f^{\mu_o}(b) - f^{\mu_o}(a)}{f^{\mu_o}(c) - f^{\mu_o}(a)}, \frac{b-a}{c-a}\right) \geq \frac{1}{2} f(\exp r). \tag{3.4}$$

Now the mapping w_{μ_o} can be extended by reflection to be a quasiconformal homeomorphism of $\mathbb{C}$. The complex dilatation of w_{μ_o} in U^* is $*\mu_o$, where

$$*\mu_o(z) = \overline{\mu_o(\bar{z})}, \quad z \in U^*.$$

Hence f^{μ_o} is again a global quasiconformal map and $f^{\mu_o} = w^{*\mu_o}$ (note that $*\mu_o$ is supported in U^*). Gardiner [8] has shown that

$$f^{\mu_o} \circ j = j \circ w^{\nu_o},$$

where j is the reflection in $\mathbb{R}$ ($j(z) = \bar{z}$), and ν_o is defined by $w_{\mu_o} \circ w_{\nu_o} = id$ (w_{ν_o} is the Teichmüller inverse of w_{μ_o}). For each triple of distinct points $a,b,c \in \mathbb{R}$ we define two maps

$$w_{abc}(\mu) = \frac{w^{\mu}(b) - w^{\mu}(a)}{w^{\mu}(c) - w^{\mu}(a)} \tag{3.5}$$

and

$$*w_{abc}(\mu) = \overline{\frac{w^{*\mu}(b) - w^{*\mu}(a)}{w^{*\mu}(c) - w^{*\mu}(a)}}.$$

First of all note that $w_{abc}(\mu)$ and $*w_{abc}(\mu)$ depends only on $\Phi(\mu)$. Further w_{abc} is a holomorphic function, so is $*w_{abc}$ (since $\mu \mapsto *\mu$ is an anti-holomorphic mapping). Thus we have defined two holomorphic functions from T into $\mathbb{C} \setminus \{0,1\}$. Since T is contractible (Earle-Eells [7]), these functions lift to holomorphic functions into the unit disk Δ (= universal covering space of $\mathbb{C} \setminus \{0,1\}$). In particular, let $\pi : \Delta \to \mathbb{C} \setminus \{0,1\}$ be a holomorphic universal covering map. There exist holomorphic mappings W_{abc}, $*W_{abc}$ from T into Δ such that $\pi \circ W_{abc} = w_{abc}$ and $\pi \circ (*W_{abc}) = *w_{abc}$. Since π is distance decreasing with respect to ρ and σ, we have for all $\alpha,\beta \in T$

$$\begin{aligned}\sigma(w_{abc}(\alpha), w_{abc}(\beta)) &= \sigma(\pi \circ W_{abc}(\alpha), \pi \circ W_{abc}(\beta)) \\ &\leq \rho(W_{abc}(\alpha), W_{abc}(\beta)),\end{aligned}$$

with a similar inequality for $*w_{abc}$. If (3.3) holds, then the above clearly implies that $c_T(0,\varphi) \geq \frac{1}{2} f(\exp r)$. Now let $\psi = \Phi(\nu_o)$. Since $\sigma(\bar{z}_1,\bar{z}_2) = \sigma(z_1,z_2)$, condition (3.4) implies that $c_T(0,\psi) \geq \frac{1}{2} f(\exp r)$. Thus to finish our proof of the lemma, we establish the following

<u>Lemma 3.</u> If $w_{\mu_o} \circ w_{\nu_o} = id$ for $\mu_o,\nu_o \in M$ and $\varphi = \Phi(\mu_o)$, $\psi = \Phi(\nu_o)$, then

$$c_T(0,\psi) = c_T(0,\varphi).$$

Proof. Right translation by ν_o:

$$M \ni \mu \xmapsto{R_{\nu_o}} \nu \in M,$$

where $w_\nu = w_\mu \circ w_{\nu_o}$, projects to a right translation map $R_\psi : T \to T$. Since R_ψ is a bi-holomorphic map, it is an isometry with respect to $c_T(\cdot,\cdot)$. Thus

$$c_T(0,\psi) = c_T(R_\psi(\varphi),\ R_\psi(0)) = c_T(\varphi,0) = c_T(0,\varphi).$$

§4. Holomorphic convexity of Teichmüller spaces. We have been studying a fairly standard family of holomorphic functions on Teichmüller space. Bers-Ehrenpreis [4] considered the family $\mathcal{F}$ of holomorphic function defined on $T(\Gamma)$ by

$$\mathcal{F} = \{w_{abc};\ (a,b,c) \in \mathbb{R}^3,\ a < b < c\},$$

where w_{abc} is defined by (3.5). We have already remarked that

$$w_{abc} : T(\Gamma) \to \mathbb{C} \setminus \{0,1\}$$

(and hence can be lifted to a bounded holomorphic function).

A slight strengthening of the Bers-Ehrenpreis [4] result is contained in the following two theorems that should be considered expository in nature (rather than new results).

Theorem A. Let Γ be trivial or a finitely generated Fuchsian group of the first kind. If $X \subset T(\Gamma)$ is a Teichmüller bounded set, then so is $\hat{X}$ the $\mathcal{F}$-hull of X.

We recall that

$$\hat{X} = \{\varphi \in T(\Gamma);\ |w(\varphi)| \leqq \sup_{\psi \in X} |w(\psi)|;\ \text{all } w \in \mathcal{F}\}.$$

We do not have a complete generalization of Theorem A to arbitrary Fuchsian groups. We can, however, prove

Theorem B. Let Γ be an arbitrary Fuchsian group. If $X \subset T(\Gamma) \hookrightarrow B_2(U^*,\Gamma)$ is Teichmüller bounded, then the closure of $\hat{X}$ in $B_2(U^*,\Gamma)$ is contained in $T(\Gamma)$, and every $\varphi \in \hat{X}$ is at positive distance from $\partial T(\Gamma)$, the boundary of $T(\Gamma)$ in $B_2(U,\Gamma^*)$.

Remark. Theorem B implies Theorem A for finitely generated Fuchsian groups of the first kind. We shall, however, give an independent (quite standard) proof of Theorem A since we are interested in specific bounds relating the diameter of $\hat{X}$ to the one for X. It will be obvious to the reader that except for one step (passing from the trivial group to Γ), we do have the possibility of obtaining such bounds. It would be interesting to see if a bound can be obtained (which is independent of Γ).

Proof of Theorem A. We begin by recalling for the reader some very important results of Ahlfors [1]. Let C be a closed Jordan curve. A quasiconformal reflection about C is an automorphism $f : \mathbb{C} \cup \{\infty\} \to \mathbb{C} \cup \{\infty\}$ such that $f^2 = 1$, $f \mid C = 1$, and $\overline{f}$ is quasiconformal. We define $K(C)$ to be the infimum of $K(\overline{f})$ over all quasiconformal reflections about C ($K(C) = +\infty$ if and only if no such quasiconformal reflection about C exists). It is quite easy to see that there exists a quasiconformal reflection about C if and only if C is a quasicircle; that is, if and only if $C = w^{\mu}(\mathbb{R})$ for some $\mu \in M = M(\{1\})$. Furthermore $K(C)$ and $K(w^{\mu})$ can be estimated in terms of each other.

A very non-trivial characterization of quasicircles is provided by the following criterium (Ahlfors [1]): Let C be an oriented Jordan curve with $\infty \in C$. Let $\zeta_1, \zeta_2, \zeta_3$ be three arbitrary points on C which follow each other in the given order. Set

$$\tilde{K}(C) = \sup \left|\frac{\zeta_2 - \zeta_1}{\zeta_3 - \zeta_1}\right|.$$

Then $\tilde{K}(C) < +\infty$ if and only if $K(C) < +\infty$ and $\tilde{K}(C)$ and $K(C)$ can be estimated one in terms of the other.

We are now ready to proof Theorem A. Let $X \subset T(\Gamma)$ be Teichmüller bounded. Constants depending only on X and Γ will be denoted by capital A's with subscripts. Thus

$$t_{T(\Gamma)}(\varphi, 0) \leqq A_1 \quad \text{all } \varphi \in X.$$

For each $\varphi \in X$ we can hence choose a $\nu \in M(\Gamma)$ such that

$$K(w^{\nu}) \leqq A_2$$

(A_2 can be chosen as $(\exp 2A_1) + 1$). Each point $\varphi \in X$ determines a quasicircle $C^{\nu} = w^{\nu}(\mathbb{R})$ with

$$K(C^{\nu}) \leqq A_3.$$

Thus by the Ahlfors result explained above

$$\tilde{K}(C^{\nu}) \leqq A_4.$$

In particular,

$$|w_{abc}(\varphi)| \leqq A_4, \quad \text{all } \varphi \in X. \tag{4.1}$$

Hence we conclude that (4.1) holds for all $\varphi \in \hat{X}$. Thus, because (4.1) holds for all a,b,c for each $\varphi \in K$, there is a $\nu \in M$ such that $\varphi = \varphi^{\nu}$

$$K(w^{\nu}) \leqq A_5.$$

If we knew that $\nu \in M(\Gamma)$ we would be done. However, Ahlfors' result only guarantees the weaker conclusion $\nu \in M$. We also know that for each $\varphi \in \hat{X}$ there is a $\tilde{\nu} \in M(\Gamma)$ such that $\varphi = \varphi^{\tilde{\nu}}$. Assume that the corresponding set

$$\{K(w^{\tilde{\nu}});\ \tilde{\nu} \in M(\Gamma) \text{ of mininal norm, } \tilde{\nu} \text{ corresponding to some } \varphi \in \hat{X}\}$$

is not bounded. Then there exists a sequence $\tilde{\nu}_n$ such that

$$\lim_n K(w^{\tilde{\nu}_n}) = +\infty.$$

We can now choose $\nu_n \in M$ such that

$$K(w^{\nu_n}) \leqq A_5 \quad \text{and} \quad \varphi^{\nu_n} = \varphi^{\tilde{\nu}_n}.$$

Using the compactness properties of quasiconformal mappings, we can not extract a convergent subsequence (assumed to be the entire sequence without loss of generality) of the sequence $\{\nu_n\}$ such that

$$w^{\nu_n} \to w^{\nu}, \text{ uniformly on compact subsets of } \mathbb{C}, \tag{4.2}$$

for some $\nu \in M$. Since $K(w^{-1}) = K(w)$, we may also assume that

$$(w^{\nu_n})^{-1} \to (w^{\nu})^{-1}, \text{ uniformly on compact subsets of } \mathbb{C}. \tag{4.3}$$

Further, we know that

$$w^{\nu_n} \circ \gamma \circ (w^{\nu_n})^{-1},\ \gamma \in \Gamma,$$

is a Möbius transformation. Hence we conclude that $\varphi^{\nu} \in T \cap B_2(U^*)$. (We have, as usual, supressed the trivial group from our notation.) By Kra [10], this means that there is a $\tilde{\nu} \in M(\Gamma)$ such that $\varphi^{\tilde{\nu}} = \varphi^{\nu}$.

Now dim $T(\Gamma) < \infty$. Hence (4.2) shows that

$$\varphi^{\nu_n} \to \varphi^{\tilde{\nu}} \quad \text{in} \quad T(\Gamma) \hookrightarrow B_2(U^*,\Gamma).$$

Bers [3] has constructed a local inverse τ of $\Phi : M \to T$ at every point $\varphi_o \in T$. This local inverse has the property that: $\varphi \in T(\Gamma)$ implies that $\tau\varphi \in M(\Gamma)$. Thus for n sufficiently large, we can choose a $\tilde{\nu}_n \in M(\Gamma)$ such that $K(w^{\tilde{\nu}_n})$ is arbitrarily close to $K(w^{\tilde{\nu}})$. This contradiction establishes the theorem.

Proof of Theorem B. Assume that there exists a sequence $\varphi_n \in \hat{X}$ such that

$$\lim_{n\to\infty} \varphi_n = \varphi \in \partial T(\Gamma).$$

We shall need the following

Lemma 4. Let $\varphi_n \in T(\Gamma)$ with $\lim_{n\to\infty} \varphi_n = \varphi \in \partial T(\Gamma)$. There exists a subsequence $\{\varphi_{n_k}\}$ of $\{\varphi_n\}$ and a sequence $\{f_k\} \subset \mathcal{F}$ such that

$$\lim_{k\to\infty} |f_k(\varphi_{n_k})| = \infty. \tag{4.4}$$

Assume for the moment that the lemma has been established. By passing to a subsequence we can take $n_k = k$ and assume the existence of functions satisfying (4.4). We know that we have a uniform bound for functions in $\mathcal{F}$ on X, and hence also on $\hat{X}$. But the limit (4.4) shows that for sufficiently large n, $\varphi_n \notin \hat{X}$.

Proof of Lemma 4. Assume that $\mathcal{F}$ is bounded on the sequence $\{\varphi_n\}$. Write $\varphi_n = \varphi^{\nu_n}$ with $\nu_n \in M$ and $\|\nu_n\| \leqq k < 1$ (possible by Ahlfors' argument). (We, of course, also have $\varphi_n = \varphi^{\tilde{\nu}_n}$, $\tilde{\nu}_n \in M(\Gamma)$.) Now, by the usual compactness properties, by passing to a subsequence we may assume that (4.2) holds. But then

$$\varphi_n = \varphi^{\nu_n} \to \varphi^{\nu}, \text{ uniformly on compact subsets of } U^*. \tag{4.5}$$

Since we also know that $\varphi_n \to \varphi$ in $B_2(U^*,\Gamma)$, we conclude that the convergence in (4.5) is $B_2(U^*,\Gamma)$-convergence and that $\varphi^{\nu} = \varphi$. In particular φ belongs to the component of the origin of $T \cap B_2(U^*,\Gamma) = T(\Gamma)$. This contradiction establishes the lemma.

Remark. A lemma very similar to the above was used by Krushkal [12].

References

[1] Ahlfors, L. V.: Quasiconformal reflections. Acta math. 109 (1963), 291 - 301.

[2] Ahlfors, L. V.: Lectures on quasiconformal mappings. Van Nostrand, Princeton N. J. (1966).

[3] Bers, L.: A non-standard integral equation with applications to quasiconformal mappings. Acta math. 116 (1966), 113 - 134.

[4] Bers, L., Ehrenpreis, L.: Holomorphic convexity of Teichmüller spaces. Bull. Amer. math. Soc. 70 (1964), 761 - 764.

[5] Beurling, A., Ahlfors, L. V.: The boundary correspondence under quasiconformal mappings. Acta math. 96 (1956), 125 - 142.

[6] Earle, C. J.: On the Carathéodory metric in Teichmüller spaces, in "Discontinuous groups and Riemann surfaces". Princeton University Press, Princeton, N. J. (1974), 99 - 103.

[7] Earle, C. J., Eells, J.: On the differential geometry of Teichmüller spaces. J. Analyse math. 19 (1967), 35 - 52.

[8] Gardiner, F.: An analysis of the group operation in universal Teichmüller space. Trans. Amer. math. Soc. 132 (1968), 471 - 486.

[9] Kobayashi, S.: Hyperbolic manifolds and holomorphic mappings. M. Dekker, New York (1970).

[10] Kra, I.: On Teichmüller spaces for finitely generated Fuchsian groups. Amer. J. Math. 91 (1969), 67 - 74.

[11] Kra, I.: On Teichmüller's theorem on the quasi-invariance of the cross ratios. Israel J. Math. 30 (1978), 152 - 158.

[12] Krushkal, S. L.: Two theorems of Teichmüller spaces. Soviet Math., Doklady 17 (1976), 704 - 707. Translation of Doklady Akad. Nauk SSSR Tom 228 (1976), No. 2.

[13] Reed, T. J.: Quasiconformal mappings with given boundary values. Duke math. J. 33 (1966), 459 - 464.

State University of New York
Stony Brook, NY 11794
U.S.A.

ON HOLOMORPHIC CONTINUABILITY OF QUASIREGULAR FUNCTIONS

Julian Ławrynowicz*

Introduction

Almost nothing is known about holomorphic continuability of plane quasiregular functions into the space $\mathbb{C}^2$ of two complex variables [10].

A natural approach is to consider the function in question as a solution of the Beltrami differential equation and to express this function as the superposition of a plane quasiconformal mapping and a nonconstant holomorphic function (Section 1). Then, in view of the theorem on existence and uniqueness of quasiconformal mappings with a given complex dilatation we may assume that the function is defined in a standard domain. Taking a rectangle as such a domain we can extend our function effectively by the natural reflections to the whole plane $\mathbb{R}^2$ as a 2-periodic function preserving the quasiregularity and its order.

If the complex dilatation is real-analytic, so is the quasiregular function, and by a theorem of Wiener its Fourier series can be decomposed into four or even three addends holomorphically continuable by the same formulae to some domains in $\mathbb{C}^2$ (Section 2). The formulae are effective, but in the general case almost nothing can be asserted at the moment on the intersection of the domains in question.

The procedure is in principle valid also for higher dimensions m (Section 2). The first difference is that the Beltrami equation has to be replaced by an overdetermined system of nonlinear homogeneous elliptic partial differential equations (Section 1). Therefore we are unable to conclude the real-analyticity of the quasiregular mapping in question from the real-analyticity of the corresponding positive definite real symmetric matrix functions which play the role of a complex dilatation. The second difference, strictly connected with the first one, is that we cannot, in general, modify the initial mapping to an n-periodic mapping of $\mathbb{R}^m$. In particular, there exists no quasiregular m-periodic mapping in $\mathbb{R}^m$.

Another natural approach is to prove that, given a plane quasiregular function defined on $\mathbb{R}^2$ (or, more generally, in an arbitrary domain

---

* During the preparation of this paper the author was supported by an Italian C.N.R. fellowship.

$D \subset \mathbb{R}^2$) whose complex dilatation κ is real-analytic, for any $x_o \in \mathbb{R}^2$ (or $x_o \in D$) there exists a positive number ε depending only on κ, such that the mapping in question is holomorphically continuable to $\{z \in \mathbb{C}^2 : ||z - z_o|| < \varepsilon\}$ (Section 3). Hence it follows that _if κ is a real-analytic plane complex-valued function defined in D with $||\kappa||_\infty < 1$, then there exists a neighbourhood of D in $\mathbb{C}^2$, onto which any solution of the Beltrami equation corresponding to κ is holomorphically continuable_. This result seems to be the most interesting. In order to prove it we construct a suitable Fréchet space of all solutions of the equation and apply a theorem of Banach [3] (p. 38, Theorem 3).

The method gives no effective information about the domain of holomorphy. The theorem can be reformulated for the plane positive difinite real symmetric matrix functions that describe the infinitesimal ellipsoids corresponding to κ, but the method fails in higher dimensions because the system of equations considered is, in general, nonlinear.

Finally, as a consequence of results obtained by Lelong [14], Kiselman [10], and Jarnicki [9] we describe effectively the open sets of holomorphy for plane quasiregular functions with complex dilatation constant on concentric annuli centred at a fixed point (Section 4). Under the additional assumptions that the functions in question are one-to-one and transform the corresponding family of concentric circles onto another family of concentric circles, we give formulae for the biholomorphic continuation and derive the precise open sets of biholomorphy. These classes of functions are strictly related to the classes studied in [11].

In this place the author should like to thank his Colleagues Profs. A. Andreotti (Pisa) and C. O. Kiselman (Uppsala) for helpful discussions.

1. Preliminaries

Let D be a domain in the finite plane $\mathbb{R}^2$. A continuous function $u : D \to \mathbb{R}^2$ is said to be _quasiregular_ if:

1° for any rectangle D_o contained with its closure in D, with sides parallel to the coordinate axes it is absolutely continuous on almost all line segments in D_o, parallel to the coordinate axes (in other words: parallel to either side of D_o),

2° there exists a constant Q, $1 \leqq Q < +\infty$, such that

$$\max_\alpha |\partial_\alpha u(x)|^2 \leq Q|J_u(x)| \quad \text{for almost every} \quad x \in D, \tag{1.1}$$

where $\partial_\alpha u$ denotes the directional derivative of u in the direction α and $|J_u|$ is the Jacobian of u.

When indicating the <u>order</u> Q we speak about Q-<u>quasiregularity</u>. Without essential loss of generality we may suppose that the determinant $J_u(x)$ is nonnegative at almost every $x \in D$. We have [18]:

<u>Remark 1.1.</u> Under the above assumption, if u is non-constant, it is discrete, open, and sense-preserving.

<u>Remark 1.2.</u> In [13] such a function is called Q-<u>quasiconformal</u>. The present author follows [15] and [7], but in his lectures the term Q-<u>quasiholomorphic</u> is used.

The mapping $\underline{D}u(x) : T_x\mathbb{D} \to T_{u(x)}u[\mathbb{D}]$, the formal derivative of u (where $T_x\mathbb{D}$ is the tangent space of $\mathbb{D}$ at x, $\mathbb{D}$ being the domain D considered as a Riemannian manifold with the metric inherited from the euclidean metric of $\mathbb{R}^2$) transforms for almost every $x \in D$ the unit circle onto an ellipse centred at the origin, characterized by the following properties:

(a) the ratio of the lengths of the major and minor of its semiaxes is equal to $(1 + |\kappa_u(x)|)/(1 - |\kappa_u(x)|)$,

(b) the angle between the major semiaxis in the positive direction and the u^1-axis is equal to $\frac{1}{2}\arg\kappa_u(x) + \arg(\partial/\partial x_c)u_c(x)$, where $\|\kappa_u\|_\infty \leqq (Q - 1)/(Q + 1)$ and

$$\frac{\partial}{\partial \overline{x_c}} u_c(x) = \kappa_u(x) \frac{\partial}{\partial x_c} u_c(x) \quad \text{for almost every} \quad x \in D \qquad (1.2)$$

with $x_c = x^1 + ix^2$, $u_c = u^1 + iu^2$, $\frac{\partial}{\partial x_c} = \frac{1}{2}(\frac{\partial}{\partial x^1} - i\frac{\partial}{\partial x^2})$, $\frac{\partial}{\partial \overline{x_c}} = \frac{1}{2}(\frac{\partial}{\partial x^1} + i\frac{\partial}{\partial x^2})$.

The equation (1.2) due to Beltrami [4] is called the <u>Beltrami differential equation</u> and the function κ_u - the <u>complex dilatation</u> of u. A homeomorphic Q-quasiregular function is called a Q-<u>quasiconformal mapping</u>. We mention the following basic result (cf. e.g. [13], pp. 239 - 248):

<u>Proposition 1.1.</u> Any non-constant plane Q-quasiregular function u (with $J_u \geqq 0$ almost everywhere) can be expressed as the composition of a plane Q-quasiconformal mapping v and a non-constant holomorphic function $f : u = f \circ v$.

In view of the theorem on existence and uniqueness of quasiconformal mappings with a given complex dilatation (cf. e.g. [13], pp. 200 - 205) without any loss of generality we may consider them in standard domains. In the case of simply connected domains we may take as such a domain the unit disc or a rectangle $\{x \in \mathbb{R}^2 : 0 < x^q < \frac{1}{2}p^q,\ q = 1,2\}$. In the case of doubly connected domains we may take as these domains the annuli $\{x \in \mathbb{R}^2 : r < |x| < R\}$ with a radial cut and with help of the logarithm map them onto some rectangles. Finally any Q-quasiconformal mapping in the unit disc or in a rectangle in question can be homeomorphically extended onto the closure

$$\Delta = \{x \in \mathbb{R}^2 : |x| \leq 1\} \text{ resp. } E_{\frac{1}{2}p^1, \frac{1}{2}p^2} = \{x \in \mathbb{R}^2 : 0 \leq x^q \leq \tfrac{1}{2}p^q,\ q = 1,2\},$$

and it is natural to call the extended mapping also Q-quasiconformal. By Proposition 1.1 the same concerns Q-quasiregular functions.

Any plane Q-quasiregular function defined in $E_{\frac{1}{2}p^1, \frac{1}{2}p^2}$ can obviously be extended by reflection in the lines $\{x \in \mathbb{R}^2 : x^1 = \frac{1}{2}k_1 p^1\}$, $k_1 = 0, 1, -1, 2, -2, \ldots$, to the uniquely determined Q-quasiregular function $\tilde{u}$ defined in $\{x \in \mathbb{R}^2 : 0 \leq x^2 \leq \frac{1}{2}p^2\}$ which is 1-periodic with a period p^1. If there are no smaller periods, p^1 is called the primitive period of $\tilde{u}$ and $\operatorname{int} E_{p^1, \frac{1}{2}p^2}$ - its fundamental domain, where $\operatorname{int} E$ denotes the interior of the set E. Similarly $\tilde{u}$ can obviously be extended by reflection in the lines $\{x \in \mathbb{R}^2 : x^2 = \frac{1}{2}k_2 p^2\}$, $k_2 = 0, 1, -1, 2, -2, \ldots$, to the uniquely determined Q-quasiregular function u^* defined on $\mathbb{R}^2$ which is 2-periodic with periods p^1 and $p^2 i$. If there are no smaller periods, p^1 and $p^2 i$ are called the primitive periods of u^* and $\operatorname{int} E_{p^1, p^2}$ - its fundamental domain.

More generally, let D be a domain in $\mathbb{R}^m$, $m \geq 2$, considered as a Riemannian manifold $\mathbb{D}$ with the metric inherited from the euclidean metric of $\mathbb{R}^m$. A continuous function $u : D \to \mathbb{R}^m$ is said to be a quasi-regular mapping if:

1° for any m-dimensional parallelepiped D_o contained with its closure in D, with sides parallel to the coordinate axes it is absolutely continuous on almost all line segments in D_o, parallel to the coordinate axes,

2° the partial derivatives of u are locally L^m-integrable,

3° there exists a constant Q, $1 \leq Q < +\infty$, such that

$$(1/Q)\,\|\underline{D}u(x)\|^m \leq |J_u(x)| \leq Q[a(x)]^m \quad \text{for almost every } x \in D, \tag{1.3}$$

where $\|\underline{D}u(x)\| = \sup|\underline{D}u(x)h|$ with the supremum taken over $|h| = 1$, $\underline{D}u$ denotes the formal derivative of u and $|J_u|$ the Jacobian, and $a(x)$ is the smallest axis of the ellipsoid onto which the unit sphere in $T_x\mathbb{D}$, the tangent space of $\mathbb{D}$ at x, is transformed by $\underline{D}u(x)$.

Remark 1.3. By [5] and Proposition 1.1 in the case $m = 2$ the condition 2° follows from 1° and 3°, whereas the second inequality in (1.3) is a consequence of the first one and of the relation $|J_u(x)|$

$|J_u(x)| = \max_\alpha |\partial_\alpha u(x)| \min_\alpha |\partial_\alpha u(x)|$.

Suppose that the semiaxes of an ellipsoid $\mathcal{E}$ in $T_x\mathbb{D}$ centred at x are $\lambda_1(x)a(x),\ldots,\lambda_{m-1}(x)a(x),a(x)$, where $\lambda_1(x) \geqq \cdots \geqq \lambda_{m-1}(x) \geqq 1$ and consider the uniparametric family of similar ellipsoids with the semiaxes $\lambda_1(x)t,\ldots,\lambda_{m-1}(x)t,t$ which may be described by the differential equation (let $\underline{D}u(x)[\mathcal{E}]$ be a given ellipsoid):

$$\langle G(x,u)\xi,\xi\rangle = t^2[\lambda_1(x)\cdots\lambda_{m-1}(x)]^{2/m} \quad \text{for} \quad \xi \in T_x\mathbb{D}, \tag{1.4}$$

where $\langle\ ,\ \rangle$ denotes the scalar product in $\mathbb{D}$ and G is a positive definite real symmetric matrix, satisfying the condition $\det G(x,u) = 1$. The family of ellipsoids given by (1.4) will be called the infinitesimal ellipsoid corresponding to $G(x,u)$. If $\lambda_1(x) = \cdots = \lambda_{m-1}(x) = 1$, the family in question is called an infinitesimal sphere. In the case $n = 2$ we speak about infinitesimal ellipses and infinitesimal circles, respectively.

The condition that the derivative $\underline{D}u(x) : T_x\mathbb{D} \to T_{u(x)}u[\mathbb{D}]$ maps the infinitesimal ellipsoid corresponding to $G(x,u)$ onto an infinitesimal ellipsoid corresponding to $H(u,x)$ is expressed by the equation

$$\langle H(u,x)Du(x)\xi,\ Du(x)\xi\rangle = |J_u(x)|^{2/m}\langle G(x,u)\xi,\xi\rangle \quad \text{for} \quad \xi \in T_x\mathbb{D} \tag{1.5}$$

for almost every $x \in D$ or, in a more concise notation,

$$[Du(x)]^*H(u,x)Du(x) = |J_u(x)|^{2/m}G(x,u) \quad \text{for almost every} \quad x \in D. \tag{1.6}$$

Remark 1.4. The system of equations (1.5) is an m-dimensional analogue of the Beltrami differential equation (1.2).

For the convenience of a reader we give the calculations required for Remark 1.4.

Let $\eta = Du(x)\xi \in T_{u(x)}u[\mathbb{D}]$, $m = 2$, and $(\eta^1)^2 + (\eta^2)^2 = 1$. Then

$$(u^1_{|1}\xi^1 + u^1_{|2}\xi^2)^2 + (u^2_{|1}\xi^1 + u^2_{|2}\xi^2)^2 = 1, \text{ where } u^k_{|j} = (\partial/\partial x_j)u^k, \tag{1.7}$$

whenever x is a point of differentiability of u. Hence

$$\begin{aligned}[(u^1_{|1})^2 + (u^2_{|1})^2](\xi^1)^2 + 2(u^1_{|1}u^1_{|2} + u^2_{|1}u^2_{|2})\xi^1\xi^2 \\ + [(u^1_{|2})^2 + (u^2_{|2})^2](\xi^2)^2 = 1.\end{aligned} \tag{1.8}$$

Let us set $\kappa_u = -\frac{\lambda - 1}{\lambda + 1}e^{2i\theta}$. Then, by (1.2), we have

$$u^1_{|1} \equiv \frac{1}{2}\left(\frac{\partial}{\partial x_c}u_c + \frac{\partial}{\partial \overline{x_c}}u_c + \frac{\partial}{\partial x_c}\overline{u_c} + \frac{\partial}{\partial \overline{x_c}}\overline{u_c}\right)$$

$$= \frac{1}{2}(1 - \frac{\lambda - 1}{\lambda + 1} e^{2i\theta}) \frac{\partial}{\partial x_c} u_c + \frac{1}{2}(1 - \frac{\lambda - 1}{\lambda + 1} e^{-2i\theta}) \frac{\partial}{\partial \overline{x_c}} \overline{u_c},$$

$$u^2_{|1} \equiv \frac{1}{2i}\left(\frac{\partial}{\partial x_c} u_c + \frac{\partial}{\partial \overline{x_c}} u_c - \frac{\partial}{\partial x_c} \overline{u_c} - \frac{\partial}{\partial \overline{x_c}} \overline{u_c}\right)$$

$$= \frac{1}{2i}(1 - \frac{\lambda - 1}{\lambda + 1} e^{2i\theta}) \frac{\partial}{\partial x_c} u_c - \frac{1}{2i}(1 - \frac{\lambda - 1}{\lambda + 1} e^{-2i\theta}) \frac{\partial}{\partial \overline{x_c}} \overline{u_c}.$$

Consequently,

$$(u^1_{|1})^2 + (u^2_{|1})^2 = |1 - \frac{\lambda - 1}{\lambda + 1} e^{2i\theta}|^2 |\frac{\partial}{\partial x_c} u_c|^2 = \frac{|(\lambda+1)e^{-i\theta} - (\lambda-1)e^{i\theta}|^2}{(\lambda + 1)^2} |\frac{\partial}{\partial x_c} u_c|^2$$

$$= (\cos^2\theta + \lambda^2 \sin^2\theta) \frac{4}{(\lambda + 1)^2} |\frac{\partial}{\partial x_c} u_c|^2$$

$$= (\lambda \sin^2\theta + \frac{1}{\lambda}\cos^2\theta) \frac{(\lambda + 1)^2 - (\lambda - 1)^2}{(\lambda + 1)^2} |\frac{\partial}{\partial x_c} u_c|^2 = (\lambda \sin^2\theta + \frac{1}{\lambda}\cos^2\theta) J_u.$$

Let us set

$$G_{11} J_u \equiv (u^1_{|1})^2 + (u^2_{|1})^2 = (\lambda \sin^2\theta + \frac{1}{\lambda}\cos^2\theta) J_u.$$

Similarly, we have

$$G_{12} J_u \equiv G_{21} J_u \equiv u^1_{|1} u^1_{|2} + u^2_{|1} u^2_{|2} = -(\lambda - \frac{1}{\lambda}) \cos\theta \sin\theta \, J_u,$$

$$G_{22} J_u \equiv (u^1_{|2})^2 + (u^2_{|2})^2 = (\lambda \cos^2\theta + \frac{1}{\lambda}\sin^2\theta) J_u,$$

so that (1.8) becomes

$$\lambda(\xi^1 \sin\theta - \xi^2 \cos\theta)^2 J_u + \frac{1}{\lambda}(\xi^1 \cos 0 + \xi^2 \sin\theta) J_u = 1.$$

This means that $Du(x)$ <u>transforms our ellipse</u> whose ratio of the lengths of the major and minor of its semiaxes is $\lambda(x)$ and the angle between the major semiaxis in the positive direction and the ξ^1-axis is $\theta + \frac{1}{2}\pi$ <u>onto the unit circle</u> whenever x is a point of differentiability of u and $J_u(x) \neq 0$. Replacing in the above consideration the unit circle by the uniparametric family of circles described by the equation $(\eta^1)^2 + (\eta^2)^2 = \lambda J_u t^2$, we arrive at (1.4) with $m = 2$, where $\det G(x,u) \equiv \det[G_{jk}] = 1$ indeed.

Let us observe now that the relation (1.2) with $\kappa_u = -\frac{\lambda - 1}{\lambda + 1} e^{2i\theta}$ is equivalent to

$$\frac{\partial}{\partial \overline{x_c}} u_c(x) = -\kappa^*_u(x) \overline{\frac{\partial}{\partial x_c} u_c(x)} \quad \text{for almost every } x \in D \tag{1.9}$$

with $\kappa^*_u = -\frac{\lambda^* - 1}{\lambda^* + 1} e^{2i\theta}$, where $\lambda^* = \lambda$ and $\theta^* = \theta + \arg \frac{\partial}{\partial x_c} u_c$.

Arguing as before, we check that the equation $(\xi^1)^2 + (\xi^2)^2 = 1$ for $\xi \in T_x \mathbb{D}$ leads to

$$\lambda^*(\eta^1\sin\theta^* - \eta^2\cos\theta^*)^2 + \frac{1}{\lambda^*}(\eta^1\cos\theta^* + \eta^2\sin\theta^*)^2 = J_u$$

whenever x is a point of differentiability of u. This means that $\underset{\sim}{D}u(x)$ transforms the unit circle onto an ellipse whose ratio of the lengths of the major and minor of its semiaxes is $\lambda^*(x)$ and the angle between the major semiaxis in the positive direction and the η^1-axis is θ^* whenever x is a point of differentiability of u and $J_u(x) \neq 0$.

The corresponding equations (1.5) are of the form

$$(\eta^1)^2 + (\eta^2)^2 = J_u[G_{11}(\xi^1)^2 + 2G_{12}\xi^1\xi^2 + G_{22}(\xi^2)^2] \text{ with } \det[G_{jk}] = 1 \quad (1.10)$$

in the case of (1.2) and

$$H_{11}(\eta^1)^2 + 2H_{12}\eta^1\eta^2 + H_{22}(\eta^2)^2 = J_u[(\xi^1)^2 + (\xi^2)^2] \text{ with } \det[H_{jk}] = 1 \quad (1.11)$$

in the case of (1.9). In the case of a differential equation

$$\frac{\partial}{\partial \overline{x_c}}u_c = -\frac{\lambda^*(\lambda^2-1)}{(\lambda\lambda^*+1)(\lambda+\lambda^*)}e^{2i\theta}\frac{\partial}{\partial x_c}u_c + \frac{\lambda(\lambda^{*2}-1)}{(\lambda\lambda^*+1)(\lambda+\lambda^*)}e^{2i\theta^*}\overline{\frac{\partial}{\partial x_c}u_c}$$

$$\text{for almost every } x \in D \quad (1.12)$$

which, as shown in [6], is an unessential generalization of (1.2), and - on the other hand - can be written in the form of a system

$$\begin{aligned} G_{22}u^1_{|1} + (G_{12} + H_{12})u^1_{|2} &= G_{11}u^2_{|1} \\ (G_{12} - H_{12})u^1_{|1} + G_{11}u^1_{|2} &= -G_{11}u^2_{|2} \end{aligned} \quad \text{almost everywhere in } D$$

with $\det[G_{jk}] = \det[H_{jk}] = 1$, the corresponding equations (1.5) are

$$H_{11}(\eta^1)^2 + 2H_{12}\eta^1\eta^2 + H_{22}(\eta^2)^2 = J_u[G_{11}(\xi^1)^2 + 2G_{12}\xi^1\xi^2 + G_{22}(\xi^2)^2] \quad (1.13)$$

and any ellipse characterized by (the so-called <u>characteristics</u>) λ, $\theta + \frac{1}{2}\pi$ is transformed by $\underset{\sim}{D}u(x)$ onto an ellipse characterized by λ^*, θ^* whenever x is a point of differentiability of x and $J_u(x) \neq 0$.

The system (1.5) for $m = 2,3,\ldots$ is a system of nonlinear homogeneous partial differential equations for the vector function $u = (u^1,\ldots,u^m)$. In view of the condition $\det G(x,u) = 1$ the number of independent equations is $\frac{1}{2}m(m+1) - 1$. Hence for $m > 2$ the system (1.5) is overdetermined. For $m = 2$ the system reduces to a system linear in $\underset{\sim}{D}u$. A detailed study of the system (1.5) is given in [1] and [7].

Without essential loss of generality we may suppose that the determinant $J_u(x)$ of the quasiregular mapping u in question is nonnegative at almost every $x \in D$. We have [18]:

Remark 1.5. Under the above assumption, if u is non-constant, it is discrete, open, and sense-preserving.

An open, discrete, and sense-preserving mapping $u : \mathbb{R}^m \to \mathbb{R}^m$ is periodic with a period $p \neq 0$, if $u(x + p) = u(x)$ for all $x \in \mathbb{R}^m$. The discreteness of u implies that the set of all periods of u is spanned by n independent periods $p^q e_q$, $p^q > 0$, $e_q \in \mathbb{R}^m$, $|e_q| = 1$, $q = 1,\dots,n$, $1 \leqq n \leqq m$, i.e. that the function u is n-periodic. The vectors $p^1 e_1,\dots,p^n e_n$ are called the primitive periods of u. In the case $m = 2$ we often identify e_1 with 1 and e_2 with i.

Let Γ be the group generated by the n translations $x \mapsto x + p^q e_q$, $q = 1,\dots,n$, and let pr denote the projection of $\mathbb{R}^m$ onto the linear space spanned by $p^1 e_1,\dots,p^n e_n$. Then the set

$$\operatorname{int} E_{p^1,\dots,p^n}, \quad \text{where} \quad E_{p^1,\dots,p^n} = \{x \in \mathbb{R}^m : pr\, x = \sum_{q=1}^{n} \xi^q p^q e_q,\ 0 \leqq \xi^q \leqq 1\},$$

is called the fundamental domain of u. Thereafter we always assume that

$$e_q^k = 1 \quad \text{for} \quad k = q \quad \text{and} \quad e_q^k = 0 \quad \text{for} \quad k \neq q,\ k = 1,\dots,m,$$

and in the case $m = 2$ we identify e_1 with 1 and e_2 with i.

Concerning quasiregular n-periodic mappings in $\mathbb{R}^m$, where $1 \leqq n \leqq m$, including examples and effictive constructions, we refer to [15]. Here we only quote the following

Proposition 1.2. There exists no quasiregular m-periodic mapping in $\mathbb{R}^m$, $m > 2$.

2. Remarks on decomposition of quasiregular n-periodic mappings in $\mathbb{R}^m$ into functions holomorphically continuable to $\mathbb{C}^n \times \mathbb{R}^{m-n}$

Let us consider first plane quasiregular functions. We need the following extension of Lemma V.4.1 in [13], p. 208:

Lemma 2.1. Any plane quasiregular function $D \ni x \mapsto u(x)$, whose complex dilatation κ_u coincides almost everywhere with a function real-analytic in x^1 and x^2, is itself real-analytic.

Proof. It is well known ([17], [8], p. 178 or [16], p. 271) that if $P(x,\underline{D})$ is a linear elliptic differential operator in the domain D in question whose coefficients are real-analytic and u is the generalized solution of $P(x,\underline{D})u = f$, where f is also real-analytic in D, then u is real-analytic in D as well. Therefore, by the definition of the ellipticity of $P(x,\underline{D})$, it is sufficient to verify that $\frac{1}{2}\xi_c - \frac{1}{2}\kappa_u(x)\overline{\xi_c} \neq 0$

for $\xi \in \mathbb{R}^2 \setminus \{0\}$. This, however, is an immediate consequence of the assumption that $\|\kappa_u\|_\infty < 1$.

Let us consider then a plane quasiregular 2-periodic function $u = (u^1, u^2)$ of $\mathbb{R}^2$ into $\mathbb{R}^2$, whose complex dilatation κ_u is real-analytic, with primitive periods p^1 and p^2. Since $|u|^2$ is integrable over the fundamental domain, by Bessel's inequality it can be expanded into the Fourier series:

$$u_c(x) = \sum_{k_1,k_2=-\infty}^{+\infty} a_{k_1,k_2} \exp 2\pi i \left(\frac{k_1}{p^1}x^1 + \frac{k_2}{p^2}x^2\right), \quad x = (x^1,x^2) \in \mathbb{R}^2, \tag{2.1}$$

with $u_c = u^1 + iu^2$ and

$$a_{k_1,k_2} = \frac{1}{4\pi^2} \int_0^{p^2}\int_0^{p^1} u_c(t)\exp\{-2\pi i(\frac{k_1}{p^1}t^1 + \frac{k_2}{p^2}t^2)\}dt^1dt^2,$$

absolutely and uniformly convergent on compact subsets of $\mathbb{R}^2$. We have

Lemma 2.2. The function (2.1) can be expressed as the sum of four functions: $u_c = u_c^1 + u_c^2 + u_c^3 + u_c^4$, given by the formulae

$$u_c^j(x) = -\frac{3}{4}a_{0,0}$$

$$+ \sum_{k_1,k_2=0}^{+\infty} a_{(-1)^j k_1,(-1)^{[j/2]}k_2} \exp 2\pi i\left\{(-1)^j \frac{k_1}{p^1}x^1 + (-1)^{[j/2]}\frac{k_2}{p^2}x^2\right\},$$

$x \in \mathbb{R}^2$,

$[s]$ denoting the greatest integer not exceeding s, which admit the unique holomorphic continuations

$$U_c^j(x,y) = -\frac{3}{4}a_{0,0}$$

$$+ \sum_{k_1,k_2=0}^{+\infty} a_{(-1)^j k_1,(-1)^{[j/2]}k_2} \exp 2\pi i\left\{(-1)^j \frac{k_1}{p^1}z^1 + (-1)^{[j/2]}\frac{k_2}{p^2}z^2\right\}$$

with $z^q = x^q + iy^q$ onto

$$D^j \supset \mathbb{R}^2 \times \{y^1 \in \mathbb{R} : (-1)^j y^1 \geqq 0\} \times \{y^2 \in \mathbb{R} : (-1)^{[j/2]} y^2 \geqq 0\}. \tag{2.2}$$

Proof. The lemma is an immediate consequence of a theorem of Wiener (cf. e.g. [20], vol. I, p. 245) according to which if to every point $z_0 = (x_0, y_o)$ of D^j there corresponds a neighbourhood $E^j_{z_o}$, $z_o \in E^j_{z_o} \subset D^j$, and a function $E^j_{z_o} \ni z \to U_c^j(z;z_0)$ such that the Fourier series of $U_c(\ ;z_0)$ converges absolutely and $U_c^j(\ ;z_0) = U_c^j \mid E^j_{z_o}$, then the Fourier series of U_c^j converges absolutely in D^j.

It is clear that the number of addends in Lemma 2.2 can be decreased. Namely, let us divide the plane into three disjoint sectors Δ_j, $j = 1,2,3$, with vertex at the origin, according to some three fixed straight lines (passing through the origin) so that within the unit disc Δ the area of each sector is the same; e.g. let

$$\Delta_j = \{k_1 + ik_2 \in \mathbb{C} : -\frac{5}{3}\pi + \frac{2}{3}j\pi < \arg(k_1 + ik_2) \leqq -\pi + \frac{2}{3}j\pi\}.$$

Then from Lemma 2.2 we obtain

Proposition 2.1. The function (2.1) can be expressed as the sum of three functions: $u_c = \tilde{u}_c^1 + \tilde{u}_c^2 + \tilde{u}_c^3$, given by the formulae

$$\tilde{u}_c^j(x) = -\frac{2}{3}a_{o,o} + \sum_{k_1+ik_2\in\Delta_j} a_{k_1,k_2} \exp 2\pi i\,(\frac{k_1}{p^1}x^1 + \frac{k_2}{p^2}x^2),\ x \in \mathbb{R}^2,$$

which admit the unique biholomorphic continuations

$$\tilde{U}_c^j(x,y) = -\frac{2}{3}a_{o,o} + \sum_{k_1+ik_2\in\Delta_j} a_{k_1,k_2} \exp 2\pi i\,(\frac{k_1}{p^1}z^1 + \frac{k_2}{p^2}z^2)$$

with $z^q = x^q + iy^q$ onto (2.2).

Next let us consider a plane quasiregular 1-periodic function $u = (u^1,u^2)$ of the strip $\{x \in \mathbb{R}^2 : 0 < x^2 < p^2\}$, whose complex dilatation κ_u is real-analytic, with primitive period p^1. It can again be expanded into the Fourier series:

$$u_c(x) = \sum_{k_1=-\infty}^{+\infty} a_{k_1}(x^2)\exp 2\pi i\frac{k_1}{p^1}x^1,\ x = (x^1,x^2) \in \mathbb{R}\times(0;p^2), \tag{2.3}$$

with $u^c = u^1 + iu^2$ and

$$a_{k_1}(x^2) = \frac{1}{2\pi}\int_0^{p^1} u_c(t^1,x^2)\exp(-2\pi i\frac{k_1}{p^1}t^1)dt^1$$

absolutely and uniformly convergent on compact subsets of $\mathbb{R}^1$. In analogy to Lemma 2.2 and Proposition 2.1 we have

Lemma 2.3. The function (2.3) can be expressed as the sum of two functions: $u_c = u_c^1 + u_c^2$, given by the formulae

$$u_c^j(x) = -\frac{1}{2}a_{o,o} \sum_{k_1=0}^{+\infty} a_{(-1)^j k_1}(x^2)\exp 2\pi i\frac{k_1}{p^1}x^1,\ x \in \mathbb{R}\times(0;p^2),$$

which for every $x^2 \in (0;p^2)$ admit the unique holomorphic continuations

$$U_c^j(x,y) = -\frac{1}{2}a_{o,o} + \sum_{k_1=0}^{+\infty} a_{(-1)^j k_1}(x^2)\exp 2\pi i\frac{k_1}{p^1}z^1$$

with $z^1 = x^1 + iy^i$ onto

$$D^j \supset \mathbb{R} \times \{y^1 \in \mathbb{R} : (-1)^j y^1 \geq 0\}. \tag{2.4}$$

Let us consider now m-dimensional quasiregular mappings. Since in this case the corresponding differential operator $P(x,\underline{D})$ is, in general, no longer linear, we have to assume that the mappings in question are real-analytic.

Let $u = (u^1,\ldots,u^m)$ be such a mapping. We suppose that it is n-periodic in the infinite m-dimensional parallelepiped $\{x \in \mathbb{R}^m : 0 < x^q < p^q, q = n+1,\ldots,m\}$, $1 \leq n \leq m-1$, and with primitive periods $p^1 e_1,\ldots, p^n e_n$. It can again be expanded into the Fourier series

$$u_c(x) = \sum_{k_1,\ldots,k_n=-\infty}^{+\infty} a_{k_1,\ldots,k_n}(x^{n+1},\ldots,x^m)\exp 2\pi i \sum_{q=1}^{n} \frac{k_q}{p^q} x^q,$$

$$x = (x^1,\ldots,x^m) \in \mathbb{R}^m, \tag{2.5}$$

with

$$u_c = (u^1 + iu^{1+m/2},\ldots,u^{m/2} + iu^m) \quad \text{for } m \text{ even},$$

$$u_c = (u^1 + iu^{(m+1)/2},\ldots,u^{(m-1)/2} + iu^{m-1},u^m) \quad \text{for } m \text{ odd},$$

and

$$a_{k_1,\ldots,k_n}(x^{n+1},\ldots,x^m)$$

$$= \frac{1}{(4\pi)^{2n}} \int_0^{p^n}\cdots\int_0^{p^1} u_c(t^1,\ldots,t^n,x^{n+1},\ldots,x^m)\exp(-2\pi i \sum_{q=1}^{n} \frac{k_q}{p^q} t^q)\,dt^1\cdots dt^n$$

absolutely and uniformly convergent on compact subsets of $\mathbb{R}^n$. In analogy to Lemmas 2.2 and 2.3 we have

<u>Lemma 2.4.</u> The mapping (2.5) can be expressed as the sum of 2^n mappings $u_c = u_c^1 + \cdots + u_c^q$, $q = 2^n$, given by the formulae

$$u_c^j(x) = -\frac{2^n-1}{2^n} a_{o,o} + \sum_{k_1,\ldots,k_n=0}^{+\infty} a_{(-1)^{[2^{-0}j]}k_1,\ldots,(-1)^{[2^{1-n}j]}k_n}(x^{n+1},$$

$$\ldots,x^m)\exp 2\pi i \sum_{q=1}^{n} (-1)^{[2^{1-q}j]} \frac{k_q}{p^q} x^q, \quad x \in \mathbb{R}^n \times \mathop{\mathsf{X}}_{q=n+1}^{m} (0;p^q),$$

$[s]$ denoting the greatest integer not exceeding s, which for every $(x^{n+1},\ldots,x^m) \in (0;p^{n+1}) \times \cdots \times (0;p^m)$ admit the unique holomorphic continuations

$$U_c^j(x,y) = -\frac{2^n-1}{2^n} a_{o,o} + \sum_{k_1,\ldots,k_n=0}^{+\infty} a_{(-1)^{[2^{-0}j]}k_1,\ldots,(-1)^{[2^{1-n}j]}k_n}(x^{n+1},$$

$$\ldots,x^m)\exp 2\pi i \sum_{q=1}^{n} (-1)^{[2^{1-q}j]}\frac{k_q}{p^q} z^q$$

with $z^q = x^q + iy^q$ onto

$$D^j \supset \mathbb{R}^n \times \mathop{X}_{q=1}^{n} \{y^q \in \mathbb{R} : (-1)^{[2^{1-q}j]} y^q \geqq 0\}. \tag{2.6}$$

It is clear that the number of addends in Lemma 2.4 can be decreased. Namely, let us divide the space $\mathbb{R}^n$ into $n + 1$ disjoint sectors Δ_j^n, $j = 1,\ldots,n + 1$, with vertex at the origin, according to some $\frac{1}{2}n(n + 1)$ fixed $(n - 1)$-dimensional hyperplanes (passing through the origin) so that within the unit ball (centred at the origin) the area of each sector is the same. Then from Lemma 2.4 we obtain

Proposition 2.2. The mapping (2.5) can be expressed as the sum of $n + 1$ mappings: $u_c = \tilde{u}_c^1 + \cdots + \tilde{u}_c^{n+1}$, given by the formulae

$$\tilde{u}_c^j(x) = -\frac{n}{n+1}a_{o,o} + \sum_{(k_1,\ldots,k_n)\in\Delta_j^n} a_{k_1,\ldots,k_n}(x^{n+1},\ldots,x^m)$$

$$\times \exp 2\pi i \sum_{q=1}^{n} \frac{k_q}{p^q}x^q,\ x \in \mathbb{R}^n \times \mathop{X}_{q=n+1}^{m} (0;p^q),$$

which for every $(x^{n+1},\ldots,x^m) \in (0;p^{n+1}) \times \cdots \times (0;p^m)$ admit the unique holomorphic continuations

$$\tilde{U}_c^j(x,y) = -\frac{n}{n+1}a_{o,o}$$

$$+ \sum_{(k_1,\ldots,k_n)\in\Delta_j^n} a_{k_1,\ldots,k_n}(x^{n+1},\ldots,x^m)\exp 2\pi i \sum_{q=1}^{n} \frac{k_q}{p^q}z^q$$

with $z^q = x^q + iy^q$ onto (2.6).

Better estimates of the sets D^j appearing in (2.2), (2.4), and (2.6) follow in particular cases from the results of Sections 3 and 4.

3. Holomorphic continuability of the plane quasiregular functions to a neighbourhood of $\mathbb{R}^2$ in $\mathbb{C}^2$

Let us consider plane quasiregular functions. By Section 1 the most important case is when the functions in question are defined on $\mathbb{R}^2$. We need the following

Lemma 3.1. Let u be a quasiregular function defined on $\mathbb{R}^2$, whose complex dilatation κ_u is real-analytic. Then, for any $x_o \in \mathbb{R}^2$ there exists a positive number ε depending only on κ_u, such that u is holomorphically continuable onto the ball

$$\{z \in \mathbb{C}^2 : ||z - x_0|| < \varepsilon\}. \tag{3.1}$$

Proof. By Lemma 2.1 u is real-analytic. Let A_n denote the set of all generalized solutions $\tilde{u}$ of the Beltrami equation (1.2), where $D = \mathbb{R}^2$, with the following property:

$\tilde{u}$ is holomorphically continuable onto $\{z \in \mathbb{C}^2 : ||z - x_0|| < 1/n\}$.

It is clear that A_n is contained in $H(\mathbb{R}^2)$, the space of all solutions of (1.2) in $D = \mathbb{R}^2$. Moreover

$$H(\mathbb{R}^2) = \bigcup_{n=1}^{+\infty} A_n.$$

Now, on $H(\mathbb{R}^2)$ we can consider the seminorms

$$||\tilde{u}||_K = \sup_K |\tilde{u}| \quad \text{on compact sets} \quad K \subset \mathbb{R}^2.$$

In such a way we obtain the Fréchet topology on $H(\mathbb{R}^2)$.

Let us observe next that A_n as a subspace of $H(\mathbb{R}^2)$ is linear. Moreover, for some n_o the set A_{n_o} is of the second category because the space is Fréchet. Remark that also A_{n_o} is a Fréchet space with the seminorm

$$||\tilde{u}||_K = \sup_K |\tilde{u}| \quad \text{on compact sets} \quad K \subset \mathbb{R}^2 \cup \{z \in C^2 : ||z - x_0|| < 1/n_o\}.$$

Finally we observe that the inclusion mapping

$$r : A_{n_o} \to H(\mathbb{R}^2)$$

is linear continuous and that the image of r equals A_{n_o} and is of the second category. Therefore the conclusion needed is a consequence of the following theorem of Banach ([3], p. 38, Theorem 3; cf. also [19], p. 75 and [2], p. 44): Suppose that A and H are complete metric topological vector spaces, $r : A \to H$ is a continuous linear mapping, and the image $r[A]$ is a set of the second category in H. Then r is surjective and open, and $r[A] = H$.

From Lemmas 2.1 and 3.1 we immediately obtain

Proposition 3.1. Let u be a quasiregular function defined on $\mathbb{R}^2$, whose complex dilatation κ_u is real-analytic. Then there exists a neighbourhood of $\mathbb{R}^2$ in $\mathbb{C}^2$, depending only on κ_u, onto which u is holomorphically continuable.

It is clear that Proposition 3.1 describes a property of the Beltrami equation (1.2) and not of its particular solution u, so it may be formulated as follows:

Theorem 3.1. If κ is a real-analytic complex-valued function defined on $\mathbb{R}^2$ with $\|\kappa\|_\infty < 1$, then there exists a neighbourhood of $\mathbb{R}^2$ in $\mathbb{C}^2$, onto which any generalized solution of

$$\frac{\partial}{\partial \bar{x}_c}u_c = \kappa\frac{\partial}{\partial x_c}u_c \quad \text{almost everywhere in } \mathbb{R}^2 \tag{3.2}$$

is holomorphically continuable.

An analogous remark applies to Lemma 3.1.

By Remark 1.4 the equation (3.2) in Theorem 3.1 may be replaced by the system of equations (1.5) with suitable G and H specified there, so that we arrive at the following version of Proposition 3.1:

Theorem 3.2. If G and H are real-analytic positive definite real symmetric 2×2-matrices satisfying the conditions $\det[G_{jk}] = \det[H_{jk}] = 1$ on $\mathbb{R}^2$, then there exists a neighbourhood of $\mathbb{R}^2$ in $\mathbb{C}^2$ onto which any generalized solution of (1.5) is holomorphically continuable.

An analogous remark applies to Lemma 3.1. Of course (1.5) is considered only almost everywhere.

Remark 3.1. It is easily seen that in all results of this section we may replace $\mathbb{R}^2$ by an arbitrary domain $D \subset \mathbb{R}^2$.

In the case of m-dimensional quasiregular mappings with $m > 2$, the system (1.5) is, in general, nonlinear, so our argument for establishing Lemma 3.1 clearly fails.

4. Quasiregular functions in the plane with complex dilatation constant on concentric annuli centred at the origin

As a particular case of plane quasiregular functions we are going to study in detail a quasiregular function $u : \Delta \to \mathbb{R}^2$ whose complex dilatation κ is constant almost everywhere in concentric annuli centred at the origin, e.g.

$$\kappa(x) = a_k \quad \text{almost everywhere for} \quad r_k < |x| < r_{k-1},\ k - 1,2,\ldots,$$

where $r_o = 1$ (cf. e.g. [12], Sections 19, 24, and 27). Let $x = (x^1,x^2) \in \mathbb{R}^2$, $a \in \mathbb{C}$, and

$$A(r,R) = \{x_c = x^1 + ix^2 \in \mathbb{C} : r < |x_c| < R\} \quad \text{for} \quad 0 \leq r < R \leq +\infty,$$

$$B_a(r,R) = \{z = (x^1 + iy^1,\ x^2 + iy^2) \in \mathbb{C}^2,$$

where

$$r < \{|x_c + a\overline{x_c}|^2 + |y_c + a\overline{y_c}|^2 - 2|\operatorname{im}[(x_c + a\overline{x_c})(\overline{y_c} + \bar{a}y_c)]|\}^{1/2},$$

$$\{|x_c + a\overline{x_c}|^2 + |y_c + a\overline{y_c}|^2 + 2|\operatorname{im}[(x_c + a\overline{x_c})(\overline{y_c} + \bar{a}y_c)]|\}^{1/2} < R\},$$

$$B_\kappa = \bigcup_{k=1}^{+\infty} B_{a_k}(r_k, r_{k-1}), \quad \operatorname{cl} \bigcup_{k=1}^{+\infty} A(r_k, r_{k-1}) = \Delta_c,$$

where $\operatorname{cl} E$ denotes the closure of the set E. In addition to the results of Sections 2 and 3 we have

Theorem 4.1. A. There exists a function U holomorphic in B_κ such that $U(x) = u(x)$ for $r_k < |x| < r_{k-1}$, $k = 1,2,\ldots$

B. There exists a function $u^* : \Delta \to \mathbb{R}^2$ whose complex dilatation κ^* satisfies

$$\kappa^*(x) = a_k \quad \text{for} \quad r_k < |x| < r_{k-1}, \; k = 1,2,\ldots,$$

such that u^* cannot be continued holomorphically to any larger open set $D \supset B_\kappa$.

Proof. The problem can be considered separately on each of the annuli $A(r_k, r_{k-1})$ for $k = 1,2,\ldots$, since they are disjoint as well as the Lie rings $B_a(R_k, R_{k-1})$ with $a = a_k$ for $k = 1,2,\ldots$, provided these rings are a_k-quasiconformal envelopes of holomorphy for $A(r_k, r_{k-1})$, respectively. This requirement is equivalent to the requirement for a Lie ring $B_1(r,R)$, $0 \leqq r < R \leqq +\infty$, to be the harmonic envelope of holomorphy for the annulus $A(r,R)$, and this is ensured by [9].

Suppose now that the function u in question is one-to-one and transforms the annuli $A(r_k, r_{k-1})$ onto $A(R_k, R_{k-1})$, $k = 1,2,\ldots$, respectively, where $R_o = 1$ and $\operatorname{cl}[A(R_1,R_o) \cup A(R_2,R_1) \cup \cdots] = \Delta_c$. In other words we assume that f is a plane quasiconformal selfmapping of Δ which transforms the concentric circles $\{x \in \mathbb{R}^2 : |x| = r_k\}$ onto $\{x \in \mathbb{R}^2 : |x| = R_k\}$, $r_{k+1} < r_k$, $k = 0,1,\ldots$, with $R_o = r_o = 1$, and within the k-th annulus formed in such a way it has a constant complex dilatation a_k almost everywhere. Let further $L_{a_k}(r_k, r_{k-1})$ consist of all $(x,y) \in \mathbb{R}^4$, satisfying

$$z_c + a_k\overline{z}_c \in A(r_k, r_{k-1}) \quad \text{and} \quad \overline{z}_c + a_k\overline{\overline{z}_c} \in A(r_k, r_{k-1}), \; k = 1,2,\ldots, \tag{4.1}$$

where $z_c = z^1 + iz^2$ and, obviously, $\overline{z}_c = \overline{z}^1 + i\overline{z}^2$, $\overline{\overline{z}_c} = \overline{z}^1 - i\overline{z}^2$. Then Theorem 4.1 may be completed as follows:

Theorem 4.2. Under the above assumptions u admits the unique biholomorphic continuation

$$f(x,y) = (U(x,y),\ V(x,y));\ U(x,y) \in \Delta,\ V(x,y) \in \Delta, \tag{4.2}$$

onto $L_\kappa = \bigcup_{k=1}^{+\infty} L_{a_k}(r_k, r_{k-1})$, where

$$U_c(x,y) = \tfrac{1}{2}[u_c([z_c + a_k\bar{z}_c]_{\bar{c}}) + \overline{u_c([\bar{z}_c + a_k\overline{z_c}]_{\bar{c}})}] \quad \text{whenever } x_c \in A(r_k, r_{k-1}), \tag{4.3}$$

$$V_c(x,y) = -\tfrac{1}{2}i[u_c([z_c + a_k\bar{z}_c]_{\bar{c}}) - \overline{u_c([\bar{z}_c + a_k\overline{z_c}]_{\bar{c}})}] \quad \text{whenever } x_c \in A(r_k, r_{k-1}), \tag{4.4}$$

and $s_{\bar{c}} = (\text{re } s, \text{im } s)$ for $s \in C$. Moreover, $f[L_\kappa] = L'_\kappa \equiv \bigcup_{k=1}^{+\infty} L_{a_k}(r_k, r_{k-1})$.

Proof. The proof is essentially contained in [9]. Namely, it is clear that the relations (4.3) and (4.4) together with (4.1) and $s_{\bar{c}} = (\text{re } s,\ \text{im } s)$ for $s \in C$ determine the unique biholomorphic continuation (4.2) of u. With the notation $(U + iV)_{\bar{c}} \equiv (U_c, V_c)$, by (4.1) and $s_{\bar{c}} = (\text{re } s,\ \text{im } s)$ for $s \in C$, we have

$$(U + iV)_{c\bar{c}} = (U + iV)^{1}_{\bar{c}} + i(U + iV)^{2}_{\bar{c}} = U_c + iV_c.$$

This, together with (4.2) and (4.3), yields $(U + iV)_{c\bar{c}}(x,y) = u_c(z_{c\bar{c}})$. Similarly,

$$[\overline{(U + iV)_{\bar{c}}}]_c(x,y) = \overline{U_c(x,y)} + i\overline{V_c(x,y)} = u_c(\bar{z}_{c\bar{c}}).$$

Hence, if $(x,y) \in L_\kappa$, we have not only (4.1), but also similar relations

$$[(U + iV)_c]_c(x,y) \in A(R_k, R_{k-1}) \quad \text{and}$$

$$[\overline{(U + iV)_{\bar{c}}}]_c(x,y) \in A(R_k, R_{k-1}),\ k = 1,2,\ldots,$$

so $f(z) \in L'_\kappa$, i.e. $f[L_\kappa] \in L'_\kappa$.

On the other hand, if $(\xi,\eta) \in L'_\kappa$, there exist points λ and μ in L_κ such that

$$\zeta_c = u_c(\lambda) \quad \text{and} \quad \bar{\zeta}_c = u_c(\mu), \text{ where } \zeta = \xi + i\eta.$$

Hence

$$\xi = U_c((\lambda + i\mu)^{c\bar{c}}),\ \eta = V_c((\lambda + i\mu)^{c\bar{c}}),$$

where $(\lambda,\mu) \to (\lambda + i\mu)^{c\bar{c}}$ is the inverse mapping of $(x,y) \to (x + iy)_{c\bar{c}}$, namely,

$$(\lambda + i\mu)^{c\bar{c}} = (\tfrac{1}{2}(\mu + \lambda)_{\bar{c}},\ \tfrac{1}{2}(i\mu - i\lambda)_{\bar{c}}).$$

Therefore $f^{-1}(\zeta) \in L_K$, i.e. $f[L_K] \supset L'_K$.

Finally we construct for the mapping u^{-1} the corresponding holomorphic continuation $\varphi : L'_K \to L_K$ and observe that both

$$f \circ \varphi | \bigcup_{k=1}^{+\infty} \Lambda(R_k, R_{k-1}) \quad \text{and} \quad \varphi \circ f | \bigcup_{k=1}^{+\infty} \Lambda(r_k, r_{k-1})$$

are identity mappings, so $\varphi = f^{-1}$, as desired.

References

[1] Ahlfors, L. V.: Conditions for quasiconformal deformations in several variables, in "Contributions to analysis". Academic Press, New York and London (1974), 19 - 25.

[2] Andreotti, A., Tomassini, G.: Spazi vettoriali topologici. Pitagora Editrice, Bologna (1978).

[3] Banach, S.: Théorie des opérations linéaires. Z subwencji Funduszu Kultury Narodowej, Warszawa (1932).

[4] Beltrami, E.: Delle variabili complesse sopra una superficie qualunque. Ann. Mat. Pura appl., II. Ser. 1 (1867/8), 329 - 336.

[5] Bers, L.: On a theorem of Mori and the definition of quasiconformality. Trans. Amer. math. Soc. 84 (1957), 78 - 84.

[6] Bojarski, B.: Generalized solutions of a system of differential equations of the first order and elliptic type with discontinuous coefficients (Russian). Math. Sbornik n. Ser. 43 (1957), 451 - 503.

[7] Bojarski, B., Iwaniec, T.: Topics in quasiconformal theory in several variables, in "Proceedings of the first Finnish-Polish summer school in complex analysis at Podlesice, Part II". Uniwersytet Łódzki, Łódź (1978), 21 - 44.

[8] Hörmander, L.: Linear partial differential operators. Springer-Verlag, Berlin - Göttingen - Heidelberg (1963).

[9] Jarnicki, M.: Analytic continuation of harmonic functions. Zeszyty nauk. Uniw. Jagielloński., Prace mat. 17 (1975), 93 - 104.

[10] Kiselman, C.O.: Prolongement des solutions d'une équation aux dérivées partielles à coefficients constants. Bull. Soc. math. France 97 (1969), 329 - 356.

[11] Ławrynowicz, J.: On the class of quasi-conformal mappings with invariant boundary points I-II. Ann. Polon. math. 21 (1969), 309 - 347.

[12] Ławrynowicz, J., Krzyż, J.: The parametrical method for quasiconformal mappings in the plane. Lecture notes, to appear.

[13] Lehto, O., Virtanen, K. I.: Quasiconformal mappings in the plane. Springer-Verlag, Berlin - Heidelberg - New York (1973).

[14] Lelong, P.: Prolongement analytique et singularités complexes des fonctions harmoniques. Bull. Soc. math. Belgique 7 (1954), 10 - 23.

[15] Martio, O., Srebro, U.: Periodic quasimeromorphic mappings in R^n. J. Analyse math. 28 (1975), 20 - 40.

[16] Morrey, C. B., Jr.: Multiple integrals in the calculus of variations. Springer-Verlag, Berlin - Heidelberg - New York (1966).

[17] Petrovski, I. G.: Sur l'analyticité des solutions des systèmes d'équations différentielles. Mat. Sbornik n. Ser. 5 (1939), 3 - 68.

[18] Rešetnjak, Ju. G.: A condition for boundedness of the index for mappings with bounded distortion (Russian). Sibir. mat. Žurn. 9 (1968), 368 - 374.

[19] Yosida, K.: Functional analysis. Springer-Verlag, Berlin - Heidelberg - New York (1968).

[20] Zygmund, A.: Trigonometric series I-II. Cambridge University Press, Cambridge (1959).

Institute of Mathematics
Polish Academy of Sciences
The Łódź Branch
ul. Kilińskiego 86
PL-90-012 Łódź
Poland

DIFFERENTIAL EQUATIONS ASSOCIATED WITH HARMONIC SPACES

Fumi-Yuki Maeda

Introduction. One of the main aims of axiomatic potential theory is to give potential theoretic methods of treating an extended class of elliptic and some parabolic second order partial differential equations. Thus, as soon as the theory of harmonic spaces has been established as an axiomatic potential theory, the following question naturally arises: given a structure of harmonic space on a Euclidean domain, can we associate to it a second order partial differential operator L so that harmonic functions are exactly the solutions of $Lu = 0$?

This type of question had been already treated by J.L. Tautz [7], [8] long before the theory of harmonic spaces took its shape by the works of M. Brelot. Later J.-M. Bony [1] gave the following result, which may be considered as a generalization of the result in [8] (though Bony made no reference to it).

Theorem of J.-M. Bony. Let Ω be an open set in the euclidean space R^k $(k \geq 1)$ and suppose a harmonic sheaf $\mathcal{H}$ is given on Ω. Assume that regular domains (with respect to $\mathcal{H}$) form a base of the topology of Ω. If all harmonic functions are C^2 (or, if there are sufficiently many C^2-harmonic functions; cf. [2]), then there is an open dense subset Ω_o of Ω and there are continuous functions a_{ij}, b_j, c on Ω_o $(i,j = 1,\ldots,k)$ such that (a_{ij}) is symmetric positive semi-definite everywhere on Ω_o and, for any open set $U \subset \Omega_o$, a C^2-function u is superharmonic on U if and only if

$$\sum a_{ij} \frac{\partial^2 u}{\partial x_i \partial x_j} + \sum b_j \frac{\partial u}{\partial x_j} + cu \leq 0$$

on U.

As is shown by an example in [1], Ω_o cannot be taken to be the whole space Ω, in general, in order that the coefficients a_{ij}, b_j, c are continuous on Ω_o or have reasonable regularity on Ω_o.

On the other hand, the author attacked the problem from entirely different direction in [6]. Let us recall here a few definitions in [6]. Let Ω be a harmonic space in the sense of Constantinescu-Cornea [3]. We denote by $\mathcal{R}$ the sheaf of real functions on Ω which are locally expressed as differences of continuous superharmonic functions. Let $\mathcal{M}$ be the sheaf of signed (Radon) measures on Ω. A sheaf homomorphism $\sigma : \mathcal{R} \to \mathcal{M}$ is called a measure representation if for any open set U in Ω and for any $f \in \mathcal{R}(U)$, $\sigma(f) \geq 0$ on U if and only if f is superharmonic on U. Now our result can be stated as follows (cf. [6; Theorem 4]):

Theorem A. Let Ω be an open set in R^k and suppose a structure of harmonic space in the sense of Constantinescu-Cornea is given on Ω. Assume that the constant function 1 and the coordinate functions x_j, $j = 1,\dots,k$, all belong to $\mathcal{R}(\Omega)$, and that a measure representation σ is given. Then, $C^2(U) \subset \mathcal{R}(U)$ for any open set $U \subset \Omega$ and there are signed measures α_{ij}, β_j, γ on Ω $(i,j = 1,\dots,k)$ such that (α_{ij}) is symmetric positive semi-definite and for any open set $U \subset \Omega$, every $f \in C^2(U)$ satisfies the equality

$$\sum \frac{\partial^2 f}{\partial x_i \partial x_j} \alpha_{ij} + \sum \frac{\partial f}{\partial x_j} \beta_j + f\gamma = -\sigma(f)$$

on U. In fact, α_{ij}, β_j, γ are given by

$$\begin{cases} \alpha_{ij} = \frac{1}{2}\{x_i\sigma(x_j) + x_j\sigma(x_i) - \sigma(x_ix_j) - x_ix_j\sigma(1)\} \\ \beta_j = -\sigma(x_j) + x_j\sigma(1) \\ \gamma = -\sigma(1). \end{cases} \tag{1}$$

Here, for a continuous function g and a signed measure μ, $g\mu$ denotes the signed measure given by $g\mu(\varphi) = \mu(g\varphi)$ for $\varphi \in C_0$ (the subscript 0 means "support compact").

In this theorem, coefficients α_{ij}, β_j, γ are defined on the whole space Ω, though they are no longer functions. The assumption on the existence of a measure representation may look very strong, but we can show its existence for a wide class of harmonic spaces (see [6]) and moreover we need something like this in order to obtain the uniqueness of the associated differential operator. In this connection, our Theorem A can be regarded as a generalization of the results in [7].

However, assumptions in Theorem A are entirely different from those in the theorem of Bony. In fact, we can easily construct examples of harmonic spaces satisfying the assumptions of Theorem A but not those in Bony's theorem, and vice versa. Thus, what we can do is simply to combine these theorems. This is our first objective of the present paper (§1).

If we restrict ourselves to self-adjoint harmonic spaces, then we can make more detailed discussions. A harmonic space is said to be self-adjoint, if it is a Brelot harmonic space with proportionality condition and if there exists a consistent system $\{G_U(x,y)\}_{U\,:\,P\text{-set}}$ of <u>symmetric</u> Green functions (for more details, see [4]). Given such a system $\{G_U(x,y)\}$, we can associate a measure representation σ in such a way that, for any open set $U \subset \Omega$ and for any $f \in \mathcal{R}(U)$, $f \mid V = \int_V G_V(\cdot,y)d\sigma(f)(y) + u_f^V$ with $u_f^V \in \mathcal{H}(V)$ for every relatively compact P-set V whose closure is contained in another P-set $\subset U$ (see [4; 2.1] or [6; §6]). This σ will be called a <u>canonical</u> measure representation.

Let α_{ij}, β_j, γ be defined as in (1) by a canonical measure representation. We remarked ([5; Remark 2]) that $u \in C^1(U) \cap \mathcal{H}(U)$ is a "weak" solution of the differential equation of the divergent form

$$\sum \frac{\partial}{\partial x_i}\left(\alpha_{ij}\frac{\partial u}{\partial x_j}\right) + \gamma u = 0$$

on U. Our second objective of this paper is to make this remark more precise and complete (§2).

<u>§1. Combination of Bony's theorem and Theorem A</u>

In this section we prove the following theorem:

<u>Theorem 1.</u> Let Ω be an open set in R^k and suppose a structure of harmonic space in the sense of Constantinescu-Cornea is given on Ω. Suppose furthermore:

(a) Regular domains form a base of the topology of Ω;

(b) There are sufficiently many C^2-harmonic functions (in the sense of Bony [2]);

(c) $1, x_1,\ldots,x_k \in \mathcal{R}(\Omega)$;

(d) There exists a measure representation σ.

Then there are an open dense subset Ω_o of Ω, a non-negative measure ν on Ω_o and continuous functions a_{ij}, b_j and c on Ω_o such that

$$\alpha_{ij} = a_{ij}\nu, \quad \beta_j = b_j\nu, \quad \gamma = c\nu$$

on Ω_o, where α_{ij}, β_j, γ are given by (1), and for any open set $U \subset \Omega_o$ and for any $f \in C^2(U)$,

$$\left\{\sum a_{ij}\frac{\partial^2 f}{\partial x_i \partial x_j} + \sum b_j \frac{\partial f}{\partial x_j} + cf\right\}\nu = -\sigma(f).$$

We need some measure theoretic preparations.

<u>Lemma 1.</u> Let Ω be a locally compact Hausdorff space, $f_1,\ldots,f_k$ be continuous functions on Ω and $\mu_1,\ldots,\mu_k$ be signed measures on Ω. If, for any open set $U \subset \Omega$, $\sum c_j f_j > 0$ on U with constants $c_1,\ldots,c_k$ implies $\sum c_j\mu_j \geq 0$ on U, then, for any open set $V \subset \Omega$,

$\sum g_j f_j > 0$ on V with continuous functions $g_1,\ldots,g_k$ implies $\sum g_j \mu_j \geq 0$ on V.

Proof. Let $g_1,\ldots,g_k$ be continuous functions on V such that $\sum g_j f_j > 0$ on V. Let $\varepsilon > 0$ be given arbitrarily. For each $x \in V$, choose a neighborhood V_x of x such that $\overline{V}_x$ is compact and contained in V. Put

$$M = \sup_{y \in V_x} \sum_j |f_j(y)| \,, \quad \delta = \inf_{y \in V_x} \sum g_j(y) f_j(y).$$

Then $M < +\infty$ and $\delta > 0$. Since g_j are continuous, there is an open neighborhood U_x of x such that $U_x \subset V_x$ and

$$|g_j(y) - g_j(x)| < \min(\varepsilon, \tfrac{\delta}{2M})$$

for all $y \in U_x$, $j = 1,\ldots,k$. Put $c_j = g_j(x)$, $j = 1,\ldots,k$. Then

$$\sum c_j f_j = \sum g_j f_j + \sum \{g_j(x) - g_j\} f_j \geq \delta - \frac{\delta}{2M} \sum |f_j| \geq \frac{\delta}{2} > 0$$

on U_x. Hence, by the assumption of the lemma, $\sum c_j \mu_j \geq 0$ on U_x, so that

$$\sum g_j \mu_j = \sum c_j \mu_j + \sum \{g_j - g_j(x)\} \mu_j \geq -\varepsilon \sum |\mu_j|$$

on U_x. Since $\bigcup_{x \in V} U_x = V$, it follows that

$$\sum g_j \mu_j \geq -\varepsilon \sum |\mu_j|$$

on V. Now the arbitrariness of $\varepsilon > 0$ implies $\sum g_j \mu_j \geq 0$ on V.

Lemma 2. Let Ω be a locally compact Hausdorff space, $f_1,\ldots,f_k$ be continuous functions on Ω such that $\sum |f_j| > 0$ on Ω and $\mu_1,\ldots,\mu_k$ be signed measures on Ω. If, for any open set $U \subset \Omega$, $\sum c_j f_j > 0$ on U with constants $c_1,\ldots,c_k$ implies $\sum c_j \mu_j \geq 0$ on U, then there is a non-negative measure ν on Ω such that $\mu_j = f_j \nu$ for all $j = 1,\ldots,k$.

Proof. Let $x \in \Omega$ be arbitrarily fixed. Without loss of generality, we may assume that $f_1(x) > 0$. Choose an open neighborhood V_x of x such that $\overline{V}_x$ is compact and $f_1 > 0$ on V_x. Put $\nu_x = \frac{1}{f_1}\mu_1$ on V_x. By our assumption, $f_1 > 0$ on V_x implies $\mu_1 \geq 0$ on V_x, so that $\nu_x \geq 0$ on V_x.

We shall show that $\mu_j = f_j \nu_x$ on V_x for all $j = 2,\ldots,k$. Let $\varepsilon > 0$ and $t \in R$ be arbitrary. Consider the functions (for fixed $j \neq 1$)

$$g_1 = \varepsilon + t\frac{f_j}{f_1} \,, \quad g_j = -t \,, \quad g_\ell = 0 \quad (\ell \neq 1, j).$$

Then

$$\sum g_\ell f_\ell = \varepsilon f_1 > 0$$

on V_x. By Lemma 1,

$$\sum g_\ell \mu_\ell \geq 0$$

on V_x. Since

$$\sum g_\ell \mu_\ell = \varepsilon\mu_1 + t\frac{f_j}{f_1}\mu_1 - t\mu_j = \varepsilon\mu_1 + t(f_j\nu_x - \mu_j),$$

we have

$$t(f_j\nu_x - \mu_j) \geq -\varepsilon\mu_1$$

on V_x. Since $\varepsilon > 0$ and $t \in R$ are arbitrary, it follows that $\mu_j = f_j\nu_x$ on V_x.

If $x \neq x'$ and $V_x \cap V_{x'} \neq \emptyset$, then we easily see that $\nu_x \mid V_x \cap V_{x'} = \nu_{x'} \mid V_x \cap V_{x'}$. Hence there is a non-negative measure ν on Ω such that $\nu \mid V_x = \nu_x$ for each $x \in \Omega$, and this ν is the required measure.

Proof of Theorem 1. By Bony's theorem, there are an open dense subset Ω_0 of Ω and continuous functions a_{ij}, b_j, c $(i,j = 1,\dots,k)$ on Ω_0 such that $a_{ij} = a_{ji}$ and, for any open set $U \subset \Omega_0$, a C^2-function u is superharmonic on U if and only if

$$\sum a_{ij}\frac{\partial^2 u}{\partial x_i \partial x_j} + \sum b_j \frac{\partial u}{\partial x_j} + cu \leq 0$$

on U.

Now, let U be any open set in Ω_0 and suppose

$$\sum p_{ij}a_{ij} + \sum q_j b_j + rc > 0 \quad \text{on} \quad U \tag{2}$$

with constants p_{ij}, q_j, r $(i,j = 1,\dots,k)$. We shall show that

$$\sum p_{ij}\alpha_{ij} + \sum q_j\beta_j + r\gamma \geq 0 \quad \text{on} \quad U. \tag{3}$$

Then Lemma 2 implies the existence of a non-negative measure ν on Ω_0 such that $\alpha_{ij} = a_{ij}\nu$, $\beta_j = b_j\nu$ and $\gamma = c\nu$ on Ω_0, and our Theorem 1 follows from Theorem A.

To show (3), let $\varepsilon > 0$ be arbitrarily given. Fix an arbitrary $x^* \in U$ and consider the function

$$u(x) = \frac{1}{2}\sum p_{ij}(x_i - x_i^*)(x_j - x_j^*) + \sum q_j(x_j - x_j^*) + r$$

for $x \in U$, where $x = (x_1,\dots,x_k)$ and $x^* = (x_1^*,\dots,x_k^*)$. Then u is a C^2-function and

$$\sum a_{ij}(x^*) \frac{\partial^2 u}{\partial x_i \partial x_j}(x^*) + \sum b_j(x^*) \frac{\partial u}{\partial x_j}(x^*) + c(x^*)u(x^*)$$

$$= \sum p_{ij}a_{ij}(x^*) + \sum q_j b_j(x^*) + rc(x^*) > 0$$

by (2). By continuity, we find an open neighborhood V_{x^*} of x^* such that $V_{x^*} \subset U$, $|x - x^*| < \varepsilon$ for $x \in V_{x^*}$ and

$$\sum a_{ij} \frac{\partial^2 u}{\partial x_i \partial x_j} + \sum b_j \frac{\partial u}{\partial x_j} + cu \geq 0$$

on V_{x^*}. Then $-u$ is superharmonic on V_{x^*}, so that $\sigma(u) \leq 0$ on V_{x^*}. Hence by Theorem A,

$$\sum \frac{\partial^2 u}{\partial x_i \partial x_j} \alpha_{ij} + \sum \frac{\partial u}{\partial x_j} \beta_j + u\gamma \geq 0$$

on V_{x^*}. Computing the left hand side, we have

$$\sum p_{ij}\alpha_{ij} + \sum q_j\beta_j + r\gamma + \frac{1}{2}\sum (p_{ij} + p_{ji})(x_i - x_i^*)\beta_j$$
$$+ \{\frac{1}{2}\sum p_{ij}(x_i - x_i^*)(x_j - x_j^*) + \sum q_j(x_j - x_j^*)\}\gamma \geq 0$$

on V_{x^*}. Hence, denoting the left hand side of (3) by μ, we obtain

$$\mu \geq -\frac{\varepsilon}{2} \sum_{i,j} |p_{ij} + p_{ji}| \cdot |\beta_j| - \frac{\varepsilon^2}{2}\{\sum_{i,j} |p_{ij}|\}|\gamma| - \varepsilon\{\sum_j |q_j|\}|\gamma| \quad (4)$$

on V_{x^*}. Since both sides of (4) are independent of x^* and since $\bigcup_{x^* \in U} V_{x^*} = U$, we see that (4) holds on U. Now the arbitrariness of $\varepsilon > 0$ implies $\mu \geq 0$, i.e., (3).

§2. The case of self-adjoint harmonic spaces

What we prove in this section is the following

Theorem 2. Let Ω be an open set in R^k and suppose a structure of self-adjoint harmonic space is given on Ω. Let σ be a canonical measure representation. Assume furthermore $1, x_1, \ldots, x_k \in \mathcal{R}(\Omega)$ (or $\mathcal{D}(\Omega)$; see below). Then, for any open set $U \subset \Omega$ and for any $f \in C^1(U) \cap \mathcal{R}(U)$ and $\varphi \in C_0^1(U)$,

$$\sum \int_U \frac{\partial f}{\partial x_i} \frac{\partial \varphi}{\partial x_j} d\alpha_{ij} - \int_U f\varphi\, d\gamma = \int_U \varphi\, d\sigma(f),$$

where α_{ij} and γ are given by (1) in terms of σ. In particular, $u \in C^1(U)$ belongs to $\mathcal{H}(U)$ if and only if

$$\sum \int_U \frac{\partial u}{\partial x_i} \frac{\partial \varphi}{\partial x_j} d\alpha_{ij} - \int_U u\varphi\, d\gamma = 0$$

for all $\varphi \in C_o^1(U)$.

To prove this theorem, we use many results established in [4] and [5]. First, let us recall the definition of the space $\mathcal{D}_{BC,loc}(U)$, which we shall denote by $\mathcal{D}(U)$ for simplicity: $f \subset \mathcal{D}(U)$ if and only if there is a sequence $\{f_n\}$ in $\mathcal{R}(U)$ such that $f_n \to f$ locally uniformly on U and $\delta_{f_n-f_m}(K) \to 0$ $(n,m \to \infty)$ for each compact set $K \subset U$. For $f,g \in \mathcal{D}(U)$, δ_f and $\delta_{[f,g]}$ are defined in a natural manner, so that, in particular, $\delta_{f_n-f}(K) \to 0$ in the above situation. The following lemma, which may be regarded as Green's formula, is essential.

Lemma 3. Let U be an open set in a self-adjoint harmonic space Ω such that $1 \in \mathcal{R}(U)$. If $f \in \mathcal{R}(U)$, $g \in \mathcal{D}(U)$ and g has compact support in U, then

$$\delta_{[f,g]}(U) + \int_U fg\, d\sigma(1) = \int_U g\, d\sigma(f). \tag{5}$$

Proof. First assume that $g \in \mathcal{R}(U)$ and $\mathrm{Supp}\, g$ is compact in U. Using [3; Theorem 2.3.1] and [6; Proposition 1], we can find a finite number of PC-domains (see [4; 2.1]) V_n, $n = 1,\dots,\ell$, and functions $h_n \in \mathcal{R}(U)$, $n = 1,\dots,\ell$, such that $\overline{V}_n \subset U$, $\bigcup_{n=1}^{\ell} V_n \supset \mathrm{Supp}\, g$, $\mathrm{Supp}\, h_n \subset V_n$ and $\sum_{n=1}^{\ell} h_n = 1$ on $\mathrm{Supp}\, g$. It is easy to see that $(h_n g) \mid V_n \subset \mathcal{P}_{BC}(V_n) - \mathcal{P}_{BC}(V_n)$ in the notation of [4], and, by virtue of [4; Lemma 2.8], $f \mid V_n \in \mathcal{H}_{BE}(V_n) + \mathcal{P}_{BC}(V_n) - \mathcal{P}_{BC}(V_n)$. Hence, by [4; Propositions 2.2 and 2.3] (or also cf. [4; Theorem 6.3]),

$$\delta_{[f,h_n g]}(V_n) + \int_{V_n} fh_n g\, d\sigma(1) = \int_{V_n} h_n g\, d\sigma(f).$$

Since $\mathrm{Supp}\,(h_n g) \subset V_n$, we see that

$$\delta_{[f,h_n g]}(U) + \int_U fh_n g\, d\sigma(1) = \int_U h_n g\, d\sigma(f).$$

Summing up in n and noting that $(\sum h_n)g = g$, we obtain (5) in this case.

Next, suppose $g \in \mathcal{D}(U)$ and $\mathrm{Supp}\, g$ is compact in U. By definition, there is a sequence $\{g_n\} \subset \mathcal{R}(U)$ such that $g_n \to g$ locally uniformly on U and $\delta_{g-g_n}(K) \to 0$ $(n \to \infty)$ for each compact set K in U. As above, we can find a $\varphi \in \mathcal{R}(U)$ such that $\varphi = 1$ on $\mathrm{Supp}\, g$ and $\mathrm{Supp}\,\varphi$ is compact in U. Put $K' = \mathrm{Supp}\,\varphi$. Then $\varphi g_n \to \varphi g = g$ uniformly on K' and we see by [5; Corollary to Theorem 2] that

$$\begin{aligned}\delta_{\varphi g_n - g}(U) &= \delta_{\varphi(g_n-g)}(K')\\ &= \int_{K'} \varphi^2\, d\delta_{g_n-g} + 2\int_{K'} \varphi(g_n-g)\, d\delta_{[\varphi,g_n-g]} + \int_{K'} (g_n-g)^2\, d\delta_\varphi \to 0 \quad (n\to\infty).\end{aligned}$$

Since $\varphi g_n \in \mathcal{R}(U)$ and $\mathrm{Supp}(\varphi g_n)$ is compact in U, the result above shows that

$$\delta_{[f,\varphi g_n]}(U) + \int_U f\varphi g_n \, d\sigma(1) = \int_U \varphi g_n \, d\sigma(f)$$

for each n. Hence letting $n \to \infty$ we obtain (5).

Proof of Theorem 2. Let $f \in C^1(U) \cap \mathcal{R}(U)$ and $\varphi \in C_0^1(U)$. By [5; Theorem 5], we know that $\varphi \in \mathcal{D}(U)$. Hence by the above lemma

$$\delta_{[f,\varphi]}(U) + \int_U f\varphi \, d\sigma(1) = \int_U \varphi \, d\sigma(f).$$

By [5; Theorem 5] again,

$$\delta_{[f,\varphi]}(U) = \int_U \sum \frac{\partial f}{\partial x_i} \frac{\partial \varphi}{\partial x_j} \, d\alpha_{ij}.$$

Hence, noting that $\gamma = -\sigma(1)$, we obtain the theorem.

References

[1] Bony, J.-M.: Détermination des axiomatiques de théorie du potentiel dont les fonctions harmoniques sont différentiables. Ann. Inst. Fourier 17, 1 (1967), 353 - 382.

[2] Bony, J.-M.: Opérateurs elliptiques dégénérés associés aux axiomatiques de la théorie du potentiel, in "Potential theory", Edizioni Cremonese, Roma (1970), 69 - 119.

[3] Constantinescu, C., Cornea, A.: Potential theory on harmonic spaces. Springer-Verlag, Berlin - Heidelberg - New York (1972).

[4] Maeda, F-Y.: Dirichlet integrals of functions on a self-adjoint harmonic space. Hiroshima math. J. 4 (1974), 685 - 742.

[5] Maeda, F-Y.: Dirichlet integrals of product of functions on a self-adjoint harmonic space. Ibid. 5 (1975), 197 - 214.

[6] Maeda, F-Y.: Dirichlet integrals on general harmonic spaces. Ibid. 7 (1977), 119 - 133.

[7] Tautz, G.L.: Zum Umkehrungsproblem bei elliptischen Differentialgleichungen I. Arch. der Math. 3 (1952), 232 - 238.

[8] Tautz, G.L.: Zum Umkehrungsproblem bei elliptischen Differentialgleichungen II and Bemerkungen. Ibid. 3 (1952), 239 - 250 and 361 - 365.

Department of Mathematics, Faculty of Science
Hiroshima University
Hiroshima 730, Japan
and
Mathematisches Institut der
Universität Erlangen-Nürnberg
D-852 Erlangen
BR Deutschland

PROBLEMS IN THE THEORY OF CLOSED RIEMANN SURFACES

Henrik H. Martens

At the Scandinavian Mathematical Congress in Oslo 10 years ago I gave a survey [36] of the classical theory of Jacobian varieties and discussed some problems from the theory of closed Riemann surfaces. Today I should like to update that survey by discussing some of the progress made since then. The selection of topics is a personal one and I make no pretense at being complete.

1. The Jacobian varieties provide an effective and natural tool for the study of certain problems in the theory of closed Riemann surfaces. Their classical construction is very simple:

Let X be a closed Riemann surface of genus $g \geq 1$. Let $\omega^1,\dots,\omega^g$ be a basis for the holomorphic differentials, and let $\gamma_1,\gamma_2,\dots,\gamma_{2g}$ be a basis for the first integral homology group, on X. The associated <u>period matrix</u> $\Pi = (\pi^i_j)$ is defined by

$$\pi^i_j = \int_{\gamma_j} \omega^i$$

Its columns are $\mathbb{R}$-linearly independent and generate a group of translations Γ of $\mathbb{C}^g$. The complex torus $J(X) = \mathbb{C}^g/\Gamma$ will be referred to as a <u>Jacobian variety</u> of X.

Let P be a fixed reference point of X. For an arbitrary point $Q \in X$ the point $u(Q) \in \mathbb{C}^g$ with coordinates

$$u^i(Q) = \int_P^Q \omega^i$$

is defined modulo periods. This defines the so-called <u>canonical embedding</u> $\kappa : X \to J(X)$. It can be extended to a map of divisors by setting $\kappa(D) = \sum d_i\kappa(Q_i)$ when $D = \sum d_iQ_i$.

An easy calculation establishes the <u>universal mapping property</u>:

Given a holomorphic map $\Phi : X \to X'$ of closed Riemann surfaces, there exists a uniquely determined holomorphism $h : J(X) \to J(X')$ such that the diagram

$$\begin{array}{ccc} X & \xrightarrow{\kappa} & J(X) \\ \phi \downarrow & & \downarrow h \\ X' & \xrightarrow{\kappa'} & J(X') \end{array}$$

commutes when the canonical embeddings are chosen so that the reference points correspond.

A much deeper result is <u>Abel's theorem</u>:

Two positive divisors of the same degree on X are linearly equivalent if and only if they are mapped on the same point in $J(X)$ by κ.

2. With these results a number of questions regarding divisors and holomorphic maps of closed Riemann surfaces can be studied naturally in the context of Jacobian varieties. Here are some simple examples:

2.1. Let $\phi : X \to X'$ be a holomorphic map of X onto X'. If $D = \sum d_i Q_i$ is a divisor on X, let $\phi(D) = \sum d_i \phi(Q_i)$. Combining Abel's theorem with the mapping property we immediately see that linearly equivalent divisors on X are mapped on linearly equivalent divisors on X'. It follows [35] that if X admits a meromorphic function with polar divisor of order n, so does X'. In particular, if X is hyperelliptic then so is X' (unless its genus is < 2). In a recent paper the last observation is proved three times for three special cases.

It would be interesting to know if Abel's theorem, which gives conditions for the existence of holomorphic maps onto the Riemann sphere, has generalizations to the case of holomorphic maps onto Riemann surfaces of positive genus. Unpublished results of Accola and Landman apparently show that, in the above context, if X is 1-hyperelliptic then so is X'. The notion of p-hyperellipticity (double covering of a surface of genus p) has recently been studied by Accola [3].

2.2. Hurwitz [25] showed that an automorphism of a closed Riemann surface of genus $g \geq 2$ is completely determined by the homomorphism it induces on the first homology group. This result receives a very natural proof when examined in the light of the universal mapping property of the Jacobian variety, and admits an immediate generalization [37]:

Let $X \to X'$ and $X \to X''$ be given maps of X onto surfaces X' and X'' of the same genus. Assume that, with respect to given bases for homology, the induced homomorphisms of the first integral homology groups are the same (i.e. expressed by the same matrix). Then there is an isomorphism $X' \to X''$ inducing the natural identification of homology and commuting with the given maps. Thus the induced homomorphism of homology determines the map $X \to X'$ in the strong sense that the conformal structure of X' is determined. (Obvious modifications must be made when X' is of genus 1).

Other generalizations of Hurwitz' result are due to Accola [1] and Gilman [24]. It would be interesting to try and interpret these in the present context.

2.3. de Franchis [18] asserted, in the language of his time, that a curve can admit at most a finite number of involutions of genus ≥ 2. As I understand it, this means that only a finite number of surfaces X' of genus $g' \geq 2$, and only a finite number of surjective maps $X \to X'$, can occur for a fixed X. His argument is that the occurrence of such a map is an algebraic condition, and that the number of possible maps is at most denumerable. The last result is attributed to Humbert and Castelnuovo, and is also a consequence of (2.2.) above.

A proof of the special case of de Franchis' theorem, the finiteness of the number of maps $X \to X'$ with X and X' fixed, is given in [37]. The general result of de Franchis should be related to the unique factorization theorem for Abelian varieties modulo isogeny.

These results are capable of substantial generalization, see [8] and the survey [31].

2.4. The question of when a given homomorphism of homology groups is induced by a holomorphic map of Riemann surfaces appears to be open except for the case of isomorphisms. As is shown in [37], such a homomorphism defines a topological map of the Jacobian varieties, and an obvious condition is that the map must be holomorphic and preserve polarization. In the case of isomorphisms this is also sufficient modulo a reflection in the Jacobian variety.

3. I now turn to some questions of divisors. Recall that the dimension of a positive divisor $D = \sum d_i Q_i$ is the projective dimension of the linear space of meromorphic functions with poles at most as allowed by the d_i.

3.1. Clifford [9] showed that if D has degree $n = \sum d_i \leq 2g - 2$ and dimension r, then $2r \leq n$. This reduces to a simple statement about the impossibility of inclusion of a subvariety of $J(X)$ in one of lower dimension [36], and this seems to be Clifford's argument. A strong form of Clifford's theorem stated without reference by Severi [50] asserts that when $0 < n < 2g - 2$ equality can occur if and only if the surface is hyperelliptic.

By virtue of Abel's theorem the set of linear equivalence classes of positive divisors of given degree and dimension are represented in a 1-1 fashion by certain subsets of the Jacobian varieties. Let G_n^r denote the subset of $J(X)$ representing positive divisors of degree n and dimension $\geq r$. Clifford's theorem states that when $0 < 2r < 2g - 2$ $G_{2r}^r \neq \emptyset$ if and only if the surface is hyperelliptic. In [38] I showed that G_n^r is an analytic subset of $J(X)$ satisfying

a) $\dim(G_n^r) \leq n - 2r$

b) $\dim(G_n^r) \geq (r + 1)(n - r) - rg$, if $G_n^r \neq \emptyset$

c) $\dim(G_n^r) = n - 2r$, for $0 < r \leq n - r < g - 1$ if and only if X is hyperelliptic.

The first inequality was apparently observed by A. Mayer in his thesis. The second inequality corresponds to classical assertions going back to Riemann, see [29] for comments on the early history. Statement c) constitutes a very strong generalization of Clifford's theorem and provides a new proof of a crucial lemma for the proof of Noether's theorem about the generation of quadratic differentials, as was shown in [41].

3.2. The inequality b) above was established on the assumption $G_n^r \neq \emptyset$. The question of whether G_n^r indeed is non-empty whenever the dimensional formula gives a non-negative bound was an outstanding open problem in 1968. It had been answered affirmatively by Meis [43] only for the case $r = 1$. Kleiman and Laksov finally solved it affirmatively in full generality in [28] and [29], basing themselves on work by Kempf [27]. Gunning [19] gave an interesting proof for the case $r = 1$ in the context of Jacobian varieties.

3.3. It is natural to suppose that $\dim(G_n^r) = (r + 1)(n - r) - rg$, generically, but this is still an open, and apparently difficult question. The case $r = 1$ can be settled easily by a result of Farkas [14] and the characterization of G_n^1 given in [38]. When $2n < g + 2$, the affirmative result is already in Farkas [13] and [14], and the rest is implicit in [41]. In fact, the proof of (3.3) of that paper shows the stronger result that for generic X $G_n^1 \setminus G_n^2$ cannot have a tangent space of dimension $> 2(n - 1) - g$ at any point. An explicit but inconclusive treatment of the cases $r = 1,2,3$ was given by Lax [33], [34], and an extensive discussion with many interesting partial results is given in [30].

4. The problem of characterizing the matrices which can occur as period matrices of a closed Riemann surface is one of long standing. Riemann showed that by choosing a so-called canonical homology basis on the surface, the period matrix may always be normalized to the form $\Pi = (E,Z)$ where Z is a symmetric matrix whose imaginary part is positive definite. He also observed that there are more such matrices than Riemann surfaces to accommodate them.

4.1. For a long time the only tangible result in this area was an ill-understood formula of Schottky [49], giving a necessary condition in genus 4, and an asymptotic formula of Poincaré (see [11]). One of the most spectacular developments of the last decade in this area has been the clarification of Schottky's results by Farkas and Rauch [15],[16]. They showed the connection of Schottky's results with a general theory of so-called theta-relations associated with unramified double coverings, and found means of generating period relations in higher genera. This theory has been further extended to a revival of Wirtinger's theory of Prym varieties [54] by Mumford [44] and others, (see e.g. Accola [3], Fay [17], Beauville [7]).

The sufficiency of the Schottky conditions in genus four is an open question despite tantalizing promises in [26] and [44].

4.2. Another very exciting development in this connection was the work of Andreotti and Mayer [5] which I mentioned in my Oslo survey. They found a way of characterizing period matrices by the dimension of the singular locus of the theta-divisor on the Jacobian variety. Again the conditions are necessary, but maybe not sufficient. The relation to the Schottky conditions is discussed in [11] and [12]. In the case of genus 4 the conditions are not sufficient, and I stated in Oslo that they included at least 136 unwanted components. This is true if you look at the Siegel upper half plane. If you divide out the action of the symplectic group (as you should) Beauville [6] has shown that the unwanted components collapse to one. He shows, in fact, that the period matrix of a generic curve of genus 4 is completely characterized by the presence of a singularity on the associated theta-divisor not arising from the vanishing of an even theta-constant.

This result would undoubtedly have pleased Wirtinger who stated it in 1938 ([53], p. 419) and claimed it was an easy consequence of Riemann's vanishing theorems.

4.3. A theorem of Torelli [51] (see also [4], [52], [42], [40]) asserts that the isomorphism class of a closed Riemann surface is completely determined by any one of its canonical period matrices. (The assertion can be given a much more precise form in terms of the canonical embedding in the Jacobian variety [40].) As a consequence it is reasonable to investigate how properties of the Riemann surface are reflected in and characterized by properties of the matrix. Such questions are most often treated via the theta functions defined by the matrix.

It was, for instance, long known that hyperellipticity gives rise to vanishing theta-constants. A classical question was whether hyperellipticity could be characterized by the vanishing of $(g-2)$ even theta constants. The conjecture holds for $g \leq 4$, but Accola [2] observed that Farkas [13a] had given a counterexample in genus 5. He has also discussed variants of the conjecture [2],[3]. Necessary and sufficient conditions for hyperellipticity can be given for all genera in terms of vanishing theta-constants (including derivatives) but they are dimensionally redundant. Farkas [13a] and Accola [3] have also tried to characterize p-hyperellipticity by vanishing theta-constants.

If the matrix Z in the canonical period matrix is a direct sum, it cannot be a canonical period matrix of a Riemann surface [40]. The associated theta function then becomes a product of lower-dimensional ones, and will have vanishing even constants. Can such period matrices be characterized by vanishing properties? The answer is yes when $g = 2,3$.

4.4. The fact that the canonical period matrix cannot split into a direct sum does not mean that a closed Riemann surface cannot admit non-canonical period matrices satisfying the Riemann relations and splitting. This situation was discussed for the case $g = 2$ by Hayashida and Nishi [20], [21], [22], [23], who found conditions for splitting. Their work also gives examples of complex tori which serve as Jacobian varieties for several, conformally distinct, Riemann surfaces. These examples are interesting in that they also show that a given complex torus can carry infinitely many distinct polarizations as Jacobian variety of the same surface (see [39]). Lange [32] has shown that the Hayashida - Nishi cases do not exhaust all possibilities in genus 2, and has completed the treatment. Earle [10] has interesting results both in genus 2 and higher, to be presented at this conference.

Hayashida and Nishi showed that in their examples, a given torus never could be the Jacobian variety of more than a finite number of surfaces. Recent work by M. S. Narasimhan and Nori [46] shows that this is true in all dimensions.

Some of the developments discussed above have already found their way into books. I mention particularly Fay [17], Gunning [19], Mumford [45] and Rauch and Farkas [48].

5. In conclusion I should like to draw attention to what may well be the most amazing development of the last few years. I refer to the discovery of the relevance of the theory of Jacobian varieties and theta-functions for certain inverse spectral problems of mathematical physics. For details I refer to [55], [56], [57], [58] and literature there cited This development grew out of the remarkable numerical experiments on the Korteweg de Vries equation by Zabusky and Kruskal [59]. The connection with theta-functions has apparently been a surprise to all workers in the field except Sonya Kowalewsky.

References

[1] Accola, R. D. M.: Automorphisms of Riemann surfaces. J. Analyse math. 18 (1967), 1 - 5.

[2] Accola, R. D. M.: Some loci of Teichmüller space for genus five defined by vanishing theta nulls, in "Contributions to analysis". Academic Press, New York and London (1974), 11 - 18.

[3] Accola, R. D. M.: Riemann surfaces, theta functions, and abelian automorphism groups. Lecture Notes in Mathematics 483, Springer-Verlag, Berlin - Heidelberg - New York (1975).

[4] Andreotti, A.: On a theorem of Torelli. Amer. J. Math. 80 (1958), 801 - 828.

[5] Andreotti, A., Mayer, A.: On period relations for abelian integrals on algebraic curves. Ann. Sc. norm. super. Pisa, Cl. Sci., IV. Ser. 21 (1967), 189 - 238.

[6] Beauville, A.: Prym varieties and the Schottky problem. Inventiones math. 41 (1977), 149 - 196.

[7] Beauville, A.: Variétés de Prym et jacobiennes intermédiaires. Ann. Sci. École norm. sup., IV. Sér. 10 (1977), 309 - 391.

[8] Borel, A., Narasimhan, R.: Uniqueness conditions for certain holomorphic mappings. Inventiones math. 2 (1967), 247 - 255.

[9] Clifford, W. K.: On the classification of loci. Philos. Trans. roy. Soc. London, Ser. A Part II (1878), 663 - 681, reprinted in: Mathematical Papers, Chelsea Publishing Company, Bronx, N.Y. (1968), 305 - 329.

[10] Earle, C.: Some jacobian varieties which split. See these Proceedings.

[11] Ehrenpreis, L., Farkas, H. M.: Some refinements of the Poincaré period relation, in "Discontinuous groups and Riemann surfaces". Princeton Univ. Press, Princeton, N.J. (1974).

[12] Ehrenpreis, L., Farkas, H. M., Martens, H. H., Rauch, H. E.: On the Poincaré relation, in "Contributions to analysis". Academic Press, New York and London (1974), 125 - 132.

[13] Farkas, H. M.: Special divisors on compact Riemann surfaces. Proc. Amer. math. Soc. 19 (1968), 315 - 318.

[13a] Farkas, H. M.: Automorphisms of Riemann surfaces and the vanishing of theta constants. Bull. Amer. math. Soc. 73 (1967), 231 - 232.

[14] Farkas, H. M.: Special divisors and analytic subloci of Teichmueller space. Amer. J. Math. 88 (1966), 881 - 901.

[15] Farkas, H. M.: On the Schottky relation and its generalization to arbitrary genus. Ann. of Math., II. Ser. 92 (1970), 56 - 81.

[16] Farkas, H. M., Rauch, H. E.: Period relations of Schottky type on Riemann surfaces. Ann. of Math., II. Ser. 92 (1970), 434 - 461.

[17] Fay, J.: Theta functions on Riemann surfaces. Lecture Notes in Mathematics 352, Springer-Verlag, Berlin - Heidelberg - New York (1973).

[18] Franchis, M. De.: Un teorema sulle involuzioni irrazionali. Rend. Circ. mat. Palermo, II. Ser. 36 (1913), 368.

[19] Gunning, R. C.: Lectures on Riemann surfaces: Jacobi varieties. Princeton University Press, Princeton, N.J. (1972).

[20] Hayashida, T.: A class number associated with the product of an elliptic curve with itself. J. math. Soc. Japan 20 (1968), 26 - 43.

[21] Hayashida, T.: A class number associated with a product of two elliptic curves. Natur Sci. Rep. Ochanomizu Univ. 16 (1965), 9 - 19.

[22] Hayashida, T., Nishi, M.: Existence of curves of genus two on a product of two elliptic curves. J. math. Soc. Japan 17 (1965), 1 - 16.

[23] Hayashida, T., Nishi, M.: On certain type of jacobian varieties of dimension 2. Natur Sci. Rep. Ochanomizu Univ. 16 (1965), 49 - 57.

[24] Gilman, J.: A matrix representation for automorphisms of compact Riemann surfaces. Linear Algebra and Appl. 17 (1977), 139 - 147.

[25] Hurwitz, A.: Über algebraischen Gebilde mit eindeutiegen Transformationen in sich. Math. Ann. 41 (1893), 403 - 442.

[26] Igusa, J.-I.: Geometric and analytic methods in the theory of theta-functions, in "Algebraic geometry". Oxford Univ. Press, London (1969), 241 - 253.

[27] Kempf, G.: On the geometry of a theorem of Riemann. Ann. of Math., II. Ser. 98 (1973), 178 - 185.

[28] Kleiman, S. L., Laksov, D.: On the existence of special divisors. Amer. J. Math. 94 (1972), 431 - 436.

[29] Kleiman, S. L., Laksov, D.: Another proof of the existence of special divisors. Acta math. 132 (1974), 163 - 176.

[30] Kleiman, S. L., Laksov, D.: r-Special subchemes and an argument of Severi's. To appear.

[31] Kobayashi, S.: Intrinsic distances, measures and geometric function theory. Bull. Amer. math. Soc. 82 (1976), 357 - 416.

[32] Lange, H.: Produkte elliptischer Kurven. Nachr. Akad. Wiss. Göttinge II. math. - phys. Kl. 8 (1975), 95 - 108.

[33] Lax, R. F.: On the dimension of varieties of special divisors. Trans. Amer. math. Soc. 203 (1975), 141 - 159.

[34] Lax, R. F.: On the dimension of varieties of special divisors, II. Illinois J. Math. 19 (1975), 318 - 324.

[35] Martens, H. H.: A remark on Abel's theorem and the mapping of linear series. Commentarii math. Helvet. 52 (1977), 557 - 559.

[36] Martens, H. H.: From the classical theory of jacobian varieties. Lecture Notes in Mathematics 118, Springer-Verlag, Berlin - Heidelberg - New York (1970), 74 - 98.

[37] Martens, H. H.: Observations on morphisms of closed Riemann surfaces Bull. London math. Soc. 10 (1978), 209 - 212.

[38] Martens, H. H.: On the varieties of special divisors on a curve. J. reine angew. Math. 227 (1967), 111 - 120.

[39] Martens, H. H.: Riemann matrices with many polarizations, in "Complex analysis and its applications, Vol. III". International Atomic Energy Agency, Vienna (1976), 35 - 48.

[40] Martens, H. H.: Torelli's theorem and a generalization for hyperelliptic surfaces. Commun. pure appl. Math. 16 (1963), 97 - 109.

[41] Martens, H. H.: Varieties of special divisors on a curve, II. J. reine angew. Math. 233 (1968), 89 - 100.

[42] Matsusaka, T.: On a theorem of Torelli. Amer. J. Math. 80 (1958), 784 - 800.

[43] Meis, T.: Die minimale Blätterzahl der Konkretizierung einer kompakten Riemannsche Fläche. Schr. Math. Inst. Münster (1960).

[44] Mumford, D.: Prym varieties I, in "Contributions to analysis". Academic Press, New York and London (1974), 325 - 350.

[45] Mumford, D.: Curves and their jacobians. The University of Michigan Press, Ann Arbor, Michigan (1975).

[46] Narasimhan, M. S., Nori, M. V.: Polarizations on an abelian variety.

[47] Noether, M.: Über die invariante Darstellung algebraischer Funktionen Math. Ann. 17 (1880), 263.

[48] Rauch, H. E., Farkas, H. M.: Theta functions with applications to Riemann surfaces. Williams and Wilkins Co., Baltimore, Md. (1974).

[49] Schottky, F.: Zur Theorie der Abelschen Functionen von vier Variabeln. J. reine angew. Math. 102 (1888), 304 - 352.

[50] Severi, F.: Vorlesungen über algebraische Geometrie. B. G. Teubner, Leipzig (1921).

[51] Torelli, R.: Sulle varietá di Jacobi. Rend. Reale Acc. Lincei, 22 (1913), 98 - 103.

[52] Weil, A.: Zum Beweis des Torellischen Satzes. Nachr. Akad. Wiss. Göttingen, II. math. phys. Kl. II (1957), 33 - 53.

[53] Wirtinger, W.: Lie's Translationsmannigfaltigkeiten und Abelsche Integrale. Monatsh. Math. Phys. 46 (1938), 384 - 431.

[54] Wirtinger, W.: Untersuchungen über Thetafunctionen. B. G. Teubner, Leipzig (1895).

[55] Airault, H., McKean, H. P., Moser, J.: Rational and elliptic solutions of the Korteweg-deVries equation and a related many-body problem. Commun. pure appl. Math. 30 (1977), 95 - 148.

[56] Dubrovin, B. A., Matveev, V. B., Novikov, S. P.: Non-linear equations of Korteweg-deVries type, finite zone linear operators and abelian varieties. Russian Math. Surveys 31 (1976), 59 - 146.

[57] Kricever, I. M.: Integration of nonlinear equations by the methods of algebraic geometry (Russian). Funkcional'. Analiz Prilozenija 11 (1977), 15 - 31.

[58] McKean, H. P., van Moerbekr, P.: Sur le spectre de quelques opérateurs et les varietes de Jacobi. Lecture Notes in Mathematics 567, Springer-Verlag, Berlin - Heidelberg - New York (1977).

[59] Zabusky, N. J., Kruskal, M. D.: Interaction of "solitons" in a collisionless plasma and the recurrence of initial states. Phys. Rev. Lett. 15 (1965), 240 - 243.

Institutt for Matematikk
Norges Tekniske Høgskole
N-7034 Trondheim-NTH
Norway

CONTINUATION OF QUASICONFORMAL MAPPINGS

O. Martio

Let D be a bounded domain in Euclidean n-space R^n, $n \geq 2$, and let E be a subset of D such that $D - E$ is a domain. The set E is called quasiconformally removable if every quasiconformal mapping of $D - E$ can be extended to a quasiconformal mapping of D. A weaker removability condition is obtained by requiring that every quasiconformal mapping of $D - E$ can be extended to a continuous mapping of D.

In the case $n = 2$ the quasiconformally removable sets were characterized by Ahlfors and Beurling [1] as the O_{AD}-sets or, equivalently, as the NED-sets. That NED-sets are quasiconformally removable for all $n \geq 2$ was shown by Aseev and Syčev [2]. A slightly weaker result was obtained earlier by Väisälä [6]. In the plane continuous extension of quasiconformal mappings has also been extensively studied. There exists a characterization, due to Grötzsch [4], Renggli [5] and others, which states that every quasiconformal mapping of $D - E$ can be extended to a continuous mapping of D if and only if

$$M(\Gamma_b) = \infty$$

for each point b in E, where $M(\Gamma_b)$ denotes the conformal modulus of the family of all paths in D with non-zero winding number with respect to b.

In this note two new sufficient conditions, obtained by R. Näkki and the author, for continuous extension are presented, valid for all $n \geq 2$.

Theorem 1. Every quasiconformal mapping of $D - E$ can be extended to a continuous mapping of D if, for each point b in E, there exists a sequence of domains D_i, each containing b, such that ∂D_i lies in $D - E$, $\operatorname{dia}(D_i) \to 0$, and

$$\limsup M(\Gamma_i) < \infty ,$$

where $M(\Gamma_i)$ denotes the n-modulus of the family Γ_i of all paths joining ∂D_i to E in D_i.

Proof. Fix $b \in E$. We may assume that $M(\Gamma_i) \leq M < \infty$ for all i. Write $G_i = D_i - E$. Then G_i is a domain, thus the cluster set $C(f,b)$ of f at b is a continuum. Suppose that $C(f,b)$ is non-degenerate. By the quasiconformality of f

$$\infty > K(f)\,M \geq K(f)\,M(\Gamma_i) \geq M(f\Gamma_i) \geq M(\Gamma_i^*)$$

where Γ_i^* is the family of all paths joining $C(f,b)$ to $f\partial D_i$ in R^n. But $\cap fG_i = C(f,b)$ implies $M(\Gamma_i^*) \to \infty$ because $C(f,b)$ is non-degenerate. This contradiction proves the theorem.

Theorem 2. Every quasiconformal mapping of $D - E$ can be extended to a continuous mapping of D if

$$\int_{F_b} \frac{dr}{r} = \infty$$

for each point b in E, where F_b is the set of all numbers r on $(0,1)$ for which the Hausdorff $(n-2)$-measure of $E \cap S^{n-1}(b,r)$ is zero.

Outline of the proof. Fix a point $b \in E$. We may assume that $b = 0$, $B^n \subset D$ and $f(D - E) \subset B^n$. Observe that for a.e. $r \in F_o$, say in a set B, the map $f_r = f|S^{n-1}(r) - E$ is ACL^n (not necessarily continuous) on $S^{n-1}(r)$ when $S^{n-1}(r)$ is realized as R^{n-1} via a stereographic projection. By Sobolev's imbedding theorem there is a map F_r continuous on $S^{n-1}(r)$ with $F_r|S^{n-1}(r) - E = f_r$ for $r \in B$.

Next it is not difficult to show, using algebraic topology, that $F_r(S^{n-1}(r))$ separates R^n. Note that F_r need not be an imbedding of $S^{n-1}(r)$, however, the set where F_r fails to be injective is thin.

The rest of the proof is based on the well-known oscillation lemma, see [3, Lemma 1]. This lemma yields

$$\int_E \frac{u(r)^n}{r}\,dr \leq C \int_E \int_{S^{n-1}(r)} |\nabla f|^n\,dS\,dr \leq C\,K(f) \int_{B^n(\alpha)\cap D} J(x,f)\,dm(x)$$

$$\leq C\,K(f)\,m(B^n)\,,$$

where $E = B \cap (0,\alpha)$ and $u(r) = \operatorname{osc}|F_r|$ over $S^{n-1}(r)$. Since

$$\int_E \frac{dr}{r} = \infty\,,$$

the above inequality gives a sequence $r_i \in B$, $r_i \to 0$, with $u(r_i) \to 0$. Let U_i be the unbounded component of $R^n - F_{r_i}(S^{n-1}(r_i))$. Now $C(f,0) \subset \cap(R^n - U_i)$. Since $\operatorname{dia}(R^n - U_i) = u(r_i) \to 0$, $C(f,0)$ is a single point. The theorem follows.

The integral condition in Theorem 2 is sharp in the following sense: For each sequence of pairwise disjoint concentric spherical rings

$R_i = \{x : a_i < |x| < b_i\}$ with $a_i \to 0$ and

$$\sum_{i=1}^{\infty} \int_{a_i}^{b_i} \frac{dr}{r} < \infty ,$$

there exists a domain D, containing each R_i, and a quasiconformal mapping f of D such that f has no limit at the origin.

If the Hausdorff (n-1)-measure of E is zero, then the condition of Theorem 2 is satisfied and consequently every quasiconformal mapping of D - E has a continuous extension to E. By a theorem of Väisälä [6] the set E is even quasiconformally removable. On the other hand, it is possible to produce examples of sets E of positive n-measure such that all quasiconformal mappings of D - E have continuous extensions to E. Note that for n = 2 such sets E are never quasiconformally removable.

References

[1] Ahlfors, L., Beurling, A.: Conformal invariants and function theoretic null-sets. Acta math. 83 (1950), 101 - 129.

[2] Aseev, V., Syčev, A.: On sets which are removable for quasiconformal space mappings (Russian). Sibir. mat. Žurn. 15 (1974), 1213 - 1227.

[3] Gehring, F.: Rings and quasiconformal mappings in space. Trans. Amer. math. Soc. 103 (1962), 353 - 393.

[4] Grötzsch, H.: Eine Bemerkung zum Koebeschen Kreisnormierungsprinzip. Berichte über die Verhandlungen der Sächsischen Academie der Wissenschaften zu Leipzig 87 (1963), 319 - 324.

[5] Renggli, H.: Quasiconformal mappings and extremal lengths. Amer. J. Math. 86 (1964), 63 - 69.

[6] Väisälä, J.: Removable sets for quasiconformal mappings. J. Math. Mech. 19 (1969), 49 - 51.

Department of Mathematics
University of Helsinki
SF-00100 Helsinki 10
Finland

A THEOREM FOR ENTIRE FUNCTIONS OF INFINITE ORDER

Joseph Miles

In this note we are concerned with the relationship between the rate of growth of an entire function f, the arguments of its zeros, and the Nevanlinna deficiency of 0, defined to be

$$d(0,f) = 1 - \limsup_{r\to\infty} \frac{N(r,0)}{T(r,f)} .$$

For some time conditions on the zeros of f sufficient to imply $d(0,f) > 0$ have been known. Of particular interest here is the following theorem, which is a combination of results in [1] and [3].

Theorem A. Suppose f is an entire function of finite order whose zeros lie on the rays $\arg z = \alpha_j$, $1 \leq j \leq k$. Associated with $\{\alpha_1, \alpha_2, \ldots, \alpha_k\}$ is a function $B : [0,\infty) \to [0,1)$ with $B(\rho) \to 1$ as $\rho \to \infty$ such that if f has order ρ then $d(0,f) \geq B(\rho)$.

A natural question [2, problem 1.12] arising from Theorem A is whether 0 is necessarily a deficient value of an entire function of infinite order with zeros restricted to a finite number of rays through the origin. We answer this question and provide additional information with the following two theorems.

Theorem 1. Suppose f is an entire function of infinite order with zeros restricted to the rays $\arg z = \alpha_j \in [0,2\pi)$, $1 \leq j \leq k$.

(i) For every strongly convex function φ,

$$\lim_{r\to\infty} \frac{1}{2\pi} \int_0^{2\pi} \varphi\left(\frac{\log^+|f(re^{i\theta})|}{N(r,0)}\right) d\theta = \infty \tag{1}$$

and in particular for $p > 1$

$$N(r,0) = o\left(\left(\frac{1}{2\pi} \int_0^{2\pi} (\log^+|f(re^{i\theta})|)^p d\theta\right)^{1/p}\right) . \tag{2}$$

(ii) There exists $E \subset [1,\infty)$ with zero logarithmic density such that

$$\lim_{r\to\infty,\, r\notin E} \frac{N(r,0)}{T(r,f)} = 0 . \tag{3}$$

Theorem 2. If $\psi(r) \to \infty$ as $r \to \infty$, there exists an entire function of infinite order with only positive zeros satisfying $d(0,f) = 0$ and

$$\frac{\log T(r,f)}{\log r} < \psi(r), \quad r > R_0 .$$

Thus we answer the above question in the negative even for functions with zeros on a single ray and arbitrarily slow infinite rate of growth; yet from Theorem 1 we note that the behavior of f which might be expected on the basis of Theorem A in fact does occur if small exceptional sets of r-values are omitted or if $N(r,0)$ is compared to a slightly larger function than $T(r,f)$.

The proof of both theorems is based on a consideration of the Fourier series of $\log|f(re^{i\theta})|$. The formulas for the Fourier coefficients $c_m(r,f)$ of $\log|f(re^{i\theta})|$, apparently first noticed in [5], are for $m \geq 1$ with the normalization $f(0) = 1$

$$c_m(r,f) = \frac{1}{2\pi}\int_{-\pi}^{\pi} e^{-im\theta} \log|f(re^{i\theta})|\,d\theta = \frac{\alpha_m}{2} r^m + \frac{1}{2m} \sum_{|z_\nu|<r} \left\{\left(\frac{r}{z_\nu}\right)^m - \left(\frac{\overline{z_\nu}}{r}\right)^m\right\} \qquad (4)$$

where $\{z_\nu\}$ is the zero set of f and $\log f(z) = \sum_{m=1}^{\infty} \alpha_m z^m$ near 0. Clearly $c_m(r,f) = \overline{c_{-m}(r,f)}$ for $m \leq -1$ and $c_0(r,f) = N(r,0)$.

To prove (1), it is certainly sufficient to restrict our attention to a sequence $r_n \to \infty$ for which

$$\limsup_{n\to\infty} \frac{N(r_n,0)}{T(r_n,f)} > 0 .$$

It can be shown from (4) using integration by parts twice and the hypothesis of infinite order that such a sequence possesses a subsequence (still denoted by r_n) such that for some set of non-negative numbers η_j, $1 \leq j \leq k$, with sum 1

$$\lim_{n\to\infty} \frac{c_m(r_n,f)}{N(r_n,0)} = \sum_{j=1}^{k} \eta_j\, e^{-im\alpha_j}$$

for every integer m. The density of the trigonometric polynomials in the continuous periodic functions then implies that the sequence of measures

$$\frac{\log|f(r_n e^{i\theta})|}{2\pi\, N(r_n,0)}\, d\theta$$

converges weakly to the positive measure with point mass η_j at α_j. Assuming with no loss in generality that $\eta_1 > 0$, we have for all sufficiently small $\delta > 0$ and all $n > n_0(\delta)$,

$$\frac{1}{2\pi}\int_{\alpha_1-\delta}^{\alpha_1+\delta}\frac{\log^+|f(r_ne^{i\theta})|}{N(r_n,0)}\,d\theta > \eta_1/2 > 0 ,$$

implying $\{\log^+|f(r_ne^{i\theta})|/N(r_n,0)\}$ is not a uniformly integrable family and thus establishing (1). The choice $\varphi(t) = t^p$ in (1) establishes (2).

To prove (3), it is sufficient by Nevanlinna's first fundamental theorem to show

$$N(r,0) = o(\max_m |c_m(r,f)|),\ r \notin E .$$

To simplify the exposition, let us only consider the case where f has positive zeros $\{z_\nu\}$ with $z_1 < z_2 < z_3 < \dots$. Writing

$$f(z) = e^{g(z)}\prod_{\nu=1}^{\infty} E(\frac{z}{z_\nu}, \nu - 1) ,$$

we obtain from (4) after two integrations by parts

$$c_m(r,f) = \beta_m r^m + N(r,0) - \frac{m}{2}\int_0^r \left(\frac{t}{r}\right)^m \frac{N(t,0)}{t}\,dt + \frac{m}{2}\int_{z_m}^r \left(\frac{r}{t}\right)^m \frac{N(t,0)}{t}\,dt$$

where $|\beta_m|^{1/m} \to 0$ as $m \to \infty$. We need only consider the case in which $N(t,0)$ has infinite order; in this case it is elementary that there exists $b_m \to \infty$ (not necessarily monotonically) such that for $m > m_0$

$$\operatorname{Re}\beta_m = \frac{m}{2}\int_{b_m}^{z_m}\frac{N(t,0)}{t^{m+1}}\,dt .$$

Thus it suffices to prove for an arbitrary $b_m \to \infty$ that

$$N(r,0) = o(\max_{m>m_0}|\frac{m}{2}\int_{b_m}^r \left(\frac{r}{t}\right)^m \frac{N(t,0)}{t}\,dt|),\ r \notin E ,$$

for some set E of zero logarithmic density. The details of the proof of this rather technical growth lemma are to be found in [4].

Theorem 2 is proved by constructing an entire function f with positive zeros such that for a sequence $r_n \to \infty$ and a sequence $\beta_n \uparrow 1$

$$\left|\left|\log|f(r_ne^{i\theta})| - \operatorname{Re}\left(\frac{1+\beta_ne^{i\theta}}{1-\beta_ne^{i\theta}}\right)N(r_n,0)\right|\right|_2 = o(N(r_n,0)) . \qquad (5)$$

The positivity of the Poisson kernel implies $m(r_n,1/f) = o(N(r_n,0))$, yielding $d(0,f) = 0$. The function f has zeros distributed in such a way that $\log N(t,0)$ is approximately a piecewise-linear convex function of $\log t$ whose derivative has large jumps near $\log r_n$. If

h is any convergent product with these zeros, then f may be taken to be

$$f(z) = e^{g(z)}h(z)$$

where g is an arbitrary entire function. The Maclaurin coefficients of g, whose effect on the Fourier coefficients of $\log|f(re^{i\theta})|$ is given explicitly in (4), are chosen so that Parseval's theorem yields (5). The essential difference between the finite order and infinite order situations is that in the latter case infinitely many coefficients of g are available to be chosen arbitrarily and can in fact be so chosen as to achieve (5); no such freedom is available in the former case. The lengthy details of the construction appear in [4].

References

[1] Edrei, A., Fuchs, W. H. J., Hellerstein, S.: Radial distribution of deficiencies of the values of a meromorphic function. Pacific J. Math. 11 (1961), 135 - 151.

[2] Hayman, W. K.: Research problems in function theory. Athlone Press, London (1967).

[3] Hellerstein, S., Shea, D. F.: Minimal deficiencies for entire functions with radially distributed zeros. Proc. London math. Soc., III. Ser. 37 (1978), 35 - 55.

[4] Miles, J.: On entire functions of infinite order with radially distributed zeros. To appear in Pacific J. Math.

[5] Nevanlinna, F.: Bemerkungen zur Theorie der ganzen Funktionen endlicher Ordnung. Commentationes Phys.-Math., Soc. Sci. Fennica II. 4 (1923).

University of Illinois
Urbana, IL 61801
U.S.A.

QUASIREGULAR MAPPINGS

Ruth Miniowitz

The theory of quasiregular mappings in n-space is apparently a natural generalization of that of analytic functions of one complex variable. Therefore it is natural to ask, which distortion theorems known for analytic functions may be generalized.

Notation and terminology are in general as in [4]. A continuous mapping $f : D \to R^n$ in a domain D, $D \subset R^n$ is said to be quasiregular (qr) if

i) f is ACL^n.

ii) There exists a constant $1 \leq K < \infty$ such that:

$$|f'(x)|^n \leq K \cdot J(x,f) \quad \text{a.e. in } D.$$

Here $f' = \left(\frac{\partial f_i}{\partial x_j}\right)_{i,j=1}^n$ is the formal derivative of f, $|f'(x)| = \max\limits_{|h|=1} |f'(x) \cdot h|$, and $J(x,f) = \det f'(x)$ is the Jacobian of f at x.

A mapping $f : D \to R^n$ is said to be quasiconformal (qc) if f is qr and injective. We denote by $K_I(f)$, $K_0(f)$ and $K(f)$, respectively, the inner, outer, and maximal dilatation of f; see [4].

Global distortion theorems that one can obtain are the following:

1) For $f : B^n \to R^n \setminus \{0\}$ K - qr and of bounded degree, we have:

$$\frac{|f(0)|}{A}\left(\frac{1-|x|}{1+|x|}\right)^{\beta} \leq |f(x)| \leq A|f(0)|\left(\frac{1+|x|}{1-|x|}\right)^{\beta};\ x \in B^n$$

where $A = e^{8\beta}$ and $\beta = 2^{3n-1}K_I(f)N^{n+1}$.

Proof. The proof is based on a theorem of Rickman's on path liftings [6]. That theorem gives a lower bound for the modulus of a certain path family Γ_R with respect to a sphere $S^{n-1}(R)$. We define a new path family $\Gamma = \bigcup\limits_{R_1 \leq R \leq R_2} \Gamma_R$, where $R_1 = |f(0)|$ and $R_2 = |f(x)|$, and get from that estimate a lower bound for $M(\Gamma)$. If Γ^* is a lifted family of the family Γ, then:

$$M(\Gamma) \leq K_I M(\Gamma^*) .$$

An upper bound for the modulus of the lifted family Γ^* can be obtained from [1]. A simple computation will lead to the above inequality.

2) For $n \geq 3$ $f : B^n \to R^n \setminus \{0\}$ locally K - qc one obtains:

$$\frac{|f(0)|}{R}\cdot \exp\{-(\frac{1}{1-|x|})^{(n+1)/(n-1)}\} \le |f(x)|$$

$$|f(x)| \le |f(0)|\cdot R\cdot \exp\{(\frac{1}{1-|x|})^{(n+1)/(n-1)}\},\ x\in B^n$$

where R is a positive constant depending only on n and K.

Proof. Using [5, 2.3] one can get an estimate for $N(r) = N(f,B^n(r))$. We construct a path family as in [6, 4.11] and the respective lifting. Using the lower bound in [7, 10.12] for the modulus of the lifted family, and the upper bound in [7, 7.5] for the modulus of the path family, in conjunction with the inequality:

$$M(\Gamma) \leqq K_0(f)\cdot N\cdot M(f\Gamma)\ ,$$

one can obtain the desired distortion theorem.

3) For $f: B^n \to R^n \setminus E$ K - qr where $E\subset R^n$ is a set satisfying: $0\in E$ and $E\cap S^{n-1}(R) \ne \emptyset$ for every $R > 0$, we have

$$(*)\quad \frac{|f(0)|}{C}(\frac{1-|x|}{1+|x|})^{\alpha} \le |f(x)| \le |f(0)|\cdot C\cdot(\frac{1+|x|}{1-|x|})^{\alpha};\ x\in B^n$$

where $\alpha = 2^{n-1}K_I(f)$ and $C = 2^{8\alpha}$.

Proof. Let $x^*\in B^n$ be such that $|x| = |x^*|$ and $|f(x^*)| = \underset{|z|=r}{\mathrm{Max}}|f(z)|$. Let I be the line segment that connects the origin and x^*. Let $A = B^n(M(r,f)) \setminus B^n(|f(0)|)$ where $M(r,f) = \underset{|z|=r}{\mathrm{Max}}|f(z)|$, $F = f(I)\cap A$, $\Gamma' = \Gamma(E,F,A)$, $\tilde{\Gamma} = (F,\partial fB^n,A)$, and Γ a lifting of the family $\tilde{\Gamma}$ in B^n, then:

$$M(\Gamma') \leqq K_I\cdot M(\Gamma)\ .$$

Using the lower bound for $M(\Gamma')$ in [7, 10.12], and the upper bound for $M(\Gamma)$ in [1] we get the inequality (*).

4) If f is K - qc in B^n and does not vanish then f maps B^n onto a domain D such that $CD\cap S^{n-1}(R) \ne \emptyset$ for every $R > 0$ and therefore (*) holds.

In the plane the generalization of the theory of p-valent functions is the theory of circumferentially mean p-valent, see [2]. The corresponding generalization for $n \ge 3$ may be the following. Let $f: D\to R^n$ be sense-preserving, discrete and open. Denote $n(y,f,D)$ the number of roots of the equation $f(x) = y$ in D with their multiplicity and:

$$p(R) = \frac{1}{\omega_{n-1}R^{n-1}}\int_{S^{n-1}(R)} n(y,f,D)\,d\Lambda$$

where $\omega_{n-1} = m_{n-1}(S^{n-1})$, and $d\Lambda$ is an element of spherical measure of $S^{n-1}(R)$.

We say that f is spherically mean p-valent $(p > 0)$ if $p(R) \leqq p$ for every $R > 0$.

A generalization to the "Length-Area" theorem of Ahlfors-Cartwright, see [2, Theorem 2.1] is the following "Volume-Area" theorem.

5) Let $f : D \to R^n$ be K - qr, $m_n(D) < \infty$,

$$\sigma(R) = H^{n-1}\{x \in D \setminus B_f;\ |f(x)| = R\}$$

where H^{n-1} is the (n - 1)-Hausdorff measure, and B_f is the branch set, then:

$$\int_0^\infty \left[\frac{\sigma(R)^n}{\omega_{n-1}R^{n-1}p(R)}\right]^{1/(n-1)} dR \leqq K_I(f)m_n(D) .$$

The proof is based on the possibility to compose D as a sum of cubes, such that f is a K - qc mapping in a domain containing each of the cubes. Then estimating $\sigma(R)$ by [3, Lemma 6] and using the appropriate generalizations to the ideas of the proof in [2, Theorem 2.1].

One can get distortion theorems and boundary behavior of qr spherically mean p-valent mappings in certain conditions by using the last theorem.

References

[1] Anderson, G. D.: Extremal rings in n-space for fixed and varying n. Ann. Acad. Sci. Fenn., Ser. A I 575 (1974).

[2] Hayman, W. K.: Multivalent functions. Cambridge University Press, Cambridge (1967).

[3] Gehring, F. W.: Lower dimensional absolute continuity properties of quasiconformal mappings. Math. Proc. of the Cambridge Philosophical Soc. 78 (1975), 81 - 93.

[4] Martio, O.,Rickman, S., Väisälä, J.: Definitions for quasiregular mappings. Ann. Acad. Sci. Fenn., Ser. A I 448 (1969).

[5] Martio, O., Rickman, S., Väisälä, J.: Distortion and singularities of quasiregular mappings. Ibid 465 (1970).

[6] Rickman, S.: A path lifting construction for discrete open mappings with application to quasimeromorphic mappings. Duke math. J. 42 (1975), 797 - 809.

[7] Väisälä, J.: Lectures on n-dimensional quasiconformal mappings. Lecture Notes in Mathematics 229, Springer-Verlag, Berlin - Heidelberg - New York (1971).

Current Address:

Department of Mathematics
University of Michigan
Ann Arbor, MI 48109
U.S.A.

SUR LA THEORIE DES FONCTIONS FINEMENT HOLOMORPHES (II)

Nguyen-Xuan-Loc *

§ 0. Introduction.

Ce travail fait suite à [4b]; dans [4b] étaient étudiées des extensions éventuelles des opérateurs ∂_z et $\partial_{\bar{z}}$ sur un ouvert fin du plan complexe: étant donné un domaine fin U du plan complexe C on a montré l'existence du différentiel complexe fin ∂_U sur un sous-algèbre de l'algèbre $\mathcal{H}_f(U)$ des fonctions finement holomorphes dans U et on a posé le problème d'extension du domaine de définition de ∂_U sur tout l'espace $H_f(U)$ des fonctions complexes finement harmoniques dans U (voir [4b], 2.4).

On montre dans cet article que ∂_U (resp. $\bar{\partial}_U$) est effectivement un opérateur linéaire de domaine de définition $H_f(U)$ et à valeurs dans $\mathcal{H}_f^q(U)$, l'algèbre des fonctions quasi-finement holomorphes dans U (resp. $\overline{\mathcal{H}}_f^q(U)$, l'algèbre des fonctions quasi-finement anti-holomorphes dans U). Comme applications de ce résultat on donne une réponse positive à la conjecture suivante de Fuglede:

<u>Principe d'unicité fine:</u> Soit $f : U \to C$ une fonction complexe finement harmonique dans un domaine fin $U \subset C$. Alors f est identiquement nulle dans U si et seulement si elle est nulle au voisinage (fin) d'un certain point x_o de U.

§ 1. Rappel.

Nous gardons les terminologies et notations de [4b]. Notons toujours par

$$Z := \{(\Omega, \mathcal{F}, (\mathcal{F}_t), (Z_s = X_s + iY_s), P^x(x \in C))\}$$

où $X = (X_s)$ et $Y = (Y_s)$ sont deux processus de Wiener standards, réels et indépendantes, le mouvement brownien standard du plan complexe C. Pour tout ouvert fin $U \subset C$ de temps de sortie τ (pour le brownien Z) on dénote par $\sum_x$ le système de probabilité:

$$\textstyle\sum_x := \{(\Omega, \tilde{\mathcal{F}}, (\tilde{\mathcal{F}}_t), P^x, \tau)\} \ (x \in U) \text{ où } \tilde{\mathcal{F}} = \mathcal{F}_\tau \text{ et où } \tilde{\mathcal{F}}_t = \mathcal{F}_{\tau\wedge t} \ (t \geq 0),$$

---

* Une partie de ce travail est réalisée pendant le séjour de l'auteur à l'Université de Joensuu (Finlande). L'auteur tient à remercier Mr. le Professeur I. Laine de son invitation et les collègues du Département de Mathématiques et Physiques de l'Université de Joensuu de leur chaleureuse hospitalité.

et par Z^τ (resp. $\bar{Z}^\tau$) le mouvement brownien Z stoppé au temps τ:

$$Z^\tau := \{(\Omega,\tilde{\mathcal{F}},(\tilde{\mathcal{F}}_t),(Z_s = X_{\tau\wedge s} + iY_{\tau\wedge s}),P^x(x \in U))\},$$

resp.

$$\bar{Z}^\tau := \{(\Omega,\tilde{\mathcal{F}},(\tilde{\mathcal{F}}_t),(\bar{Z}_s = X_{\tau\wedge s} - iY_{\tau\wedge s}),P^x(x \in U))\}.$$

Soit $\mathcal{M}_x^2[0,\tau[$ l'espace des martingales réelles du système $\sum_x$, continues et bornées dans $L^2(\Omega,\tilde{\mathcal{F}},P^x)$, de temps de vie τ et nulles au temps nul. Notons par

$$H_x^2[0,\tau[= \mathcal{M}_x^2[0,\tau[+ i\mathcal{M}_x^2[0,\tau[,$$

alors pour tout couple d'éléments $M = M_1 + iM_2$ et $N = N_1 + iN_2$ de H_x^2 on peut associer un processus réel de variation bornée comme suit:

$$<M,N> = <M_1,N_1> + <M_2,N_2>,$$

où $<M_1,N_1>$ et $<M_2,N_2>$ sont respectivement des processus de variation bornée associés avec des couples de martingales de carré intégrable (M_1,N_1) et (M_2,N_2). On peut déduire facilement à partir du cas réel (voir par exemple [2], Theorem 2.2) l'estimation à priori suivante: Soit $0 < p < \infty$, ils existent deux constantes universelles c_p et C_p telles que,

$$c_p E_x(\sup_{s\leq t}|M_s|^{2p}) \leq E_x(<M,M>_t) \leq C_p E_x(\sup_{s\leq t}|M_s|^{2p}) \tag{1}$$

pour tout $t \geq 0$ et pour tout $M = (M_s) \in H_x^2[0,\tau[$.

L'espace hilbertien des processus holomorphes (resp. anti-holomorphes, harmoniques complexes) du système $\sum_x$ est noté par $\{\mathcal{H}_x[0,\tau[,((\ ,\ ))_x\}$ (resp. par $\{\bar{\mathcal{H}}_x[0,\tau[,((\ ,\ ))_x\}$, $\{\mathcal{H}_x^1[0,\tau[,((\ ,\ ))_x\}$). Rappelons que tous ces espaces sont contenus dans $H_x^2[0,\tau[$ et que $\mathcal{H}_x^1[0,\tau[$ est la somme directe de $\mathcal{H}_x[0,\tau[$ et $\bar{\mathcal{H}}_x[0,\tau[$. Tout élément de $\mathcal{H}_x[0,\tau[$ (resp. de $\bar{\mathcal{H}}_x[0,\tau[$, $\mathcal{H}_x^1[0,\tau[$) est de la forme $W\cdot Z^\tau = \{(W\cdot Z^\tau)_s\}$ (resp. $W\cdot\bar{Z}^\tau = \{(W\cdot\bar{Z}^\tau)_s\}$, $W_1\cdot Z^\tau + W_2\cdot\bar{Z}^\tau$), où $W = \{(U_s + iV_s)\}$ (resp. W_1, W_2) est un élément de l'espace des intégrands complexes $L_x^2(<Z>_\tau)$. L'intégrale stochastique complexe $W\cdot Z^\tau$ (resp. $W\cdot\bar{Z}^\tau$) est par définition,

$$W\cdot Z^\tau = (U\cdot X^\tau - V\cdot Y^\tau) + i(V\cdot X^\tau + U\cdot Y^\tau),$$

resp.

$$W\cdot\bar{Z}^\tau = (U\cdot X^\tau + V\cdot Y^\tau) + i(V\cdot X^\tau - U\cdot Y^\tau),$$

où $U\cdot X^\tau$, $V\cdot Y^\tau$,..., sont des intégrales stochastiques réels, et le produit scalaire de deux éléments $W\cdot Z^\tau$ et $W'\cdot Z^\tau$ (resp. $W\cdot\bar{Z}^\tau$ et $W'\cdot\bar{Z}^\tau$)

de $\mathcal{H}_x[0,\tau[$ (resp. $\overline{\mathcal{H}}_x[0,\tau[)$ est par définition,

$$((W\cdot Z^\tau, W'\cdot Z^\tau))_x = E_x(<W\cdot Z^\tau, W'\cdot Z^\tau>_\infty) = 2E_x\left(\left\{\int_0^\infty (U_s U_s' + V_s V_s')d(s\wedge\tau)\right\}\right),$$

resp. (2)

$$((W\cdot \overline{Z}^\tau, W'\cdot \overline{Z}^\tau))_x = E_x(<W\cdot \overline{Z}^\tau, W'\cdot \overline{Z}^\tau>_\infty).$$

Etant donné un élément W de $L^2_x(<Z>_\tau)$ et un entier naturel k on définit l'intégral stochastique itéré $W\cdot Z^\tau_k$ du système Σ_x d'une manière inductive comme suit:

$$W\cdot Z^\tau_1 := W\cdot Z^\tau,$$

$$W\cdot Z^\tau_i = (W\cdot Z^\tau_{i-1})\cdot Z^\tau, \quad 2 \leq i \leq k, \tag{3}$$

pourvu que

$$W\cdot Z^\tau_{i-1} \in L_x(<Z>_\tau) \cap \mathcal{H}_x[0,\tau[, \quad 2 \leq i \leq k-1.$$

On peut définir de la même facon les intégrales stochastiques itérés de la forme $W\cdot \overline{Z}^\tau_k$ ou de la forme $W\cdot \overline{Z}^\tau_k \cdot Z^\tau_{k'}$. Il est intéressant à noter que les intégrales stochastiques itérés peuvent être calculés explicitement par un système de polynomes de Hermite (voir par exemple [5], p. 36). On dénote enfin par ∂ et $\overline{\partial}$ les opérateurs $\frac{1}{2}(\partial/\partial x - i\partial/\partial y)$ et $\frac{1}{2}(\partial/\partial x + i\partial/\partial y)$, par ∂^k et $\overline{\partial}^k$ les opérateurs $(\partial)^k$ et $(\overline{\partial})^k$ où k est un entier ≥ 2.

L'algèbre des fonctions finement holomorphes (resp. anti-holomorphes, l'espace des fonctions complexes finement harmoniques) dans un ouvert fin $U \subset \mathbb{C}$ est noté par $\mathcal{H}_f(U)$ (resp. $\overline{\mathcal{H}}_f(U)$, $H_f(U)$). Pour tout $f \in H_f(U)$ et tout $x \in U$ il existe un triplet $\{(f_n),x,V_x\}$ associé à f au point x. Rappelons que, par définition, cela veut dire qu'il existe un voisinage fin compact V_x de x dans U et une suite de fonctions (f_n), harmoniques dans un voisinage de V_x, qui converge uniformément sur V_x vers f. L'algèbre des fonctions holomorphes (resp. anti-holomorphes, l'espace des fonctions complexes harmoniques) dans un voisinage de V_x est noté par $R(V_x)$ (resp. $\overline{R}(V_x)$, $H_o(V_x)$). Rappelons que pour tout élément f de $R(V_x)$ (resp. $\overline{R}(V_x)$, $H_o(V_x)$) et pour tout entier k on a la représentation suivante du processus $\{f(Z^\tau) - f(Z^\tau_o)\}$ (τ dénote le temps de sortie de V_x) dans $\mathcal{H}_x[0,\tau[$ (resp. dans $\overline{\mathcal{H}}_x[0,\tau[$, $\mathcal{H}^1_x[0,\tau[)$:

$$\begin{cases} f(Z^\tau) - f(Z^\tau_o) = \sum_{i=1}^{k} (\partial^i f(Z^\tau_o)\cdot Z^\tau_i) + \partial^{k+1} f(Z^\tau)\cdot Z^\tau_{k+1} \\ \text{resp.} \\ f(Z^\tau) - f(Z^\tau_o) = \sum_{i=1}^{k} (\overline{\partial}^i f(Z^\tau_o)\cdot \overline{Z}^\tau_i) + \overline{\partial}^{k+1} f(Z^\tau)\cdot \overline{Z}^\tau_{k+1}, \end{cases} \tag{4}$$

$$- \sum_{i=1}^{k} (\partial^i f(Z_0^\tau)\cdot Z_i^\tau + \bar\partial^i f(Z_0^\tau)\cdot \bar Z_i^\tau) + \partial^{k+1} f(Z^\tau)\cdot Z_{k+1}^\tau + \bar\partial^{k+1} f(Z^\tau)\cdot \bar Z_{k+1}^\tau .$$

§ 2. Existence et unicite de ∂_U et $\bar\partial_U$ sur l'espace $H_f(U)$.

Lemme 1. Soit $f : U \to C$ une fonction complexe finement harmonique dans un ouvert fin $U \subset C$ et soit $\{(f_n),x,V_x\}$ un triplet associé à la fonction f au point x de U. Alors sont équivalents:

(a) Pour tout entier k et pour le point a de l'intérieur fin $\overset{\circ f}{V}_x$ de V_x la suite $\{\partial^k f_n(a)\}$ est de Cauchy dans C.

(b) La suite d'intégrales stochastiques itérés $\{\partial^{k+1} f(Z^\tau)\cdot Z_{k+1}^\tau\}$ est de Cauchy dans $\mathcal{H}_a[0,\tau[$ (τ dénote le temps de sortie de V_x).

De plus on a:

Pour tout $k \in N$ et pour q.p.(*) point $a \in \overset{\circ f}{V}_x$ la suite $\{\partial^k f_n(a)\}$ est de Cauchy et la fonction $\partial_U^k f$ définie ponctuellement (q.p.) comme $\partial_U^k f(y) = \lim_m \partial^k f(y)$ q.p. dans $\overset{\circ f}{V}_x$ est indépendant du triplet $\{(f_n),x,V_x\}$. (5)

Preuve. Traitons d'abord le cas $k = 1$. Notons par $\mathcal{N}$ le filtre sur N^2 défini par l'ordre naturel: $\alpha = (m,n)$ et $\alpha' = (m',n')$ sont deux éléments de $\mathcal{N}$ alors $\alpha \leq \alpha'$ si et seulement si $m \leq m'$ et $n \leq n'$. En posant pour tout $\alpha \in \mathcal{N}$, $f_\alpha = (f_m - f_n)$ alors $\{f_\alpha | \alpha \in \mathcal{N}\}$ est un filtre dénombrable dans $H_o(V_x)$ qui converge vers la fonction nulle. Fixons un point a arbitraire de $\overset{\circ f}{V}_x$ alors en appliquant les formules (4) pour les fonctions f_α et pour le système $\sum_a$, on a:

$$f_\alpha(Z^\tau) - f_\alpha(Z_0^\tau) = \partial f_\alpha(Z^\tau) + \bar\partial f_\alpha(Z^\tau)\cdot \bar Z^1 . \tag{6}$$

D'après l'inégalité a priori (1):

$$\|f_\alpha(Z^\tau) - f_\alpha(Z_0^\tau)\|_a = E_a(\langle f_\alpha(Z^\tau) - f_\alpha(Z_o^\tau), f_\alpha(Z^\tau) - f_\alpha(Z_o^\tau)\rangle_\infty) \leqq C|f_\alpha|_{\sup}$$

où C est une constante et où $|\cdot|_{\sup}$ dénote la norme supremum. Ce implique que les filtres $\{f_\alpha(Z^\tau) - f_\alpha(Z_0^\tau) | \alpha \in \mathcal{N}\}$ (resp. $\{\partial f_\alpha(Z^\tau)\cdot Z^\tau | \alpha\in\mathcal{N}\}$ $\{\bar\partial f_\alpha(Z^\tau)\cdot \bar Z^\tau | \alpha \in \mathcal{N}\}$) converges vers zéro dans $\mathcal{H}_a^1[0,\tau[$ (resp. $\mathcal{H}_a[0,\tau[$ $\overline{\mathcal{H}}_a[0,\tau[$). D'autre part puisque les fonctions ∂f_α appartiennent à $R(V_x)$ on a d'après les mêmes formules (4):

$$\partial f_\alpha(Z^\tau) = \partial f_\alpha(Z_0^\tau)\cdot Z^\tau + \partial^2 f_\alpha(Z^\tau)\cdot Z_2^\tau \quad \text{dans } \mathcal{H}_a[0,\tau[. \tag{7}$$

(a) $\Longleftrightarrow$ (b). Calculons les normes dans $\mathcal{H}_a[0,\tau[$ les deux membres de (7):

$$\|f_\alpha(Z^\tau)\cdot Z^\tau\|_a^2 = \|f_\alpha(Z_0^\tau)\cdot Z^\tau\|_a^2 + \|\partial^2 f_\alpha(Z^\tau)\cdot Z_\alpha^\tau\|^2 + 2((\partial f_\alpha(Z_0^\tau)\cdot Z^\tau, \partial^2 f_\alpha(Z^\tau)\cdot Z_2^\tau))_a \tag{8}$$

(*) L'abréviation q.p. signifie quasi-partout, càd, sauf sur un ensemble polaire.

En remarquant que $\partial f_\alpha(Z_0^\tau) = \partial f_\alpha(a)$ p.s.P^a. on a:

$$\|\partial f_\alpha(Z_0^\tau)\cdot Z^\tau\|_a^2 = |\partial f_\alpha(a)|^2\cdot E_a(\tau)$$

et

$$((\partial f_\alpha(Z_0^\tau)\cdot Z^\tau, \partial^2 f_\alpha(Z^\tau)\cdot Z_2^\tau))_a^2 \leq |\partial f_\alpha(a)|^2 E_a(\tau)\|\partial^2 f_\alpha(Z^\tau)\cdot Z_2^\tau\|_a^2$$

(d'après l'inégalité de Schwarz).

En faisant tendre α vers ∞ dans les deux membres de (8) et en remarquant que $E_a(\tau)$ est strictement positive on a l'équivalence de (a) et (b) pour le cas $k = 1$.

Nous montrons maintenant l'assertion (5) pour le cas $k = 1$. Posons

$$\partial^2 f_\alpha(Z^\tau)\cdot Z^\tau = Y_\alpha^1 + iY_\alpha^2$$

où $Y_\alpha^1 = (Y_\alpha^1(s))$ et $Y_\alpha^2 = (Y_\alpha^2(s))$ sont des processus réels alors on a d'après les formules (5):

$$\|\partial^2 f_\alpha(Z^\tau)\cdot Z^\tau\|_a = 2E_a\left(\int_0^\infty (Y_\alpha^1(s)^2 + Y_\alpha^2(s)^2)d(s\wedge\tau)\right) = 2E_a\left(\int_0^\infty |Y_\alpha(s)|^2 d(s\wedge\tau)\right)$$

ce qui entraine si on pose $\partial f_\alpha(a) = r_\alpha(a)e^{i\theta_\alpha}$,

$$((\partial f_\alpha(Z_0^\tau)\cdot Z^\tau, \partial^2 f_\alpha(Z^\tau)\cdot Z_2^\tau))_a = E_a(<\partial f_\alpha(Z_0^\tau)\cdot Z^\tau, \partial^2 f_\alpha(Z^\tau)\cdot Z_2^\tau>_\infty)$$

$$= 2r_\alpha(a)\cdot E_a\left(\int_0^\infty (\cos\theta_\alpha\cdot Y_\alpha^1(s) + \sin\theta_\alpha\cdot Y_\alpha^2(s))d(s\wedge\tau)\right).$$

L'égalité (8) prend donc la forme:

$$2\|\partial f_\alpha(Z^\tau)\cdot Z^\tau\|_a^2 = r_\alpha(a)^2\cdot E_a(\tau) + $$
$$+ 2r_\alpha(a)\cdot E_a\left(\int_0^\infty (\cos\theta_\alpha\, Y_\alpha^1(s) + \sin\theta_\alpha Y_\alpha^2(s))d(s\wedge\tau)\right) + E_a\left(\int_0^\infty |Y_\alpha(s)|^2 d(s\wedge\tau)\right). \qquad (8')$$

Supposons qu'il existe un point $a \in V_x^{Of}$ tel que ou bien le filtre $\{\partial f_\alpha(a)\}$ n'a pas de limite ou bien il converge vers un nombre complexe différent de zéro, alors en prenant un sous-filtre on peut supposer toujours que

$$\lim \partial f_\alpha(a) = r(a)e^{i\theta} \neq 0 \text{ et } \infty.$$

En tenant compte de (7) on peut aussi supposer que le filtre $\{Y_\alpha | \alpha \in \mathcal{N}\}$ converge vers un élément $Y = (Y^1 + iY^2)$ différent de zéro dans $L_a^2(<Z>_\tau)$. Par conséquent on a lorsque α tend vers l'infini:

$$0 = r(a)^2 E_a(\tau) + 2r(a)\cdot E_a\left(\int_0^\infty (\cos\theta\, Y^1(s) + \sin\theta\, Y^2(s))d(s\wedge\tau)\right) +$$
$$+ E_a\left(\int_0^\infty |Y(s)|^2 d(s\wedge\tau)\right). \qquad (8")$$

En remarquant que la quantité $\cos\theta\, Y^1(s) + \sin\theta\, Y^2(s)$ n'est d'autre que la projection du vecteur aléatoire $Y(s)$ sur la direction fixe $e^{i\theta}$ on en déduit d'après l'inégalité de Schwarz que le descriminant Δ' de l'equation (8") est non-positif:

$$\Delta' = E_a\left\{\left({}_0\int^{\infty}(\cos\theta\, Y^1(s) + \sin\theta\, Y^1(s))d(s\wedge\tau)\right)\right\}^2 - E_a(\tau)E_a\left({}_0\int^{\infty}|Y(s)|^2 d(s\wedge\tau)\right)$$

$$\leqq E_a\left({}_0\int^{\infty}\sqrt{Y^1(s)^2 + Y^2(s)^2}^{\,2}\, d(s\wedge\tau)\right)^2 - E_a\left({}_0\int^{\infty}d(s\wedge\tau)\right)E_a\left({}_0\int^{\infty}|Y(s)|^2 d(s\wedge\tau)\right) \leqq 0$$

Si $\Delta' < 0$ l'équation (8") est impossible mais si $\Delta' = 0$ le vecteur aléatoire $Y(s)$ reste toujours colinéaire avec le vecteur fixe $e^{i\theta}$, ce qui est absurde puisque ce fait implique que $Y = 0$.

Il nous reste à examiner le cas où $\lim_n \partial f_n(a)$ est égale à ∞ dans C. Or en tenant compte de l'équivalence de (a) et (b) on a dans ce cas:

$$\lim_n |\partial f_n(a)| = +\infty \quad \text{et} \quad \lim_n \|\partial^2 f_n(Z^\tau)\cdot Z_2^\tau\|_a = +\infty. \qquad (8''')$$

Ainsi l'ensemble $\overset{\circ}{V}{}^f_x$ admet une partition en deux ensembles presque-boreliens disjoints: $\overset{\circ}{V}{}^f_x = U_o \cup A_\infty$ où

$$U_o := \{a \in \overset{\circ}{V}{}^f_x \mid \lim_n \partial f_n(a) \text{ existe et } \neq +\infty\},$$

$$A_\infty := \{a \in \overset{\circ}{V}{}^f_x \mid \lim_n \partial f_n(a) \text{ existe et } = +\infty\}.$$

Montrons maintenant que A_∞ est polaire. En effet si A_∞ est non-polair ils existent d'une part d'un sous-ensemble compact, non-polaire K de A_∞ et d'autre part une composante connexe de $\overset{\circ}{V}{}^f_x$ dont l'intersection avec K est non-polaire. Puisque U_o est finement dense dans $\overset{\circ}{V}{}^f_x$ (voir [7]) on peut fixer d'un point a de U_o dans cette composante connexe. D'une part on a d'après [4a], Théorème 2.3:

$$P^a\{T < \tau\} > 0$$

où T dénote le temps d'entrée de K pour Z et d'autre part d'après la propriété de Markov forte:

$$\|\partial^2 f_n(Z^\tau)\cdot Z_2^\tau\|_a^2 = 2E_a\left({}_0\int^{\infty}|\partial^2 f_n(Z^\tau)\cdot Z^\tau|^2(s)d(s\wedge\tau)\right)$$

$$\geqq 2E_a\left\{E_a\left({}_0\int^{\infty}|\partial^2 f_n(Z^\tau)\cdot Z^\tau|^2(s)d(s\wedge\tau)\mid \mathcal{F}_T\right)\cdot\chi_{\{T<\tau\}}\right\}$$

$$\geqq 2E_a\left\{E_{Z_T}\left({}_0\int^{\infty}|\partial^2 f(Z^\tau)\cdot Z^\tau|^2(s)d(s\wedge\tau)\right);\ T < \tau\right\}.$$

En tenant compte du fait que Z_T appartient p.s.P^a à K pour les ω de l'événement $\{T < \tau\}$ on arrive à une contradiction en faisant tendre n vers $+\infty$ dans les deux membres de l'inégalité ci-dessus car d'une part le premier membre reste toujours borné puisque $a \in U_o$ tandis le second membre est non-borné à cause de (8"').

La deuxième partie de l'assertion (5) est déjà montrée dans [4b], Lemme 7.b: la fonction $\partial_U f$ définie ponctuellement sur l'ouvert fin U_o par $\lim_n \partial f_n(y)$, $y \in U_o$, est indépendante du triplet $\{(f_n),x,V_x\}$ et est finement continue dans U_o, càd, finement continue q.p. dans $\overset{of}{V}_x$.

Le cas général: Fixons un point a de U_o et faisons tendre α vers l'infini dans (7) nous obtenons,

$$\lim \partial^2 f_\alpha(Z^\tau)\cdot Z_2^\tau = 0 \quad \text{dans} \quad \mathcal{H}_a[0,\tau[,$$

puisqu'on a d'après la première partie $\lim \partial f_\alpha(a) = 0$. D'autre part puisque $\{\partial^2 f_\alpha | \alpha \in \mathcal{N}\}$ est aussi un filtre dans $R(V_x)$, on a pour tout α:

$$\partial^2 f_\alpha(Z^\tau)\cdot Z_2^\tau = \partial^2 f_\alpha(Z_o^\tau)\cdot Z_2^\tau + \partial^3 f_\alpha(Z^\tau)\cdot Z_3^\tau .$$

En répétant les arguments de la démonstration du cas où $k = 1$ on obtient l'equivalence de (a) et (b) et l'assertion (5) pour le cas $k = 2$, les autres cas s'en déduisent facilement par récurrence.

Remarque. L'exemple donné par Debiard et Gaveau dans [6], p. 119, montre qu'ils existent des fonctions de $H_f(U)$ telles que les ensembles polaires A_∞ associés ne sont pas vides. Il est intéressant de se demander si A_∞ est vide lorsque f est de plus finement holomorphe (resp. anti-holomorphe), voir aussi remarque du théorème 4.

Définition 2. A toute fonction f de $H_f(U)$ on peut associer une et une seule suite $\{(\partial_U^k f)\}$ de fonctions quasi-finement continues dans U telle que

$$\lim_n \partial^k f_n(y) = \partial_U^k f(y) \quad \text{q.p. } y \in \overset{of}{V}_x \tag{9}$$

pour tout triplet $\{(f_n),x,V_x\}$ associé à f au point x de U et pour tout entier $k \in \mathbb{N}$. La fonction $\partial_U^k f$ s'appelle la dérivée ponctuelle d'ordre k de f.

Corollaire 3. Soit $f : U \to C$ une fonction complexe finement harmonique dans un ouvert fin $U \subset C$ et soit $\{(f_n),x,V_x\}$ un triplet associé à f au point x de U. Alors pour tout entier $k \geq 2$ et pour q.p. point $a \in \overset{of}{V}_x$ il existe une sous-suite (f_{n_i}) de (f_n) (dépendante de k et a) telle que:

$$\lim_{n_i} \partial^k f_{n_i}(Z^\tau)\cdot Z_{k-1}^\tau = \partial_U^k f(Z^\tau)\cdot Z_{k-1}^\tau \quad \text{dans} \quad \mathcal{H}_a[0,\tau[. \tag{10}$$

Preuve. Nous traitons le cas $k = 2$, les autres cas s'en déduisent par récurrence. Montrons d'abord le résultat assez général suivant: Soit

(g_n) une suite de fonctions définies dans un voisinage de V_x telle que $(g_n(Z^\tau))$ est une suite de Cauchy dans $L^2_a(<Z>_\tau)$ et telle que (g_n) converge ponctuellement vers une fonction finement continue g, alors:

$$\lim_n g_n(Z^\tau)\cdot Z^\tau = g(Z^\tau)\cdot Z^\tau \quad \text{dans} \quad \mathcal{H}_a[0,\tau[. \tag{11}$$

En effet puisque $(g_n(Z^\tau))$ est de Cauchy dans $L^2_a(<Z>_\tau)$ il existe alors un processus $Y = (Y(s)) \in L^2_a(<Z>_\tau)$ qu'on peut choisir de telle facon que ses parties imaginaire et réelle sont continues à droite, tel que:

$$\lim_n E_a\left(\int_0^\infty |g_n(Z^\tau) - Y|^2(s)d(s\wedge\tau)\right) = 0 .$$

Il existe donc un évenement $\Omega_1 \subset \Omega$ de mesure P^a égale à 1 et une sous-suite (g_{n_i}) de (g_n) tels que:

$$\lim_{n_i} \int_0^\infty |g_{n_i}(Z^\tau) - Y|^2(\omega,s)d(s\wedge\tau(\omega)) = 0, \text{ pour chaque } \omega \in \Omega_1.$$

D'autre part pour chaque $\omega \in \Omega_1$ fixé en posant $F_{n_i}(s) = |g_{n_i}(Z^\tau)|(\omega,0)$ $(s < \tau(\omega))$, alors il existe une sous-suite (F_{n_j}) de (F_{n_i}) telle que:

$$Y(\omega,s) = \lim_{m_j} F_{m_j}(\omega,s) = \lim_{m_j} g_{m_j}(Z^\tau_s(\omega)) = g(Z^\tau_s(\omega))$$

p.s. pour s sur $[0,\tau(\omega)[$. La fonction $g = g_1 + ig_2$ est par hypothèses finement continue dans $\overset{\text{of}}{V_x}$ il s'ensuit qu'il existe un évenement $\Omega_2 \subset \Omega_1$ de mesure pour P^a égale à 1 tel que pour tout $\omega \in \Omega_2$ les fonctions $s \to g_i(Z^\tau_s(\omega))$ $(i = 1,2)$ sont continues sur $[0,\tau(\omega)[$, d'autre part les parties réelle et imaginaire de la fonction $s \to Y(\omega,s)$ sont continues à droite sur le même interval $[0,\tau(\omega)[$ on obtient finalement:

$$Y(\omega,s) = g(Z^\tau_s(\omega)) \quad \text{sur} \quad [0,\tau(\omega)[\quad \text{pour tout} \quad \omega \in \Omega_2,$$

càd, l'assertion (11).

Pour montrer (10) avec $k = 1$ il suffit d'après ce qui précède de montrer qu'il existe une sous-suite (f_{n_i}) de (f_n) telle que:

$$\{\partial^2 f_{n_i}(Z^\tau)\} \quad \text{est de Cauchy dans} \quad L^2_a(<Z>_\tau). \tag{12}$$

D'après le lemme 1 le filtre $\{\partial^2 f_\alpha(Z^\tau)\cdot Z^\tau_2 | \alpha \in \mathcal{N}\}$ converge vers 0 dans $\mathcal{H}_a[0,\tau[$ pour tout $a \in \overset{\text{of}}{V_x}$, càd:

$$\lim E_a(<\partial^2 f_\alpha(Z^\tau)\cdot Z^\tau_2, \partial^2 f_\alpha(Z^\tau)\cdot Z^\tau_2>_\infty) = 2\lim E_a\left(\int_0^\infty |\partial^2 f_\alpha(Z^\tau)\cdot Z^\tau|^2(s)d(s\wedge\tau)\right)$$

$$= 2\lim E_a\left(\int_0^\infty (<\partial^2 f_\alpha(Z^\tau)\cdot Z^\tau, \partial^2 f_\alpha(Z^\tau)\cdot Z^\tau>_s + H^\alpha_s)d(s\wedge\tau)\right) = 0,$$

où $H = (H^\alpha_s)$ est une martingale uniformément intégrable.

En remarquant que

$$\lim E_a\left(\int_0^\infty H_s^\alpha d(s\wedge\tau)\right) = \lim E_a(H^\tau\cdot\tau) = 0$$

on a:

$$\lim E_a\left(\int_0^\infty \langle\partial^2 f_\alpha(Z^\tau)\cdot Z^\tau, \partial^2 f_\alpha(Z^\tau)\cdot Z^\tau\rangle_s d(s\wedge\tau)\right) = 0.$$

D'autre part il existe un événement $\Omega_2 \subset \Omega$ de mesure pour P^a égale à 1 tel que pour tout $\omega \in \Omega'$ and pour tout $\alpha \in \Omega_2$ les deux fonctions non-négatives $\langle\partial^2 f_\alpha(Z^\tau)\cdot Z^\tau, \partial^2 f_\alpha(Z^\tau)\cdot Z^\tau\rangle_s(\omega)$ et $(s\wedge\tau(\omega))$ sont croissantes sur $[0,\tau(\omega)[$. L'intégration parties des intégrales de Stieltjes sur $[0,\tau(\omega)[$ donne:

$$\int_0^\infty \langle\partial^2 f_\alpha(Z^\tau)\cdot Z^\tau, \partial^2 f_\alpha(Z^\tau)\cdot Z^\tau\rangle_s(\omega)d(s\wedge\tau(\omega)) =$$

$$\langle\partial^2 f_\alpha(Z^\tau)Z^\tau, \partial^2 f_\alpha(Z^\tau)Z^\tau\rangle_\infty(\omega)\tau(\omega) - \int_0^\infty (s\wedge\tau(\omega))d\langle\partial^2 f_\alpha(Z^\tau)Z^\tau, \partial^2 f_\alpha(Z^\tau)Z^\tau\rangle_s(\omega).$$

Ce qui entraine:

$$\lim E_a(\langle\partial^2 f_\alpha(Z^\tau)Z^\tau, \partial^2 f_\alpha(Z^\tau)Z^\tau\rangle_\infty) = \lim E_a\left(\left(\int_0^\infty |\partial^2 f_\alpha(Z^\tau)|^2(s)d(s\wedge\tau)\right)\cdot\tau\right) = 0.$$

Il s'ensuit que le filtre de variables aléatoires,

$$\left\{\left(\int_0^\infty |\partial^2 f_\alpha(Z^\tau)|^2(s)d(s\wedge\tau)\right)\tau \,\middle|\, \alpha \in \mathcal{N}\right\}$$

converge en probabilité vers 0 pour P^a, d'autre part puisque τ, étant le temps de sortie de $\overset{\circ}{V}{}^f_x$, est p.s. positive on en déduit qu'il existe un sous-filtre $\mathcal{N}'$ de $\mathcal{N}$ tel que:

$$\lim_{\mathcal{N}'} \int_0^\infty |\partial^2 f_\alpha(Z^\tau)|^2(s)d(s\wedge\tau(\cdot)) = 0 \quad \text{p.s.}P^a. \tag{13}$$

En évoquant l'inégalité à priori (1) on a d'autre part:

$$E_a(\sup_{s<\infty}|\partial^2 f_\alpha(Z^\tau)\cdot Z|^2(s)) \leqq C_2 E_a(\langle\partial^2 f_\alpha(Z^\tau)\cdot Z^\tau\rangle_\infty) < +\infty$$

ce qui montre que les termes du premier membre de (13) sont bornés par une variable aléatoire intégrable. On a d'après le théorème de convergence bornée de Lebesgue:

$$\lim_{\mathcal{N}'} E_a\left(\int_0^\infty |\partial^2 f_\alpha(Z^\tau)|^2(s)d(s\wedge\tau)\right) = 0,$$

càd, (12) est démontrée.

Remarques. Au lieu de considérer la partie holomorphe $\{\partial^k f_m(Z^\tau)\cdot Z_k^\tau\}$ de la suite de processus harmoniques $\{f_m(Z^\tau) - f_m(Z_o^\tau)\}$ nous pouvons considérer sa partie anti-holomorphe $\{\bar\partial^k f_m(Z^\tau)\cdot\bar Z_k^\tau\}$ pour étudier les suites $\{\bar\partial^k f_m\}$. Avec des arguments analogues à ceux du lemme 1 et du corollaire 3 nous pouvons énoncer les résultats suivants:

<u>Lemme 1'.</u> Sous les mêmes hypothèses que celles du lemme 1, sont équivalents:

(a) Pour tout entier k et pour le point a de $\overset{\circ f}{V_x}$ la suite $\{\bar\partial^k f_n(a)\}$ est de Cauchy dans $\mathbb{C}$.

(b) La suite d'intégrales stochastiques itérés $\{\bar\partial^{k+1} f(Z^\tau)\cdot \bar Z^\tau_{k+1}\}$ est de Cauchy dans $\mathcal{H}_a[0,\tau[$.

De plus on a: Pour tout k et pour q.p. $a \in \overset{\circ f}{V_x}$ la suite $\{\bar\partial^k f_n(a)\}$ est de Cauchy et la fonction $\partial^k_U f$ définie ponctuellement quasi-partout comme $\partial^k_U(y) = \lim_n \partial^k f(y)$, $y \in U_o$, est indépendant du triplet $\{(f_n), x, V_x\}$.

<u>Definition 2'.</u> A toute fonction f de $H_f(U)$ on peut associer une et une seule suite $\{\bar\partial^k_U f\}$ de fonctions quasi-finement continues dans U telle que: Pour tout triplet $\{(f_n), x, V_x\}$ associé à f au point x de U et pour tout entier $k \in \mathbb{N}$, $\lim_n \bar\partial^k f_n(y) = \bar\partial^k_U f(y)$ q.p. La fonction $\bar\partial^k_U f$ s'appelle la <u>dérivée à bord ponctuelle d'ordre k</u> de f.

<u>Corollaire 3'.</u> Sous les mêmes hypothèses que celles du lemme 1, alors pour tout entier $k \geq 2$ et pour q.p. point $a \in \overset{\circ f}{V_x}$ il existe une sous-suite (f_{n_i}) de (f_n) telle que:

$$\lim_{n_i} \bar\partial^k f_n(Z^\tau)\cdot \bar Z^\tau_{k-1} = \partial^k_U f(Z^\tau)\cdot \bar Z^\tau_{k-1} \quad \text{dans } \mathcal{H}_a[0,\tau[.$$

<u>Théorème 4.</u> a) Pour tout entier $k \in \mathbb{N}$ la dérivée ponctuelle d'ordre k $\partial^k_U f$ (resp. $\bar\partial^k_U f$) d'une fonction finement harmonique f $(\in H_f(U))$ est quasi-finement holomorphe dans U (resp. anti-holomorphe dans U)

b) Pour tout entier $k \in \mathbb{N}$ la fonction f admet un développement de Taylor d'ordre k de la forme: Pour tout $x \in U$ il existe un voisinage fin compact V_x de x dans U tel que,

$$f(Z^\tau) - f(Z^\tau_o) = \sum_{i=1}^{k} (\partial^i_U f(Z^\tau_o) Z^\tau_i + \bar\partial^i_U f(Z^\tau_o)\bar Z^\tau_i) + \partial^{k+1}_U f(Z^\tau) Z^\tau_{k+1} + \bar\partial^{k+1}_U f(Z^\tau)\bar Z^\tau_{k+1} \tag{14}$$

dans $\mathcal{H}^1_a[0,\tau[$ pour q.p. point $a \in \overset{\circ f}{V_x}$. De plus la représentation (14) est unique.

<u>Preuve.</u> Nous traitons le cas $k = 1$, les autres cas s'en déduisent par récurrence. Fixons un point $a \in \overset{\circ f}{V_x}$ alors on a pour tout $n \in \mathbb{N}$:

$$f_n(Z^\tau) - f_n(Z^\tau_o) = \partial f_n(Z^\tau) Z^\tau + \bar\partial f_n(Z^\tau)\bar Z^\tau$$

et

$$f_n(Z^\tau) - f_n(Z^\tau_o) = \partial f_n(Z^\tau_o) Z^\tau + \bar\partial f_n(Z^\tau_o)\bar Z^\tau + \partial^2 f_n(Z^\tau) Z^\tau_2 + \bar\partial^2 f_n(Z^\tau)\bar Z^\tau_2 \tag{15}$$

(voir (4)). En utilisant la formule (11) pour les suites $\{\partial f_n(Z^\tau) Z^\tau\}$ et $\{\bar\partial f_n(Z^\tau)\bar Z^\tau\}$ et la formule (12) pour les suites $\{\partial^2 f_n(Z^\tau) Z^\tau\}$ et $\{\bar\partial^2 f_n(Z^\tau)\bar Z^\tau\}$ on obtient en tendant n vers l'infini dans (15):

Pour q.p. $a \in \overset{\circ f}{V}_x$,

$$f(Z^\tau) - f(Z_0^\tau) = \partial_U f(Z^\tau) Z^\tau + \overline{\partial}_U f(Z^\tau)\overline{Z}^\tau \quad \text{dans } \mathcal{H}_a^1[0,\tau[$$

et (15')

$$f(Z^\tau) - f(Z_0^\tau) = \partial_U f(Z_0^\tau) Z^\tau + \overline{\partial}_U f(Z_0^\tau)\overline{Z}^\tau + \partial_U^2 f(Z^\tau) Z_2^\tau + \overline{\partial}_U^2 f(Z^\tau)\overline{Z}_2^\tau .$$

D'autre part puisque $\mathcal{H}_a^1[0,\tau[$ est la somme directe de $\mathcal{H}_a[0,\tau[$ et $\overline{\mathcal{H}}_a[0,\tau[$ les représentations dans (15') sont uniques, on a donc:

$$\partial_U f(Z^\tau) - \partial_U f(Z_o^\tau) = \partial_U^2 f(Z^\tau) Z^\tau \quad \text{dans } \mathcal{H}_a[0,\tau[$$

et (16)

$$\overline{\partial}_U f(Z^\tau) - \overline{\partial}_U f(Z_o^\tau) = \overline{\partial}_U^2 f(Z^\tau)\overline{Z}^\tau \quad \text{dans } \overline{\mathcal{H}}_a[0,\tau[.$$

Mais (16) n'est autre que la condition (b) du théorème de Cauchy-Riemann pour les fonctions finement holomorphes (resp. finement anti-holomorphes) ce qui entraine que $\partial_U f \in \mathcal{H}_f^q(U)$ (resp. $\overline{\partial}_U f \in \overline{\mathcal{H}}_f^q(U)$), [4], Théorème 5.

Remarques. 1) Dans une communication privée T. J. Lyons nous a informé qu'en utilisant la critère de continuité pour les mesures de Keldych et les techniques d'approximation rationnelle il a pu démontrer que la dérivée d'une fonction finement holomorphe est finement holomorphe (voir [3]).

2) Dans [4b] nous avons annoncé que la suite $\{\partial^2 f_m(Z^\tau)Z_2^\tau\}$ est elle même de Cauchy dans $\mathcal{H}_a[0,\tau[$ si f est finement holomorphe, mais d'après le corollaire 3 nous ne savons pas encore que la propriété ci-dessus est vraie ou non. En fait nous avons démontré essentiellement dans la preuve du corollaire 3 le fait suivant: Soit $\{(Y_n\cdot Z^\tau)Z^\tau\}$ une suite d'intégrales stochastiques itérés qui est de Cauchy dans $\mathcal{H}_a[0,\tau[$ alors il existe une sous-suite (Y_{n_i}) telle que la suite $\{Y_{n_i}\cdot Z^\tau\}$ est de Cauchy dans $L_a^2(\langle Z\rangle_\tau) \cap \mathcal{H}_a[0,\tau[$. Je tiens à remercier T. J. Lyons de m'avoir signalé l'erreur ci-dessus. Quelque auteur a utilisé la propriété ci-dessus pour les suites $\{\partial^k f_m(Z^\tau)Z_k^\tau\}$ comme les 'hypothèses a priori' pour étudier les dérivées des fonctions finement harmoniques réelles!

Corollaire 5. (Principe d'unicité fine) Soit $f : U \to C$ une fonction complexe finement harmonique dans un domaine fin $U \subset C$. Alors f est identiquement nulle dans U si et seulement si f est nulle dans un voisinage fin d'un certain point x_0 de U.

Preuve. D'après le théorème 4 les dérivées $\partial_U f$ et $\overline{\partial}_U f$ de f sont respectivement quasi-finement holomorphe et finement anti-holomorphe dans U, elles sont donc des transformations harmoniques fines dans C (voir [1]). D'autre part il est clair que $\partial_U f$ et $\overline{\partial}_U f$ sont aussi nulles dans un voisinage fin du point donné x_0 de U, ils sont donc identiquement nulles car $\{U \setminus e_f\}$ est finement ouvert et finement connexe. D'après la première représentation de f dans la formule (15) la fonction f est nulle sur un ensemble finement dense de U, elle est donc identiquement nulle dans U car elle y est finement continue. Remarquons que l'ensemble polaire $\{e_f\}$ depend de fonction f et que nous avons utilisé la propriété: $\{U \setminus e_f\}$ est un domaine fin si U l'est (voir [4a]).

References

[1] Fuglede, B.: Finely harmonic mappings and finely holomorphic functions. Ann. Acad. Sci. Fenn., Ser. A I 2 (1976), 113 - 127.

[2] Getoor, R. K., Sharpe, M. J.: Conformal martingales. Inventiones math. 16 (1972), 271 - 308.

[3] Lyons, T. J.: A theorem in fine potential theory and applications to finely holomorphic functions. Manuscript.

[4a] Nguyen-Xuan-Loc, Watanabe, T.: Characterization of fine domains for certain class of Markov processes. Z. Wahrscheinlichkeitstheorie verw. Gebiete 21 (1972), 167 - 178.

[4b] Nguyen-Xuan-Loc: Sur la théorie des fonctions finement holomorphes. Bull. Sci. math., II. Sér. 102 (1978), 337 - 364.

[5] McKean, H. P.: Stochastic integrals. Academic Press, New York - London (1969).

[6] Debiard, A. - Gaveau, B.: Différentiabilité des fonctions finement harmoniques. Inventiones math. 29 (1975), 111 - 123.

Mathématiques (Bat. 425)
Université Paris-Sud
F-91405 Orsay
France

INTEGRAL MEANS AND THE THEOREM OF HAMILTON, REICH AND STREBEL

Marvin Ortel*

0. Introduction

Let H denote the Banach space of functions, f, analytic on the upper half plane, U, with

$$||f|| = \iint_U |f|\,dxdy < \infty.$$

A dilatation, κ, is a complex function, bounded and measurable on U, with $||\kappa||_\infty < 1$; as usual

$$||\kappa||_\infty = \text{ess sup } \{|\kappa(z)| : z \in U\}.$$

We say that κ is an extremal dilatation if

$$\sup_{||f||=1} \left|\iint_U f(z)\kappa(z)\,dxdy\right| = ||\kappa||_\infty. \qquad (0.1)$$

This terminology is motivated by a remarkable connection, discovered by Richard Hamilton, Edgar Reich and Kurt Strebel, between the linear extremal problem (0.1) and the theory of quasiconformal mapping.

Theorem [2, 5, 6]. Let F be a quasiconformal homeomorphism of U with complex dilatation κ. Then F is extremal for its boundary values if and only if κ satisfies condition (0.1).

For dilatations in the angular class (defined below), there are explicit conditions equivalent to (0.1). In the following, R and $\hat{R}$ denote the real line and the extended real line respectively; $r(\theta)$ is a strictly positive continuous function on $[0,\pi]$; a V-sequence, $\langle f_n \rangle$ (we also write $\langle f_n \rangle \in V$), is a sequence of functions from H, each of unit norm, for which $\lim_{n\to\infty} f_n(z) = 0$ uniformly on compact subsets of U.

Definition (0.1). Let κ be a dilatation and let $\{\lambda_x : x \in \hat{R}\}$ be an indexed family of functions, each λ_x bounded and measurable on

---

* This research was partially supported by grant NRC 2057-04, under the administration of Professor Walter Schneider, at Carleton University, Ottawa.

$(0,\pi)$. Suppose for every $\varepsilon > 0$ and every $x \in \hat{R}$, we may choose $\delta(\varepsilon,x) > 0$ so that, if $0 < r(\theta) < \delta(\varepsilon,x)$ for $0 < \theta < \pi$, we have

$$\begin{aligned}\overline{\lim_{n\to\infty}} \left| \int_0^\pi \int_0^{r(\theta)} f_n(x + re^{i\theta})(\kappa(x + re^{i\theta}) - \lambda_x(\theta))rdrd\theta \right| \\ \leq \varepsilon \overline{\lim_{n\to\infty}} \int_0^\pi \int_0^{r(\theta)} |f_n(x + re^{i\theta})| rdrd\theta\end{aligned} \tag{0.2}$$

$$\begin{aligned}\overline{\lim_{n\to\infty}} \left| \int_0^\pi \int_{r(\theta)^{-1}}^\infty f_n(re^{i\theta})(\kappa(re^{i\theta}) - \lambda_\infty(\theta))rdrd\theta \right| \\ \leq \varepsilon \overline{\lim} \int_0^\pi \int_{r(\theta)^{-1}}^\infty |f_n(re^{i\theta})| rdrd\theta,\end{aligned} \tag{0.3}$$

for any V-sequence $\langle f_n \rangle$. Then κ is said to be <u>angular with associated</u> functions $\{\lambda : x \in \hat{R}\}$.

In the primary examples, $\{\lambda_x : x \in \hat{R}\}$ is formed from limits of κ; it is easy to see that κ is angular if the following simple conditions hold:

for each $x \in R$,

$$\lim_{r\to 0} \kappa(x + re^{i\theta}) = \lambda_x(\theta), \text{ uniformly for } 0 < \theta < \pi, \tag{0.4}$$

$$\lim_{r\to\infty} \kappa(re^{i\theta}) = \lambda_\infty(\theta), \text{ uniformly for } 0 < \theta < \pi. \tag{0.5}$$

A wide variety of quasiconformal maps have dilatations satisfying (0.4) and (0.5). These examples, as well as the proof of the next result, will appear in [4].

<u>Theorem (0.1).</u> Let κ be angular with associated functions $\{\lambda_x : x \in \hat{R}\}$. Then

$$\max_{\langle f_n\rangle \in V} \left[\overline{\lim_{n\to\infty}} \left| \iint_U f_n \kappa dxdy \right| \right] = \sup_{x\in\hat{R}} \left| \frac{1}{\pi} \int_0^\pi \lambda_x(0) e^{-2i\theta} d\theta \right| \tag{0.6}$$

Also, κ is an extremal dilatation if and only if either

$$\kappa(z) \equiv \|\kappa\|_\infty \overline{f_o(z)} / |f_o(z)|, \text{ some } f_o \text{ in } H \tag{0.7}$$

or

$$\sup_{x\in\hat{R}} \left| \frac{1}{\pi} \int_0^\pi \lambda_x(\theta) e^{-2i\theta} d\theta \right| = \|\kappa\|_\infty. \tag{0.8}$$

Equality (0.6) is the central fact in Theorem 0.1. The characterization of extremals, (0.7) and (0.8), is a consequence of (0.6).

Several questions are suggested with regard to conditions (0.2) and (0.3), and the concrete special conditions (0.4) and (0.5). Suppose we drop the uniformity restriction in (0.4) and (0.5). Is κ still an angular dilatation with associated family $\{\lambda_x : x \in \hat{R}\}$? In section 1, we give a counterexample; the construction also suggests a positive result which is proved in section 4. Section 2 contains a result on integral means of analytic functions, due to Hardy, Ingham and Polya, which is used in the proof. In section 3, we obtain a new condition, which applies to an arbitrary dilatation, and is necessary for extremality.

1. Example

The existence of limits (0.4) and (0.5), without the assumption of uniformity, does not imply that a dilatation is angular. We construct a dilatation, κ, and a V-sequence, $\langle f_n \rangle$, with the following properties:

1) $\kappa(z) = 0$ for all z not in a set $E \subset U$; $\overline{E} \cap \hat{R} = \{0\}$;

2) $\lim\limits_{r \to 0} \kappa(re^{i\theta}) = 0$, for all θ, $0 < \theta < \pi$, but the limits are not uniform in θ;

3) $\lim\limits_{n \to \infty} \iint\limits_U f_n \kappa dxdy = \|\kappa\|_\infty$;

4) $\lim\limits_{n \to \infty} \int_{\theta_1}^{\theta_2} \int_0^\infty |f(re^{i\theta})| r dr d\theta = 0$ if θ_1 and θ_2 are fixed, $0 < \theta_1 < \theta_2 < \pi$.

By property 3), κ is an extremal dilatation. Also, by 1) and 2) we may form a family $\{\lambda_x : x \in \hat{R}\}$ by taking $\lambda_x(\theta) = \lim\limits_{r \to 0} \kappa(x + re^{i\theta})$, $\lambda_\infty(\theta) = \lim\limits_{r \to \infty} \kappa(re^{i\theta})$ for $0 < \theta < \pi$; then $\lambda_x(\theta) = 0$ for all x and θ. By equation (0.6) of Theorem (0.1) and property 3), we see that κ is not angular with the associated family $\{\lambda_x : x \in \hat{R}\}$. We will not give the details, but property 4) and the proof of the equation (0.6), imply that κ is not angular in conjunction with any associated family. The breakdown is caused by the non-uniformity of the limits when $x = 0$.

Construction: For $0 < s < t < 1$ and $0 < \theta < \phi < \pi$, we set

$$C(s,t,\theta,\phi) = \{z : s < |z| < t,\ \theta < \arg z < \phi\}.$$

We construct a sequence $C_n = C(s_n, t_n, \theta_n, \phi_n)$ and a V-sequence, $\langle f_n \rangle$:

the sets C_n are pairwise disjoint; $\phi_n \searrow 0$, $t_n \searrow 0$; $\iint_{C_n} |f_n| dxdy \geq 1 - 1/n$ (recall that $\|f_n\| = 1$). Once this is accomplished, we set

$$\kappa(z) = \begin{cases} \frac{1}{2}\, \overline{f_n(z)}/|f_n(z)|, & z \in C_n \\ 0, & z \notin \bigcup_n C_n \end{cases}$$

Properties 1 through 4 follow immediately from this construction.

The functions f_n are formed by translating members of the family

$$g_\rho(z) = -B_\rho z^{\rho-2}(z+i)^{-\rho-2}, \qquad 0 < \rho < 1;$$

here, $B_\rho > 0$ is chosen so that $\iint_U |g_\rho| dxdy = 1$. It is not difficult to show, for fixed x and fixed $r > 0$, that

$$\lim_{\rho \to 0} \iint_{|z-x|<r} |g_\rho(z-x)| dA(z) = 1.$$

Thus, take $\varepsilon > 0$ and choose $x > 0$, $r > 0$ so that $0 < x - r < x + r < \varepsilon$ and so that $\arg z < \varepsilon$ for all z satisfying $|z - x| < r$. By (1.1), we may choose $\rho > 0$ so that $\iint_{|z-x|<r} |g_\rho(z-x)| dA(z) > 1 - \varepsilon/2$. We may choose s, t, θ, ϕ, subject to $x - r < s < t < x + r$ and $0 < \theta < \phi < \varepsilon$, so that

$$\iint_{C(s,t,\theta,\phi)} |g_\rho(z-x)| dA(z) > 1 - \varepsilon/2 - \varepsilon/2.$$

Now, if $f_1, f_2, \ldots, f_n$ and $C_1, C_2, \ldots, C_n$ have been constructed, we choose $\varepsilon < s_n, \theta_n$ and set $f_{n+1} = g_\rho(z-x)$, $C_{n+1} = C(s,t,\theta,\phi)$ as above. This completes the construction.

In contrast to property 5, <u>no</u> V-sequence, $\langle f_n \rangle$, satisfies

$$\lim_{n\to\infty} \int_{\theta_1}^{\theta_2} \int_0^\infty |f_n(re^{i\theta})| r dr d\theta = 1$$

when θ_1 and θ_2 are fixed, $0 < \theta_1 < \theta_2 < \pi$ ([1]).

2. Integral means of functions in H

For $0 < \theta < \pi$, denote the ray $\ell(\theta) : r \mapsto re^{i\theta}$, $0 < r < \infty$, and for f in H, set

$$L(f,\theta) = \int_{\ell(\theta)} |f(z)z| |dz|.$$

<u>We show that</u> $L(f,\theta)$ <u>is a convex function of</u> θ <u>on</u> $(0,\pi)$.

For $\varepsilon \geq 0$, set $S(\varepsilon) : \varepsilon < \operatorname{Im} \omega < \pi - \varepsilon$ and, for ω in $S(0)$, form

$g(\omega) = f(e^{\omega})e^{2\omega}$. For $0 < \theta < \pi$. write $J(g,\theta) = \int_{-\infty}^{\infty} |g(u + i\theta)|du$. Then

$$\int_0^{\pi} J(g,\theta)d\theta = \iint_{S(0)} |g|dudv = \|f\|, \tag{2.1}$$

and

$$J(g,\theta) = L(f,\theta), \qquad 0 < \theta < \pi; \tag{2.2}$$

so it suffices to show that $J(g,\theta)$ is convex on $0 < \theta < \pi$. If $\varepsilon > 0$, <u>then</u> g <u>is bounded on</u> $S(\varepsilon)$. This follows easily from the area form of the Mean Value Theorem and the fact that $\iint_{S(0)} |g|dudv$ is finite. If $J(g,\theta)$ is finite for $\theta = \varepsilon$ and $\theta = \pi - \varepsilon$, then, by a result of Hardy, Ingham and Polya ([3]), $J(g,\theta)$ is convex on $(\varepsilon, \pi - \varepsilon)$. However, by (2.1), the means $J(g,\theta)$ are finite for almost every θ. Thus $J(g,\theta)$ is convex on $(0,\pi)$.

3. A necessary condition

<u>Lemma.</u> Let ϕ, μ be non-negative functions on $(0,\pi)$. We assume that μ is bounded and measurable and we assume that ϕ is convex and integrable. Then

$$\int_0^{\pi} \phi(t)\mu(t)dt \leqq C(\mu),$$

where

$$C(\mu) = \sup_{0<h<\pi} \left\{\frac{1}{h}\int_0^h \mu(t)dt, \frac{1}{h}\int_{\pi-h}^{\pi} \mu(t)dt\right\}.$$

<u>Proof.</u> We may express ϕ as a convex combination of two non-negative monotone functions and approximate each of these by a step function. Thus it suffices to assume that ϕ is a non-negative, non-decreasing step function on $(0,\pi)$. We then have

$$\phi(t) = \sum_{j=1}^{n} c_j\phi_j(t),$$

where $\phi_j(t) = \begin{cases} 0 & 0 < t < h_j \\ (\pi - h_j)^{-1} & h_j < t < \pi \end{cases}$, and $0 \leqq h_1 < h_2 < \cdots < h_n < \pi$.

Since ϕ is a non-decreasing, each c_j is non-negative; also $\sum_{j=1}^{n} c_j = \int_0^{\pi} \phi(t)dt$. Therefore

$$\int_0^{\pi} \phi(t)\mu(t)dt \leqq \left[\max_{j=1,2,\ldots,n} \int_0^{\pi} \phi_j(t)\mu(t)dt\right] \cdot \sum_{j=1}^{n} c_j$$

$$\leq \left[\sup_{0<h<\pi} \frac{1}{h} \int_{\pi-h}^{\pi} \mu(t)dt\right] \cdot \int_0^{\pi} \phi(t)dt \leq C(\mu) \int_0^{\pi} \phi(t)dt,$$

as required in the Lemma.

Now, for a dilatation κ, set

$$\mu(\kappa)(\theta) = \sup\{|\kappa(re^{i\theta})| : 0 < r < \infty\}.$$

For f in Π, it is clear that

$$\left|\iint_U f\kappa dxdy\right| \leq \int_0^{\pi} L(f,\theta)\mu(\kappa)(\theta)d\theta$$

and, since $L(f,\theta)$ is a convex on $(0,\pi)$ and $\int_0^{\pi} L(f,\theta)d\theta = \|f\|$, the Lemma gives the estimate

$$\left|\iint_U f\kappa dxdy\right| \leq C(\mu(\kappa))\,\|f\|. \tag{*}$$

Since $C(\mu(\kappa)) \leq \|\kappa\|_\infty$, (*) immediately leads to the following necessary condition.

<u>Theorem (3.1).</u> Let κ be an extremal dilatation. Then $C(\mu(\kappa)) = \|\kappa\|_\infty$.

Theorem (3.1) relates to the construction in section 1. For $0 < \delta < 1$, let $\Theta(\kappa,\delta)$ denote the set of arguments, θ, for which $\mu(\kappa)(\theta) \leq \delta\|\kappa\|_\infty$. How "thick" can $\Theta(\kappa,\delta)$ be near 0 and π? The following answer follows directly from Theorem (3.1).

<u>Corollary.</u> Let κ be extremal and let $0 < \delta < 1$. Then $\Theta(\kappa,\delta)$ has metric density zero at one of the points 0, π.

4. The uniformity condition

Consider the following dilatation

$$\kappa(x,y) = \begin{cases} 1/2, & 0 < x < 1, \quad 0 < y < \sqrt{x} \\ -1/2, & -1 < x < 0, \; 0 < y < \sqrt{-x}; \\ 0, & \text{otherwise} \end{cases}$$

also, let $\{\lambda_x : x \in \hat{R}\}$ be the associated family formed from the limiting values of κ along lines crossing R. At $x = 0$, for instance,

$$\lambda_0(\theta) = \begin{cases} 1/2 & 0 < \theta < \pi/2 \\ -1/2 & \pi/2 < \theta < \pi \end{cases}$$

We have

$$\lim_{r\to 0} \kappa(re^{i\theta}) = \lambda_o(\theta), \qquad 0 < \theta < \pi,$$

but these limits are not assumed uniformly. So it is not clear that κ is angular. However, the uniformity only breaks down on an interior interval (θ_1,θ_2), $0 < \theta_1 < \theta_2 < \pi$. This is in contrast to the example of section 1 and, in the present case, we have a positive result.

Theorem (4.1). Let κ be a dilatation. Suppose that, for each $x \in R$

$$\lim_{r\to 0} \kappa(x + re^{i\theta}) = \lambda_x(\theta), \qquad 0 < \theta < \pi$$

and

$$\lim_{r\to\infty} \kappa(re^{i\theta}) = \lambda_\infty(\theta), \qquad 0 < \theta < \pi.$$

Moreover, assume that, for each x, the corresponding limits are uniform, in θ, on intervals $(0,\theta_1(x))$ and $(\theta_2(x),\pi)$, where $0 < \theta_1(x) < \theta_2(x) < \pi$. Then κ is angular with associated functions $\{\lambda_x : x \in \hat{R}\}$.

Proof. We verify condition (0.2) only in the case $x = 0$. The proof when $x \neq 0$ is the same. Set $x = 0$, $\lambda_o(\theta) = \lambda(\theta)$, $\theta_1(0) = \theta_1$, $\theta_2(0) = \theta_2$ let $\varepsilon > 0$ and let $<f_n>$ be an arbitrary V-sequence. We choose $\delta(\varepsilon)$ so that $|\kappa(re^{i\theta}) - \lambda(\theta)| < \varepsilon/2$ for $0 < \theta < \theta_1$, $\theta_2 < \theta < \pi$ and $0 < r < \delta(\varepsilon)$. Then if $0 < r(\theta) < \delta(\varepsilon)$, $0 < \theta < \pi$, we have

$$\begin{aligned} &\left|\left(\int_0^{\theta_1} + \int_{\theta_2}^{\pi}\right) \int_{r=0}^{r(\theta)} f_n(re^{i\theta})(\kappa(re^{i\theta}) - \lambda(\theta))\,r\,dr\,d\theta\right| \\ &\quad \leq \varepsilon/2\left(\int_0^{\theta_1} + \int_{\theta_2}^{\pi}\right) \int_{r=0}^{r(\theta)} |f_n(re^{i\theta})|\,r\,dr\,d\theta \end{aligned} \tag{4.1}$$

for all n.

To estimate the part of the integral for arguments $\theta_1 < \theta < \theta_2$, note that

$$M_n(\theta) \equiv \int_0^{r(\theta)} |f_n(re^{i\theta})(\kappa(re^{i\theta}) - \lambda(\theta))|\,r\,dr \leq 2L(f_n,\theta), \qquad 0 < \theta < \pi.$$

Since $L(f_n,\theta)$ is convex on $(0,\pi)$ and $\int_0^\pi L(f_n,\theta)d\theta = \|f\| = 1$, there is a uniform bound $L(f_n,\theta) \leq K$ which holds for $\theta_1 \leq \theta \leq \theta_2$ and all n. Hence $M_n(\theta) \leq 2K$ for $\theta_1 \leq \theta \leq \theta_2$ and all n. Also, as $<f_n>$

vanishes on compact subsets of U and $\lim\limits_{r\to 0} \kappa(re^{i\theta}) = \lambda(\theta)$, it follows that $\lim\limits_{n\to\infty} M_n(\theta) = 0$ for <u>all</u> θ, $0 < \theta < \pi$. By the Bounded Convergence Theorem, we conclude

$$\lim_{n\to\infty} \int_{\theta_1}^{\theta_2} M_n(\theta)\,d\theta = 0.$$

Therefore

$$\lim_{n\to\infty} \int_{\theta_1}^{\theta_2} \int_{0}^{r(\theta)} |f_n(re^{i\theta})(\kappa(re^{i\theta}) - \lambda(0))|\,r\,dr\,d\theta = \lim_{n\to\infty} \int_{\theta_1}^{\theta_2} M_n(\theta)\,d\theta = 0.$$

This, combined with (4.1) completes the proof.

References

[1] Belna, C., Ortel, M.: Extremal quasiconformal mappings: necessary conditions. J. Analyse math. 33 (1978), 1 - 11.

[2] Hamilton, R.S.: Extremal quasiconformal mappings with prescribed boundary values. Trans. Amer. math. Soc. 138 (1969), 399 - 406.

[3] Hardy, G.H., Ingham, A.E., Polya, G.: Theorems concerning mean values of analytic functions. Proc. roy. Soc., Ser. A 113 (1927), 542 - 569.

[4] Ortel, M.: Quasiconformal maps extremal for their boundary values. To appear.

[5] Reich, E., Strebel, K.: Extremal quasiconformal maps with given boundary values, in "Contributions to Analysis". Academic Press, New York and London (1974), 375 - 391.

[6] Strebel, K.: On quadratic differentials and extremal quasiconformal mappings, in "Proceedings of the International Congress of Mathematicians, Vancouver 1974, Vol. 2", Canadian Mathematical Congress (1975), 223 - 227.

Department of Mathematics
University of Hawaii
2565 The Mall
Honolulu, HI 96822
U.S.A.

INNER FUNCTIONS WITH A LEVEL-SET OF INFINITE LENGTH

George Piranian

1. Introduction

Corresponding to every meromorphic function f in the unit disk D and every positive number R, let $\mathcal{L}E(R,f)$ denote the length of the R-level-set

$$E(R,f) = \{z \in D : |f(z)| = R\} .$$

In 1974, A. Weitsman asked whether a bounded holomorphic function can have a level set of infinite length, and in 1978, he and I showed that if $\{R_j\}$ is a countable set in the interval $(0,1)$, then there exists an outer function f such that $\mathcal{L}E(R_j,f) = \infty$ for each index j (see [6]). In a more recent paper [1], C. Belna and I proved that there exists a Blaschke product B such that $\mathcal{L}E(1/e,B) = \infty$. About the same time, Peter W. Jones gave an alternate proof (private communication). Later, Jones constructed a bounded holomorphic function f such that each nonempty R-level-set of f has infinite length [4].

The present paper gives an analogue for singular inner functions to the result on Blaschke products, and it states some open problems.

2. The singular inner function

Theorem 1. If $A(z) = \exp(z+1)/(z-1)$ in D, then $\mathcal{L}(1/e, A \circ A) = \infty$.

Proof. Let γ denote the circle $|w| = 1/e$. The image of γ under the mapping A^{-1} is the circle γ^* with center $1/2$ and radius $1/2$. The inverse of γ^* under exponentiation consists of a family of congruent curves $\delta_n (n = 0, \pm 1, \ldots)$ such that the closure of δ_n passes through the point at infinity and the point $2n\pi i$. Consequently, the function A^{-1} maps the circle γ^* onto a family $\{\delta_n^*\}$, where δ_n^* denotes a curve whose closure passes through the two points 1 and $\frac{2n\pi i + 1}{2n\pi i - 1} = 1 - i/n\pi + O(n^{-2})$ $(n \neq 0)$. Because for $n \neq 0$ the length of

---

The author gratefully acknowledges support from the National Science Foundation.

δ_n^* has the order of magnitude $1/|n|$, the length of $E(1/e, A \circ A)$ is infinite. This concludes the proof.

3. The order of normality

Ch. Pommerenke [7, p. 3] has defined the order of normality of a meromorphic function f in D as the quantity

$$\sup_{z \in D} (1 - |z|^2) \frac{|f'(z)|}{1 + |f(z)|^2}$$

(f is normal if and only if its order of normality is finite). S. Dragosh [2, Theorems 2 and 3] showed that if f is meromorphic in D and some level-set $E(R,f)$ contains a sequence of Koebe arcs, then the order of normality of f is not less than the solution C^* of the equation

$$1 + \sqrt{1 + C^2} = C \exp\sqrt{1 + C^2} ,$$

and the assertion fails if the constant C^* is replaced with a larger number. The value of C^* is slightly less than 2/3.

Theorem 2. The order of normality of the atomic function is C^*.

It is convenient to prove slightly more, namely, that if $b > 0$ and $f(z) = \exp b(z + 1)/(z - 1)$ in D, then the order of normality of f is C^*. The formula

$$f'(z) = -2bf(z)/(z - 1)^2$$

implies that

$$(1 - |z|^2) \frac{|f'(z)|}{1 + |f(z)|^2} = (1 - |z|^2) \frac{|f'(z)/f(z)|}{|f(z)| + 1/|f(z)|}$$

$$= \frac{1 - |z|^2}{|1 - z|^2} \cdot \frac{2b}{|f(z)| + 1/|f(z)|} .$$

On the circle C_r with center r $(0 < r < 1)$ and radius $1 - r$, the real part of $(z + 1)/(z - 1)$ is constant, and because the point $2r - 1$ lies on C_r, the constant value is $-r/(1 - r)$. It follows that

$$|f(z)| + 1/|f(z)| = 2 \cosh br/(1 - r)$$

on C_r. Also, if $z = r + (1 - r)e^{i\phi}$, then

$$|z|^2 = r^2 + (1 - r)^2 + 2r(1 - r)\cos \phi ,$$

and therefore

$$1 - |z|^2 = 2r(1 - r)(1 - \cos\phi) .$$

Moreover,

$$|1 - z|^2 = [2(1 - r)|\sin\phi/2|]^2 = 2(1 - r)^2(1 - \cos\phi) .$$

Consequently,

$$(1 - |z|^2)|f'(z)|/(1 + |f(z)|^2) = \frac{br}{1 - r} /\cosh \frac{br}{1 - r} .$$

We deduce that regardless of the value of b, the order of normality of our function is the maximum value of $\lambda/\cosh \lambda$. By differential calculus, the maximum occurs at the point where $\lambda = \coth \lambda$, and the maximum value is

$$\frac{\lambda}{\cosh \lambda} = \frac{\coth \lambda}{\cosh \lambda} = 1/\sinh \lambda .$$

Now, if $\lambda = \coth \lambda$ and $C = 1/\sinh \lambda$, then

$$1 + \sqrt{1 + C^2} = 1 + \coth \lambda = \frac{\sinh \lambda + \cosh \lambda}{\sinh \lambda} = C \exp \lambda = C \exp(\coth \lambda)$$
$$= C \exp\sqrt{1 + C^2} .$$

In other words, the order of normality of each of the functions $\exp b(z + 1)/(z - 1)$ is the constant C^* in Dragosh's theorem.

By different methods, J. S. Hwang has obtained similar results [3].

The measure $d\mu$ in the Herglotz representation of the function $A \circ A$ consists of point masses. It is easy to see that in the neighborhood of each corresponding point of the unit circle, $A \circ A$ has the local order of normality C^*, and it follows immediately that the order of normality of $A \circ A$ is not less than C^*. The following theorem, kindly supplied by D. Campbell, implies that the order of normality of $A \circ A$ is not greater than C^*.

Theorem 3. If f is normal and g is a holomorphic mapping of D into D, then the function $f \circ g$ is normal, and its order of normality is not greater than that of f.

Proof. The order of $f \circ g$ is the supremum over D of the quantity

$$\frac{(1 - |z|^2)|f'(g) \cdot g'|}{1 + |f(g)|^2} = \frac{(1 - |z|^2)|g'(z)|}{1 - |g(z)|^2} \cdot \frac{(1 - |g(z)|^2)|f'(g)|}{1 + |f(g)|^2} .$$

The first factor on the right is not greater than 1 because the mapping g does not increase hyperbolic length. The second factor obviously can not exceed the order of normality of f.

4. Unsolved problems

Problem 1. A Blaschke product B is <u>indestructible</u> if for each α in D the function $(\alpha - B)/(1 - \bar{\alpha}B)$ is also a Blaschke product [5, p. 697]. Can an indestructible Blaschke product have a level-set of infinite length?

Problem 2 (private communication from D. Campbell). Each known level-set of infinite length of a bounded holomorphic function consists of infinitely many components, each of finite length. Can a bounded holomorphic function in D have constant modulus on an arc of infinite length?

Problem 3 (raised by a participant in the Conference at Joensuu). Does there exist a continuous singular measure $d\mu$ such that the corresponding inner function has a level-set of infinite length?

Problem 4 (A. Weitsman). Can a univalent holomorphic function have a level-set of infinite length? My conjecture: If f is holomorphic and univalent in D, then $\mathcal{L}E(R,f) < \pi^2$ for all R.

Problem 5. If $\{R_j\}$ is a countable set in $(0,1)$, does there exist a Blaschke product or a singular inner function each of whose R_j-level-sets has infinite length?

Problem 6. For various function classes (such as the family of Blaschke products or of indestructible Blaschke products) determine the infimum of the order of normality of functions possessing a level-set of infinite length.

Problem 7. If f is holomorphic in D, then the set of values R for which $\mathcal{L}E(R,f) = \infty$ is of type G_δ. If M is a set of type G_δ on the positive real axis, does there exist a holomorphic function f in D such that $\mathcal{L}E(R,f) = \infty$ if and only if $R \in M$?

References

[1] Belna, C., Piranian, G.: A Blaschke product with a level-set of infinite length. To appear in "Memorial Volume for Professor P. Turán"

[2] Dragosh, S.: Koebe sequences of arcs and normal meromorphic functions. Trans. Amer. math. Soc. 190 (1974), 207 - 222.

[3] Hwang, J. S.: On the O.K. constant of Doob and Seidel class. Preprint (1978).

[4] Jones, P. W.: Bounded holomorphic functions with all level sets of infinite length. To appear in Michigan math. J. 26 (1979).

[5] McLaughlin, R.: Exceptional sets for inner functions. J. London math. Soc., II. Ser. 4 (1972), 696 - 700.

[6] Piranian, G., Weitsman, A.: Level sets of infinite length. Commentarii math. Helvet. 53 (1978), 161 - 164.

[7] Pommerenke, Ch.: Estimates for normal meromorphic functions. Ann. Acad. Sci. Fenn., Ser. A I 476 (1970).

The University of Michigan
Ann Arbor, MI 48109
U.S.A.

ON THE UNIQUENESS PROBLEM FOR EXTREMAL QUASICONFORMAL MAPPINGS WITH PRESCRIBED BOUNDARY VALUES

Edgar Reich*

0. Section 1 will be devoted to a survey of a class of extremal problems for plane quasiconformal mappings. This will serve the double purpose of presenting a historical review as well as motivating, in Section 2, a contribution to the question of when the extremal mappings are unique.

1. Quasiconformal mappings were introduced by Grötzsch [4] who considered mappings of an oriented rectangle $R = \{0 \le x \le a,\ 0 \le y \le 1\}$ onto an oriented rectangle $R' = \{0 \le u \le a',\ 0 \le v \le 1\}$, $a' > a$, with correspondence of the vertices. He showed that for this class of mappings the affine stretch

$$w = u + iv = A_k(z) = Kx + iy, \quad (z = x + iy,\ K = \frac{a'}{a})$$

is extremal in the sense that the maximal dilatation is minimized. Furthermore, A_k is the unique extremal mapping in this case. In a far-reaching generalization Teichmüller [16] considered quasiconformal mappings of a Jordan domain Ω with n $(n \ge 4)$ distinguished boundary points, $z_j \in \partial\Omega$, $j = 1,2,\ldots,n$, onto a Jordan domain Ω', for which $f(z_j)$, $j = 1,2,\ldots,n$, are specified, and showed that there is a unique extremal mapping $f^*(z)$. The complex dilatation of $f^*(z)$ is of the form

$$\kappa(z) = \frac{f^*_{\bar z}}{f^*_z} = k\,\frac{\overline{\varphi(z)}}{|\varphi(z)|} \tag{1.1}$$

where $0 < k < 1$, and where $\varphi(z)$ belongs to the Banach space $\mathcal{B}(\Omega)$ of functions analytic in Ω for which

$$||\varphi|| = \iint_{\Omega} |\varphi(z)|\,dx\,dy < \infty. \tag{1.2}$$

A mapping with complex dilatation of the form (1.1) for which $\varphi \in \mathcal{B}(\Omega)$ is referred to as a Teichmüller mapping with finite norm.

More generally, let Ω be an arbitrary region of the complex plane

---

* Work done with support of the National Science Foundation, and the Forschungsinstitut für Mathematik, E.T.H., Zürich.

C, $\Omega \neq C$. For convenience we will assume that Ω is simply connected, although this is not essential. Let g be a quasiconformal mapping of Ω, and Q_g the class of all quasiconformal mappings of Ω whose boundary values, on $\partial\Omega$, (in the sense of prime ends) agree pointwise with those of g. The question of whether g is an extremal mapping in the class Q_g, that is, whether g possesses minimal maximal dilatation among all mappings with the same pointwise boundary values as g, was also formulated by Teichmüller [16]. Important progress on this question was due to Strebel [14, 15] who showed [15] that if g is a Teichmüller mapping with finite norm then g is an extremal mapping in the class Q_g, and moreover it is the <u>unique</u> <u>extremal</u> element of Q_g.

If Ω has <u>finite</u> area, and $g(z) = A_K(z)$, $z \in \Omega$, then the foregoing remark shows that A_K is the <u>unique</u> <u>extremal</u> mapping in the class Q_g. If Ω has <u>infinite</u> area, on the other hand, then, as the following examples show, more information on Ω is needed to determine whether A_K is extremal and/or uniquely extremal.

<u>Example 1.1</u> [14] $\Omega = \{z : c < \arg z < c + \lambda\}$, $0 < \lambda < 2\pi$. In this case A_K is <u>not</u> extremal in the class Q_g.

<u>Example 1.2</u> [14] $\Omega = \{z = x + iy : -1 < x < 1,\ y > 0\}$. In this case A_K is <u>uniquely</u> <u>extremal</u> in the class Q_g.

<u>Example 1.3</u> $\Omega = S \cup H$, where $S = \{z = x + iy : 0 < x \leq 1,\ 0 < y < x^2\}$, $H = \{z : \operatorname{Re} z > 1\}$. (This is a modification of Strebel's classical "chimney" example [14, pg. 316].) We assert that A_K is <u>extremal</u> for the boundary values it induces on $\partial\Omega$, but not <u>uniquely</u> <u>extremal</u>. To prove extremality we use a general necessary [5] and sufficient [8, 9] condition which can be formulated as follows: If g is a quasiconformal mapping of Ω, and $\kappa(z) = g_{\bar{z}}/g_z$, let Λ_κ be the linear functional

$$\Lambda_\kappa[\varphi] = \iint_\Omega \kappa(z)\varphi(z)\,dx\,dy, \quad \varphi \in \mathcal{B}(\Omega). \tag{1.3}$$

Then g <u>is extremal in the class</u> Q_g if and only if*

$$\|\Lambda_\kappa\| = \|\kappa\|_\infty. \tag{1.4}$$

Since A_K has complex dilatation $\kappa(z) = (K - 1)/(K + 1)$, condition (1.4), which is necessary and sufficient for extremality of A_K, becomes

---

* For a corresponding necessary and sufficient condition for extremality in case the admissible class of mappings is defined by a point-dependent dilatation bound see [10].

$$\|\Lambda_\Omega\| = 1, \tag{1.5}$$

where

$$\Lambda_\Omega[\varphi] = \iint_\Omega \varphi(z)\,dx\,dy, \quad \varphi \in \mathcal{B}(\Omega). \tag{1.6}$$

To show that (1.5) is satisfied in Example 1.3, let $F(z)$ be any function analytic at $z = 0$ and belonging to $\mathcal{B}(\Omega)$, and such that $F(0) = 1$ (for example, $F(z) = (z + 1)^{-3}$). Let

$$\varphi_n(z) = \frac{n^3}{2} e^{-nz} F(z), \quad z \in \Omega.$$

As is easily verified,

$$\lim_{n\to\infty} \|\varphi_n\| = \lim_{n\to\infty} \Lambda_\Omega[\varphi_n] = 1.$$

Thus, (1.5) holds, implying that A_K is extremal.

To see that A_K is not uniquely extremal, consider

$$f(z) = \begin{cases} A_K(z) & , \quad z \in S \\ K - 1 + z, & z \in H \end{cases}.$$

f has the same boundary values on $\partial\Omega$ as A_K; furthermore,

$$\operatorname*{ess\,sup}_{z \in \Omega} |f_{\bar z}/f_z| = \frac{K-1}{K+1}.$$

Therefore f is also extremal.

<u>Example 1.4</u> $\Omega_\alpha = \{x + iy : y > |x|^\alpha\}$, $\alpha > 1$. It is known [2] that A_K is <u>extremal</u> for any α, $\alpha > 1$. It has been shown [2, 6] that A_K is <u>uniquely extremal</u> if and only if $\alpha \geq 3$. (For a generalization involving geometric criteria cf. [7])

2. For reference we recall the following.

<u>Conjecture</u>* Suppose the quasiconformal mapping g of Ω is extremal within the class Q_g, and $\kappa(z) = g_{\bar z}/g_z$. Then g is a <u>unique</u> extremal if and only if the Hahn-Banach extension of Λ_K from $\mathcal{B}(\Omega)$ to $\mathcal{L}^1(\Omega)$ is unique**.

In the case when g is a Teichmüller mapping with finite norm verification of the conjecture is immediate. However since the unit ball

* Cf [10], last paragraph, for a somewhat more general version.

** Since g is assumed extremal condition (1.4) holds.

of $\mathcal{L}^{\infty}(\Omega)$ is not strictly convex the simplest standard test [3] for uniqueness of the Hahn-Banach extension fails to yield any information in general.

Our current contribution to the question of whether the conjecture holds is the following. First of all, we restrict ourselves in what follows to the case of affine mappings A_K of simply-connected regions Ω. Under a certain restriction on Ω, of a geometric nature, we succeed in establishing the truth of the conjecture in one direction. Technically Ω could have either finite or infinite area, but it is of course only in the latter case that a problem exists.

We will say that Ω is <u>expansive</u> if the set

$$\{w = A_K(z) : z \in \Omega\} \quad (K \geq 1)$$

increases as K increases.

<u>Theorem.</u> Suppose Ω is expansive, $\overline{\Omega} \neq C$. If for some K, $(K > 1)$, the mapping $A_K(z)$, $z \in \Omega$, is uniquely extremal for its boundary values, then the linear functional Λ_Ω possesses a unique Hahn-Banach extension from $\mathcal{B}(\Omega)$ to $\mathcal{L}^1(\Omega)$.

We shall require the following result [13, 12 (Theorem 5)]. Let $L(w)$, $w \in C$ be a complex-valued function with properties (i), (ii), (iii):

(i) $L(w)$ is continuous

(ii) L_w, $L_{\bar{w}}$ exist in the distributional sense, and as locally integrable functions

(iii) $|L_{\bar{w}}| \leq \frac{1}{2}$ a.e.

<u>Then the Löwner-type differential equation,</u>

$$\frac{dw}{dt} = L(w), \quad t \geq 0, \quad w \in C \tag{2.1}$$

<u>has a unique solution</u> $w = f(t,z)$ <u>that satisfies the initial condition</u> $f(0,z) = z$. Considered as a function of z, for fixed t, $f(t,z)$ <u>is an e^t-quasiconformal mapping of the plane</u>.

Proceeding now with the proof of the theorem we first use the fact that $\|\Lambda_\Omega\| = 1$, (where the norm is taken with respect to $\mathcal{B}(\Omega)$) since $A_K(z)$, $z \in \Omega$, is extremal. The formula

$$\tilde{\Lambda}[\varphi] = \iint_\Omega \varphi(z)\,dx\,dy, \quad \varphi \in \mathcal{L}^1(\Omega),$$

defines an extension of the linear functional Λ_Ω from $\mathcal{B}(\Omega)$ to $\mathcal{L}^1(\Omega)$. Clearly the norm (with respect to $\mathcal{L}^1(\Omega)$) has the value $\|\tilde{\Lambda}\| = 1$; i.e. $\tilde{\Lambda}$ is a Hahn-Banach extension of Λ_Ω. Suppose, contrary to the assertion that is to be verified, $\hat{\Lambda}$ is also a Hahn-Banach extension of Λ_Ω from

$\mathcal{B}(\Omega)$ to $\mathcal{L}^1(\Omega)$, $\hat{\Lambda} \neq \tilde{\Lambda}$. By Riesz's theorem, $\hat{\Lambda}$ has a representation of the form

$$\hat{\Lambda}[\varphi] = \iint_\Omega \tau(z)\varphi(z)\,dx\,dy, \quad \varphi \in \mathcal{L}^1(\Omega), \tag{2.2}$$

where $\tau \in \mathcal{L}^\infty(\Omega)$, $\|\tau\|_\infty = \|\hat{\Lambda}\| = \|\Lambda_\Omega\| = 1$, but where

$$\operatorname{meas}\{z : \tau(z) \neq 1\} > 0, \tag{2.3}$$

since $\hat{\Lambda} \neq \tilde{\Lambda}$.

Choosing $a \in \operatorname{Ext}\Omega$ let $h(z)$ be defined by

$$h(z) = \begin{cases} \dfrac{\tau(z) - 1}{z - a}, & z \in \Omega \\ 0, & z \in C \setminus \Omega. \end{cases}$$

Clearly $h(z)$ is bounded; moreover $h \in \mathcal{L}^p(C)$, $p > 2$. Let Ph denote the Pompeiu operation,

$$\begin{aligned}(Ph)(z) &= -\frac{1}{\pi}\iint_C h(\zeta)\left[\frac{1}{\zeta - z} - \frac{1}{\zeta - a}\right]d\xi\,d\eta \\ &= -\frac{(z - a)}{\pi}\iint_\Omega \frac{\tau(\zeta) - 1}{(\zeta - z)(\zeta - a)^2}\,d\xi\,d\eta.\end{aligned} \tag{2.4}$$

We shall define the function $L(w)$ as

$$L(w) = \frac{1}{2}[w + \bar{w} + (w - a)(Ph)(w)], \quad w \in C. \tag{2.5}$$

Conditions (i) and (ii) are satisfied [1, Chapter V]. Moreover, for a.a.w,

$$L_{\bar{w}} = \frac{1}{2}[1 + (w - a)h(w)] = \begin{cases} \tau(w)/2, & w \in \Omega \\ 1/2, & w \in C \setminus \Omega \end{cases}. \tag{2.6}$$

Therefore (iii) holds.

We claim that

$$L(w) = \frac{w + \bar{w}}{2}, \quad w \in C \setminus \Omega \tag{2.7}$$

To verify this, note that

$$\hat{\Lambda}[\varphi] = \Lambda_\Omega[\varphi], \quad \varphi \in \mathcal{B}(\Omega),$$

that is

$$\iint_\Omega [\tau(\zeta) - 1]\varphi(\zeta)\,d\xi\,d\eta = 0, \quad \varphi \in \mathcal{B}(\Omega).$$

Applying the above to the function

$$\varphi(\zeta) = \frac{1}{(\zeta - w)(\zeta - a)^2},$$

with w chosen in the complement of Ω, we find that

$$(Ph)(w) = 0, \quad w \in C \setminus \Omega.$$

Thus (2.7) follows from (2.5).

Consider now the solution $w = f(t,z)$ of (2.1), with $L(w)$ defined by (2.5), $f(0,z) = z$, $z \in C$. We assert that

$$w = f(t,z) = e^t x + iy, \quad z = x + iy \in C \setminus \Omega \qquad (2.8)$$

As seen by differentiation, the function $f(t,z)$ defined by (2.8) satisfies for $w = f(t,z)$, z fixed,

$$\frac{dw}{dt} = \frac{w + \overline{w}}{2}.$$

Since Ω is expansive we can conclude by (2.7) that the function $f(t,z)$ of (2.8) satisfies (2.1). By hypothesis $A_K(z)$ is an extremal mapping of Ω for its boundary values on $\partial\Omega$. Since $f(\log K,z)$ is K-quasiconformal, and since, by (2.7), $f(\log K,z) = A_K(z)$, $z \in \partial\Omega$, we see that the restriction of $f(\log K,z)$ to Ω is a competetive extremal mapping of Ω. In view of (2.3) and (2.6), condition (iii) holds in the stronger sense that $|L_{\overline{w}}| < \frac{1}{2}$ on a set of positive measure. It is now not difficult to verify that (2.8) could not possibly hold everywhere, i.e., $f(\log K,z)$ differs from $A_K(z)$. This contradicts the hypothesis of unique extremality of A_K, completing the proof of the Theorem.

As an application, for Example 1.4, we can conclude that the linear functional

$$\iint_{\Omega_\alpha} \varphi(z)\,dx\,dy, \quad \varphi \in \mathcal{B}(\Omega_\alpha), \qquad (2.9)$$

has a unique Hahn-Banach extension to $\mathcal{L}^1(\Omega_\alpha)$, if $\alpha \geq 3$. In fact, we have recently shown [11] by a more direct method that the condition $\alpha \geq 3$ is both necessary and sufficient for a unique Hahn-Banach extension of (2.9).

References

[1] Ahlfors, L.V.: Lectures on quasiconformal mappings. Van Nostrand, Princeton, N.J. (1966).

[2] Blum, E.: Die Extremalität gewisser Teichmüllerscher Abbildungen des Einheitskreises. Commentarii math. Helvet. 44 (1969), 319 - 340

[3] Foguel, S.R.: On a theorem by A.E. Taylor. Proc. Amer. math. Soc. 9 (1958), 325.

[4] Grötzsch, H.: Über die Verzerrung bei schlichten nichtkonformen Abbildungen und eine damit zusammenhängende Erweiterung des Picardschen Satzes. Ber. über die Verhandlungen der Sächs. Akad. Wiss., Leipzig, math.-phys. Kl. 80 (1928), 503 - 507.

[5] Hamilton, R.S.: Extremal quasiconformal mappings with prescribed boundary values. Trans. Amer. math. Soc. 138 (1969), 399 - 406.

[6] Reich, E., Strebel, K.: On the extremality of certain Teichmüller mappings. Commentarii math. Helvet. 45 (1970), 353 - 362.

[7] Reich, E.: On extremality and unique extremality of affine mappings. Lecture Notes in Mathematics 419, Springer-Verlag, Berlin - Heidelberg - New York (1974), 294 - 304.

[8] Reich, E., Strebel, K.: Extremal quasiconformal mappings with given boundary values, in "Contributions to Analysis". Academic Press, New York and London (1974), 375 - 391.

[9] Reich, E.: A generalized Dirichlet integral. J. Analyse math. 30 (1976), 456 - 463.

[10] Reich, E.: Quasiconformal mappings with prescribed boundary values and a dilatation bound. Arch. rat. Mech. Analysis 68 (1978), 99 - 112.

[11] Reich, E.: Uniqueness of Hahn-Banach extensions from certain spaces of analytic functions. To appear in Math. Z.

[12] Reimann, H.M.: Ordinary differential equations and quasiconformal mappings. Inventiones math. 33 (1976), 247 - 270.

[13] Schwartz, G.P.: Parametric representations of plane quasiconformal mappings. Dissertation, University of Minnesota (1970).

[14] Strebel, K.: Zur Frage der Eindeutigkeit extremaler quasikonformer Abbildungen des Einheitskreises. Commentarii math. Helvet. 36 (1962), 306 - 323.

[15] Strebel, K.: Zur Frage der Eindeutigkeit extremaler quasikonformer Abbildungen des Einheitskreises II. Commentarii math. Helvet. 39 (1964), 77 - 89.

[16] Teichmüller, O.: Extremale quasikonforme Abbildungen und quadratische Differentiale. Abh. Preuss. Akad. Wiss. 22 (1940), 1 - 197.

University of Minnesota
Minneapolis, MN 55455
U.S.A.

EXTENSIONS OF QUASICONFORMAL DEFORMATIONS

H. M. Reimann

A continuous function $f : R^n \to R^n$, $n \geq 1$, is a quasiconformal deformation, if there exists a constant c such that

$$\left| \frac{(a, f(x+a) - f(x))}{|a|^2} - \frac{(b, f(x+b) - f(x)}{|b|^2} \right| \leq c$$

for all $x \in R^n$ and for all $a,b \in R^n$ with $|a| = |b| \neq 0$. The smallest constant c with this property is denoted by $\|f\|_Q$. Equivalently, the continuous function $f : R^n \to R^n$, $n \geq 2$, is a quasiconformal deformation, if it has locally integrable distributional derivatives, if $f(x) = O(|x|\log|x|)$ for $|x| \to \infty$ and if

$$Sf = \frac{1}{2}(Df + {}^tDf) - \frac{1}{n}(\text{trace } Df)\, I$$

is essentially bounded. Here Df denotes the Jacobian matrix of f, which exists a.e. For the essential supremum

$$\|Sf\|_\infty = \operatorname*{ess\,sup}_{x \in R^n} |Sf(x)|$$

the inequality

$$\|Sf\|_\infty \leq \frac{n-1}{n} \|f\|_Q$$

holds (the norm of a matrix being defined by $|A| = \sup_{|z|=1} |Az|$). Conversely, $\|f\|_Q$ is bounded by a function of $\|Sf\|_\infty$ and n only (provided that $f(x) = O(|x|\log|x|)$ for $|x| \to \infty$). In fact it can be verified ([4], proof of proposition 12) that

$$\|f\|_Q \leq C(n)\, \|Sf\|_\infty$$

for some constant $C(n)$.

In the following it will be crucial, that quasiconformal deformations satisfy a "Zygmund condition":

$$|f(x+y) + f(x-y) - 2f(x)| \leq k|y| \qquad x,y \in R^n$$

with a constant k,

$$k \leq 4\, \|f\|_Q \leq 4\, C(n)\, \|Sf\|_\infty$$

The original definition for quasiconformal deformations is due to Ahlfors [1]. For the above properties see [4].

In this paper we show that quasiconformal deformations $f : R^n \to R^n$ can be extended to quasiconformal deformations $F : R^{n+1} \to R^{n+1}$ such tha

$$(F_1(x_1,\dots,x_n,0),\dots,F_{n+1}(x_1,\dots,x_n,0)) = (f_1(x_1,\dots,x_n),\dots,f_n(x_1,\dots,x_n),0$$

It may be worthwhile to point out, that this extension problem is connected with the corresponding extension problem for quasiconformal mappings $h : R^n \to R^n$. This is due to the fact (see [1], [4], [5]) that the solutions $h(t) = h(x,t)$ for a differential equation

$$\frac{dh}{dt} = f(h,t)$$

with initial condition $h(0) = x \in R^n$ are quasiconformal mappings $h(x,T) : R^n \to R^n$ (T fixed), provided that $f : R^n \to R^n$ is a quasiconformal deformation (which may depend on the parameter t). The more difficult extension problem for quasiconformal mappings can therefore be reduced to the following question: Can every quasiconformal mapping h be obtained as the solution $h(x,T)$ of a differential equation with a vector fiel $f(x,t)$, which is a quasiconformal deformation (depending on a parameter t, $0 \leq t \leq T$)? For $n = 2$ this question can be answered affirmatively. The main theorem on the existence of quasiconformal mappings with prescribed dilatation leads to parametric representations, which are in fact solutions of such differential equations. The result obtained in this paper therefore implies the extendability of quasiconformal mappings from two to three dimensions. However for this case, Ahlfors [2] has already given an explicit construction, which takes care of the extension problem for quasiconformal mappings.

It already became apparent from Carlesons discussion of the extension problem for quasiconformal mappings from three to four dimensions [3], that the main obstacle to be overcome is the construction of suitable approximations for quasiconformal mappings. The extension problem for deformations being solved, we find this point of view confirmed.

For the formulation of the theorem $B_r(x)$ denotes the n-dimensional ball with center x and radius r and $S_r(x)$ the corresponding sphere in R^n. The surface measure on $S_r(x)$ is $d\sigma$ and $\nu(\xi) = (\nu_1,\dots,\nu_n)(\xi)$ is the outer normal unit vector at $\xi \in S_r(x)$. We also set

$$\alpha_n = \int_{B_1(0)} d^n\xi, \qquad \beta_n = \int_{S_1(0)} d\sigma(\xi) = n\alpha_n$$

and for the points in R^{n+1} we write

$$(x_1,\dots,x_{n+1}) = (x,z) = (x_1,\dots,x_n,z).$$

Theorem. Every quasiconformal deformation $f : R^n \to R^n$ extends to a quasiconformal deformation $F : R^{n+1} \to R^{n+1}$ given by

$$F_i(x,z) = \frac{1}{\alpha_n |z|^n} \int_{B_{|z|}(x)} f_i(\xi)d\xi \qquad i = 1,\dots,n$$

$$F_{n+1}(x,z) = \frac{1}{\beta_n |z|^{n-1}} \int_{S_{|z|}(x)} \sum_{i=1}^{n} F_i(\xi,z)\nu_i(\xi)d\sigma(\xi)$$

The calculations show that $\|SF\|_\infty$ is bounded by a function depending on $\|Sf\|_\infty$, n and the constant k in the Zygmund condition for f. This constant k only depends on $\|Sf\|_\infty$ and n (see the introduction). Collecting the estimates and denoting the partial derivatives, which exist a.e., by

$$\frac{\partial F_i}{\partial x_j} = F_{i,j} = F_{ij}$$

we have for a.e. x:

$$\tfrac{1}{2} | F_{ji} + F_{ij} | \leqq \|Sf\|_\infty \qquad 1 = i < j = n$$

$$\tfrac{1}{2} | F_{n+1,i} + F_{i,n+1} | \leqq \frac{kn}{2} \qquad i = 1,\dots,n$$

$$| F_{ii} - \frac{1}{n} \sum_{j=1}^{n} F_{jj} | \leq \|Sf\|_\infty \qquad i = 1,\dots,n$$

$$| F_{n+1,n+1} - \frac{1}{n+1} \sum_{j=1}^{n+1} F_{jj} | = \frac{n}{n+1} | F_{n+1,n+1} - \frac{1}{n} \sum_{j=1}^{n} F_{jj} | \leqq A(k + \|Sf\|_\infty)$$

and hence for $i = 1,\dots,n$

$$| F_{ii} - \frac{1}{n+1} \sum_{j=1}^{n+1} F_{jj} | \leqq | F_{ii} - \frac{1}{n} \sum_{j=1}^{n} F_{jj} | + | \frac{1}{n(n+1)} \sum_{j=1}^{n} F_{jj} - \frac{1}{n+1} F_{n+1,n+1} |$$

$$\leq \|Sf\|_\infty + \frac{A}{n}(k + \|Sf\|_\infty)$$

where A depends on n only. Keeping in mind that

$$k \leqq 4\, C(n) \|Sf\|_\infty$$

this clearly shows that

$$\|SF\|_\infty \leqq B \|Sf\|_\infty$$

or equivalently

$$\|F\|_Q \leqq B' \|f\|_Q$$

for some constants B, B' depending on the dimension.

We remark at this point that for $n = 1$ a quasiconformal deformation is a continuous function $f : R \to R$ satisfying the Zygmund condition

$$|f(x + y) + f(x - y) - 2f(x)| \leq k|y| \qquad x,y \in R.$$

The theorem shows, that such a function extends to a quasiconformal deformation $F : R^2 \to R^2$ defined by

$$F_1(x,z) = \frac{1}{2z} \int_{x-z}^{x+z} f(\xi)d\xi$$

$$F_2(x,z) = \frac{1}{2}(F_1(x + z,z) - F_1(x - z,z)).$$

The remainder of the paper will be devoted to the proof of this theorem.

We begin with the remark that the extension is a linear operation. Referring to some standard regularization procedure, it therefore suffices to show that the inequalities hold for $z \neq 0$. At those points F is continuously differentiable. Furthermore, the extension being symmetric, we can assume $z > 0$.

The growth condition for F easily follows from the corresponding condition

$$f(x) = O(|x|\log|x|) \qquad \text{for} \quad |x| \to \infty$$

for the quasiconformal deformation f.

Let us assume that $\|f\|_Q$ is finite. As a consequence $\|Sf\|_\infty = c$ is finite and f satisfies the Zygmund condition

$$|f(x + y) + f(x - y) - 2f(x)| \leq k|y| \qquad x,y \in R^n$$

with a constant $k \leq 4 \|f\|_Q$.

The implications

$$\frac{1}{2} |F_{ji} + F_{ij}| \leq c \qquad 1 \leq i < j \leq n$$

$$|F_{ii} - \frac{1}{n} \sum_{j=1}^{n} F_{jj}| \leq c \qquad i = 1,\ldots,n$$

are immediate. It will be shown that

$$\frac{1}{2} |F_{i,n+1} + F_{n+1,i}| \leq kn(1 + \sqrt{n/2}) \qquad i = 1,\ldots,n$$

and

$$|F_{n+1,n+1} - \frac{1}{n} \sum_{j=1}^{n} F_{jj}| \leq cn + kn^{3/2} + kn\alpha_{n-1}\beta_n^{-1}.$$

1) Estimate for $F_{i,n+1}$

$$F_{i,n+1}(x,z) = \frac{-n}{\alpha_n z^{n+1}} \int_{B_z(x)} f_i d\xi + \frac{n}{z} \frac{1}{\beta_n z^{n-1}} \int_{S_z(x)} f_i d\sigma$$

$$= \frac{n}{z}\left(\frac{1}{\beta_n z^{n-1}} \int_{S_z(x)} f_i d\sigma - F_i(x,z)\right)$$

$$|F_{i,n+1}(x,z)| \leq \frac{n}{z} |f_i(x) - F_i(x,z)| + \frac{n}{z}\left|\frac{1}{\beta_n z^{n-1}} \int_{S_z(x)} f_i d\sigma - f_i(x)\right| .$$

The Zygmund condition implies

$$|f_i(x) - F_i(x,z)| = \frac{1}{\alpha_n z^n} \left| \int_{B_z(x)} (f_i(x) - f_i(\xi))d\xi \right|$$

$$= \frac{1}{2\alpha_n z^n} \left| \int_{B_z(0)} (2f_i(x) - f_i(x+\xi) - f_i(x-\xi))d\xi \right|$$

$$\leq \frac{k\beta_n}{2\alpha_n z^n} \int_0^z r^n dr = \frac{k}{2} \frac{n}{n+1} z$$

and similarly

$$\left|\frac{1}{\beta_n z^{n-1}} \int_{S_z(x)} f_i d\sigma - f_i(x)\right| \leq \frac{1}{2\beta_n z^{n-1}} \int_{S_z(0)} |f_i(x+\xi) + f_i(x-\xi) - 2f_i(x)| d\sigma(\xi)$$

$$\leq \frac{kz}{2} .$$

The two inequalities combined yield

$$|F_{i,n+1}(x,z)| \leq \frac{kn}{2}(1 + \frac{n}{n+1}) \leq kn .$$

2) Estimate for $F_{n+1,i}$.

This term can be written in the form

$$F_{n+1,i}(x,z) = \frac{1}{2\beta_n z^{n-1}} \int_{S_z(x)} \sum_{j=1}^{n} F_{ji}(\xi,z)\nu_j(\xi)d\sigma(\xi)$$

$$= \frac{1}{2\beta_n z^{n-1}} \int_{S_z(0)} \sum_{j=1}^{n} (F_{ji}(x+\xi,z) - F_{ji}(x-\xi,z))\nu_j(x+\xi)d\sigma(\xi)$$

The derivatives F_{ji}, $1 \leq i,j \leq n$, in this formula are given by

$$F_{ji}(x,z) = \frac{1}{\alpha_n z^n} \int_{S_z(x)} f_j(\xi)\nu_i(\xi)d\sigma(\xi)$$

$$= \frac{n}{z}\,\frac{1}{2\beta_n z^{n-1}} \int_{S_z(0)} (f_j(x+\xi) - f_j(x-\xi))\nu_i(\xi)\,d\sigma(\xi).$$

If ν' is the normal to the ball $B_z(x+\xi)$ at the point $x+\xi+\rho$, it follows that

$$F_{ji}(x+\xi,z) - F_{ji}(x-\xi,z)$$

$$= \frac{n}{z}\,\frac{1}{2\beta_n z^{n-1}} \int_{S_z(0)} (f_j(x+\xi+\rho) - f_j(x+\xi-\rho) - f_j(x-\xi+\rho) + f_j(x-\xi-\rho))\nu_i'\,d\sigma(\rho)$$

The Zygmund condition implies

$$|f_j(x+\xi+\rho) + f_j(x-\xi-\rho) - f_j(x+\xi-\rho) - f_j(x-\xi+\rho)| \le k|\xi+\rho| + k|\xi-\rho| \le 2kz\sqrt{2}$$

hence

$$|F_{ji}(x+\xi,z) - F_{ji}(x-\xi,z)| \le nk\sqrt{2}$$

and finally

$$|F_{n+1,i}(x,z)| \le \frac{1}{2\beta_n z^{n-1}} \int_{S_z(0)} \Big(\sum_{j=1}^{n}|F_{ji}(x+\xi,z)-F_{ji}(x-\xi,z)|^2\Big)^{1/2}\Big(\sum_{j=1}^{n}\nu_j^2\Big)^{1/2} d\sigma(\xi)$$

$$\le nk\sqrt{n/2}\ .$$

3) An expression for $F_{n+1,n+1} - \frac{1}{n}\sum_{i=1}^{n} F_{ii}$.

Inserting $\xi = x + z\nu$, $|\nu| = 1$, in the definition for F_{n+1} we obtain successively

$$F_{n+1}(x,z) = \frac{1}{\beta_n} \int_{S_1(0)} \sum_{i=1}^{n} F_i(x+z\nu,z)\nu_i\,d\sigma(\nu)$$

$$F_{n+1,n+1}(x,z) = \frac{1}{\beta_n} \int_{S_1(0)} \sum_{i=1}^{n} F_{i,n+1}(x+z\nu\,,\,z)\nu_i\,d\sigma(\nu)$$

$$+ \frac{1}{\beta_n} \int_{S_1(0)} \sum_{j=1}^{n}\sum_{i=1}^{n} F_{ij}(x+z\nu,z)\nu_i\nu_j\,d\sigma(\nu)$$

$$= \frac{1}{\beta_n z^{n-1}} \int_{S_z(x)} \sum_{i=1}^{n} F_{i,n+1}(\xi,z)\nu_i\,d\sigma$$

$$+ \frac{1}{\beta_n z^{n-1}} \int_{S_z(x)} \sum_{i,j} F_{ij}(\xi,z)\nu_i\nu_j\,d\sigma(\nu).$$

This leads to a representation as a sum of three integrals:

$$F_{n+1,n+1} - \frac{1}{n}\sum_{i=1}^{n} F_{ii} = I_1 + I_2 + I_3$$

with

$$I_1(x,z) = \frac{1}{\beta_n z^{n-1}} \int_{S_z(x)} \sum_{i=1}^{n} F_{i,n+1}(\xi,z)\nu_i d\sigma$$

$$I_2(x,z) = \frac{1}{2\beta_n z^{n-1}} \int_{S_z(x)} \sum_{i\neq j} (F_{ij}(\xi,z) + F_{ji}(\xi,z))\nu_i \nu_j d\sigma$$

$$I_3(x,z) = \frac{1}{\beta_n z^{n-1}} \int_{S_z(x)} \sum_{i=1}^{n} F_{ii}(\xi,z)\nu_i^2 d\sigma - \frac{1}{n}\sum_{i=1}^{n} F_{ii}(x,z).$$

4) Estimates for I_1, I_2 and I_3.

Given the estimate

$$|F_{i,n+1}(x,z)| \leqq kn$$

the integral I_1 is easily seen to be bounded:

$$|I_1(x,z)| \leqq \frac{1}{\beta_n z^{n-1}} \int_{S_z(x)} \Big(\sum_{i=1}^{n} |F_{i,n+1}(\xi,z)|^2\Big)^{1/2} \Big(\sum_{i=1}^{n} \nu_i^2\Big)^{1/2} d\sigma \leqq kn^{3/2}.$$

Similarly it can be concluded from

$$\frac{1}{2}|F_{ij}(x,z) + F_{ji}(x,z)| \leqq c \qquad\qquad 1 \leqq i < j \leqq n$$

that I_2 is bounded:

$$|I_2(x,z)| \leqq c\sqrt{n^2 - n}\,\frac{1}{\beta_n z^{n-1}} \int_{S_z(x)} \sum_{i\neq j} \nu_i^2\nu_j^2 d\sigma \leqq c\sqrt{n^2 - n} \leqq cn\,.$$

Observe that

$$\sum_{i\neq j} \nu_i^2\nu_j^2 = \sum_{i=1}^{n} \nu_i^2(1 - \nu_i^2) < 1\,.$$

If use is made of the identity

$$\frac{1}{n} = \frac{1}{\beta_n} \int_{S_1(0)} \nu_i^2 d\sigma$$

the integral I_3 can be rewritten as

$$I_3(x,z) = \frac{1}{\beta_n z^{n-1}} \int_{S_z(x)} \sum_{i=1}^{n} (F_{ii}(\xi,z) - F_{ii}(x,z))\nu_i^2 d\sigma(\xi)$$

$$= \frac{1}{2\beta_n z^{n-1}} \int_{S_z(0)} \sum_{i=1}^{n} (F_{ii}(x+\xi,z) + F_{ii}(x-\xi,z) - 2F_{ii}(x,z))\nu_i^2 d\sigma .$$

For $i = 1,\dots,n$ the derivatives F_{ii} are given by

$$F_{ii}(x,z) = \frac{1}{\alpha_n z^n} \int_{S_z(x)} f_i(\rho)\nu_i(\rho)d\sigma(\rho) .$$

Therefore

$$|F_{ii}(x+\xi,z) + F_{ii}(x-\xi,z) - 2F_{ii}(x,z)|$$

$$\leqq \frac{n}{z}\frac{1}{\beta_n z^{n-1}} \int_{S_z(0)} |f_i(x+\xi+\rho) + f_i(x-\xi+\rho) - 2f_i(x+\rho)|\,|\nu_i|d\sigma(\rho)$$

$$\leqq nk \frac{1}{\beta_n z^{n-1}} \int_{S_z(0)} |\nu_i|d\sigma = nk\frac{2\alpha_{n-1}}{\beta_n} .$$

This leads to the following inequality for the third integral:

$$|I_3(x,z)| \leqq \frac{1}{2\beta_n} nk \frac{2\alpha_{n-1}}{\beta_n} \int_{S_z(0)} \sum_{i=1}^{n} \nu_i^2 d\sigma = nk\alpha_{n-1}\beta_n^{-1} .$$

The final estimate is

$$|F_{n+1,n+1}(x,z) - \frac{1}{n}\sum_{i=1}^{n} F_{ii}(x,z)| \leqq cn + kn^{3/2} + nk\alpha_{n-1}\beta_n^{-1} .$$

References

[1] Ahlfors, L.V.: Quasiconformal deformations and mappings in R^n. J. Analyse math. 30 (1976), 74 - 97.

[2] Ahlfors, L.V.: Extension of quasiconformal mappings from two to three dimensions. Proc. nat. Acad. Sci. USA 51 (1964), 768 - 771.

[3] Carleson, L.: The extension problem for quasiconformal mappings, in "Contributions to analysis". Academic Press, New York and London (1974), 39 - 47.

[4] Reimann, H.M.: Ordinary differential equations and quasiconformal mappings. Inventiones math. 33 (1976), 247 - 270.

[5] Semenov, V.I.: One parameter groups of quasiconformal mappings in euclidean space (Russian). Sibir. mat. Žurn. 17 (1976), 177 - 193.

Universität Bern
Mathematisches Institut
Sidlerstraße 5
CH-3000 Bern
Switzerland

REMOVABLE SINGULARITIES OF ANALYTIC AND MEROMORPHIC FUNCTIONS OF SEVERAL COMPLEX VARIABLES

Juhani Riihentaus

1. Introduction

Shiffman [5, Lemma 3] has obtained the following results concerning removable singularities of analytic functions of several complex variables.

Let G be a domain of $\mathbb{C}^n$. Let $E \subset G$ be closed in G. Let $f : G \setminus E \to \mathbb{C}$ be an analytic function. Then f can be extended to a unique analytic function $f^* : G \to \mathbb{C}$ if one of the following conditions is satisfied:

(1) f can be extended to a continuous function $f^* : G \to \mathbb{C}$, and $H^{2n-1}(E) < \infty$.

(2) f is bounded and $H^{2n-1}(E) = 0$.

(3) $H^{2n-2}(E) = 0$.

Harvey-Polking [2, Remark after Theorem 1.1] announces that in view of the fact that the envelope of meromorphy is the same as the envelope of holomorphy the above result (3) of Shiffman implies the corresponding result for meromorphic functions:

(4) Let G be a domain of $\mathbb{C}^n$. Let $E \subset G$ be closed in G. Let f be meromorphic in $G \setminus E$. If $H^{2n-2}(E) = 0$, then f has a unique meromorphic extension to G.

Above H^β is the β-dimensional Hausdorff measure.

In section 3 we generalize the above results (1)-(3) of Shiffman. To be more precise our result (Theorem 3.1 below) gives an answer to the following problem.

Let G be a domain of $\mathbb{C}^n$. Let $E \subset G$ be closed in G. Let $E_1, E_2 \subset E$ (not necessarily closed in G). Let $f : G \setminus E_1 \to \mathbb{C}$ be a continuous function such that $f \mid G \setminus E$ is analytic. How large can the sets E, E_1, E_2 be and what must be assumed of the boundary behavior of f in $E_1 \setminus E_2$ in order to ensure that f has an analytic extension to G?

Our result is sharp when Hausdorff measure is used (Remark 3.4 below). Moreover, our result contains as special cases the results of [3, Theorem 3.1, Theorem 3.5].

Our method of proof is first to reduce the problem considered to the case of one complex variable, then to use a slight generalization of a result of Wohlhauser [8, Satz III.1] and finally to construct the desired extension using normal family arguments.

Theorem 3.6 below gives a corresponding generalization to the above result (4) of Harvey-Polking. In addition, this result contains as a special case the result of [4, Theorem 3.1]. The proof is based on Theorem 3.1, on a lemma, which is essentially based on Wohlhauser's result, and on Levi's extension theorem.

We use mainly the same notation and terminology as in [3] and [4].

For the theory of analytic and meromorphic functions of several complex variables we refer e.g. to [7]. However, for the sake of completeness we recall here the following facts.

Let G be a domain of $\mathbb{C}^n$. Let f be a meromorphic function in G. Then there is $N(f) \subset G$ closed in G and $I(f) \subset N(f)$ closed in G such that f is analytic in $G \setminus N(f)$ and has a spherically continuous extension f^* to $G \setminus I(f)$. In the following we will speak simply of the meromorphy of $f^* : G \setminus I(f) \to \mathbb{C}^*$ in G. Moreover, $N(f)$ is of σ-finite $(2n-2)$-dimensional Hausdorff measure and $I(f)$ is of σ-finite $(2n-4)$-dimensional Hausdorff measure (see e.g. [6, pp. 13-15]).

2. Lemmas

We begin with the following results of Federer and Shiffman. See [5, Corollary 4 and Lemma 2].

2.1. Lemma. Let A be an arbitrary subset of $\mathbb{C}^n$. Let α be an arbitrary non-negative number and let k be an integer, $0 \leq k \leq n-1$.

(1) If $H_*^{2n-1}(A) < \sigma_\infty$, then $H_*^1(A \cap p^{-1}\{x\}) < \sigma_\infty$ for H^{2n-2}-almost all $x \in \mathbb{C}^{n-1}$, and $H_*^{2n-3}(A \cap q^{-1}\{y\}) < \sigma_\infty$ for H^2-almost all $y \in \mathbb{C}$.

(2) If $H^{2n-2+\alpha}(A) = 0$, then $H^\alpha(A \cap p^{-1}\{x\}) = 0$ for H^{2n-2}-almost all $x \in \mathbb{C}^{n-1}$, and $H^{2n-4+\alpha}(A \cap q^{-1}\{y\}) = 0$ for H^2-almost all $y \in \mathbb{C}$.

(3) If $a \in \mathbb{C}^n$, $\alpha > 0$, and $H^{2k+\alpha}(A) = 0$, then there is a complex $(n-k)$-plane P through the point a such that $H^\alpha(A \cap P) = 0$.

We recall that for an arbitrary set $B \subset \mathbb{C}^n$, $H^0(B)$ is the number of points of B.

Next we give the results from the theory of analytic and meromorphic functions of one complex variable we need in section 3. The first lemma is just a slight generalization of Wohlhauser's result [8, Satz III.1].

2.2. Lemma. Let G be a domain of $\mathbb{C}$. Let $E \subset G$ be closed in G and let $H^1(E) < \sigma_\infty$. Let $E_1 \subset E$ (not necessarily closed in G) be such that $H^1(E_1) = 0$. Let $f : G \setminus E_1 \to \mathbb{C}$ be a continuous function such that $f \mid G \setminus E$ is analytic and such that for each $z_0 \in E_1$

$$(z - z_o)f(z) \to 0,\ z \in G \setminus E,\ z \to z_o.$$

Then there is a unique analytic function $f^* : G \to \mathbb{C}$ such that

$$f^* \mid G \setminus E_1 = f.$$

Proof. Since $H^1(E) < \sigma_\infty$, int $E = \emptyset$. Hence it is sufficient to show the existence of a local extension.

Take $z_o \in E$ arbitrarily. From the assumption it follows that

$$(z - z_o)f(z) \to 0,\ z \in G \setminus E,\ z \to z_o. \qquad (2.2.1)$$

Using (2.2.1) and the continuity of f we find a neighborhood U_{z_o} of z_o in G such that

$$|(z - z_o)f(z)| < 1$$

whenever $z \in U_{z_o} \setminus E_1$. Define $g : U_{z_o} \setminus E_1 \to \mathbb{C}$,

$$g(z) = (z - z_o)f(z).$$

Then g is bounded and continuous. Moreover, $g \mid U_{z_o} \setminus E$ is analytic. By Wohlhauser's result [8, Satz III.1] g has a unique analytic extension $g^* : U_{z_o} \to \mathbb{C}$. Consequently $f^* : U_{z_o} \to \mathbb{C}^*$,

$$f^*(z) = \frac{g^*(z)}{z - z_o},$$

provides a meromorphic extension to $f \mid U_{z_o} \setminus E_1$. To conclude the proof we just notice that (2.2.1) implies that f^* is in fact analytic.

Proceeding in a way similar to the above proof we get a corresponding result for meromorphic functions:

2.3. Lemma. Let G be a domain of $\mathbb{C}$. Let $E \subset G$ be closed in G and let $H^1(E) < \sigma_\infty$. Let $E_1 \subset E$ (not necessarily closed in G) be such that $H^1(E_1) = 0$. Let $f : G \setminus E_1 \to \mathbb{C}^*$ be a spherically continuous function such that $f \mid G \setminus E$ is analytic and such that for each $z_o \in E_1$ there is $n(z_o) \in \mathbb{N}$ for which

$$(z - z_o)^{n(z_o)} f(z) \to 0,\ z \in G \setminus E,\ z \to z_o.$$

Then there is a unique meromorphic function $f^* : G \to \mathbb{C}^*$ such that

$$f^* \mid G \setminus E_1 = f.$$

The following lemmas are well-known.

<u>2.4. Lemma.</u> Let G be a domain of $\mathbb{C}$. Let W be a family of analytic functions $f : G \to \mathbb{C}$. If W is locally uniformly bounded, then W is equicontinuous.

<u>2.5. Lemma.</u> Let G be a domain of $\mathbb{C}$. Let $H \subset G$ be dense in G. Let the sequence of functions $f_k : G \to \mathbb{C}$, $k = 1,2,\dots$, converge in H. If the set $\{f_k \mid k = 1,2,\dots\}$ is equicontinuous, then the sequence f_k converges c-uniformly.

<u>2.6. Lemma.</u> Let G be a domain of $\mathbb{C}$. If the sequence of analytic functions $f_k : G \to \mathbb{C}$, $k = 1,2,\dots$, converges c-uniformly to a function $f : G \to \mathbb{C}$, then f is analytic.

For the proof of the following simple topological lemma see [3, Lemma 2.6].

<u>2.7. Lemma.</u> Let S be a metric space. Let $S_1 \subset S$ be dense in S. Let G be a domain of $\mathbb{C}$. Let $f : S_1 \times G \to \mathbb{C}$ be a function such that for each $x \in S_1$ the function $g_x : G \to \mathbb{C}$,

$$g_x(y) = f(x,y),$$

is continuous. Then there is a unique continuous function $f^* : S \times G \to \mathbb{C}$ such that

$$f^* \mid S_1 \times G = f$$

if and only if the following condition is fullfilled: For each $x_0 \in S$ and for each sequence $x_k \to x_0$, $x_k \in S_1$, $k = 1,2,\dots$, the sequence g_{x_k} converges c-uniformly.

The last lemma is Levi's extension theorem. For the proof see e.g. [2, Theorem 2.1 (b)].

<u>2.8. Lemma.</u> Let $z_0 = (x_0;y_o) \in \mathbb{C}^n$ have a neighborhood $D^n(z_0,r) = D^{n-1}(x_0,s) \times B^2(y_o,t)$. Let t' be such that $0 < t' < t$. Let $f : D^{n-1}(x_0,s) \times (B^2(y_o,t) \setminus \overline{B}^2(y_o,t')) \to \mathbb{C}$ be an analytic function such that for H^{2n-2}-almost all $x \in D^{n-1}(x_0,s)$ the analytic function $g_x : B^2(x_0,t) \setminus \overline{B}^2(y_o,t') \to \mathbb{C}$,

$$g_x(y) = f(x;y),$$

has a meromorphic extension to $B^2(y_o,t)$. Then f has a unique meromorphic extension to $D^n(z_0,r)$.

3. Singularities

We begin with a generalization to the above results (1)-(3) of Shiffman:

3.1. Theorem. Let G be a domain of $\mathbb{C}^n$. Let $E \subset G$ be closed in G and let $H^{2n-1}(E) < \sigma_\infty$. Let $E_1 \subset E$ (not necessarily closed in G) be such that $H^{2n-1}(E_1) = 0$. Let $E_2 \subset E$ (not necessarily closed in G) be such that $H^{2n-2}(E_2) = 0$. Let $f : G \setminus E_1 \to \mathbb{C}$ be a continuous function such that $f \mid G \setminus E$ is analytic and for each $z_0 \in E_1 \setminus E_2$ and for each complex line $T(z_0, z_0')$,

$$T(z_0, z_0') = \{z_0 + t(z_0' - z_0) \mid t \in \mathbb{C}\},$$

through the point z_0,

$$|z - z_0| f(z) \to 0, \ z \in (G \setminus E) \cap T(z_0, z_0'), \ z \to z_0.$$

Then there is a unique analytic function $f^* : G \to \mathbb{C}$ such that

$$f^* \mid G \setminus E_1 = f.$$

Proof. We give an induction proof. If $n = 1$, the theorem is just Lemma 2.2. For the case $n > 1$, assume that the theorem holds for $m = n - 1$.

Since $H^{2n-1}(E) < \sigma_\infty$, int $E = \emptyset$. Hence it is sufficient to show the existence of a local extension.

Let $z_0 = (x_0; y_o) = (z_1^o, \ldots, z_{n-1}^o, z_n^o) \in E$ be arbitrarily given. Since $H^{2n-1}(E_1) = 0$, there is by Lemma 2.1 (3) a complex line P through the point z_0 such that $H^1(E_1 \cap P) = 0$. Making an affine coordinate transformation, if necessary, we may suppose that $z_0 = 0$ and $P = \{0\} \times \mathbb{C}$. Since $H^1(E_1 \cap P) = 0$, there is $t > 0$ such that

$$\{0\} \times \bar{B}^2(t) \subset G,$$

$$\{0\} \times S^1(t) \subset G \setminus E_1.$$

Since $\{0\} \times S^1(t) \subset G \setminus E_1$ is compact and f is continuous,

$$M = \sup \{|f(0,y)| \mid y \in S^1(t)\}$$

is finite. Using the continuity of f and the triangle inequality we see that there is an open set U in $\mathbb{C}^n$ such that $U \setminus E_1$ is a neighborhood of $\{0\} \times S^1(t)$ in $G \setminus E_1$ and

$$|f(z)| < M + 1 \tag{3.1.1}$$

whenever $z \in U \setminus E_1$. Since $\{0\} \times S^1(t) \subset G \setminus E_1$ is compact, there is $s = (r_1,\dots,r_{n-1})$ such that $\bar{D}^{n-1}(s) \times \bar{B}^2(t) \subset G$ and $\bar{D}^{n-1}(s) \times S^1(t) \subset U$.

Denote

$$B_1 = \{x \in D^{n-1}(s) \mid H^1((\{x\} \times B^2(t)) \cap E) < \sigma_\infty\},$$

$$B_2 = \{x \in D^{n-1}(s) \mid H^1((\{x\} \times B^2(t)) \cap E_1) = 0\},$$

$$B_3 = \{x \in D^{n-1}(s) \mid (\{x\} \times B^2(t)) \cap E_2 = \emptyset\},$$

$$B = B_1 \cap B_2 \cap B_3.$$

By Lemma 2.1 (1) and (2), $H^{2n-2}(D^{n-1}(s) \setminus B) = 0$. Hence B is dense in $D^{n-1}(s)$. For each $x \in B$ denote

$$E_n(x) = \{y \in B^2(t) \mid (x;y) \in E\},$$

$$E_n^1(x) = \{y \in B^2(t) \mid (x;y) \in E_1\}.$$

Then $E_n(x) \subset B^2(t)$ is closed in $B^2(t)$ and $H^1(E_n(x)) < \sigma_\infty$. Respectively, $E_n^1(x) \subset E_n(x)$ and $H^1(E_n^1(x)) = 0$. For each $x \in B$ define $g_x : B^2(t) \setminus E_n^1(x) \to \mathbb{C}$,

$$g_x(y) = f(x;y).$$

Then g_x is continuous, and $g_x \mid B^2(t) \setminus E_n(x)$ is analytic. Moreover, by assumption for each $y_o' \in E_n^1(x)$

$$|y - y_o'| g_x(y) \to 0, \; y \in B^2(t) \setminus E_n(x), \; y \to y_o.$$

By Lemma 2.2 g_x has a unique analytic extension $g_x^* : B^2(t) \to \mathbb{C}$.

Since $\bar{D}^{n-1}(s) \times S^1(t) \subset U$ is compact and U is open in $\mathbb{C}^n$, there is t', $0 < t' < t$, such that $\bar{D}^{n-1}(s) \times S^1(t') \subset U$. Since $H^1(E_n^1(x)) = 0$ for each $x \in B$, it follows from (3.1.1), the continuity of g_x^*, and the maximum principle that the family of analytic functions

$$g_x^{**} = g_x^* \mid B^2(t') : B^2(t') \to \mathbb{C}, \; x \in B, \tag{3.1.2}$$

is uniformly bounded by $M + 1$.

We next show that the function $f \mid (D^{n-1}(s) \times B^2(t')) \setminus E_1$ has a continuous extension to $D^{n-1}(s) \times B^2(t')$.

By Lemma 2.7, Lemma 2.5 and Lemma 2.4 it is sufficient to show that for each $x' \in D^{n-1}(s)$ and for each sequence $x_k \to x'$, $x_k \in B$, $k = 1,2,\dots$, the sequence of analytic functions $g_{x_k}^{**}$ converges in a dense subset of $B^2(t')$. To see this we proceed as follows.

Denote

$$C_1 = \{y \in B^2(t') \mid H^{2n-3}((D^{n-1}(s) \times \{y\}) \cap E) < \sigma_\infty\},$$

$$C_2 = \{y \in B^2(t') \mid H^{2n-3}((D^{n-1}(s) \times \{y\}) \cap E_1) = 0\},$$

$$C_3 = \{y \in B^2(t') \mid H^{2n-4}((D^{n-1}(s) \times \{y\}) \cap E_2) = 0\},$$

$$C = C_1 \cap C_2 \cap C_3.$$

By Lemma 2.1 (1) and (2), $H^2(B^2(t') \setminus C) = 0$. Hence C is dense in $B^2(t')$. For each $y \in C$ denote

$$E_1(y) = \{x \in D^{n-1}(s) \mid (x;y) \in E\},$$

$$E_1^1(y) = \{x \in D^{n-1}(s) \mid (x;y) \in E_1\},$$

$$E_1^2(y) = \{x \in D^{n-1}(s) \mid (x;y) \in E_2\}.$$

Then $E_1(y) \subset D^{n-1}(s)$ is closed in $D^{n-1}(s)$ and $H^{2n-3}(E_1(y)) < \sigma_\infty$. Respectively, $E_1^1(y) \subset E_1(y)$, $E_1^2(y) \subset E_1(y)$ and $H^{2n-3}(E_1^1(y)) = 0$, $H^{2n-4}(E_1^2(y)) = 0$. For each $y \in C$ define $h_y : D^{n-1}(s) \setminus E_1^1(y) \to \mathbb{C}$,

$$h_y(x) = f(x;y).$$

Then h_y is continuous, and $h_y \mid D^{n-1}(s) \setminus E_1(y)$ is analytic. Moreover, by the assumption for each $x' \in E_1^1(y) \setminus E_1^2(y)$ and for each complex line $T_1(x',x")$ in $\mathbb{C}^{n-1}$,

$$T_1(x',x") = \{x' + t(x" - x') \mid t \in \mathbb{C}\}$$

through the point x',

$$|x - x'|h_y(x) \to 0,\ x \in (D^{n-1}(s) \setminus E_1(y)) \cap T_1(x',x"),\ x \to x'.$$

By the induction hypothesis the function h_y has a unique analytic extension $h_y^* : D^{n-1}(s) \to \mathbb{C}$.

Denote

$$E_n^1 = \left(\bigcup_{k=1}^{\infty} E_n^1(x_k)\right) \cap B^2(t').$$

For each $y \in C \cap (B^2(t') \setminus E_n^1)$ and for each $k = 1,2,\ldots,$

$$g_{x_k}^{**}(y) = f(x_k;y) = h_y^*(x_k).$$

Since h_y^* is continuous, the convergence of the sequence $g_{x_k}^{**}$ in $C \cap (B^2(t') \setminus E_n^1)$ follows. Since $C \cap (B^2(t') \setminus E_n^1)$ is dense in $B^2(t')$, $f \mid (D^{n-1}(s) \times B^2(t')) \setminus E_1$ has a continuous extension

$f^* : D^{n-1}(s) \times B^2(t') \to \mathbb{C}$.

Using Lemma 2.7 and Lemma 2.6 we see that f^* is analytic, concluding the proof.

3.2. Corollary. (See [3, Theorem 3.1 and Theorem 3.5].) Let G be a domain of $\mathbb{C}^n$. Let $E \subset G$ be closed in G and let $H^{2n-1}(E) < \sigma_\infty$. Let $E_1 \subset E$ (not necessarily closed in G). Let $f : G \setminus E_1 \to \mathbb{C}$ be a continuous function such that $f \mid G \setminus E$ is analytic. Then f can be extended to a unique analytic function $f^* : G \to \mathbb{C}$ if one of the following conditions is satisfied:

(1) For each $z_0 \in E_1$ and for each complex line $T(z_0, z_0')$,

$$T(z_0, z'_0) = \{z_0 + t(z_0' - z_0) \mid t \in \mathbb{C}\},$$

through the point z_0,

$$|z - z_0| f(z) \to 0, \; z \in (G \setminus E) \cap T(z_0, z_0'), \; z \to z_0,$$

and $H^{2n-1}(E_1) = 0$.

(2) $H^{2n-2}(E_1) = 0$.

3.3. Corollary. [5, Lemma 3 (i)-(iii)]. Let G be a domain of $\mathbb{C}^n$. Let $E \subset G$ be closed in G. Let $f : G \setminus E \to \mathbb{C}$ be an analytic function. Then f can be extended to a unique analytic function $f^* : G \to \mathbb{C}$ if one of the following conditions is satisfied:

(1) f can be extended to a continuous function on G, and $H^{2n-1}(E) < \sigma_\infty$.

(2) f is bounded, and $H^{2n-1}(E) = 0$.

(3) $H^{2n-2}(E) = 0$.

3.4. Remark. It is easy to see that the result of Theorem 3.1 is sharp in the following sense:

(1) If we assume that $H^{2n-1}(E) < \sigma_\infty$ and $H^{2n-1}(E_1) = 0$, then the condition $H^{2n-2}(E_2) = 0$ cannot be replaced by the condition $H^{2n-2}(E_2) < \infty$. This can be seen e.g. from the example given in [3, Remark 3.8].

(2) If we assume that $H^{2n-1}(E) < \sigma_\infty$ and $H^{2n-2}(E_2) = 0$, then the condition $H^{2n-1}(E_1) = 0$ cannot be replaced by the condition $H^{2n-1}(E_1) < \infty$. This can be seen e.g. from the example given in [3, Remark 3.3].

(3) If we assume that $H^{2n-1}(E_1) = 0$ and $H^{2n-2}(E_2) = 0$, then the condition $H^{2n-1}(E) < \sigma_\infty$ cannot for any $\alpha > 0$ be replaced by the condition $H^{2n-1+\alpha}(E) = 0$. This can be seen e.g. from the example given in [3, Remark 3.4].

3.5. Remark. In Theorem 3.1 the boundary behavior assumption: For each $z_0 \in E_1 \setminus E_2$ and for each complex line $T(z_0, z_0')$ through the point z_0

$$|z - z_0| f(z) \to 0, \; z \in (G \smallsetminus E) \cap T(z_0,z_0'), \; z \to z_0, \qquad (3.5.1)$$

can be replaced by neither of the following two conditions:

(1) For each $z_0 \in E_1 \smallsetminus E_2$ and for each complex line $T(z_0,z_0')$ through the point z_0

$$|z - z_0| f(z)$$

is bounded in $(G \smallsetminus E) \cap T(z_0,z_0')$.

(2) For each $z_0 = (z_1^o,\ldots,z_n^o) \in E_1 \smallsetminus E_2$

$$(z_1 - z_1^o) \ldots (z_n - z_n^o) f(z) \to 0, \; z \in G \smallsetminus E, \; z \to z_0.$$

This can be seen just from the example $f : \mathbb{C}^n \smallsetminus (\mathbb{C}^{n-1} \times \{0\}) \to \mathbb{C}$,

$$f(x;y) = \frac{1}{y}.$$

However, if we replace in Theorem 3.1 the condition (3.5.1) by either the condition (1) or (2) above we get a meromorphic extension. See Theorem 3.6 and Corollary 3.7 below.

Next we give an analogue of Theorem 3.1 for meromorphic functions.

3.6. Theorem. Let G be a domain of $\mathbb{C}^n$. Let $E \subset G$ be closed in G and let $H^{2n-1}(E) < \sigma_\infty$. Let $E_1 \subset E$ (not necessarily closed in G) be such that $H^{2n-1}(E_1) = 0$. Let $E_2 \subset E$ (not necessarily closed in G) be such that $H^{2n-2}(E_2) = 0$. Let $f : G \smallsetminus E_1 \to \mathbb{C}^*$ be a spherically continuous function such that $f \mid G \smallsetminus E$ is analytic and for each $z_0 \in E_1 \smallsetminus E_2$ and for each complex line $T(z_0,z_0')$,

$$T(z_0,z_0') = \{z_0 + t(z_0' - z_0) \mid t \in \mathbb{C}\},$$

through the point z_0, there is $n(z_0,z_0') \in \mathbb{N}$ such that

$$|z - z_0|^{n(z_0,z_0')} f(z) \to 0, \; z \in (G \smallsetminus E) \cap T(z_0,z_0'), \; z \to z_0.$$

Then f has a unique meromorphic extension to G.

Proof. Since $H^{2n-1}(E) < \sigma_\infty$, int $E = \emptyset$. Hence it is sufficient to show that each point $z_0 \in E$ has a neighborhood U_{z_0} in G such that $f \mid U_{z_0} \smallsetminus E_1$ has a meromorphic extension to U_{z_0}.

We consider first the case $z_0 \in E \smallsetminus E_1$. Suppose first that $f(z_0) \neq \infty$. Using the spherical continuity of f at the point z_0 and Theorem 3.1 we find a neighborhood U_{z_0} of z_0 in G such that $f \mid U_{z_0} \smallsetminus E_1$ has a unique analytic extension $g : U_{z_0} \to \mathbb{C}$.

If $f(z_0) = \infty$, it similarly follows that there is a neighborhood U_{z_0} of z_0 in G such that $1/f \mid U_{z_0} \setminus E_1$ has a unique analytic extension $h : U_{z_0} \to \mathbb{C}$. Therefore $h^* = 1/h : U_{z_0} \to \mathbb{C}^*$ gives the desired meromorphic extension of $f \mid U_{z_0} \setminus E_1$ to U_{z_0}. Moreover, h^* is spherically continuous.

To treat the case $z_0 = (x_0;y_o) \in E_1$ we proceed as follows. Denote

$$E_1' = (G \setminus \bigcup_{z \in E \setminus E_1} U_z) \cap E,$$

where U_z is as above. Then $E_1' \subset G$ is closed in G. Since $E_1' \subset E_1$, $H^{2n-1}(E_1') = 0$. Define $f^* : G \setminus E_1' \to \mathbb{C}^*$,

$$f^*(z) = \begin{cases} g(z), \text{ when } z \in U_{z_0'} \text{ for some } z_0' \in E \setminus E_1 \text{ such that } f(z_0') \neq \infty, \\ h^*(z), \text{ when } z \in U_{z_0'} \text{ for some } z_0' \in E \setminus E_1 \text{ such that } f(z_0') = \infty, \\ f(z), \text{ when } z \in G \setminus E, \end{cases}$$

where g and h^* are as above.

Since $\operatorname{int} E = \emptyset$, f^* is well defined. Moreover, f^* is meromorphic and spherically continuous. Hence there is a set $N(f^*) \subset G \setminus E_1'$ closed in $G \setminus E_1'$ such that $H^{2n-2}(N(f^*)) < \sigma_\infty$ and $f^* \mid (G \setminus E_1') \setminus N(f^*)$ is analytic. Denote $E_1'' = E_1' \cup N(f^*)$. Then $E_1'' \subset G$ is closed in G and $H^{2n-1}(E_1'') = 0$. By Lemma 2.1 (3) there is a complex line P through the point z_0 such that $H^1(E_1'' \cap P) = 0$. Making an affine coordinate transformation, if necessary, we may suppose that $z_0 = 0$ and $P = \{0\} \times \mathbb{C}$.

Since $H^1(E_1'' \cap (\{0\} \times \mathbb{C})) = 0$, there is $t > 0$ such that $\{0\} \times \bar{B}^2(t) \subset G$ and $\{0\} \times S^1(t) \subset G \setminus E_1''$. Since $\{0\} \times S^1(t)$ is compact in $G \setminus E_1''$ and E_1'' is closed in G, there are $s = (r_1,\dots,r_{n-1})$, $r_j > 0$, $j = 1,\dots,n-1$, and t_1, t_2, $0 < t_1 < t < t_2$, such that $D^{n-1}(s) \times B^2(t_2) \subset G$ and $D^{n-1}(s) \times (B^2(t_2) \setminus \bar{B}^2(t_1)) \subset G \setminus E_1''$. Hence $f^* \mid D^{n-1}(s) \times (B^2(t_2) \setminus \bar{B}^2(t_1))$ is analytic.

Denote

$$B_1 = \{x \in D^{n-1}(s) \mid H^1((\{x\} \times B^2(t_2)) \cap E) < \sigma_\infty\},$$

$$B_2 = \{x \in D^{n-1}(s) \mid H^1((\{x\} \times B^2(t_2)) \cap E_1'') = 0\},$$

$$B_3 = \{x \in D^{n-1}(s) \mid (\{x\} \times B^2(t_2)) \cap E_2 = \emptyset\},$$

$$B = B_1 \cap B_2 \cap B_3.$$

By Lemma 2.1 (1) and (2), $H^{2n-2}(D^{n-1}(s) \setminus B) = 0$. For each $x \in B$ denote

$$E_n(x) = \{y \in B^2(t_2) \mid (x;y) \in E\},$$

$$E_n^1(x) = \{y \in B^2(t_2) \mid (x;y) \in E_1''\}.$$

Then $E_n(x) \subset B^2(t_2)$ is closed in $B^2(t_2)$ and $H^1(E_n(x)) < \sigma_\infty$. Respectively, $E_n^1(x) \subset E_n(x)$ is closed in $B^2(t_2)$ and $H^1(E_n^1(x)) = 0$. For each $x \in B$ define $g_x : B^2(t_2) \setminus E_n^1(x) \to \mathbb{C}$,

$$g_x(y) = f^*(x;y).$$

Then g_x is analytic. Moreover, for each $y_o' \in E_n^1(x)$ there is $n(y_o') \in \mathbb{N}$ such that

$$|y - y_o'|^{n(y_o')} g_x(y) \to 0, \ y \in B^2(t_2) \setminus E_n(x), \ y \to y_o'. \tag{3.6.1}$$

If $(x;y_o') \in E_1'$, then (3.6.1) follows from the assumption, since $E_1' \subset E_1$. If $(x;y_o') \in N(f^*)$, then (3.6.1) follows from the facts that f^* is meromorphic and shperically continuous. By Lemma 2.3 g_x has a unique meromorphic extension $g_x^* : B^2(t_2) \to \mathbb{C}^*$. By Lemma 2.8 $f^* \mid D^{n-1}(s) \times (B^2(t_2) \setminus \bar{B}^2(t_1))$ has a unique meromorphic extension to $D^{n-1}(s) \times B^2(t_2)$, concluding the proof.

3.7. Corollary. Let G be a domain of $\mathbb{C}^n$. Let $E \subset G$ be closed in G and let $H^{2n-1}(E) < \sigma_\infty$. Let $E_1 \subset E$ (not necessarily closed in G) be such that $H^{2n-1}(E_1) = 0$. Let $E_2 \subset E$ (not necessarily closed in G) be such that $H^{2n-2}(E_2) = 0$. Let $f : G \setminus E_1 \to \mathbb{C}^*$ be a spherically continuous function such that $f \mid G \setminus E$ is analytic and for each $z_o \in E_1 \setminus E_2$ there is a neighborhood U_{z_o} of z_o in G and an analytic function $g_{z_o} : U_{z_o} \to \mathbb{C}$ not identically zero such that

$$g_{z_o}(z) f(z) \to 0, \ z \in G \setminus E, \ z \to z_o.$$

Then f has a unique meromorphic extension to G.

Proof. Proceeding as in the proof of Theorem 3.6 we see that each $z_o \in E \setminus E_1$ has a neighborhood U_{z_o} in G such that $f \mid U_{z_o} \setminus E_1$ has a unique spherically continuous and meromorphic extension $f_{z_o} : U_{z_o} \to \mathbb{C}^*$.

Using the boundary behavior assumption and Theorem 3.1 we see that each $z_o \in E_1 \setminus E_2$ has a neighborhood U_{z_o} in G such that $f \mid U_{z_o} \setminus E_1$ has a unique meromorphic extension $h_{z_o}^* = h_{z_o}/g_{z_o}$ to U_{z_o}.

Denote

$$E_2' = (G \setminus \bigcup_{z \in (E \setminus E_1) \cup (E_1 \setminus E_2)} U_z) \cap E,$$

where U_z is as above. Then $E_2' \subset G$ is closed in G. Since $E_2' \subset E_2$, $H^{2n-2}(E_2') = 0$. Define f^*,

$$f^*(z) = \begin{cases} f_{z_o}(z), \text{ when } z \in U_{z_o} \text{ for some } z_o \in E \setminus E_1, \\ h^*_{z_o}(z), \text{ when } z \in U_{z_o} \text{ for some } z_o \in E_1 \setminus E_2 \text{ and } h^*_{z_o}(z) \text{ is defined,} \\ f(z), \text{ when } z \in G \setminus E \end{cases}$$

where f_{z_o} and $h^*_{z_o}$ are as above.

Since int $E = \emptyset$, f^* is well defined. Moreover, f^* is meromorphic in $G \setminus E_2'$. Hence there are sets $N(f^*)$, $I(f^*) \subset G \setminus E_2'$ closed in $G \setminus E_2'$ such that $H^{2n-2}(N(f^*)) < \sigma_\infty$, $H^{2n-4}(I(f^*)) < \sigma_\infty$ and $f^* \mid (G \setminus E_2') \setminus N(f^*)$ is analytic and $f^* \mid (G \setminus E_2') \setminus I(f^*)$ is spherically continuous. Denote $E' = E_2' \cup N(f^*)$, $E_2'' = E_2' \cup I(f^*)$. Then $E' \subset G$ is closed in G and $H^{2n-1}(E') = 0$. Similarly, $E_2'' \subset G$ is closed in G and $H^{2n-2}(E_2'') = 0$. Then $f^* : G \setminus E_2'' \to \mathbb{C}^*$ is spherically continuous and $f^* \mid G \setminus E'$ is analytic. By Theorem 3.6 f^* has a unique meromorphic extension f^{**} to G, concluding the proof.

<u>3.8. Corollary.</u> Let G be a domain of $\mathbb{C}^n$. Let $E \subset G$ be closed in G and let $H^{2n-1}(E) = 0$. Let $E_1 \subset E$ (not necessarily closed in G) be such that $H^{2n-2}(E_1) = 0$. Let f be meromorphic in $G \setminus E$. Then f can be extended to a unique meromorphic function f^* in G if one of the following conditions is satisfied:

(1) For each $z_o \in E \setminus E_1$ and for each complex line $T(z_o, z_o')$,

$$T(z_o, z_o') = \{z_o + t(z_o' - z_o) \mid t \in \mathbb{C}\},$$

through the point z_o, there is $n(z_o, z_o') \in \mathbb{N}$ such that

$$|z - z_o|^{n(z_o, z_o')} f(z) \to 0, \; z \in ((G \setminus E) \setminus N(f)) \cap T(z_o, z_o'), \; z \to z_o.$$

(2) For each $z_o \in E \setminus E_1$ there is a neighborhood U_{z_o} of z_o in G and an analytic function $g_{z_o} : U_{z_o} \to \mathbb{C}$ not identically zero such that

$$g_{z_o}(z) f(z) \to 0, \; z \in (G \setminus E) \setminus N(f), \; z \to z_o.$$

<u>3.9. Corollary.</u> Let G be a domain of $\mathbb{C}^n$. Let $E \subset G$ be closed in G. Let f be meromorphic in $G \setminus E$. Then f can be extended to a unique meromorphic function f^* in G if one of the following conditions is satisfied:

(1) For each $z_o \in E$ and for each complex line $T(z_o, z_o')$,

$$T(z_o, z_o') = \{z_o + t(z_o' - z_o) \mid t \in \mathbb{C}\},$$

through the point z_o, there is $n(z_o, z_o') \in \mathbb{N}$ such that

$$|z - z_0|^{n(z_0,z_0')} f(z) \to 0,\ z \in ((G \setminus E) \setminus N(f)) \cap T(z_0,z_0'),\ z \to z_0$$

and $H^{2n-1}(E) = 0$.

(2) For each $z_0 \in E$ there is a neighborhood U_{z_0} of z_0 in G and an analytic function $g_{z_0} : U_{z_0} \to \mathbb{C}$ not identically zero such that

$$g_{z_0}(z)f(z) \to 0,\ z \in (G \setminus E) \setminus N(f),\ z \to z_0,$$

and $H^{2n-1}(E) = 0$.

(3) [2, Remark after Theorem 1.1] $H^{2n-2}(E) = 0$.

3.10. Remark. It is easy to see that the result of Theorem 3.6 is sharp in the following sense:

(1) If we assume that $H^{2n-1}(E) < \sigma_\infty$ and $H^{2n-1}(E_1) = 0$, then the condition $H^{2n-2}(E_2) = 0$ cannot be replaced by the condition $H^{2n-2}(E_2) < \infty$. This can be seen e.g. from the example given in [4, Remark 3.4].

(2) If we assume that $H^{2n-1}(E) < \sigma_\infty$ and $H^{2n-2}(E_2) = 0$, then the condition $H^{2n-1}(E_1) = 0$ cannot be replaced by the condition $H^{2n-1}(E_1) < \infty$. This can be seen e.g. from the example given in [3, Remark 3.3].

(3) If we assume that $H^{2n-1}(E_1) = 0$ and $H^{2n-2}(E_2) = 0$, then the condition $H^{2n-1}(E) < \sigma_\infty$ cannot for any $\alpha > 0$ be replaced by the condition $H^{2n-1+\alpha}(E) = 0$. This can be seen e.g. from the example given in [3, Remark 3.4].

References

[1] Federer, H.: Geometric measure theory. Springer-Verlag, Berlin - Heidelberg - New York (1969).

[2] Harvey, R., Polking, J.: Extending analytic objects. Commun. pure appl. Math. 28 (1975), 701 - 727.

[3] Riihentaus, J.: Removable singularities of analytic functions of several complex variables. Math. Z. 158 (1978), 45 - 54.

[4] Riihentaus, J.: An extension theorem for meromorphic functions of several variables. Ann. Acad. Sci. Fenn., Ser. A I 4 (1978), 145 - 149.

[5] Shiffman, B.: On the removal of singularities of analytic sets. Michigan math. J. 15 (1968), 111 - 120.

[6] Stolzenberg, G.: Volumes, limits, and extensions of analytic varieties. Lecture Notes in Mathematics 19, Springer-Verlag, Berlin - Heidelberg - New York (1966).

[7] Whitney, H.: Complex analytic varieties. Addison-Wesley, Reading - Menlo Park - London - Don Mills (1972).

[8] Wohlhauser, A.: Hebbare Singularitäten quasikonformer Abbildungen und lokal beschränkter holomorpher Funktionen. Math. Z. 124 (1972), 37 - 42.

Kosteperänkatu 2 B 84
SF-90100 Oulu 10
Finland

AHLFORS' TRIVIAL DEFORMATIONS AND LIOUVILLE'S THEOREM IN R^n

Jukka Sarvas

1. Introduction

1.1. Let G be a domain in R^n, $n \geq 2$, and $f : G \to R^n$ a continuous mapping with locally integrable distributional partial derivatives $D_i f$, $i = 1,2,\ldots,n$. Let $Df(x)$ be the Jacobian matrix of f at almost every $x \in G$. According to Ahlfors we define

$$Sf(x) = \frac{1}{2}[Df(x) + \widetilde{Df(x)}] - \frac{1}{n}\operatorname{tr}(Df(x))I$$

where I is the unit matrix, $\tilde{A}$ denotes the transpose and $\operatorname{tr}(A)$ the trace of an $n\times n$-matrix A. The mapping f is called a <u>trivial deformation</u> if $Sf(x) = 0$ for almost every $x \in G$.

If $n = 2$, Sf corresponds to the complex derivative $D_{\bar{z}}f = (1/2)(\partial f/\partial x + i\partial f/\partial y)$, $z = x + iy$, and $Sf = 0$ is the Cauchy-Riemann equation. Therefore, $f : G \to R^2$ is an analytic mapping if and only if f is a trivial deformation. If $n \geq 3$, the following theorem of Ahlfors [1; Prop. 1] shows that there are rather few trivial deformations:

<u>1.2. Theorem</u> (Ahlfors). If $f : G \to R^n$ is a trivial deformation in a domain $G \subset R^n$ and $n \geq 3$, then f is of the form

$$f(x) = a + Bx + 2(c \cdot x)x - |x|^2 c, \quad x \in G,$$

where $a,c \in R^n$, $B : R^n \to R^n$ is a linear mapping with $SB = 0$ and $c \cdot x$ denotes scalar product in R^n.

The proof of this theorem is short and rather easy. First, by a standard smoothing procedure, one can reduce the general case to a C^3-case. For a C^3-function f the proof only involves elementary methods.

Theorem 1.2 resembles Liouville's theorem for conformal mappings in R^n, $n \geq 3$. We state the C^2-version of Liouville's theorem here. If $f : G \to R^n$ is differentiable and $\det Df(x) \neq 0$, $x \in G$, we write

$$Xf(x) = |\det Df(x)|^{-1/n} Df(x) \quad \text{and} \quad Zf(x) = \widetilde{Xf(x)}Xf(x).$$

A mapping $f : G \to R^n$ is <u>conformal</u> if $\det Df(x) \neq 0$ and $Zf(x) = I$ for all $x \in G$. Let $\bar{R}^n$ be the one point compactification of R^n. By a <u>Möbius transformation</u> $f : \bar{R}^n \to \bar{R}^n$ we mean a mapping which is a composition of a finite number of reflections in (n-1)-planes or inversions in (n-1)-spheres in $\overline{R}^n$. A Möbius transformation can always be written in one of the following forms: $f(x) = Tx + a$ or $f(x) = J(Tx + a)$ where $T : R^n \to R^n$ is a linear conformal mapping, $a \in R^n$ and $J : \overline{R}^n \to \overline{R}^n$ is the inversion $J(x) = |x|^{-2}x$, $x \neq 0$, $J(0) = \infty$.

1.3. Theorem (Liouville). If $f : G \to R^n$ is a conformal C^2-function, $G \subset R^n$ a domain and $n \geqq 3$, then f is a restriction of a Möbius transformation.

The purpose of this paper is to show that, in fact, Theorems 1.2 and 1.3 are equivalent in the sense that either can rather easily be derived from another. This also gives a new easy proof for the C^2-version of Liouville's theorem. Our method is based on some elementary considerations on ordinary differential equations and Möbius transformations.

2. Trivial deformations and conformal mappings

2.1. The close relation between trivial deformations and conformal mappings is demonstrated by an ordinary differential equation. Let $G \subset R^n$ be open, $n \geqq 2$, and $f : G \to R^n \in C^1$. If $U \subset G$ is open and $\overline{U}$, the closure of U in R^n, is a compact subset of G, then there is $s > 0$ and a mapping $\varphi : (-s,s) \times U \to R^n$ so that φ is the unique solution of the differential equation

$$\dot{\varphi}(t,x) = f(\varphi(t,x)),\ \varphi(0,x) = x, \quad \text{for } |t| < s,\ x \in U, \tag{2.2}$$

where $\dot{\varphi}(t,x) = (\partial/\partial t)\varphi(t,x)$. The mapping $x \to \varphi_t(x) = \varphi(t,x)$, $x \in U$, is a C^1-diffeomorphism onto $\varphi_t U$ and it is called a flow generated by f. Also (here $D = \partial/\partial x$),

$$(D\varphi_t(x))^{\bullet} = D\dot{\varphi}_t(x) = Df(\varphi_t(x))D\varphi_t(x) \quad \text{for } |t| < s,\ x \in U.$$

If $t \in (-s,s)$, the solution $\varphi(u,x)$ extends in a unique way to all (u,x) for which u, $u + t \in (-s,s)$ and $x \in \varphi_t U$, and for such u and t it holds $\varphi_u \circ \varphi_t(x) = \varphi_{u+t}(x)$ for $x \in U$.

The starting point of our reasoning is the following lemma:

2.3. Lemma. Let $f : G \to R^n \in C^1$ and $\varphi_t : U \to R^n$, $|t| < s$, be as in (2.2). If $\varphi_t : U \to R^n$ is conformal for all $t \in (-s,s)$, then $Sf(x) = 0$ for all $x \in U$. Conversely, if $Sf(x) = 0$ for all $x \in G$, then every $\varphi_t : U \to R^n$ is conformal for $|t| < s$.

Proof. We apply the differentation formula $(\det A(t))^{\bullet} = \det A(t) \operatorname{tr}(A(t)^{-1}\dot{A}(t))$ where the matrix valued function $t \to A(t)$

is differentiable and $\det A(t) \neq 0$. Since $\det D\varphi_t(x) > 0$ for $(t,x) \in (-s,s) \times U$, we obtain

$$(X\varphi_t(x))^{\bullet} = [(\det D\varphi_t(x))^{-1/n} D\varphi_t(x)]^{\bullet}$$
$$= -\frac{1}{n}(\det D\varphi_t(x))^{-1-1/n}\det D\varphi_t(x)\,\mathrm{tr}[(D\varphi_t(x))^{-1}(D\varphi_t(x))^{\bullet}]D\varphi_t(x)$$
$$+ (\det D\varphi_t(x))^{-1/n}(D\varphi_t(x))^{\bullet}.$$

Now, $(D\varphi_t(x))^{\bullet}_{t=0} = (D\dot{\varphi}_t(x))_{t=0} = Df(x)$ for $x \in U$. Also $D\varphi_o(x) = I$ for $x \in U$. Therefore we have

$$(X\varphi_t(x))^{\bullet}_{t=0} = -(1/n)\mathrm{tr}(Df(x))I + Df(x),\ x \in U.$$

Because $(\widetilde{X\varphi_t})^{\bullet} = [(X\varphi_t)^{\bullet}]^{\sim}$ and $(Z\varphi_t)^{\bullet} = (\widetilde{X\varphi_t}X\varphi_t)^{\bullet} = (\widetilde{X\varphi_t})^{\bullet}X\varphi_t + (\widetilde{X\varphi_t})(X\varphi_t)^{\bullet}$, we obtain

$$(Z\varphi_t(x))^{\bullet}_{t=0} = -\frac{2}{n}\mathrm{tr}(Df(x)) + Df(x) + \widetilde{Df(x)} = 2Sf(x) \tag{2.4}$$

for $x \in U$. This is Ahlfors' formula [2; (0.1)]. Actually, it is true for all $x \in G$, since every $x \in G$ has a neighborhood where the flow φ_t is defined for small $|t|$.

Now, if $\varphi_t : U \to R^n$ is conformal for all $t \in (-s,s)$, then $Z\varphi_t(x) = I$ for all $t \in (-s,s)$ and so $Sf(x) = (1/2)(Z\varphi_t(x))^{\bullet}_{t=0} = 0$ for all $x \in U$. Conversely, let $Sf(x) = 0$ for all $x \in G$. Let $t \in (-s,s)$. Because $\varphi_{u+t} = \varphi_u \circ \varphi_t$ for small $|u|$ and $Z(\varphi_u \circ \varphi_t)(x) = \widetilde{X\varphi_t}(x)[Z\varphi_u(\varphi_t(x))]X\varphi_t(x)$, we get by (2.4)

$$(Z\varphi_t(x))^{\bullet} = (\frac{\partial}{\partial u}Z\varphi_{u+t}(x))_{u=0} = \widetilde{X\varphi_t}(x)[\frac{\partial}{\partial u}Z\varphi_u(\varphi_t(x))]_{u=0}X\varphi_t(x)$$
$$= 2\widetilde{X\varphi_t}(x)Sf(\varphi_t(x))X\varphi_t(x) = 0$$

for $|t| < s$ and $x \in U$. Therefore, $Z\varphi_t(x) = I + {}_0\int^t (Z\varphi_u(x))^{\bullet}du = I$ for $x \in U$. So $\varphi_t : U \to R^n$ is conformal for all $t \in (-s,s)$. The lemma is proved.

3. Theorem 1.2 implies Theorem 1.3

3.1. Note that for the inversion $J : \bar{R}^n \to \bar{R}^n$ we have

$$DJ(x) = |x|^{-2}(I - 2Q_x),\ x \neq 0,$$

where $Q_x : R^n \to R^n$ is the orthogonal projection

$$Q_x y = |x|^{-2}(y \cdot x)x,\ y \in R^n, \tag{3.2}$$

from R^n onto the 1-dimensional subspace $L_x = \{\lambda x : \lambda \in R^1\}$. Thus

$2Q_x - I$ is the reflection in the axis L_x. Note also that $J^{-1} = J$ and $(I - 2Q_x)^{-1} = I - 2Q_x$.

3.3. Proof of Theorem 1.3. Let $h : G \to G' \in C^2$ be conformal where G and G' are domains in R^n and $n \geq 3$. By performing preliminary Möbius transformations we may assume that $0 \in G$, $h(0) = 0$ and $Dh(0) = I$. Then

$$h(x) = x + |x|\varepsilon(x) \quad \text{with} \quad \varepsilon(x) \to 0 \quad \text{as} \quad x \to 0. \tag{3.4}$$

Put $g = J \circ h$. Now $Dg(x) = |h(x)|^{-2}(I - 2Q_{h(x)})Dh(x)$ and so

$$(Dg(x))^{-1} = |h(x)|^2(Dh(x))^{-1}(I - 2Q_{h(x)}), \; x \neq 0. \tag{3.5}$$

Choose an arbitrary $w \in R^n$, $w \neq 0$. By (3.2)

$$(Dg(x))^{-1}w = |h(x)|^2(Dh(x))^{-1}w - 2(h(x) \cdot w)h(x). \tag{3.6}$$

Define $f : G \to R^n$ by $f(x) = (Dg(x))^{-1}w$ if $x \neq 0$ and $f(0) = 0$. Then $f \in C^1$ by (3.6). It is easy to see that for every $y \in G$ there is a neighborhood U_y of y in G and $s_y > 0$ so that the mapping

$$\varphi_t(x) = g^{-1}(g(x) + tw) = (h^{-1} \circ J)((J \circ h)(x) + tw)$$

is defined for $|t| < s_y$ and $x \in U_y$. Differentation shows that

$$\dot{\varphi}_t(x) = [Dg(\varphi_t(x))]^{-1}w = f(\varphi_t(x)), \; \varphi_o(x) = x,$$

for $t \in (-s_y, s_y)$ and $x \in U_y$. Because every $\varphi_t : U_y \to R^n$ is conformal as a composed mapping of conformal mappings, Lemma 2.3 implies that $Sf(x) = 0$ for every $x \in U_y$. Hence, $Sf(x) = 0$ for all $x \in G$. Then by Theorem 1.2 f is of the form

$$f(x) = a + Bx + 2(c \cdot x)x - |x|^2c, \; x \in G, \tag{3.7}$$

where $a, c \in R^n$ and B is a linear mapping. The condition $f(0) = 0$ implies $a = 0$. Let $e \in R^n$, $|e| = 1$, be arbitrary. Inserting $x = se$ into (3.7) we get for small $s > 0$ by (3.4) and (3.5)

$$\begin{aligned} sBe = B(se) &= -s^2[2(c \cdot e)e - c] + f(se) \\ &= -s^2[2(c \cdot e)e - c] + s^2|e + \varepsilon(se)|^2(Dh(se))^{-1}(I - 2Q_{h(se)})w. \end{aligned}$$

Dividing by s and letting $s \to 0$ gives $Be = 0$. This yields $B = 0$. Therefore $f(x) = 2(c \cdot x)x - |x|^2c$ with $c \neq 0$. Putting $x = sc$ for small $s > 0$ we obtain

$$s^2|c|^2c = f(sc) = s^2|c + |c|\varepsilon(sc)|^2(Dh(sc))^{-1}(I - 2Q_{h(sc)})w.$$

If $s \to 0$, then $Q_{h(sc)}w \to Q_c w$ by (3.4), and so the above implies $c = (I - 2Q_c)w$, and this yields $w = (I - 2Q_c)c = -c$. Therefore

$$(Dg(x))^{-1}w = -(2(w \cdot x)x - |x|^2 w) = |x|^2(I - 2Q_x)w$$

for all $w \neq 0$ and $x \in G \setminus \{0\}$. This implies $(Dg(x))^{-1} = |x|^2(I - 2Q_x)$, or equivalently, $Dg(x) = |x|^{-2}(I - 2Q_x) = (DJ)(x)$, $x \neq 0$. Then for some $d \in R^n$ we have $(J \circ h)(x) = g(x) = J(x) + d$, $x \in G$, $x \neq 0$, and so $h(x) = J(J(x) + d)$ for all $x \in G$. This proves Theorem 1.3.

4. Proof of Theorem 1.2 by Theorem 1.3

4.1. Let $f : G \to R^n$ be a trivial deformation in a domain $G \subset R^n$ and $n \geq 3$. We have to show that f is of the form

$$f(x) = a + Bx + 2(c \cdot x)x - |x|^2 c, \ x \in G, \tag{4.2}$$

where $a, c \in R^n$ and $B : R^n \to R^n$ is linear with $SB = 0$. Fix an arbitrary $x_0 \in G$. Note that for any $u, v \in R^n$ the linear mapping $Ax = (x \cdot u)v - (x \cdot v)u$, $x \in R^n$, is skew-symmetric, i.e. $A + \tilde{A} = 0$, and so $SA = 0$. This and a short computation show that we may assume $x_0 = 0 = f(0)$.

Suppose next that $f \in C^1$. Let $U \subset G$ be any domain so that $0 \in U$ and $\bar{U}$ is a compact subset of G. There is $s > 0$ so that f generates a flow $\varphi_t : U \to R^n$, $|t| < s$, i.e.

$$\dot{\varphi}_t(x) = f(\varphi_t(x)), \ \varphi_o(x) = x \quad \text{for } x \in U \text{ and } |t| < s.$$

By Lemma 2.3 and Theorem 1.3 for each $t \in (-s,s)$ the mapping $\varphi_t : U \to R^n$ is a restriction of a Möbius transformation. The condition $f(0) = 0$ implies $\varphi_t(0) = 0$, and so φ_t has to be of the form

$$\varphi_t = T_t \circ R_{w(t)}, \ |t| < s,$$

where $T_t : R^n \to R^n$ is a conformal linear mapping and $R_{w(t)}(x) = J(J(x) + w(t))$, $w(t) \in R^n$. Since $DR_{w(t)}(0) = I$, we have $D\varphi_t(0) = T_t$, and because $f \in C^1$, we get

$$\dot{T}_t = (D\varphi_t)^{\bullet}(0) = D\dot{\varphi}_t(0) = Df(\varphi_t(0))D\varphi_t(0) = Df(0)T_t,$$

for $|t| < s$. Also $T_o = D\varphi_o(0) = I$. Then by Lemma 2.3 $SA = 0$ with $A = Df(0)$. Therefore $S(f - A)(x) = 0$ for all $x \in G$, and we may suppose that $Df(0) = 0$. Then the above reasoning shows that $T_t = I$ for $|t| < s$, and so $\varphi_t = R_{w(t)}$ for $|t| < s$. Choose an arbitrary $x \in U$, $x \neq 0$. Then $\varphi_t(x) \neq 0$ for $|t| < s$, and we get $w(t) = J(\varphi_t(x)) - J(x)$.

Therefore $\dot{w}(t)$ exits for all $t \in (-s,s)$, and we have

$$\dot{w}(t) = (DJ)(\varphi_t(x))\dot{\varphi}_t(x) = (DJ)(\varphi_t(x))f(\varphi_t(x)).$$

This yields $\dot{w}(0) = (DJ)(x)f(x) = |x|^{-2}(I - 2Q_x)f(x)$. Thus $f(x) = |x|^2(I - 2Q_x)\dot{w}(0) = 2(c \cdot x)x - |x|^2 c$ with $c = -\dot{w}(0)$. This proves the theorem for $f \in C^1$.

If f is not a C^1-function, we employ a standard smoothing procedure. Let $\theta : R^n \to R^1$ be a non-negative C^∞-function with $\theta(x) = 0$ for $|x| \geqq 1$ and $\int_{R^n} \theta(x)dx = 1$. Set $\theta_\varepsilon(x) = \varepsilon^{-n}\theta(\varepsilon^{-1}x)$ for $\varepsilon > 0$, and define

$$f_\varepsilon(x) = \int_{R^n} \theta_\varepsilon(y)f(x - y)dy \quad \text{for} \quad x \in G_\varepsilon$$

where $G_\varepsilon = \{x \in G : B^n(x,\varepsilon) \subset G\}$ and $B^n(x,\varepsilon) = \{y \in R^n : |y - x| < \varepsilon\}$. Then $f_\varepsilon \to f$ uniformly in compact subsets of G as $\varepsilon \to 0$. Also $f_\varepsilon \in C^\infty$ and $Df_\varepsilon(x) = \int_{R^n} \theta_\varepsilon(y)Df(x - y)dy$. Therefore

$$Sf_\varepsilon(x) = \int_{R^n} \theta_\varepsilon(y)Sf(x - y)dy = 0 \quad \text{for} \quad x \in G_\varepsilon.$$

Then the first part of the proof yields that f_ε is of the form (4.2). Because $f_\varepsilon \to f$ uniformly in compact subsets of G, it is not difficult to see that f also has to be of the form (4.2).

References

[1] Ahlfors, L.: Conditions for quasiconformal deformations in several variables, in "Contributions to analysis". Academic Press, New York and London (1974), 19 - 25.

[2] Ahlfors, L.: Quasiconformal deformations and mappings in R^n. J. Analyse math. 30 (1976), 74 - 97.

Department of Mathematics
University of Helsinki
SF-00100 Helsinki 10
Finland

AN APPLICATION OF THE CALCULUS OF VARIATIONS FOR GENERAL FAMILIES OF QUASICONFORMAL MAPPINGS

M. Schiffer and G. Schober*

1. Introduction.

It is customary to obtain closed families of plane quasiconformal (q.c.) mappings by restrictions on the complex dilatation. For example, if $\mu_n = (f_n)_{\bar{z}}/(f_n)_z$ are the complex dilatations of q.c. mappings f_n that converge locally uniformly to a q.c. mapping f and if

$$|\mu_n| \leqq k \quad \text{a.e.} \tag{1}$$

for a constant $k < 1$, then the complex dilatation $\mu = f_{\bar{z}}/f_z$ also satisfies

$$|\mu| \leqq k \quad \text{a.e.} \tag{2}$$

In addition, K. Strebel [8] has shown that $\limsup_{n \to \infty} |\mu_n| \leqq |\mu|$ a.e. As a result, in (1) and (2) one may replace the constant k by any nonnegative measurable function with essential supremum less than one. Such families have been studied by R. Kühnau [1], the authors [5, 7], and others.

More generally, let D be a plane domain and let us consider families $\mathcal{F}$ of q.c. mappings $f : D \to \mathbb{C}$ whose complex dilatations are restricted by a relation of the form

$$F(\mu(z),z) \leqq 0 \quad \text{for a.e. } z \in D. \tag{3}$$

Some conditions will be imposed on $F(t,z)$ in Section 2, and we shall give examples of closed families of q.c. mappings generated this way in Section 3.

Very recently, the authors [6] developed a calculus of variations for general families of q.c. mappings defined by relations of the form (3). It is valuable from two points of view. On the one hand, one can pose interesting extremal problems over such families of functions. Then the variational method leads to partial differential equations satisfied

* This work was supported in part by grants MCS-75-23332-A03 and MCS-77-01831-A01 from the National Science Foundation.

by an extremal function. In cases where these equations are sufficiently simple and restrictive, one is led to an explicit solution of the extremal problem and, consequently, to interesting inequalities. We applied this method in [6] to solve a coefficient problem. In Section 4 of this article we shall give a further application to obtain new inequalities of "Grunsky-type".

A valuable second point of view, on the other hand, is obtained by posing extremal problems that lead to fundamental solutions for the partial differential equations arising from the variational procedure. This program was followed in [5, 6]. The many possibilities for relations in (3) give rise to different partial differential equations. We shall comment on this aspect at the end of Section 2.

2. Calculus of variations.

Let $\mathcal{F}$ be the family of q.c. mappings $f : D \to \mathbb{C}$ whose complex dilatations $\mu = f_{\bar{z}}/f_z$ satisfy the relation (3). We shall now impose conditions on the functions $F(t,z)$ and a normalization on the family $\mathcal{F}$.

Assume that $F(t,z)$ is defined in $\{t : |t| < 1\} \times D$ and that

(i) $F(t,z)$ is measurable in z for each fixed t, and

(ii) $F(t,z)$ is continuous in t for a.e. fixed z.

These are the so-called Carathéodory conditions, which imply that $F(\mu(z),z)$ is a measurable function of z whenever μ is measurable, so that relation (3) is meaningful.

We shall also assume that $F(t,z)$ has the following differentiability property.

(iii) Except for a set N of area zero, the complex partial derivative $F_t(t,z)$ exists in $\mathcal{D} = \{t : |t| < 1\} \times (D \setminus N)$, is uniformly bounded on compact subsets of $\mathcal{D}$, and

$$F(\tau,z) = F(t,z) + 2\mathrm{Re}\{(\tau - t)F_t(t,z)\} + o(|\tau - t|)$$

as $\tau \to t$, where the term $o(|\tau - t|)$ is uniform on compact subsets of $\mathcal{D}$.

In addition, we shall assume that

(iv) relation (3) implies $\|\mu\|_\infty < 1$.

To simplify the presentation, we shall assume that

(v) the domain D contains a fixed neighborhood of ∞ in which the mappings f are analytic and normalized so that

$$f(z) = z + o(1) \quad \text{as} \quad z \to \infty.$$

In order to construct variations in the normalized family $\mathcal{F}$, let

E be a compact set and $a(w)$ a complex-valued measurable function such that $|a(w)|$ is the characteristic function of the set E. Then for each ε, $0 < \varepsilon < 1$, there exists a q.c. mapping $\Phi : \mathbb{C} \to \mathbb{C}$ that satisfies the Beltrami equation

$$\Phi_{\bar{w}} = \varepsilon a \Phi_w \quad \text{a.e.}$$

and has the normalization $\Phi(w) = w + o(1)$ as $w \to \infty$. It has the representation

$$\Phi(w) = w - \frac{\varepsilon}{\pi} \iint_E \frac{a(\zeta)}{\zeta - w}\, dm(\zeta) + O(\varepsilon^2) \tag{4}$$

where m denotes planar Lebesgue measure and the term $O(\varepsilon^2)/\varepsilon^2$ is uniformly bounded depending only on E. If f belongs to $\mathcal{F}$ and if the complex dilatation $(\Phi \circ f)_{\bar{z}}/(\Phi \circ f)_z$ satisfies (3), then $\Phi \circ f$ also belongs to $\mathcal{F}$ and provides a variation of the function f.

Now consider the problem of finding the maximum value of a functional λ over the family $\mathcal{F}$. Assume that

(vi) $\mathcal{F}$ is compact and $\lambda : \mathcal{F} \to \mathbb{R}$ is an upper semicontinuous functional.

Then there will be a (not necessarily unique) function $f \in \mathcal{F}$ for which

$$\lambda(f) = \max_{\mathcal{F}} \lambda. \tag{5}$$

Finally, assume that λ has a nonzero functional derivative at f in the following sense.

(vii) An a.e. nonvanishing, locally integrable function A exists in $f(D)$ such that

$$\lambda(\Phi \circ f) = \lambda(f) + \frac{\varepsilon}{\pi} \operatorname{Re} \iint_E A(\zeta) a(\zeta)\, dm(\zeta) + O(\varepsilon)$$

under variations of the form (4), whenever $\Phi \circ f \in \mathcal{F}$ and $\varepsilon \to 0$.

With the assumptions above, we proved in [6] that the method of variations leads to the following conclusions.

<u>Theorem 1.</u> Under the assumptions (i) through (vii), the extremal function f for the problem (5) satisfies the differential relations

$$F\left(\frac{f_{\bar{z}}(z)}{f_z(z)}, z\right) = 0 \quad \text{and} \quad \overline{F_t\left(\frac{f_{\bar{z}}(z)}{f_z(z)}, z\right)} A(f(z)) f_z(z)^2 \geqq 0 \tag{6}$$

for a.e. $z \in D$. If, in addition, the function A is analytic and nonzero in a neighborhood of a point $w_o = f(z_o)$ and if J denotes any local integral

$$J(w) = \int^{w} \sqrt{A(\zeta)}\, d\zeta,$$

then $\mathcal{J} = J \circ f$ satisfies the differential relations

$$F\left(\frac{\mathcal{J}_{\bar z}(z)}{\mathcal{J}_z(z)}, z\right) = 0 \quad \text{and} \quad \overline{F_t\left(\frac{\mathcal{J}_{\bar z}(z)}{\mathcal{J}_z(z)}, z\right)}\mathcal{J}_z(z)^2 \geqq 0 \tag{7}$$

for a.e. z in a neighborhood of z_o.

Note that the differential inequalities in (6) and (7) may be written as differential equations. For example, the latter inequality is equivalent to the equation

$$F_t(\mathcal{J}_{\bar z}/\mathcal{J}_z, z)\overline{\mathcal{J}_z} = |F_t(\mathcal{J}_{\bar z}/\mathcal{J}_z, z)|\mathcal{J}_z. \tag{8}$$

On the one hand, the relations (6) and (7) are important conditions satisfied by an extremal function for the problem (5). On the other hand, one can choose the function $F(t,z)$ used to define the family $\mathcal{F}$ in such a way as to obtain interesting equations in (6), (7), or (8). In the next section we shall give an example where equations (7) and (8) reduce to a linear system.

It becomes important to identify functions $F(t,z)$ that give rise to compact families $\mathcal{F}$. Then the compactness of $\mathcal{F}$, in conjunction with an appropriate extremal problem, can be used to provide a q.c. mapping that implies existence of a corresponding solution of (6), (7), or (8) and even a representation for it in terms of the q.c. mapping. An application of this approach to fundamental solutions of some linear and nonlinear equations is found in [6].

3. An example.

Let c be a complex function and k a nonnegative function, both measurable in a domain D, with $\| |c| + k \|_\infty < 1$. Set

$$F(t,z) = |t - c(z)|^2 - k(z)^2. \tag{9}$$

One easily verifies that this function satisfies (i) through (iv). The corresponding family $\mathcal{F}$ consists of q.c. mappings whose complex dilatations satisfy

$$|\mu - c| \leqq k \qquad \text{a.e. in } D.$$

In [6] we showed that this family is closed in the topology of locally uniform convergence. With the normalization (v), it is compact. Therefore it provides fertile ground for the variational method.

In terms of the function (9), the equations (7) and (8) of Theorem 1 take the form

$$|\mathcal{F}_{\bar{z}}/\mathcal{F}_z - c| = k$$

$$(\mathcal{F}_{\bar{z}}/\mathcal{F}_z - c)\mathcal{F}_z = |\mathcal{F}_{\bar{z}}/\mathcal{F}_z - c|\overline{\mathcal{F}_z}.$$

Combining them, we find that

$$\mathcal{F}_{\bar{z}} = c\mathcal{F}_z + k\overline{\mathcal{F}_z}. \tag{10}$$

In terms of the real and imaginary parts of $\mathcal{F} = u + iv$, this equation is equivalent to the linear system

$$u_x = \alpha v_x + \beta v_y$$

$$-u_y = \gamma v_x + \alpha v_y$$

where $\alpha\delta = i(c - \bar{c})$, $\beta\delta = |1 + c|^2 - k^2$, $\gamma\delta = |1 - c|^2 - k^2$, and $\delta = (1 - k)^2 - |c|^2$. Elimination of v leads formally to the second order equation

$$\left(\frac{\gamma u_x + \alpha u_y}{\beta\gamma - \alpha^2}\right)_x + \left(\frac{\alpha u_x + \beta u_y}{\beta\gamma - \alpha^2}\right)_y = 0$$

for u. In [6] we obtained existence theorems for (weak) fundamental solutions of this equation and representations for them in terms of q.c. mappings.

4. Applications.

Let $\Sigma_{c,k}$ denote the family of q.c. mappings $f : \mathbb{C} \to \mathbb{C}$ that are analytic with expansion

$$f(z) = z + \sum_{n=1}^{\infty} b_n z^{-n} \quad \text{in} \quad |z| > 1 \tag{11}$$

and that have complex dilatations $\mu = f_{\bar{z}}/f_z$ satisfying

$$|\mu - c| \leq k \quad \text{a.e. in} \quad |z| < 1.$$

Here we assume that c is a complex constant and k is a nonnegative constant satisfying

$$|c| + k < 1.$$

For $c = 0$ the families $\Sigma_{0,k}$ have been studied by R. Kühnau, O. Lehto, the authors, and many others (see, e.g., [2, 3, 4, 7]).

In [6] we showed that

$$|b_1 - c| \leq k \tag{12}$$

for all f in $\sum_{c,k}$ and that equality occurs only for the functions

$$f(z) = \begin{cases} z + (c + ke^{i\alpha})/z & \text{for } |z| > 1 \\ z + (c + ke^{i\alpha})\bar{z} & \text{for } |z| \leq 1. \end{cases}$$

In this article we shall derive inequalities of "Grunsky-type" for the family $\sum_{c,k}$.

Theorem 2. Let f belong to $\sum_{c,k}$ and

$$\log \frac{f(z) - f(\zeta)}{z - \zeta} = \sum_{m,n=1}^{\infty} \gamma_{mn} z^{-m} \zeta^{-n}$$

for $|z| > 1$ and $|\zeta| > 1$. Set

$$a = \frac{1}{2k}\,[1 + k^2 - |c|^2 - \sqrt{(1 + k^2 - |c|^2)^2 - 4k^2}\,] \tag{13}$$

and

$$b = \frac{1}{2\bar{c}}\,[1 - k^2 + |c|^2 - \sqrt{(1 + k^2 - |c|^2)^2 - 4k^2}\,]. \tag{14}$$

If $k = 0$, define $a = 0$, and if $c = 0$, define $b = 0$. Then

$$\left| \sum_{m,n=1}^{N} \gamma_{mn}\lambda_m\lambda_n + (1-a^2) \sum_{n=1}^{N} \frac{b^n}{n(1-a^2|b|^{2n})}\lambda_n^2 \right| \leq a \sum_{n=1}^{N} \frac{1 - |b|^{2n}}{n(1-a^2|b|^{2n})}|\lambda_n|^2 \tag{15}$$

for all complex constants $\lambda_1, \lambda_2, \ldots, \lambda_N$.

Proof. The family $\sum_{c,k}$ is a special case of the example in Section 3. In particular, the functions $(c(z)$ and $k(z))$ of that example vanish identically in $|z| > 1$ and have constant values in $|z| \leq 1$. Consequently, $\sum_{c,k}$ is compact and satisfies conditions (i) through (v).

For fixed $\lambda_1, \ldots, \lambda_N$ let us consider

$$\lambda(f) = \text{Re} \sum_{m,n=1}^{N} \gamma_{mn}\lambda_m\lambda_n \tag{16}$$

as a functional on $\sum_{c,k}$. It is continuous, fulfilling condition (vi).

Let f be a function in $\sum_{c,k}$ for which the maximum of the functional (16) occurs. Under a variation of the form (4), we have

$$\log \frac{\Phi \circ f(z) - \Phi \circ f(\zeta)}{z - \zeta} = \log \frac{f(z) - f(\zeta)}{z - \zeta} - \frac{\varepsilon}{\pi} \iint_E \frac{a(t)\,dm(t)}{[t-f(z)][t-f(\zeta)]} + O(\varepsilon^2)$$

For t fixed and z in some neighborhood of ∞, define the Faber polynomials ϕ_n by the generating function

$$\log \frac{z}{f(z) - t} = \sum_{n=1}^{\infty} \frac{1}{n}\,\phi_n(t) z^{-n}.$$

Then the variation of the Grunsky coefficients γ_{mn} is given by

$$\gamma^*_{mn} = \gamma_{mn} - \frac{\varepsilon}{\pi mn} \iint_E \phi'_m(t)\phi'_n(t)a(t)dm(t) + O(\varepsilon^2)$$

so that

$$\lambda(\Phi \circ f) = \lambda(f) - \frac{\varepsilon}{\pi} \operatorname{Re} \iint_E [\sum_{m=1}^{N} \frac{1}{m} \lambda_m \phi'_m(t)]^2 a(t)dm(t) + O(\varepsilon^2).$$

The function

$$A(t) = -[\sum_{m=1}^{N} \frac{1}{m} \lambda_m \phi'_m(t)]^2$$

fulfills condition (vii). Consequently, by Theorem 1, the function

$$\mathcal{F} = \int^f \sqrt{A(t)}\, dt = i \sum_{m=1}^{N} \frac{1}{m} \lambda_m \phi_m \circ f$$

satisfies the differential equation (10) in $|z| < 1$ and $\mathcal{F}_{\bar z} = 0$ in $|z| > 1$.

The interesting boundary value problem that we encounter is to determine the continuous function $\mathcal{F}$, which is analytic for $|z| > 1$ and satisfies the equation (10) for $|z| < 1$. In addition, since the Faber polynomial ϕ_m is of the form (cf. [7, p. 39])

$$\phi_m \circ f(z) = z^m - m \sum_{n=1}^{\infty} \gamma_{mn} z^{-n}$$

for $|z| > 1$, the function

$$\mathcal{F}(z) = i \sum_{m=1}^{N} \lambda_m [\frac{z^m}{m} - \sum_{n=1}^{\infty} \gamma_{mn} z^{-n}] \tag{17}$$

for $|z| > 1$.

Define the auxiliary function

$$\mathcal{J}(\zeta) = \mathcal{F}(\zeta - b\bar\zeta) - a\overline{\mathcal{F}(\zeta - b\bar\zeta)}$$

where the constants a and b in (13) and (14) are chosen to satisfy the equations $abk - b + c = 0$ and $ab\bar c - a + k = 0$. Note that $0 \leq a < 1$ and $|b| < 1$. The function $\mathcal{J}$ has been constructed so that equation (10) implies $\mathcal{J}_{\bar\zeta} = 0$ in the ellipse $\mathcal{E} = \{\zeta : |\zeta - b\bar\zeta| < 1\}$.

Since $\mathcal{J}$ is analytic in $\mathcal{E}$, it follows for any analytic polynomial p that

$$0 = \int_{\partial\mathcal{E}} \mathcal{J}(\zeta)p(\zeta)d\zeta = \int_{|z|=1} [\mathcal{F}(z) - a\overline{\mathcal{F}(z)}]p\left(\frac{z + b/z}{1 - |b|^2}\right) \frac{z - b/z}{1 - |b|^2} \frac{dz}{z}.$$

For each positive integer n it is possible to choose a polynomial $p(\zeta)$ so that

$$p\left(\frac{z + b/z}{1 - |b|^2}\right) = \frac{z^n - (b/z)^n}{z - b/z}.$$

With this choice we have

$$0 = \int_{|z|=1} [\mathcal{J}(z) - a\overline{\mathcal{J}(z)}][z^n - (b/z)^n] \frac{dz}{z}, \quad n = 1,2,3,\ldots .$$

It follows that the coefficients of (17) satisfy

$$0 = -\sum_{m=1}^{N} \lambda_m \gamma_{mn} + \frac{a}{n}\overline{\lambda_n} - \frac{b^n}{n}\lambda_n + ab^n \sum_{m=1}^{N} \overline{\lambda_m \gamma_{mn}}$$

for $1 \leq n \leq N$. This equation implies

$$\sum_{m=1}^{N} \lambda_m \gamma_{mn} = \frac{-(1 - a^2)b^n\lambda_n + a(1 - |b|^{2n})\overline{\lambda_n}}{n(1 - a^2|b|^{2n})}.$$

Multiplying by λ_n and summing, we find that

$$\sum_{m,n=1}^{N} \gamma_{mn}\lambda_m\lambda_n = \sum_{n=1}^{N} \frac{-(1 - a^2)b^n\lambda_n^2 + a(1 - |b|^{2n})|\lambda_n|^2}{n(1 - a^2|b|^{2n})}$$

in the extreme situation. Due to the arbitrary nature of the complex constants $\lambda_1,\ldots,\lambda_N$, the inequality (15) follows.

By the nature of a variational argument, the inequality (15) is sharp for any choice of $\lambda_1,\ldots,\lambda_N$. Furthermore, relations determining the extremal functions could be obtained.

There are many consequences of inequalities of the form (15). We shall conclude with a few of them as corollaries. Since their proofs are well known, we refer the reader to [2, 7] for any details that are not familiar. To avoid repetition we use the notation of Theorem 2, the expansion (11), and the identities

$$\frac{a(1 - |b|^2)}{1 - a^2|b|^2} = k \quad \text{and} \quad \frac{(1 - a^2)b}{1 - a^2|b|^2} = c.$$

The inequalities (15) have the following apparent strengthening.

Corollary 1.

$$\sum_{m=1}^{N} \frac{m(1-a^2|b|^{2m})}{(1-|b|^{2m})}\left|\frac{(1-a^2)b^m\lambda_m}{m(1-a^2|b|^{2m})} + \sum_{n=1}^{N}\gamma_{mn}\lambda_n\right|^2 \leq \sum_{n=1}^{N}\frac{a^2(1-|b|^{2n})|\lambda_n|^2}{n(1-a^2|b|^{2n})}$$

As an application of Corollary 1, choose $\lambda_1 = 1$ and $\lambda_n = 0$ for $n > 1$. Also use the fact that $\gamma_{m1} = -b_m$.

Corollary 2. $$|b_1 - c|^2 + \frac{k}{a}\sum_{m=2}^{\infty} \frac{m(1 - a^2|b|^{2m})|b_m|^2}{1 - |b|^{2m}} \leq k^2$$

Observe that the earlier result (12) is also a consequence of Corollary 2.

The choices $\lambda_n = z^{-n}$ and $\lambda_n = \sqrt{6}\, nz^{-n-1}$ in (15) yield the following.

Corollary 3. If $|z| > 1$, then

$$\left|\log f'(z) + (1 - a^2) \sum_{n=1}^{\infty} \frac{b^n z^{-2n}}{n(1 - a^2|b|^{2n})}\right| \leq a \sum_{n=1}^{\infty} \frac{(1 - |b|^{2n})|z|^{-2n}}{n(1 - a^2|b|^{2n})}$$

and

$$\left|\{f,z\} + 6(1 - a^2) \sum_{n=1}^{\infty} \frac{nb^n z^{-2n-2}}{(1 - a^2|b|^{2n})}\right| \leq 6a \sum_{n=1}^{\infty} \frac{n(1 - |b|^{2n})|z|^{-2n-2}}{(1 - a^2|b|^{2n})}$$

where $\{f,z\} = 6 \lim_{\zeta \to z} \frac{\partial^2}{\partial z \partial \zeta} \log \frac{f(z) - f(\zeta)}{z - \zeta}$ is the familiar Schwarzian derivative.

References

[1] Kühnau, R.: Wertannahmeprobleme bei quasikonformen Abbildungen mit ortsabhängiger Dilatationsbeschränkung. Math. Nachr. 40 (1969), 1 - 11.

[2] Kühnau, R.: Verzerrungssätze und Koeffizientenbedingungen vom Grunskyschen Typ für quasikonforme Abbildungen. Math. Nachr. 48 (1971), 77 - 105.

[3] Lehto, O.: Schlicht functions with a quasiconformal extension. Ann. Acad. Sci. Fenn., Ser. AI 500 (1971).

[4] Schiffer, M., Schober, G.: Coefficient problems and generalized Grunsky inequalities for schlicht functions with quasiconformal extensions. Arch. rat. Mech. Analysis 60 (1976), 205 - 228.

[5] Schiffer, M., Schober, G.: Representation of fundamental solutions for generalized Cauchy-Riemann equations by quasiconformal mappings. Ann. Acad. Sci. Fenn., Ser. AI 2 (1976), 501 - 531.

[6] Schiffer, M., Schober, G.: A variational method for general families of quasiconformal mappings. To appear in J. Analyse math.

[7] Schober, G.: Univalent functions - selected topics. Lecture Notes in Mathematics 478, Springer-Verlag, Berlin - Heidelberg - New York (1975).

[8] Strebel, K.: Ein Konvergenzsatz für Folgen quasikonformer Abbildungen. Commentarii math. Helvet. 44 (1969), 469 - 475.

Stanford University
Stanford, CA 94305
U.S.A.

Indiana University
Bloomington, IN 47405
U.S.A.

BERS' TEICHMÜLLER SPACES OF KLEIN SURFACES

Mika Seppälä *

In [5] I have shown that the reduced Teichmüller space $T(X)$ of a compact non-classical Klein surface X is a real analytic manifold which can be realized as the fixed-point set of an antianalytic involution σ^* of the Teichmüller space $T(X_c)$, where X_c is the complex double of X (which is a compact Riemann surface). A similar result holds also for Bers' Teichmüller spaces of non-classical Klein surfaces. In Sections 1 and 2 we recall the necessary definitions. Section 3 summarizes the results of [5]. In Sections 4 and 5 we prove a topologica result (Theorem 2 and Lemma 2). As a consequence of Theorem 2 we show, in Section 6, that the Bers' Teichmüller space of a non-orientable surface with boundary is an infinite dimensional real-analytic Banach manifold.

1. In this lecture Z will always denote a compact non-orientable surface with a non-empty boundary ∂Z. The topological surface Z can be endowed with a dianalytic structure X, the pair (Z,X) is a Klein surface and is usually denoted by X ([1]).

The only non-trivial covering transformation of the orientable double covering $\pi_o : Z_0 \to Z$ of Z is an orientation reversing involution $\sigma_o : Z_0 \to Z_0$. The boundary of Z_0 consists of two parts which are mapped homeomorphically onto ∂Z by π_o and onto each other by σ_o. Identifying all boundary points $p \in \partial Z_0$ with the corresponding points $\sigma_o(p)$ we obtain a surface Z_c which does not have boundary. The projection π_o induces a mapping $\pi : Z_c \to Z$, and σ_o induces $\sigma : Z_c \to Z_c$. It is clear that $Z = Z_c/\sigma$ and $Z = Z_0/\sigma_o$.

If X is a dianalytic structure on Z, then we can find an analytic structure X_o on Z_0 for which $\pi_o : X_o \to X$ is dianalytic. We have two choices corresponding to different orientations of Z_0. Let us now fix an orientation on Z_0 and on Z_c. In the sequel we shall consider only those analytic structures of Z_0 (of Z_c) which agree with the given orientation. Then also the above analytic structure X_o becomes unique. The analytic structure X_o of Z_0 induces an analytic structure X_c on Z_c. The mapping $\pi : X_c \to X$ is dianalytic and the mapping

* Research supported by the Finnish Cultural Foundation.

$\sigma : X_c \to X_c$ and $\sigma_o : X_o \to X_o$ are antianalytic, i.e. σ and σ_o are antiinvolutions.

The triple (X_c, π, σ) is the complex double of X, and (X_o, π_o, σ_o) is the orienting double of X ([1], page 40).

2. Let $f : Z \to Z$ be a homeomorphism. Consider dianalytic structures X and X' of Z. If the mapping $f : X \to X'$ is quasiconformal, then let $K_f(X,X')$ denote the maximal dilatation of $f : X \to X'$; otherwise put $K_f(X,X') = +\infty$. By an XX'-extremal mapping in some class of homeomorphisms $Z \to Z$ we mean a mapping f for which the number $K_f(X,X')$ is the smallest in the class under consideration.

In the case of surfaces with boundary we have two alternative definitions for Teichmüller spaces. In order to give these definitions we will have to consider two homotopy classes of mappings of Z. Let $HI(Z)$ denote the group of those topological self-mappings of Z which are homotopic to the identity mapping, and let $HIMB(Z)$ denote the subgroup of $HI(Z)$ consisting of the mappings which are homotopic to the identity mapping modulo the boundary of Z (i.e. under a homotopy which keeps ∂Z point-wise fixed). We say that two self-mappings f and g of Z are homotopic modulo ∂Z, if $f^{-1} \circ g$ belongs to the class $HIMB(Z)$.

Let W be a fixed dianalytic structure on Z, and let $K(Z)$ denote the set of all dianalytic structures of Z. Denote by $M(W)$ the subset of $K(Z)$ consisting of those dianalytic structures X of Z for which the class $HIMB(Z)$ contains quasiconformal mappings $X \to W$.

In the set $K(Z)$ we define an equivalence relation setting $X \sim X'$ if $HI(Z)$ contains a dianalytic mapping $X \to X'$. In the set $M(W)$ we define: $X \approx X'$ if $HIMB(Z)$ contains a dianalytic mapping $X \to X'$. These relations are obviously equivalence relations. The corresponding quotient spaces

$$T(Z) = K(Z)/\sim \quad \text{and} \quad T_B(W) = M(W)/\approx$$

are the reduced and the Bers' Teichmüller spaces of Z. In the case where Z is orientable these definitions agree with the classical ones. Note that if $\partial Z = \emptyset$, $T_B(W) = T(Z)$.

In $T(Z)$ we define the Teichmüller metric d setting

$$d(p,p') = \inf \left\{ \frac{1}{2} \log K_f(X,X') \mid f \in HI(Z),\ X \in p,\ X' \in p' \right\} .$$

In $T_B(W)$ the Teichmüller metric d is defined in the same way - just replace $HI(Z)$ by $HIMB(Z)$ in the definition.

The class $HIMB(Z)$ is more restricted than $HI(Z)$. This reflects on the fact that $T_B(W)$ is harder to study than $T(Z)$. Also the dimension of $T(Z)$ will be, in our case, finite while that of $T_B(W)$ is infinite.

3. For a Klein surface X we can define the algebraic genus $g(X)$, which turns out to be the same as the genus of the Riemann surface X_c (cf.[5], Lemma 1.1, page 12). We shall consider here only the general case $g(X) \geq 2$ - the only non-orientable surface with boundary that will be excluded is the Möbius strip.

The double Z_c of Z is a compact orientable surface without boundary. Hence its Teichmüller space is a complex manifold. In [5] I have studied reduced Teichmüller spaces and proved the following theorem which can also be derived from old results of Earle ([4]).

Theorem . $T(Z)$ is a manifold with a natural real analytic structure such that the mapping $[X] \to [X_c]$ is a real analytic immersion of $T(Z)$ into the complex manifold $T(Z_c)$. The mapping $\sigma : Z_c \to Z_c$ induces an antianalytic involution σ^* of $T(Z_c)$ whose fixed point locus is the image of $T(Z)$ under the above immersion.

If X is a dianalytic structure on Z, then there exists a real analytic immersion of $T(Z)$ into the real Banach space of quadratic differentials on X.

A proof can be based on the uniqueness of the Teichmüller mapping between compact Riemann surfaces, and on the topological result ([5], page 27) which states that a mapping $f : Z \to Z$ belongs to $HI(Z)$ if and only if it has a lifting $\tilde{f} : Z_c \to Z_c$ which belongs to $HI(Z_c)$. The same reasoning cannot be applied to $T_B(W)$ because an extremal mapping in the class $HIMB(Z)$ need not be unique. A similar result can be, however, proved for the spaces $T_B(W)$. In order to do this we will first define the induced mappings π_o^* and σ_o^*.

4. Let $W_o \in K(Z_0)$ be the complex double of W. The mapping $\pi_o^* : T_B(W) \to T_B(W_o)$, $[X] \to [X_o]$ is well-defined since $\pi_o : Z_0 \to Z$ is a covering projection; π_o^* is clearly continuous and one-to-one.

Consider an analytic structure Y on Z_0, and let Y' be the analytic structure for which $\sigma_o : Y \to Y'$ is antianalytic. Setting

$$\sigma_o^*([Y]) = [Y']$$

we obtain a well-defined mapping $\sigma_o^* : T_B(W_o) \to T_B(W_o)$ which is an isometry.

Theorem 2. $\pi_o^*(T_B(W)) = \{p \in T_B(W_o) \mid \sigma_o^*(p) = p\}$.

5. We will prove the above theorem using results concerning the universal covering surfaces of Klein surfaces. Let $X \in M(W)$. Since we have assumed that $g(X) \geq 2$, we can write $X_o = (\bar{D} \cap \Omega)/G$, where G is a Fuchsian group acting in the unit disk D, Ω is the region of discontinuity of G, and $\bar{D}$ is the closure of D. A lifting $\tilde{\sigma}_o : \bar{D} \cap \Omega \to \bar{D} \cap \Omega$ of σ_o is an antianalytic mapping. It will not be an involution, but it satisfies $\tilde{\sigma}_o^2 \in G$. G together with $\tilde{\sigma}_o$ generates a reflection group $R = (G, \tilde{\sigma}_o)$ for which $X = (\bar{D} \cap \Omega)/R$.

Let $f : X \to X$ be homeomorphism. A lifting $\tilde{f} : \bar{D} \cap \Omega \to \bar{D} \cap \Omega$ can be uniquely extended to a homeomorphism $\tilde{f} : \bar{D} \to \bar{D}$ ([6], Lemma 3). In the sequel we shall assume that all liftings are self-mappings of $\bar{D}$. Let us now recall a simple fact about the class HIMB(Z).

Lemma 1. A homeomorphism $f : X \to X$ $(f : X_o \to X_o)$ belongs to the class HIMB(Z) (HIMB(Z_0)) if and only if it has a lifting $\tilde{f} : \bar{D} \to \bar{D}$ which agrees with the identity mapping on ∂D.

To prove Theorem 2 we need the following auxiliary result.

Lemma 2. Let $Y, Y' \in M(W_o)$ be analytic structures of Z_0, and let $\sigma_o : Y \to Y$, $\sigma_o' : Y' \to Y'$ be antiinvolutions which are homotopic mod ∂Z_0. There is an YY'-extremal mapping $f : Y \to Y'$ in the class HIMB(Z_0) for which

$$f \circ \sigma_o = \sigma_o' \circ f \,. \tag{1}$$

Proof. Let G be a Fuchsian group for which $Y = (\bar{D} \cap \Omega)/G$, where Ω is the region of discontinuity of G. Respectively, write $Y' = (\bar{D} \cap \Omega')/G'$. Consider a lifting $\alpha : \bar{D} \to \bar{D}$ of the identity mapping $Y \to Y'$ (i.e. α is a lifting of the identity mapping of Z_0 - Y and Y' are just different analytic structures on Z_0). Since $\sigma_o : Y \to Y$ and $\sigma_o' : Y' \to Y'$ are, as self-mappings of Z_0, homotopic mod ∂Z_0, we can, by Lemma 1, choose their liftings $\tilde{\sigma}_o : \bar{D} \to \bar{D}$ and $\tilde{\sigma}_o' : \bar{D} \to \bar{D}$ (which are antianalytic mappings) in such a way that on ∂D they satisfy

$$\alpha \circ \tilde{\sigma}_o = \tilde{\sigma}_o' \circ \alpha \,. \tag{2}$$

Let $\{p_1, p_2, p_3, \ldots\}$ be a dense subset of ∂Z_0. Denote by HI_n the set of those quasiconformal mappings $u : \bar{D} \to \bar{D}$ which satisfy:

(i) $u : \bar{D} \to \bar{D}$ is a lifting of some mapping $Y \to Y'$, and

(ii) u coincides with α on the limit set of G and at those points of ∂D which lie over the set $\{p_1, \sigma_o(p_1), p_2, \sigma_o(p_2), \ldots, p_n,$

$\sigma_o(p_n)\} \subset \partial Y$.

By the general Teichmüller extremal mapping theorem, the extremal mapping u_n in the class HI_n is unique. By (2) the mapping $\tilde{\sigma}_o'^{-1} \circ u_n \circ \tilde{\sigma}_o$ belongs to the class HI_n. Since $\tilde{\sigma}_o'^{-1}$ and $\tilde{\sigma}_o$ are antianalytic, $\tilde{\sigma}_o'^{-1} \circ u_n \circ \tilde{\sigma}_o$ is extremal in HI_n as well. From the uniqueness it follows then that

$$u_n \circ \tilde{\sigma}_o = \tilde{\sigma}_o' \circ u_n . \tag{3}$$

The maximal dilatations of the mappings u_n are bounded by the maximal dilatation K of an YY'-extremal mapping of the class $HIMB(Z_o)$. Since the family $\{u_n\}$ is normal, we can pick up a sequence $(u_{j_1}, u_{j_2}, \ldots)$ which converges to a mapping $\tilde{f} : \bar{D} \to \bar{D}$. By condition (ii) $\tilde{f}$ is a quasiconformal self-mapping of $\bar{D}$. Since each u_n is K-quasiconformal, also $\tilde{f}$ is K-quasiconformal. The mapping $\tilde{f}$ induces a K-quasiconformal mapping $f : Y \to Y'$. Since the set $\{p_1, p_2, \ldots\}$ is dense in ∂Y, it follows from condition (ii), that $\tilde{f} : \bar{D} \to \bar{D}$ coincides with α on ∂D. Hence, by Lemma 1, $f \in HIMB(Z_o)$. It follows that $f : Y \to Y'$ is YY'-extremal in $HIMB(Z_o)$. By (3) $\tilde{f}$ satisfies $\tilde{f} \circ \tilde{\sigma}_o = \tilde{\sigma}_o' \circ \tilde{f}$. Hence $f : Y \to Y'$ satisfies (1). This completes the proof. The idea to use the general Teichmüller extremal mapping theorem in the above reasoning is due to P. Tukia.

Corollary. The mapping $\pi_o^* : T_B(W) \to T_B(W_o)$ is an isometry.

Proof. Let $[X_o]$, $[X_o'] \in T_B(W_o)$ be the images of the points $[X]$, $[X'] \in T_B(W)$ under the mapping π_o^*. Denote the Teichmüller metrics in $T_B(W)$ and in $T_B(W_o)$ by d and d_o, respectively. It suffices to show that $d_o([X_o],[X_o']) \geq d([X],[X'])$, since the converse inequality is obvious. This follows, since, by Lemma 2, we can find an X_oX_o'-extremal mapping $f : X_o \to X_o'$ from $HIMB(Z_o)$ which commutes with σ_o and hence induces a mapping $g : X \to X'$ which belongs to $HIMB(Z)$ and has the same maximal dilatation as f.

To prove Theorem 2 note first that obviously $\pi_o^*(T_B(W))$ is contained in the fixed-point set of $\sigma_o^* : T_B(W_o) \to T_B(W_o)$. Let us prove the converse inclusion. Let $[Y'] \in T_B(W_o)$ be a fixed-point of σ_o^*. If $Y'' \in M(W_o)$ is the analytic structure of Z_o for which $\sigma_o : Y' \to Y''$ is antianalytic, then $\sigma_o^*([Y']) = [Y'']$, and there exists a conformal mapping $g : Y'' \to Y'$ which belongs to the class $HIMB(Z_o)$. Let $\sigma_o' = g \circ \sigma_o$. Then $\sigma_o' : Y' \to Y'$ is antianalytic. Since σ_o is an involution and since $g \in HIMB(Z_o)$, $\sigma_o'^2 \in HIMB(Z_o)$. Being analytic $\sigma_o'^2 : Y' \to Y'$ must then be the identity mapping. Hence σ_o' is an involution.

Let now $Y \in M(W_o)$ be some analytic structure of Z_o for which

$\sigma_o : Y \to Y$ is antianalytic. Then, by Lemma 2, there is a quasiconformal mapping $f : Y \to Y'$ which belongs to $HIMB(Z_o)$ and satisfies (1). Let Y_o be the analytic structure of Z_o for which $f : Y_o \to Y'$ is analytic. Since $f \in HIMB(Z_o)$, $[Y_o] = [Y']$.

On the other hand, since $\sigma_o' : Y' \to Y'$ is antianalytic such is $\sigma_o = f^{-1} \circ \sigma_o' \circ f : Y_o \to Y_o$ as well. Hence $[Y'] = [Y_o] \in \pi_o^*(T_B(W))$. The proof is complete.

6. Let X_o be the orienting double of X. Express X_o as $(\bar{D} \cap \Omega)/G$, where Ω is the region of discontinuity of G. If φ is a holomorphic function in $D^* = \{z \in \hat{C} \,|\, |z| > 1\}$ then define

$$||\varphi|| = \sup \{|\varphi(z)|(|z|^2 - 1) \,|\, z \in D^*\} \tag{4}$$

and denote

$$B_2(D^*) = \{\varphi | \varphi \text{ holomorphic in } D^* \text{ and } ||\varphi|| < \infty\} .$$

Here $\hat{C} = C \cup \{\infty\}$ denotes the one-point compactification of the complex plane C. $B_2(D^*)$ with the norm (4) is a complex Banach space. Let us denote by $B_2(D^*,G)$ the subspace of $B_2(D^*)$ consisting of those holomorphic functions φ which satisfy

$$\varphi(g(z))(\partial g(z)/\partial z)^2 = \varphi(z), \quad g \in G ,$$

where $\partial/\partial z = (\partial/\partial x - i\partial/\partial y)/2$ has the usual meaning. $B_2(D^*,G)$ is the space of quadratic differentials of G (or of X_o). Since G is of the second kind, $B_2(D^*,G)$ is infinite dimensional ([3], Theorem 7, page 111).

Let $\tilde{\sigma}_o : \bar{D} \to \bar{D}$ be a lifting of the antiinvolution $\sigma_o : X_o \to X_o$. It can be extended to an antianalytic self-mapping of the whole infinite plane $\hat{C}$. Now we can define the mapping $\sigma_o : B_2(D^*,G) \to B_2(D^*,G)$ setting

$$\sigma_o^{\#}(\varphi)(z) = \overline{\varphi(\tilde{\sigma}_o(z))\left(\frac{\partial \tilde{\sigma}_o(z)}{\partial \bar{z}}\right)^2} ,$$

where $\frac{\partial}{\partial \bar{z}} = \frac{1}{2}(\frac{\partial}{\partial x} + i\frac{\partial}{\partial y})$ and the long bar denotes complex conjugation.

Bers has constructed a homeomorphism ρ of $T_B(W_o)$ onto an open set in $B_2(D^*,G)$ ([3], Theorem 6, page 111). Identifying $T_B(W_o)$ with its image in $B_2(D^*,G)$ we can make $T_B(W_o)$ into an infinite dimensional complex Banach manifold. This structure does not depend on the choice of G, neither does it depend on the choice of $X_o \in M(W_o)$. A computation shows that the diagram

$$\begin{array}{ccc} T_B(W_o) & \xrightarrow{\rho} & B_2(D^*,G) \\ \sigma_o^* \uparrow & & \uparrow \sigma_o^{\#} \\ T_B(W_o) & \xrightarrow{\rho} & B_2(D^*,G) \end{array} \qquad (5)$$

commutes.

By Theorem 2 we may identify $T_B(W)$ with the fixed point set of σ_o^* in $T_B(W_o)$. Since the diagram (5) commutes, and since $\sigma_o^{\#}$ is conjugate-linear, σ_o^* is an antiinvolution of $T_B(W_o)$. It follows that the complex structure of $T_B(W_o)$ induces a real-analytic structure on $\pi_o^*(T_B(W)) = \{p \in T_B(W_o) | \sigma_o^*(p) = p\}$. By π_o^* we can pull this structure back on $T_B(W)$. Then the mapping $\rho \circ \pi_o^*$ is a real analytic immersion of $T_B(W)$ into the real Banach space $\{\varphi \in B_2(D^*,G) | \sigma_o^{\#}(\varphi) = \varphi\} = B_2(D^*,R)$, which is the space of quadratic differentials of the reflection group $R = (G,\tilde{\sigma}_o)$ or, which is the same thing, the space of quadratic differentials on X.

We have proved the following result.

Theorem 3. Let Z be a compact non-orientable topological surface with a non-empty boundary. Assume that Z is not the Möbius-strip. Let W be a dianalytic structure on Z. Then $T_B(W)$ is an infinite-dimensional real analytic Banach manifold. The mapping $\pi_o^* : T_B(W) \to T_B(W_o)$, $[X] \to [X_o]$, is a real analytic immersion, and given a dianalytic structure X on Z, there exists a real analytic immersion ρ' of $T_B(Z)$ into the real Banach space $B_2(D^*,R)$, where R is as above.

References

[1] Alling, N. L., Greenleaf, N.: Foundations of the theory of Klein surfaces. Lecture Notes in Mathematics 219, Springer-Verlag, Berlin - Heidelberg - New York (1971).

[2] Bers, L.: Quasiconformal mappings and Teichmüller's theorem, in "Analytic Functions". Princeton University Press, Princeton, N. J. (1960), 89 - 119.

[3] Bers, L.: Automorphic forms and general Teichmüller spaces, in "Proceedings of the Conference on Complex Analysis. Minneapolis 1964", Springer-Verlag, Berlin - Heidelberg - New York (1965), 109 - 113.

[4] Earle, C. J.: Teichmüller spaces of groups of the second kind. Acta math. 112 (1964), 91 - 97.

[5] Seppälä, M.: Teichmüller spaces of Klein surfaces. Ann. Acad. Sci. Fenn. Ser. A I, Diss. 15 (1978).

[6] Tukia, P.: Extension of boundary homeomorphisms of discrete groups

of the unit disk. Ann. Acad. Sci. Fenn. Ser. A I 548 (1973).

Helsinki School of Economics
Department of Methodological Sciences
SF-00100 Helsinki 10
Finland

GROWTH PROBLEMS FOR A CLASS OF ENTIRE FUNCTIONS VIA SINGULAR INTEGRAL ESTIMATES

Daniel F. Shea and Stephen Wainger

Let $f(z)$ be an entire function of finite nonintegral order ρ and put $n(r)$ = number of zeros of $f(z)$ in $|z| \leq r$,

$$M(r,f) = \sup_{\theta} |f(re^{i\theta})|, \quad L(r,f) = \inf_{\theta} |f(re^{i\theta})| .$$

By a classical theorem of Pôlya and Valiron [2] [4],

$$\limsup_{r\to\infty} \frac{n(r)}{\log M(r,f)} \geqq C(\rho) \tag{1}$$

where

$$C(\rho) = \frac{1}{\pi} \sin \pi\rho \quad (0 \leq \rho \leq 1)$$

$$\frac{1}{\pi} |\sin \pi\rho| \geqq C(\rho) \geqq A_0 \frac{|\sin \pi\rho|}{1 + \log\rho} \quad (1 < \rho < \infty) \tag{2}$$

for an absolute constant A_0. The upper estimate for $C(\rho)$ in (2) comes from "Lindelöf functions": these are certain well known entire functions of any finite order with all zeros regularly distributed along a single ray $\arg z$ = constant.

It has been conjectured that these functions are extremal for this problem but not even the order of magnitude of $C(\rho)$ is known, for ρ large. In this direction, we prove:

Theorem 1. Let $f(z)$ have order ρ $(1 < \rho < \infty)$ and real negative zeros only. Fix a branch of $\log f(z)$ in $|\arg z| < \pi$, and define

$$B(r,f) = \sup_{\theta} |\log f(re^{i\theta})| .$$

Then

$$\limsup_{r\to\infty} \frac{n(r)}{B(r,f)} \geqq A\, |\sin \pi\rho| \tag{3}$$

for an absolute constant A.

The proof applies also to functions with zeros on ν rays through 0, with A on the right side of (3) replaced by A/ν^{α} and $\alpha = \rho - [$

In particular, we have

$$\limsup_{r\to\infty} \frac{\log L(r,f)}{n(r)} \geqq - A^{-1} |\csc \pi\rho| \tag{4}$$

for the functions of Theorem 1.

While (1) may be valid with $C(\rho) = A\,|\sin \pi\rho|$ for arbitrary entire functions of order ρ, some restriction on the arguments of the zeros is essential for (3) and (4), as examples of Hayman show ([1], cf. the estimates on pp. 501 - 503).

We give here an outline of the proof; details will appear elsewhere. Assume for simplicity that $f(z)$ is a canonical product having order $\rho \neq$ integer, and put $q = [\rho]$. Then if we determine $\log f(z)$ so that $\log f(r)$ is real for $r > 0$,

$$B(r,f) = r^{q+1} \sup_{0<\varphi<\pi} \left| \int^{\infty} \frac{n(t)}{t^{q+1}} K(r,t,\varphi)\,dt \right| , \qquad (5)$$

$$K(r,t,\varphi) = \frac{r - t\cos\varphi - it\sin\varphi}{r^2 + t^2 - 2tr\cos\varphi} .$$

Thus $B(r,f)/r^{q+1}$ can be considered a kind of maximal function, and the proof of Theorem 1 follows from the norm estimate

$$\left\{ \int_{r_n}^{\infty} \left(\frac{B(r,f)}{r^{q+1}}\right)^p dr \right\}^{1/p} \leq \left(A \sin\left(\frac{\pi}{p}\right)\right)^{-1} \left\{ \int_{r_n}^{\infty} \left(\frac{n(r)}{r^{q+1}}\right)^p dr \right\}^{1/p} ,$$

valid for $p = (q + 1 - \rho)^{-1} + \varepsilon_n$ and suitable sequences $r_n \to \infty$, $\varepsilon_n \to 0$. This inequality is deduced from (5) and known estimates (cf.[3], pp. 7, 42, 62, 67) for the $L^p(0,\infty)$ norms of the maximal functions

$$M\varphi(r) = \sup_{\varepsilon>0} \frac{1}{2\varepsilon} \int_{|t-r|<\varepsilon} \varphi(t)\,dt ,$$

$$P^*\varphi(r) = \sup_{\varepsilon>0} \varepsilon \int_0^{\infty} \varphi(t)\,[(r - t)^2 + \varepsilon^2]^{-1} dt ,$$

$$H^*\varphi(r) = \sup_{\varepsilon>0} \left| \int_{|t-r|>\varepsilon} \frac{\varphi(t)}{t - r}\,dt \right|$$

with $\varphi(t) = n(t)/t^{q+1}$.

References

[1] Hayman, W. K.: The minimum modulus of large integral functions. Proc. London math. Soc., III. Ser. 2 (1952), 469 - 512.

[2] Pólya, G.: Bemerkungen über unendlichen Folgen und ganzen Funktionen. Math. Ann. 88 (1923), 169 - 183.

[3] Stein, E. M.: Singular integrals and differentiability properties of functions. Princeton University Press, Princeton (1970).

[4] Valiron, G.: Sur les fonctions entières d'ordre fini et d'ordre nul, et en particulier les fonctions à correspondance régulière. Ann. Fac. Sci. Univ. Toulouse, III. Sér. 5 (1913), 117 - 257.

Department of Mathematics
University of Wisconsin
Madison, WI 53706
U.S.A.

BEMERKUNG ZU EINEM SATZ VON YOSIDA

Norbert Steinmetz*

1. Einleitung.

Die Riccatischen Differentialgleichungen

$$w' = a(z) + b(z)w + c(z)w^2 \tag{1}$$

zeichnen sich unter allen Differentialgleichungen

$$w' = R(z,w),\ R \text{ rational in } z \text{ und } w\ , \tag{2}$$

dadurch aus, daß nur sie im Großen eindeutige und nichtrationale Lösungen zulassen. Dieser bekannte Satz von Malmquist [3] ist enthalten in dem

Satz von Yosida [10]. Entweder es ist jede im Großen eindeutige Lösung der binomischen Differentialgleichung

$$w'^n = R(z,w),\ R \text{ rational in } z \text{ und } w\ , \tag{3}$$

rational oder aber es liegt eine hyperriccatische Differentialgleichung

$$w'^n = a_o(z) + a_1(z)w + \cdots + a_{2n}(z)w^{2n} \tag{4}$$

vor.

Unbeantwortet läßt der Satz von Yosida die Frage, unter welchen Voraussetzungen die Gleichung (4) tatsächlich im Großen eindeutige Lösungen besitzt. Bei Riccatischen Differentialgleichungen (1) reicht dazu aus, daß die Koeffizienten Polynome sind. Die entsprechende Aussage kann im allgemeinen Fall (4) nicht richtig sein, da, wie hier gezeigt wird, für die Existenz transzendenter meromorpher Lösungen weit einschneidendere Bedingungen notwending sind als der Satz von Yosida angibt. Andererseits sind jene unter zusätzlichen Voraussetzungen nur an die Koeffizienten auch hinreichend dafür, daß jede Lösung im Großen eindeutig ist. Sämtliche Wertverteilungsgrößen der Lösungen lassen sich unmittelbar der Differentialgleichung entnehmen.

---

*Teil der Dissertationsschrift des Verfassers.

2. Bezeichnungen.

Die Differentialgleichung (3) wird als irreduzibel angenommen ($n = n_1 m$ und $R \equiv R_1^m$ ist nur für $m = 1$ möglich). Ferner wird $n \geqq 2$ vorausgesetzt und der einfach zu behandelnde Sonderfall.

$$w'^n = a(z)R(w) \tag{5}$$

ausgeschlossen ($n = 1$ erübrigt sich aufgrund des Malmquistschen Satzes; die Differentialgleichungen (5) sind bis auf den Faktor $a(z)$ aus der Theorie der elliptischen Funktionen bekannt und in diesem Zusammenhang weniger interessant).

Die Beweise basieren auf den Methoden der Wertverteilungslehre (hierzu und zu den üblichen Bezeichnungen vgl. Nevanlinna [4]). $S(r,f)$ bezeichnet wie üblich eine für $r > 0$ erklärte Funktion mit $S(r,f) = o(T(r,f))$ $(r \to \infty,\ r \notin E)$ außerhalb einer Menge E von endlichem Maß. Nach dem Satz über die Schmiegungsfunktion der logarithmischen Ableitung ist

$$m\left(r,\frac{f^{(k)}}{f^{(p)} - c}\right) = S(r,f) \quad (0 \leqq p < k,\ c \in C) .$$

3. Ergebnisse.

Satz 1. Unter den angegebenen Voraussetzungen ist entweder jede im Großen eindeutige Lösung von (3) rational oder aber es ist $n = 2$, und (3) läßt sich mittels einer linearen Substitution in die Normalform

$$w'^2 = c(z)w(w + a(z))^2 \quad (a'(z),\ c(z) \not\equiv 0) \tag{6}$$

überführen.

Satz 1 gibt wiederum keine Auskunft darüber, ob (6) im Großen eindeutige Lösungen besitzt. Zumindest kann nicht erwartet werden, daß jede Lösung im Großen eindeutig ist. Dies muß nicht einmal bei Polynomkoeffizienten der Fall sein, wie das Beispiel

$$w'^2 = 4zw(w + z)^2 \tag{7}$$

zeigt. Die Substitutionen $zw = y^2$ und $y = -v'/v$ führen (7) über in die lineare Differentialgleichung

$$v'' - \frac{1}{2z}v' + z^2 v = 0 , \tag{8}$$

und umgekehrt ist mit jeder Lösung $v \not\equiv 0$ von (8) auch $w = z^{-1}(v'/v)^2$ eine Lösung von (7). Die allgemeine Lösung von (8) ist aber $v = c_1 v_1(z) + c_2 z^{3/2} v_2(z)$ (v_j ganze Funktionen mit $v_j(0) = 1$). Somit ist mit

$\Phi = c_1^2 v_1^2 - c_2^2 z^3 v_2^2$ und $c = c_1 c_2$ das allgemeine Integral von (7)

$$w = \frac{1}{z}\left(\frac{\Phi'}{2\Phi}\right)^2 + \left(\frac{3c}{4\Phi}\right)^2 + 3\,\frac{c}{\sqrt{z}}\,\frac{\Phi'}{(2\Phi)^2}$$

mehrdeutig. Dagegen gilt

Satz 2. Jede Lösung der Differentialgleichung

$$w'^2 = 4w(a(z) + c(z)w)^2 \quad \left(\left(\frac{c(z)}{a(z)}\right)' \not\equiv 0\right) \tag{9}$$

mit Polynomkoeffizienten ist im Großen eindeutig. Die transzendenten Lösungen haben folgende Eigenschaften: (a) $\lambda(w) = 1 + \frac{1}{2}$ Grad ac, (b) $m(r,k) = O(\log r)$ $(k \in \hat{C})$ und (c) $\theta(0,w) = \theta(\infty,w) = \frac{1}{2}$.

Wichtigstes Hilfsmittel zum Beweis von Satz 1 bildet der nachfolgende Satz 3. Er enthält und verallgemeinert u.a. auch die Sätze von Malmquist und Yosida sowie Ergebnisse von Rellich [5], Wittich [7], Laine [2] und Strelitz [6].

Definition 1. Ein Differentialpolynom in einer meromorphen Funktion f ist ein Polynom in $f, f', f'', \ldots$ mit meromorphen Koeffizienten,

$$P[f] = \sum a(z) f^{j_0} f'^{j_1} \cdots f^{(n) j_n} . \tag{10}$$

$d(P) = \max\,(j_o + j_1 + \cdots + j_n)$ heißt Grad und $d^*(P) = \max\,(j_o + 2j_1 + \cdots + (n+1)j_n)$ Gewicht von P.

Definition 2. Eine meromorphe Funktion heißt zulässig für die Differentialgleichung

$$P[w] = H(z,w) = \frac{H_1(z,w)}{H_2(z,w)} \tag{11}$$

(H_j ganze Funktionen in z und w), wenn gilt:

(a) $T(r,a) = S(r,f)$ für alle in P auftretenden Koeffizienten,

(b) $T(r,H(z,\tau)) = S(r,f)$ für alle τ einer Menge $\Omega \subset C$ mit endlichem Häufungspunkt.

Bemerkung. In [2] werden ebenfalls Differentialgleichungen (11) betrachtet, wobei die rechte Seite H eine rationale Funktion in w mit Koeffizienten $b(z)$ ist. Zulässige Funktionen werden durch (a) und

(b') $T(r,b) = S(r,f)$ für alle Koeffizienten b

definiert. In diesem Sinne zulässige Funktionen sind es auch im Sinne obiger Definition. Sind alle Koeffizienten rational, so ist jede transzendente meromorphe Funktion zulässig.

Satz 3. Die Differentialgleichung (11) besitzt nur dann eine zulässige Lösung, wenn H bezüglich w ein Polynom vom Grad höchstens $d^*(P)$ ist.

Bemerkung. Besitzt (11) eine zulässige ganze Lösung oder eine zulässige Lösung mit $N(r,f) = S(r,f)$, so ist der Grad von H höchstens $d(P)$.

4. Beweis der Sätze 1 und 2.

Es sei f eine transzendente meromorphe Lösung von (3). Dann ist notwendig R ein Polynom in w vom Grad höchstens $2n$ $(= d^*(w'^n))$. Nach Übergang zur neuen Veränderlichen $(w-\tau)^{-1}$ $(R(z,\tau) \not\equiv 0)$ kann man annehmen, daß die Differentialgleichung

$$w'^n = R(z,w) = Q(z,w)Q_0(w) = a_0(z)\prod_{j=1}^{\ell}(w-\alpha_j(z))^{\lambda_j}\prod_{j=1}^{m}(w-\tau_j)^{\mu_j} \tag{12}$$

mit $\sum_{j=1}^{\ell}\lambda_j + \sum_{j=1}^{m}\mu_j = 2n$ vorliegt. Definiert man für $1 \leq k < n$ Differentialpolynome F_k und G_k durch

$$\frac{d^k}{dz^k}[f'(z)]^n = f'(z)F_k[f(z)]$$

und

$$\frac{d^k}{dz^k}[R(z,f(z))] = R_k(z,f(z)) + G_k[f(z)]$$

$(R_k(z,w) = \frac{\partial^k}{\partial z^k}R(z,w))$, so genügt f mit

$$P_k[w] := (F_k[w] - \frac{1}{w'}G_k[w])^n$$

neben (12) auch der Differentialgleichung

$$P_k[w] = \frac{R_k^n(z,w)}{R(z,w)} = H_k(z,w)\ . \tag{13}$$

Für $n = 2$ ist $P_1[w] = (2w'' - R_w(z,w))^2$ ein Differentialpolynom. Mithin ist Satz 3 anwendbar und die rechte Seite in (13) ein Polynom in w. Obwohl im allgemeinen P_k kein Differentialpolynom ist, läßt sich zeigen, daß die Aussage von Satz 3 auch für die speziellen Gleichungen (13) gilt (vgl. Nr. 5). Das hat aber $\lambda_j > k$ für alle $k < n$ und somit $\lambda_j \geq n$ zur Folge. Zwei Fälle sind noch möglich:

$$w'^n = a_0(z)[(w-\alpha(z))(w-\beta(z))]^n \tag{14}$$

und

$$w'^n = a_o(z)(w - \gamma(z))^\lambda Q_o(w) \quad (n \leq \lambda < 2n) \ . \tag{15}$$

Der erste Fall führt aber auf eine Riccatische Differentialgleichung im Widerspruch zu $n \geq 2$. Denn setzt man $P_o(z,w) = (w - \alpha(z))(w - \beta(z))$ ($\alpha + \beta$ und $\alpha\beta$ sind eindeutig!), so ist $b_o(z) = f'(z)/P_o(z,f(z))$ rational und $R(z,w) \equiv [b_o(z)P_o(z,w)]^n$.

Im zweiten Fall ist zunächst $\lambda = n$ nachzuweisen (der Fall $\lambda = 2n$ ist in (14) mit $\alpha \equiv \beta$ enthalten). Dies ist sicher richtig, wenn $f - \gamma$ unendlich viele Nullstellen $z_1, z_2, \ldots$ hat. Darunter gibt es Stellen z_ν mit $\gamma'(z_\nu) \neq 0$ und $a_o(z_\nu)Q_o(\gamma(z_\nu)) \neq 0,\infty$. Weiter ist $f'(z_\nu) = 0$ mit der Vielfachheit $s \geq 1$. Aus (15) folgt $ns = \lambda$ ($f - \gamma$ verschwindet in z_ν einfach), also $s = 1$ und $\lambda = n$. Der Fall, daß $f - \gamma$ nur endlich viele Nullstellen besitzt, kann aber gar nicht eintreten. Ansonsten ist mit einem passenden Polynom q die Funktion $g = q/(f - \gamma)$ ganz transzendent und Lösung der Differentialgleichung

$$[qy' - q'y - \gamma'y^2]^n = (-1)^n a_o q^\lambda y^{2n-\lambda} Q_o(\gamma + \frac{q}{y}) \ , \tag{16}$$

in der nur ein Term maximaler Dimension, nämlich $\gamma'^n y^{2n}$, auftritt. Nach Wittich [9: S. 64] sind aber dadurch ganze transzendente Lösungen ausgeschlossen.

In (12) ist $m \geq 2$, da sonst (14) mit $\alpha = \gamma$ und $\beta = \tau_1$ vorliegt. Die Werte τ_j werden von f unendlich oft angenommen (dies folgt aus $m(r,1/(f - \tau_j)) = S(r,f)$, vgl. Nr. 5), und zwar fast immer mit derselben Vielfachheit $p_j \geq 2$, wobei $n(p_j - 1) = \mu_j p_j$. Aus $n = \sum_{j=1}^{m} \mu_j = n \sum_{j=1}^{m} (1 - \frac{1}{p_j}) \geq n \frac{m}{2} \geq n$ folgt $m = p_1 = p_2 = 2$ und $\mu_1 = \mu_2 = \frac{n}{2}$, d.h. es ist $n = 2k$ eine gerade Zahl und $R(z,w) \equiv a_o(z)[(w - \tau_1)(w - \tau_2)(w - \gamma(z))^2]^k$. Genau wie im ersten Fall folgt $k = 1$. Zur Gleichung (6) gelangt man durch Ersetzen von w durch $(w - \tau_1)/(w - \tau_2)$

Der Beweis von Satz 2 folgt unmittelbar aus dem Zusammenhang zwischen der Differentialgleichung (9) und der Riccatischen Differentialgleichung

$$y' = a(z) + c(z)y^2 \ , \tag{17}$$

die aus (9) durch die Substitution $w = y^2$ hervorgeht. Die Aussagen von Satz 2 sind somit Folgerungen der Ergebnisse von Wittich [8] über die (sämtlich) eindeutigen Lösungen von (17).

5. Beweis von Satz 3.

Es sei f eine zulässige Lösung von (11) und

$$h(z) = P[f(z)] = H(z,f(z)). \tag{18}$$

Man darf $H(z,\tau) \not\equiv 0$ für $\tau \in \Omega$ annehmen. In einer Nullstelle von $f - \tau$ verschwindet auch $h - H(z,\tau)$ (oder h und $H(z,\tau)$ besitzen eine Polstelle, die dann mit einem Pol eines Koeffizienten a zusammenfallen muß). Die Hilfsfunktion

$$\Phi(z,\tau) = \frac{h(z) - H(z,\tau)}{f(z) - \tau} \tag{19}$$

besitzt somit in einer p-fachen Nullstelle des Nenners eine höchstens (p - 1)-fache Polstelle. Ausgehend von (19) werden Hilfsfunktionen

$$\Phi(z,\tau_1,\ldots,\tau_s) = \frac{h(z) - Q_s(z,f(z),\tau_1,\ldots,\tau_s)}{(f(z) - \tau_1)\cdots(f(z) - \tau_s)} \tag{20}$$

($\tau_j \in \Omega$, $\tau_j \neq \tau_k$ für $j \neq k$) rekursiv durch

$$\Phi(z,\tau_1,\ldots,\tau_{s-1},\rho,\sigma) = [\Phi(z,\tau_1,\ldots,\tau_{s-1},\rho) - \Phi(z,\tau_1,\ldots,\tau_{s-1},\sigma)]/(\rho - \sigma)$$

definiert. $Q_s(z,w,\tau_1,\ldots,\tau_s)$ ist ein Polynom in w vom Grad $\leqq s - 1$, die Koeffizienten sind lineare Ausdrücke in den Funktionen $H(z,\tau_j)$. Es sei nun $\Phi(z) := \Phi(z,\tau_1,\ldots,\tau_{d+1})$ $(d = d^*(P))$ und $\Psi(z) = \Phi(z)(f(z) - \tau_{d+1})$. Ist $\Phi(z) \not\equiv 0$, so folgt

$$T(r,f) = T(r,f - \tau_{d+1}) + O(1) \leqq T(r,\Phi) + T(r,\Psi) + O(1) . \tag{21}$$

Eine einfache Rechnung ergibt für die Schmiegungsfunktionen von Φ und Ψ die Abschätzung

$$m(r,\Phi) + m(r,\Psi) \leqq 2 \sum_{j=1}^{d+1} m(r,\frac{1}{f - \tau_j}) + S(r,f) . \tag{22}$$

Für die Anzahlfunktionen gilt konstruktionsgemäß

$$N(r,\Phi) + N(r,\Psi) \leqq 2 \sum_{j=1}^{d+1} N_1(r,\frac{1}{f - \tau_j}) + S(r,f) \tag{23}$$

(Polstellen treten außer in Nullstellen von $f - \tau_j$ nur noch in den Polstellen der Koeffizienten $a(z)$ und $H(z,\tau_j)$ auf, nicht aber in Polstellen von f). Aus (21), (22) und (23) folgt

$$T(r,f) \leqq 2 \sum_{j=1}^{d+1} [m(r,\frac{1}{f - \tau_j}) + N_1(r,\frac{1}{f - \tau_j})] + S(r,f)$$

oder, wenn q der Funktionen Φ nicht identisch verschwinden,

$$qT(r,f) \leq 2 \sum_{j=1}^{q(d+1)} [m(r,\frac{1}{f-\tau_j}) + N_1(r,\frac{1}{f-\tau_j})] + S(r,f) \leq 4T(r,f) + S(r,f)$$

was nur für $q \leq 4$ möglich ist.

Aus dem Verschwinden von $\Phi(z,\tau_1,\ldots,\tau_{d+1})$ etwa folgt aber, daß f neben (11) noch der Differentialgleichung

$$P[w] = Q(z,w)$$

$(Q(z,w) = Q_{d+1}(z,w,\tau_1,\ldots,\tau_{d+1}))$ genügt. Q hat in w den Grad höchstens $d = d^*(P)$. Zum Nachweis von

$$H(z,w) = Q(z,w) \tag{24}$$

und somit zum vollständigen Beweis von Satz 3 betrachte man bei festem $\tau \in \Omega$ die meromorphe Funktion

$$\chi(z) = H(z,\tau) - Q(z,\tau) .$$

Da χ in den Nullstellen von $f - \tau$ verschwindet, folgt (sofern $\chi(z) \not\equiv 0$)

$$\overline{N}(r,\frac{1}{f-\tau}) \leq N(r,\frac{1}{\chi}) + S(r,f) \leq T(r,\chi) + S(r,f) = S(r,f) ,$$

was nach dem Zweiten Haupsatz nur für zwei Werte τ möglich ist. Somit gilt für alle z und eine sich im Endlichen häufende Folge (τ_j)

$$H(z,\tau_j) - Q(z,\tau_j) \equiv 0 ,$$

also (24) nach dem Identitätssatz für Funktionen zweier Veränderlichen.

Ein Beweis von Satz 3 steht noch aus für die in der angegebenen Weise aus (3) hervorgegangene Gleichung (13). Eine genaue Durchsicht des Beweises von Satz 3 zeigt, daß von der Funktion $h(z) = P[f(z)] = H(z,f(z))$ nur die nachstehenden beiden Eigenschaften benötigt wurden, nicht aber die Tatsache, daß $P = P_k$ ein Differentialpolynom ist.

a) Es gibt eine Menge $\Omega \subset \mathbb{C}$ mit endlichem Häufungspunkt, so daß

$$m(r,\frac{h}{\prod_{j=1}^{s}(f-\tau_j)}) = S(r,f) \quad \text{für beliebige } \tau_1,\ldots,\tau_s \in \Omega .$$

b) In einer μ-fachen Polstelle von f besitzt h einen höchstens $d\mu$-fachen Pol ($d > 0$ eine absolute Konstante). Andere Polstellen treten nur in Polen der Koeffizienten von P auf. Ist $R(z,w)$ in (3) ein Polynom in w vom Grad $\gamma > n$, so folgt für jede Lösung f

$m(r,f') = \frac{\gamma}{n} m(r,f) + O(\log r)$ (vgl. [1 : S. 100]) und zusammen mit $m(r,f') \leqq m(r,f) + S(r,f)$ unmittelbar $m(r,f) = S(r,f)$. Ist $R(z,w) = (w - \tau)^{\mu} Q(z,w)$ mit $\mu < n$ und $Q(z,\tau) \not\equiv 0$, so folgt für $g = 1/(f - \tau)$ ebenso $m(r,g) = S(r,g)$, also $m(r,1/(f - \tau) = S(r,f)$ (g genügt der Differentialgleichung $y'^n = (-1)^n Q(z,\tau) y^{\gamma} + \cdots$, wobei $\gamma = 2n - \mu > n$).

Zum Nachweis von (a) sei $\Omega = \{\tau: R(z,\tau) \not\equiv 0\}$. Dann folgt zunächst nach Definition von h

$$m\left(r, \frac{h}{\prod_{j=1}^{s} (f - \tau_j)}\right) \leqq n \cdot m(r, G_k[f]/f') + S(r,f) .$$

Weiter tritt in der Darstellung $G_k[w] = \sum \gamma(z) w^{j_0} w'^{j_1} + \cdots + w^{(k) j_k}$ kein Polynom in w auf, d.h. es ist stets ein $j_\kappa > 0$ $(\kappa > 0)$, und nach der Vorbemerkung und dem Satz über die Schmiegungsfunktion der logarithmischen Ableitung ist $m(r, G_k[f]/f') = S(r,f)$.

Zum Beweis von (b) ist zu zeigen, daß sich h in fast allen Nullstellen von f' regulär verhält. Dabei kann man sich auf diejenigen Nullstellen $z = z_0$ von f' beschränken, für die entweder

(i) $f(z_0) = \tau_j$, $Q(z_0, \tau_j) \neq 0, \infty$ oder

(ii) $f(z_o) = \alpha_j(z_o)$, $\alpha_j'(z_o) \neq 0$, $\alpha_j(z_o) - \alpha_k(z_0) \neq 0, \infty$ $(k \neq j)$ und $a_o(z_0) \neq 0, \infty$ gilt.

Für Stellen der ersten Sorte folgt die Behauptung unmittelbar aus $H_k(z,w) = Q_o^{n-1}(w) [\frac{\partial^k}{\partial z^k} Q(z,w)]^n / Q(z,w)$. Im zweiten Fall sei $f'(z_0) = 0$ λ-fach. Da $f - \alpha_j$ in z_0 einfach verschwindet, folgt aus (12) $n\lambda = \lambda_j$. Somit enthält R_k den Faktor $w - \alpha_j(z)$ mindestens $\lambda_j - k \geqq \lambda$-fach, und $h(z) = H_k(z, f(z))$ ist regulär in $z = z_0$.

Literaturverzeichnis.

[1] Bieberbach, L.: Theorie der gewöhnlichen Differentialgleichungen. Springer-Verlag, Berlin - Göttingen - Heidelberg - New York (1965).

[2] Laine, I.: On the behaviour of the solutions of some first order differential equations. Ann. Acad. Sci. Fenn., Ser. A I 497 (1971).

[3] Malmquist, J.: Sur les fonctions à un nombre fini de branches définies par les équations différentielles du premier ordre. Acta math. 36 (1913), 297 - 343.

[4] Nevanlinna, R.: Eindeutige analytische Funktionen. Springer-Verlag, Berlin (1936).

[5] Rellich, F.: Über die ganzen Lösungen einer gewöhnlichen Differentialgleichung erster Ordnung. Math. Ann. 117 (1940), 587 - 589.

[6] Strelitz, S.: A remark on meromorphic solutions of differential

equations. Proc. Amer. math. Soc. 65 (1977), 255 - 261.

[7] Wittich, H.: Ganze Lösungen der Differentialgleichung $w'' = f(w)$. Math. Z. 47 (1942), 422 - 246.

[8] Wittich, H.: Einige Eigenschaften der Lösungen von $w' = a(z) + b(z)w + c(z)w^2$. Arch. der Math. 5 (1954), 226 - 232.

[9] Wittich, H.: Neuere Untersuchungen über eindeutige analytische Funktionen. Springer-Verlag, Berlin - Heidelberg - New York (1968).

[10] Yosida, K.: A generalization of a Malmquist's theorem. Jap. J. Math. 9 (1932), 253 - 256.

Mathematisches Institut I
Universität Karlsruhe (TH)
D-7500 Karlsruhe
BR Deutschland

INFLATABLE FAMILIES OF HOLOMORPHIC FUNCTIONS

Kurt Strebel

1. Introduction. The trajectory structure of a quadratic differential does not depend on a positive constant factor. On the other hand one frequently encounters the situation, that a sequence of differentials tends locally uniformly to zero where one would like to have a limit which is different from zero. This would then define the limiting trajectory structure. It is therefore natural to try to blow the sequence up by multiplying its elements with constant factors in such a way that the limit is not identical zero. In this paper a necessary and sufficient condition is given which allows this kind of inflation.

2. The definition. We say that a sequence of holomorphic functions f_n in a domain G or on a Riemann surface can be blown up or is inflatable in the proper sense, if there is a sequence of complex numbers λ_n (multipliers) such that the sequence $(\lambda_n f_n)$ converges locally uniformly in G to a function f which is not identically equal to zero.

We say that a sequence or more generally a family F of holomorphic functions f can be blown up or is inflatable, if every sequence (f_n) of functions of F has a subsequence (f_{n_i}) which is inflatable in the proper sense; in other words for which there exists a sequence of multipliers λ_i such that $\lambda_i f_{n_i} \to f \not\equiv 0$ locally uniformly in G.

The definition makes sense for holomorphic quadratic differentials as well.

3. The multipliers. Let us first take a sequence (f_n) which can be blown up in the proper sense: $\lambda_n f_n \to f \not\equiv 0$ uniformly on compact sets. If we take a sequence of numbers $\rho_n \to \rho \neq 0$, then $\lambda_n \rho_n f_n \to \rho f \not\equiv 0$, so the sequence $\mu_n = \lambda_n \rho_n$ also serves as a sequence of multipliers. Conversely, let (μ_n) be a sequence of multipliers for (f_n): $\mu_n f_n \to g \not\equiv 0$. Let $z_o \in G$ be a point such that $g(z_o) \neq 0$ and $f(z_o) \neq 0$. Then $\rho_n = \frac{\mu_n}{\lambda_n} \to \frac{g(z_o)}{f(z_o)} \neq 0$, and $\mu_n = \rho_n \lambda_n$.

Let us call two sequences (λ_n) and (μ_n) of complex numbers equivalent if there exists a convergent sequence (ρ_n) with $\lim_{n\to\infty} \rho_n = \rho \neq 0$ such that $\lambda_n = \rho_n \mu_n$. We then have: If a sequence (f_n) of holomorphic functions can be blown up in the proper sense, all its sequences of multipliers are equivalent, and conversely, a sequence of multipliers can always be replaced by an equivalent one. In particular, the limit function is determined up to a constant factor, and its zeroes are well determined.

Let $\lambda_n f_n \to f \not\equiv 0$ and assume $f(z_0) \neq 0$. Then $\lambda_n f_n(z_0)$ has only finitely many zero terms, and

$$\frac{f_n}{f_n(z_0)} = \frac{\lambda_n f_n}{\lambda_n f_n(z_0)} \to \frac{f}{f(z_0)} = g \not\equiv 0$$

locally uniformly. Therefore the sequence $(f_n(z_0))^{-1}$ serves as a sequence of multipliers for every point z_0, where $f(z_0) \neq 0$.

It is easy to see that one can always choose positive multipliers: Let $\lambda_n f_n \to f \not\equiv 0$, $\lambda_n = |\lambda_n| e^{i\vartheta_n}$. We can choose a subsequence $\vartheta_{n_j} \to \vartheta_o$. Let $\rho_j = e^{i\vartheta_{n_j}}$. Then $\rho_j \to e^{i\vartheta_o}$, and the sequence $|\lambda_{n_j}| = \lambda_{n_j}/\rho_j$ is a sequence of multipliers for the subsequence (f_{n_j}). Since, in the definition of an inflatable family, we only need subsequences which are properly inflatable, we could without loss of generality replace the definition by restricting the factors to positive numbers.

<u>4. The local property.</u> Assume that the family F can be blown up. Then, just by restriction, it has of course this property also locally: Any blown up sequence is also blown up for every neighborhood, since $f \equiv 0$ in a neighborhood would imply $f \equiv 0$ in G.

Conversely, if F is locally inflatable, it also has this property globally, and every sequence which is blown up in a neighborhood is actually blown up for the entire domain G: If $\lambda_n f_n \to f \not\equiv 0$ locally uniformly in $U \subset G$, then f can be continued analytically to all of G and $\lambda_n f_n \to f$ in G.

<u>Proof.</u> Let F be a family of holomorphic functions in a domain G and assume that every point $z \in G$ has a neighbourhood U such that the restriction of F to U can be blown up. Let $\lambda_n f_n \to f \not\equiv 0$ locally uniformly in U. Let V be a neighborhood which overlaps with U and in which the family F is also inflatable. Then, the sequence (f_n) has a subsequence (f_{n_i}) with multipliers μ_i such that $\mu_i f_{n_i} \to g \not\equiv 0$ locally uniformly in V. In $U \cap V$ we have $\lambda_{n_i} f_{n_i} \to f$, $\mu_i f_{n_i} \to g$, and hence the sequences (λ_{n_i}) and (μ_i) are equivalent, $\lambda_{n_i} = \rho_i \mu_i$, $\rho_i \to \rho \neq 0$. The sequence $\lambda_{n_i} f_{n_i} = \rho_i \mu_i f_{n_i}$ converges locally uniformly in V. Its limit coincides with f in $U \cap V$ and therefore is the analytic continuation of f in V.

We now want to show that already $(\lambda_n f_n)$ converges locally uniformly to f in $U \cup V$. In the opposite case we would have a sequence $z_k \to z \in V$

such that $|\lambda_{n_k} f_{n_k}(z_k) - f(z_k)| \geq \varepsilon$ for some positive ε and a subsequence $(\lambda_{n_k} f_{n_k})$ of $(\lambda_n f_n)$. The family F can be blown up in V, and therefore the sequence (f_{n_k}) has a subsequence $(f_{n_{k_i}})$ with $\nu_i f_{n_{k_i}} \to \tilde{f}$ locally uniformly in V. As $\lambda_{n_{k_i}} f_{n_{k_i}} \to f$ locally uniformly in $U \cap V$, the two sequences of multipliers are equivalent: $(\lambda_{n_{k_i}}) \sim (\nu_i)$. Therefore the sequence $(\lambda_{n_{k_i}} f_{n_{k_i}})$ converges locally uniformly in V and its limit is of course $\tilde{f}$. But then, $\lambda_{n_{k_i}} f_{n_{k_i}}(z_{k_i}) \to f(z_o)$, which contradicts $|\lambda_{n_k} f_{n_k}(z_k) - f(z_k)| \geq \varepsilon > 0$.

5. The logarithmic derivative. Inflatable families can be characterized by means of the logarithmic derivative. Let (f_n) be a sequence of holomorphic functions which is properly inflatable. Let $\lambda_n f_n \to f \not\equiv 0$ locally uniformly. Then, the logarithmic derivative f'/f is meromorphic (possibly identically zero) and

$$\frac{(\lambda_n f_n)'}{\lambda_n f_n} = \frac{f_n'}{f_n} \to \frac{f'}{f}$$

locally uniformly outside of the zeroes of f.

Conversely, let $E \subset G$ be a set with no accumulation point in G and let the sequence (f_n'/f_n) converge locally uniformly in $G \setminus E$ to a function g which is holomorphic in $G \setminus E$. Let $z_o \notin E$ and choose a simply connected neighbourhood U of z_o with $\bar{U} \cap E = \{\ \}$. Then, for $z \in U$,

$$\int_{z_o}^{z} \frac{f_n'(\zeta)}{f_n(\zeta)} d\zeta = \log \frac{f_n(z)}{f_n(z_o)} \to \int_{z_o}^{z} g(\zeta)\, d\zeta$$

and therefore

$$\frac{f_n(z)}{f_n(z_o)} \to e^{\int_{z_o}^{z} g(\zeta)\, d\zeta} = f(z).$$

The sequence (f_n) is thus locally inflatable outside E and hence globally, with multipliers $\lambda_n = 1/f_n(z_o)$. The logarithmic derivative of the limit function f is $f'/f = g$. It results that a family F of holomorphic functions f is inflatable if and only if the logarithmic derivatives f'/f have the following property: Every sequence (f_n'/f_n) has a subsequence (f_{n_i}'/f_{n_i}) which converges locally uniformly outside of a set E with no accumulation point in G.

6. The variation of the argument. In order to formulate a useful necessary and sufficient condition for the family F to be inflatable

we must define the variation of the argument of a holomorphic function along a piecewise smooth curve γ. If f has no zeroes on γ (γ can be closed or not) we fix $\arg f(z_o)$ at an arbitrary point $z_o \in \gamma$ and then continue it along γ. We define

$$v_\gamma(\arg f) = \max_{z\in\gamma} \arg f(z) - \min_{z\in\gamma} \arg f(z).$$

The definition is clearly independent of the chosen point z_o and the determination of $\arg f(z_o)$, and also of the orientation of γ.

If f has zeroes z_j on the (oriented) arc γ, the $\arg f$ is not defined at these points, but it is defined on the intermediate subintervals of γ. We go around the points z_j along circular arcs centered at z_j and then let these arcs shrink to zero. The $\arg f$ is then a piecewise continuous function, but the jumps depend on whether we go around a z_j on its right or on its left hand side. Let m_j be the order of the zero z_j and let α_j be the angle on the right between the two consecutive subarcs of γ meeting at z_j. The jump of the argument then will be $m_j\alpha_j$ if we pass on the right, $-m_j(2\pi - \alpha_j) = m_j\alpha_j - 2\pi m_j$ if we pass on the left. In order to get a definite number we take v_γ to be the maximum of the variations of all the finitely many resulting piecewise continuous argument functions.

<u>7. A necessary condition.</u> Let F be inflatable and let γ be a piecewise smooth curve. Then the variation of the argument of the functions $f \in F$ is uniformly bounded.

The proof is indirect. We assume that there is no such bound. Then, there is a sequence (f_n) of elements of F such that $v_\gamma(\arg f_n) \to \infty$. This sequence has a subsequence which can be blown up properly, and we can assume that the multipliers are positive real numbers. Denoting the sequence by (f_n) again rather than (f_{n_i}), we have $\lambda_n f_n \to f \not\equiv 0$ locally uniformly, with $\lambda_n > 0$ for all n. If f has no zeroes on γ, the argument of $\lambda_n f_n$, which is the same as the argument of f_n, tends to the argument of f. This leads to a contradiction.

In the case of the presence of zeroes the arc γ is subdivided into neighborhoods of the zeroes of f and the complementary arcs. If the variation of the argument of f_n is bounded for all n, along each of these pieces, we have a contradiction again.

Let γ_o be a subarc of γ which contains a zero $z_o = 0$ of f of order m. Let U be a circular neighborhood of z_o containing γ_o and on the closure of which f has no zero except z_o. As $\lambda_n f_n \to f$ uniformly in $\overline{U}$, the number of zeroes of f_n in $\overline{U}$ is equal to m for all sufficiently large n. Let z_{n_i} be these zeroes, $i = 1,\dots,m$, each counted as many times as its multiplicity shows. Then $z_{n_i} \to 0$, $\prod_i (z - z_{n_i}) = P_n(z) \to z^m$ uniformly on $\bar{U}$ and therefore, because of the maximum principle , $f_n(z)/P_n(z) \to f(z)/z^m$ uniformly in $\bar{U}$. As $f(z)/z^m$ has no zero on γ_o, $v_{\gamma_o} \arg(f_n(z)/P_n(z)) \to v_{\gamma_o} \arg(f(z)/z^m)$, and therefore $v_{\gamma_o} \arg(f_n(z)/P_n(z))$ is bounded for all n. Now

$$v_{\gamma_o} \arg f_n(z) = v_{\gamma_o} \arg \frac{f_n(z)}{P_n(z)} P_n(z) \leqq v_{\gamma_o} \arg \frac{f_n(z)}{P_n(z)} + v_{\gamma_o} \arg P_n(z)$$

$$\leqq v_{\gamma_o} \arg \frac{f_n(z)}{P_n(z)} + \sum_i v_{\gamma_o} \arg(z - z_{n_i}).$$

The $\arg(z - \zeta)$ for z running through γ_o can only increase by a finite number which is independent of $\zeta \in U$. Therefore the right hand side is bounded and we have a contradiction again.

This proof is purely local and therefore works on Riemann surfaces as well.

8. Examples. The sequence $f_n(z) = z^n$, defined in $G = \mathbb{C}$, is not inflatable in any neighbourhood $U \subset G$, since there always exists an arc γ, e.g. a subarc of $|z| = r$, on which $v_\gamma \arg f_n(z) \to \infty$. Of course, these functions have a zero $z = 0$ in G of increasing order, and the zeroes of an inflatable sequence clearly can only accumulate towards the boundary of the domain. Puncturing $\mathbb{C}$ at zero, i.e. considering the functions f_n in $G = \mathbb{C} \setminus \{0\}$, would provide an example without zeroes which cannot be blown up in any neighborhood.

The sequence $f_n(z) = e^{nz}$ has no zeroes in $G = \mathbb{C}$ and is not inflatable in any neighborhood $U \subset G$, as $v_\gamma \arg f_n(z) = n(y_2 - y_1)$ on any vertical interval with endpoints y_1 and $y_2 > y_1$.

By Runges theorem one can construct sequences which can be blown up in some neighborhoods but not globally.

9. Sufficiency of the condition. Simply connected domains. We want to show that the above condition is also sufficient in the following sense. Let F be a family of holomorphic functions in a simply connected domain G. We take G to be the unit disk, but it can just as well be the whole plane or any simply connected domain. Let there exist a sequence of piecewise smooth Jordan curves γ_k, $k = 1,2,\ldots$, tending to the boundary of G, and assume that there exist constants $M_k < \infty$ such that $v_{\gamma_k}(\arg f) \leq M_k$ for all $f \in F$ and all $k \in \mathbb{N}$. Then, the family F is inflatable.

Proof. a) Let (f_n) be a sequence of elements of F and let $\gamma_k = \gamma$ be any one of the curves. Let $D = D_k$ be the domain bounded by γ. We first assume that the functions f_n have no zeroes in $\bar{D}$. We choose a fixed point $z_0 \in D$ and that branch of the logarithm

$$g_n = \log \frac{f_n}{f_n(z_0)}$$

which is zero at z_0. Let $M_k = M$ be the bound for the variation of the argument of $f \in F$ along γ. The function g_n maps γ onto a closed curve with a variation of the imaginary part which is $\leq M$, for every n. On the other hand, the winding number of $g_n(\gamma)$ around zero is at least one, as $g_n(z_0) = 0$, and there are no poles. Therefore the image $g_n(D)$ lies in the fixed horizontal strip $|\mathrm{Im}\, w| < M$ for all n. Let h be a conformal mapping of this horizontal strip onto the unit disk. Then, the sequence $h \circ g_n = h_n$ is bounded in D and therefore forms a normal family. We can select a subsequence $h \circ g_{n_i} = h_{n_i}$ which converges locally uniformly in D, and so does therefore the subsequence $g_{n_i} = h^{-1} \circ h_{n_i}$. Let $g = \lim_{i\to\infty} g_{n_i}$. We conclude

$$\lambda_i f_{n_i} = \frac{f_{n_i}}{f_{n_i}(z_0)} = e^{g_{n_i}} \to e^g \not\equiv 0, \quad \lambda_i = \frac{1}{f_{n_i}(z_0)},$$

locally uniformly in D.

b) We now allow the functions f_n to have zeroes in $\bar{D}$. For any f_n, let $\tilde{\gamma}$ be equal to γ except near the zeroes of f_n, which we circumvent along exterior circular arcs of radius ρ, centered at the zeroes. We then have

$$\frac{1}{2\pi}\int_{\tilde{\gamma}} d \arg f_n = N_n = \text{number of zeroes of } f_n \text{ in the closure } \bar{D} \text{ of } D.$$

For $\rho \to 0$, the left hand side becomes the increment of one of the argument functions of f_n along γ,

$$\arg f_n(z^+) - \arg f_n(z^-) = 2\pi N_n \geq 0,$$

z^+ and z^- being the two replica of a point $z \in \gamma$. But this is clearly $\leq$ maximum - minimum of this particular argument function and hence $\leq v_\gamma(\arg f_n) \leq M$. The number of zeroes of any f_n in $\bar{D}$ is thus bounded by $N_n \leq M/2\pi$.

Let now P_n be the polynomial $P_n(z) = \Pi(z - z_{n_i})$, z_{n_i} being the zeroes of f_n in $\bar{D}$. Then f_n/P_n has no zeroes in $\bar{D}$ and has a bounded variation of the argument:

$$v_\gamma(\arg \frac{f_n}{P_n}) \leq v_\gamma(\arg f_n) + v_\gamma \arg P_n \leq v_\gamma(\arg f_n) + \sum v_\gamma \arg(z - z_{n_i}),$$

which clearly is bounded. We therefore can apply the previous theorem to the sequence (f_n/P_n): $\lambda_k(f_{n_k}/P_{n_k}) \to g \not\equiv 0$ locally uniformly in D. By passing to a subsequence again we may assume that the zeroes of the P_{n_k} converge. Therefore the P_{n_k} converge to a polynomial $P \not\equiv 0$, uniformly on compact sets of $\mathbb{C}$. We have

$$\lambda_k f_{n_k} \to Pg \not\equiv 0$$

locally uniformly in D.

We have seen that the sequence (f_n) can be blown up in $D = D_k$, for every k. It can therefore be blown up in G, and in fact the inflated subsequence $(\lambda_i f_{n_i})$ for D_1 already converges locally uniformly in G with limit $f \not\equiv 0$.

10. Multiply connected domains. In this section we consider an arbitrary plane domain G and a family of holomorphic functions in G. We assume that there exists an exhaustion of G by domains D_k which are bounded by finitely many piecewise smooth Jordan curves on each of which the variation of the argument of all the functions $f \in F$ is uniformly bounded. Again, by passing from local to global inflatability, it is enough to prove that F is inflatable for every domain D_k. Let $D = D_k$ be any one of these domains, with boundary curves $\gamma_o, \gamma_1, \dots, \gamma_p$, γ_o being the outer boundary.

a) Let (f_n) be an arbitrary sequence of elements of F and assume first that they have no zeroes in $\bar{D}$. We have, by assumption

$$\left|\frac{1}{2\pi}\int_{\gamma_i} d\arg f_n\right| \leq \frac{1}{2\pi} v_{\gamma_i}(\arg f_n) \leq M_i,$$

M_i being the bound for the variation of the argument on γ_i. We now pick a point z_i in the interior of γ_i, $i = 1,\dots,p$, and then m_i such that

$$\frac{1}{2\pi}\int_{\gamma_i} d\arg \frac{f_n}{(z-z_i)^{m_i}} = 0.$$

The integer exponent m_i depends on the function f_n, but it is uniformly bounded for all n of the sequence. Because of homology and the absence of zeroes we then also have

$$\int_{\gamma_o} d\arg \frac{f_n}{P_n} = 0 \quad \text{with} \quad P_n(z) = \prod_{i=1}^{p} (z-z_i)^{m_i}.$$

Fix $z_0 \in D$. The function

$$g_n = \log\left(\frac{f_n}{P_n}\frac{P_n(z_0)}{f_n(z_0)}\right)$$

is a single valued holomorphic function in $\bar{D}$ with $g_n(z_0) = 0$. At least one of the images $g_n(\gamma_i)$, $i = 0,1,\ldots,p$, of the boundary curves has a positive winding number about 0. The corresponding curve is therefore contained in a horizontal strip of width M, say, which contains zero. The other boundary curves of $g_n(D)$ also lie in horizontal strips of width M, but not necessarily containing zero. If the entire image surface $g_n(D)$ were not contained in a horizontal strip $|\mathrm{Im}\, v| \leq (p+1)M$, there would be a point $w \in g_n(D)$ such that the horizontal straight line through w does not meet the boundary of $g_n(D)$. This is impossible, since this surface is obviously bounded. We now have again a normal family (g_n) in D. There is a subsequence which converges locally uniformly in D, $g_{n_k} \to g$, and hence

$$\lambda_k \frac{f_{n_k}}{P_{n_k}} \to e^g .$$

By passing to a subsequence again we can assume $P_{n_k} \to P$ uniformly in $\bar{D}$, hence $\lambda_k f_{n_k} \to Pe^g$, where of course $P \not\equiv 0$.

b) We now let the functions f_n have zeroes in $\bar{D}$. As in the simply connected case, the number $N(f)$ of zeroes of the functions $f \in F$ in $\bar{D}$ is bounded:

$$N(f) = \frac{1}{2\pi} \sum_{i=0}^{p} \int_{\tilde{\gamma}_i} d \arg f ,$$

where the curves $\tilde{\gamma}_i$ are equal to the curves γ_i except for small exterior circular arcs around the zeroes of f which lie on the curves γ_i. Letting the radius ρ of these arcs tend to zero, we get

$$N(f) \leq \frac{1}{2\pi} \sum_{i=0}^{p} v_{\gamma_i}(\arg f) ,$$

which is bounded by assumption.

Let now P_n be the polynomial with leading coefficient one which has exactly the zeroes of f_n in $\bar{D}$. By passing to a subsequence we may assume that the zeroes converge, and hence the polynomials P_n converge uniformly on compact subsets of $\mathbb{C}$. The quotients f_n/P_n have no zeroes in $\bar{D}$, and

$$v_{\gamma_i}(\arg \frac{f_n}{P_n}) \leq v_{\gamma_i}(\arg f_n) + v_{\gamma_i}(\arg P_n)$$

is uniformly bounded. Applying the previous result we see that the sequence (f_n/P_n) is inflatable in D, and so is therefore the sequence (f_n) itself.

The family F has now been shown to be inflatable in every domain D_k of our exhaustion, hence locally inflatable in G. But then this is true for G itself and any properly blown up sequence for D_1 converges locally uniformly in G to a holomorphic function $f \not\equiv 0$.

11. Riemann surfaces. Let R be an arbitrary Riemann surface, and F a family of holomorphic functions on R. We assume that there exists an exhaustion of R by surfaces R_k which are bounded by piecewise smooth Jordan curves $\Gamma_1^{(k)}, \ldots, \Gamma_{p_k}^{(k)}$ and such that the difference $R_{k+1} \setminus R_k = \cup G_j^{(k)}$ consists of planar domains $G_j^{(k)}$. Moreover, with respect to the exhaustion (R_k), the family F should have the property that the variation of the argument of the functions $f \in F$ is bounded on each $\Gamma_i^{(k)}$, $i = 1,\ldots,p_k$, $k = 1,2,3,\ldots$:

$$V_{\Gamma_i^{(k)}}(\arg f) \leq M^{(k)}$$

for all i (we take $M^{(k)}$ to be the largest bound for all individual curves $\Gamma_i^{(k)}$) and all k. Then the family F is inflatable on R.

Proof. Because of the previous theorem on multiply connected domains the family F can be blown up in each individual $G_j^{(k)}$, for all j and k. We now cover the boundary curves $\Gamma_i^{(k)}$ by narrow annuli, having their two piecewise smooth boundary curves in the two adjacent domains $G_j^{(k)}$ respectively. By the necessity of the condition, the variation of the argument of the functions $f \in F$ is also bounded on these two curves, for every annulus. Therefore, by the sufficiency again, F is inflatable in the annuli, thus locally inflatable on R, since the domains $G_j^{(k)}$ plus the annuli form an open covering of R. We conclude that F can be blown up on R, and in fact that every properly blown up sequence $(\lambda_n f_n)$ for R_1 (or any neighborhood) converges locally uniformly on R to a function $f \not\equiv 0$.

Universität Zürich
Freie Straße 36
CH-8032 Zürich
Schweiz

INNER-OUTER FACTORIZATION ON MULTIPLY CONNECTED DOMAINS

Ion Suciu

In [2] M. B. Abrahamse and R. G. Douglas proved a Wold decomposition for the subnormal operator whose spectrum is contained in the closure of a bounded, open, and connected subset Ω of the complex plane such that $\partial\Omega$ consists of finite number of nonintersecting analytic Jordan curves, and the normal spectrum is contained in $\partial\Omega$. They proved that any such operator is a direct sum of a normal operator having spectrum in $\partial\Omega$ with an appropriate bundle shift. Since they give also a functional model for the bundle shift as the multiplication by coordinate function on a space of analytic cross sections of a fibre bundle over Ω, their Wold decomposition furnishes a quite satisfactory quantity of analyticity which can be exploited in factorization theorems for operator valued functions on Ω or $\partial\Omega$ in a similar way as the classical Wold decomposition for isometries was used in the factorization of operator valued functions on the unit disk or circle (cf. [7], [9], [14], [15], [16]).

In this paper, using Wold decomposition of Abrahamse and Douglas we shall prove a Beurling type factorization theorem for a class of operator valued analytic functions on Ω. Following [2], in section 1, we shortly describe the basic facts about bundle shifts. In section 2 we discuss the operator valued automorphic analytic functions. The factorization theorem of Beurling type is proved in section 3.

In [13] we shall discuss this problematics in connection with the Szegö - Kolmogorov - Krein type factorizations.

1. Bundle shift and Wold decomposition.

Let Ω be a bounded, open, and connected subset of the complex plane whose boundary $\partial\Omega$ consists of $n + 1$ nonintersecting analytic Jordan curves. In all what follows z_0 will be a fixed point in Ω and m the harmonic measure supported by $\partial\Omega$ of the point z_0. Denote by A the algebra of all complex valued functions which are analytic in Ω and continuous on $\partial\Omega$. Any element of A can be uniformly approximated on $\partial\Omega$ by rational functions with poles off $\overline{\Omega}$ (Mergelyan theorem). If we consider A as function algebra on $\partial\Omega$ then it is an hypo-Dirichlet algebra (of [3]) and m is the unique logmodular representing measure for the complex homomorphism of A given by the evalutions on the point

z_o.

Let $\Pi_o(\Omega)$ be the fundamental group for Ω. It is known that $\Pi_o(\Omega)$ is a free abelian group with n generators. For a separable Hilbert space $\mathcal{F}$ let $\mathrm{Hom}(\Pi_o(\Omega), \mathcal{U}(\mathcal{F}))$ be the group of all group homomorphisms of $\Pi_o(\Omega)$ into the group $\mathcal{U}(\mathcal{F})$ of unitary operators on $\mathcal{F}$.

Let $C_1,\ldots,C_n$ be n cuts in Ω such that if C is the union of C_i, $\Omega \setminus C$ is simply connected. For a function h which is holomorphic in $\Omega \setminus C$, having analytic continuations along any curves in Ω, and $A \in \Pi_o(\Omega)$ we shall denote by $(h \circ A)(z)$ the values in $\gamma(1)$ of the analytic continuation of h along the closed curve γ in A which begins and ends in $z \in \Omega \setminus C$. Let $\alpha \in \mathrm{Hom}(\Pi_o(\Omega), \mathcal{U}(\mathcal{F}))$. We say that an $\mathcal{F}$-valued function h which is holomorphic in $\Omega \setminus C$ and admits analytic continuation along any curve in Ω produces an α-automorphic multiform function on Ω if for any $A \in \Pi_o(\Omega)$ and $z \in \Omega \setminus C$ we have

$$(h \circ A)(z) = \alpha(A)h(z). \tag{1.1}$$

Form (1.1) it results that $\|h(z)\|$ is a well defined subharmonic function on Ω. We shall denote by $H^2(\alpha)$ the space of all $\mathcal{F}$-valued functions h on $\Omega \setminus C$ which produces an α-automorphic multiform function on Ω such that there exists a positive harmonic function u in Ω verifying $\|h(z)\|^2 \leq u(z)$, $z \in \Omega$. If we put for any $h \in H^2_{\mathcal{F}}(\alpha)$

$$\|h\|^2 = \inf\{u(z_o),\ u \text{ harmonic in } \Omega,\ u(z) \geq \|h(z)\|^2,\ z \in \Omega\} \tag{1.2}$$

we obtain a norm on $H^2_{\mathcal{F}}(\alpha)$ with respect to which $H^2_{\mathcal{F}}(\alpha)$ becomes an Hilbert space. It can be showed that any element $h \in H^2_{\mathcal{F}}(\alpha)$ has well-defined non-tangential boundary limits at $\partial\Omega$, almost everywhere with respect to the measure m. These limits define a function h_α in $L^2_{\mathcal{F}}(m)$ and $h \to h_\alpha$ is an isometric imbedding of $H^2_{\mathcal{F}}(\alpha)$ into $L^2_{\mathcal{F}}(dm)$:

$$\|h\|^2 = \int_{\partial\Omega} \|h_\alpha(z)\|^2 dm(z), \quad h \in H^2_{\mathcal{F}}(\alpha). \tag{1.3}$$

Whenever it is necessary, we shall consider $H^2_{\mathcal{F}}(\alpha)$ as a subspace of $L^2_{\mathcal{F}}(m)$, via above described imbedding. If α is the identity 1 of $\mathrm{Hom}(\Pi_o(\Omega), \mathcal{U}(\mathcal{F}))$ we shall write $H^2_{\mathcal{F}}(\Omega)$ instead of $H^2_{\mathcal{F}}(1)$. The elements in $H^2_{\mathcal{F}}(\Omega)$ of the form $h(z) = f(z)a$, when f runs over A and a runs over $\mathcal{F}$, span a dense subspace in $H^2_{\mathcal{F}}(\Omega)$.

In [2] M. B. Abrahamse and R. G. Douglas realized $H^2_{\mathcal{F}}(\alpha)$ as a space of analytic cross sections of a flat unitary vector bundle, with fiber $\mathcal{F}$, over Ω, canonically attached to α. They introduced the <u>bundle shift operator</u> T_α on $H^2_{\mathcal{F}}(\alpha)$ as the operator given by multiplication with the identical function z on $H^2_{\mathcal{F}}(\alpha)$. The operator T_α is uniquely determine (up to a unitary equivalence) by the unitary equivalence class of α. Moreover, using the Grauert-Bungart theorem on the triviality of analytic bundles over Ω, they proved that for any $\alpha \in \mathrm{Hom}(\Pi_o(\Omega), \mathcal{U}(\mathcal{F}))$, T_α is similar with $T_\Omega = T_1$. Each T_α is the restriction to $H^2_{\mathcal{F}}(\alpha)$ of the normal operator N_Ω defined as the multiplication by z on $L^2(m)$. T_α is pure subnormal operator having N_Ω as minimal normal extension. If T is pure subnormal operator on a separable Hilbert space H, such that $\sigma(T) \subset \bar{\Omega}$ and the spectrum $\sigma(N)$ of its minimal normal extension N is contained in $\partial\Omega$, then there exist a Hilbert space $\mathcal{F}$ and an element $\alpha \in \mathrm{Hom}(\Pi_o(\Omega), \mathcal{U}(\mathcal{F}))$ such that T is unitarily equivalent to T_α. As a consequence they obtained the Wold decomposition we state here in a form which will be convenient further.

Theorem 0. Let N be a normal operator on a separable Hilbert space $\mathcal{K}$ such that $\sigma(N) \subset \partial\Omega$. Let $\mathcal{K}_+ \subset \mathcal{K}$ be an invariant subspace for N. Suppose that N is the minimal normal extension of $N_+ = N|\mathcal{K}_+$. Then there exists a unitary representation α of $\Pi_o(\Omega)$ on a separable Hilbert space $\mathcal{F}$ (possibly 0) such that $\mathcal{K}$ can be isometrically identified with a direct sum $L^2(m) \oplus \mathcal{K}_1$ in such that N becomes a direct sum of N_Ω with a normal operator N_1 on $\mathcal{K}_1$ having spectrum in $\partial\Omega$, $\mathcal{K}_+$ becomes $H^2(\alpha) \oplus \mathcal{K}_1$ and N_+ becomes $T_\alpha \oplus N_1$.

In [2] a functional model for T_α in terms of automorphic functions on the universal covering space, was also given. The construction of the covering space for Ω produces the following:

1. A group G of linear fractional transformations that map the unit disk D onto D.
2. An open G-invariant subset Γ of ∂D of zero Lebesgue measure.
3. A simply connected open G-invariant subset D' containing $D \cup \Gamma$.
4. An open set Ω' containing $\bar{\Omega}$.
5. A holomorphic covering map π from D' onto Ω', such that $\pi(D) = \Omega$, $\pi(\Gamma) = \partial\Omega$, and G is the group of all linear fractional transformations A having the property $\pi \circ A = \pi$.

We can suppose also $\pi(0) = z_o$.

In fact the group G is isomorphic with $\Pi_o(\Omega)$. This isomorphism is given by the so called lifting to the universal cover procedure which we briefly describe. For $\lambda \in D$ and $A \in \Pi_o(\Omega)$ let γ be a representant of A which begins and ends in $z = \pi(\lambda)$. Define $A\lambda$ to be $(\pi^{-1} \circ \gamma)(z)$ where $\pi^{-1} \circ \gamma$ is the analytic continuation of π^{-1} along the curve γ.

The normalized Lebesgue measure μ on ∂D lifts to the universal cover the measure m in the sense that for any $f \in L^1(m)$ we have

$$\int_{\partial D} (f \circ \pi)(\lambda) d\mu(\lambda) = \int_{\partial\Omega} f(z) dm(z). \tag{1.4}$$

Consider now the subspace $H^2_{\mathcal{F}}(D;\alpha)$ of α-automorphic functions in $H^2_{\mathcal{F}}(D)$:

$$H^2(D;\alpha) = \{h \in H^2_{\mathcal{F}}(D) : h(A\lambda) = \alpha(A)h(\lambda),\ \lambda \in D,\ A \in G\}. \tag{1.5}$$

If $\alpha = 1$ we shall denote $H^2_{\mathcal{F}}(D;\alpha)$ by $H^2_{\mathcal{F}}(D;G)$. Let $\pi^{-1}_{z_0}$ be the holomorphic function in $\Omega \setminus C$ obtained by the analytic continuation of π^{-1} along the curves in $\Omega \setminus C$ which begin in z_0. For any $h \in H^2_{\mathcal{F}}(D;\alpha)$ the function $h \circ \pi^{-1}_{z_0}$ produces an α-automorphic multiform holomorphic function on Ω. Using (1.4) it results that $h \circ \pi^{-1}_{z_0} \in H^2_{\mathcal{F}}(\alpha)$ and $\|h\| = \|h \circ \pi^{-1}_{z_0}\|$. In fact the map $h \to h \circ \pi^{-1}_{z_0}$ is an isometric isomorphism between $H^2_{\mathcal{F}}(D;\alpha)$ and $H^2_{\mathcal{F}}(\alpha)$ which make T_α unitarily equivalent to the multiplication by π on $H^2_{\mathcal{F}}(D;\alpha)$.

In what follows we shall use freely one or the other functional model for T_α in a way which will be convenient in the context, having in mind the identifications briefly described in this section.

2. Operator valued (α,β)-automorphic analytic functions.

Let now $\mathcal{E}$, $\mathcal{F}$ be two separable Hilbert spaces. Following B. Sz.-Nagy and C. Foiaş [16] we shall denote by the triplet $\{\mathcal{E},\mathcal{F},\Theta(\lambda)\}$ an $\mathcal{L}(\mathcal{E},\mathcal{F})$-valued analytic function on D. Let α be a representation of G on $\mathcal{F}$ and β a representation of G on $\mathcal{E}$. The analytic function $\{\mathcal{E},\mathcal{F},\Theta(\lambda)\}$ on D will be called <u>(α,β)-automorphic</u> if for any $\lambda \in D$ and A in G we have

$$\Theta(A\lambda) = \alpha(A)\Theta(\lambda)\beta(A)^*. \tag{2.1}$$

We shall write α-automorphic instead of (α,β)-automorphic in case $\beta = 1$, and β^*-automorphic in case $\alpha = 1$.

The result about similarity in [2] can be formulated in terms of automorphic functions as follows: There exists a bounded analytic function $\{\mathcal{F},\mathcal{F},\Phi^\alpha(\lambda)\}$ which is α-automorphic, having as inverse a bounded analytic function $\{\mathcal{F},\mathcal{F},\Psi^\alpha(\lambda)\}$ which is α^*-automorphic, such that

$$\begin{aligned} \Phi^\alpha H^2_{\mathcal{F}}(D;G) &= H^2_{\mathcal{F}}(D;\alpha) \\ \Psi^\alpha H^2_{\mathcal{F}}(D;\alpha) &= H^2_{\mathcal{F}}(D;G) \end{aligned} \tag{2.2}$$

where Φ^α and Ψ^α are the operators of pointwise multiplications by $\Phi^\alpha(\lambda)$ and $\Psi^\alpha(\lambda)$ on $H^2_{\mathcal{F}}(D)$.

The existence of the functions $\{\mathcal{F},\mathcal{F},\Phi^\alpha(\lambda)\}$ and $\{\mathcal{F},\mathcal{F},\Psi^\alpha(\lambda)\}$ comes from the Grauert-Bungart theorem on the triviality of analytic vector bundles over Ω, and it seems to be very difficult to precise more about them. The operators Φ^α and Ψ^α realise a similarity between bundle shifts T_α and T_Ω. We shall call them an α-pair of similarity.

Remark 1. For any $a \in \mathcal{F}$ there exists $h \in H^2_{\mathcal{F}}(D;\alpha)$ such that $h(0) = a$. Indeed $\Psi^\alpha(0)a$ considered as constant function on D belongs to $H^2_{\mathcal{F}}(D;G)$. Then $h = \Phi^\alpha\Psi^\alpha(0)a$ belongs to $H^2_{\mathcal{F}}(D;\alpha)$ and $h(0) = \Phi^\alpha(0)\Psi^\alpha(0)a = a$.

If $\{\mathcal{E},\mathcal{F},\Theta(\lambda)\}$ is (α,β)-automorphic in D then it lifts to the convering space the (α,β)-automorphic multiform function defined by

$$\Theta'(z) = (\Theta \circ \pi^{-1})(z), \quad z \in \Omega. \tag{2.3}$$

Conversely, any (α,β)-automorphic multiform holomorphic function on Ω with values in $\mathcal{L}(\mathcal{E},\mathcal{F})$ gives rise by lifting procedure to an (α,β)-automorphic function analytic on D.

We shall work only with (α,β)-automorphic functions on D but using above considerations, we obtain interpretations for some results in terms of multiform functions on Ω, or in terms of bundle maps.

There are no diffculties to define the boundedness for the (α,β)-automorphic functions. But, when we want to define L^2-boundedness (in the strong sense) some difficulties arise.

In the case $\{\mathcal{E},\mathcal{F},\Theta(\lambda)\}$ is α-automorphic the corresponding α-automorphic multiform function $\Theta'(z)$ on Ω produces a subharmonic (uniform) function $z \to \|\Theta'(z)\|$ on Ω. We say that $\{\mathcal{E},\mathcal{F},\Theta(\lambda)\}$ is L^2-bounded if for any $a \in \mathcal{E}$ there exists an harmonic function u_a on Ω such that for any $z \in \Omega$

$$\|\Theta'(z)a\|^2 \leqq u_a(z)\,\|a\|^2 .$$

If we put

$$(V_\Theta a)(\lambda) = \Theta(\lambda)a, \quad \lambda \in D,\ a \in \mathcal{E}, \tag{2.4}$$

we obtain a bounded operator V_Θ from $\mathcal{E}$ into $H^2_{\mathcal{F}}(\alpha)$. Indeed

$$\|V_\Theta a\|^2_{H^2_{\mathcal{F}}(\alpha)} = \int_{\partial D}\|\Theta(\lambda)a\|^2_{\mathcal{F}}d\mu(\lambda) = \int_{\partial\Omega}\|\Theta'(\lambda)a\|^2_{\mathcal{F}}dm(z)$$
$$\leqq \int_{\partial\Omega} u_a(z)\,\|a\|^2\,dm(z) \leqq M_a\,\|a\|^2 .$$

Standard arguments imply that V_Θ is bounded.

In case $\beta \neq 1$ such arguments are not more applicable. But they suggest the following definitions.

We say that an (α,β)-automorphic function $\{\mathcal{E},\mathcal{F},\Theta(\lambda)\}$ is L^2-bounded if there exists a β-pair of similarity Φ^β, Ψ^β such that the formula

$$(V_\Theta a)(\lambda) = \Theta(\lambda)\Phi^\beta(\lambda)a \tag{2.5}$$

defines a bounded operator V_0 from $\mathcal{E}$ into $H^2_{\mathcal{F}}(\alpha)$.

If $\beta = 1$ clearly this definition is the preceding one.

In [12] we showed that denoting

$$D(\Theta_+) = \{h \in H^2_{\mathcal{E}}(\beta) : \lambda \to \Theta(\lambda)h(\lambda) \in H^2_{\mathcal{F}}(\alpha)\} \tag{2.6}$$

and setting

$$(\Theta_+ h)(\lambda) = \Theta(\lambda)h(\lambda), \quad \lambda \in D,\ h \in D(\Theta_+), \tag{2.7}$$

we obtain a dense domain closed operator from $H^2_{\mathcal{E}}(\beta)$ into $H^2_{\mathcal{F}}(\alpha)$. This operator intertwines the point evaluations determined by Φ^β on $H^2_{\mathcal{E}}(\beta)$ (see for details [12]). Any non-bounded operator Q from $H^2_{\mathcal{E}}(\beta)$ into $H^2_{\mathcal{F}}(\alpha)$ having this intertwining property is representable as Θ_+ for an appropriate L^2-bounded (α,β)-automorphic function $\{\mathcal{E},\mathcal{F},\Theta(\lambda)\}$. The function $\{\mathcal{E},\mathcal{F},\Theta(\lambda)\}$ is bounded if and only if Θ_+ is bounded, and, in this case, the intertwining property which appear is the usual intertwining of T_α with T_β property.

The correspondence between the non-bounded operator Θ_+ and the function $\{\mathcal{E},\mathcal{F},\Theta(\lambda)\}$ depends, in general, by the β-pair of similarity Φ^β, Ψ^β. In the bounded case, however, this correspondence does not depend of Φ^β. Moreover, in this case there exist boundary strong limits, in the Fatou sense, for $\Theta(\lambda)$ and Θ_+ can be realized as pointwise multiplication on $H^2_{\mathcal{E}}(\beta)$ considered as the space of functions on ∂D or $\partial\Omega$. In the non-bounded case, in general, we have not such a boundary limit for $\Theta(\lambda)$.

We say that the (α,β)-automorphic function $\{\mathcal{E},\mathcal{F},\Theta(\lambda)\}$ is <u>inner</u> if it is bounded and the corresponding Θ_+ is an isometry from $H^2_{\mathcal{E}}(\beta)$ into $H^2_{\mathcal{F}}(\alpha)$. Clearly then the boundary limits of $\Theta(\lambda)$ is μ-a.e. isometric on D (or m-a.e. on $\partial\Omega$).

The L^2-bounded (α,β)-automorphic function $\{\mathcal{E},\mathcal{F},\Theta(\lambda)\}$ is called <u>outer</u> if the corresponding Θ_+ has dense range in $H^2_{\mathcal{F}}(\alpha)$.

Since for any $f \in A$ and $a \in \mathcal{E}$, $\Phi^\beta(f\circ\pi)a \in D(\Theta_+)$ and $\Theta_+\Psi^\beta(f\circ\pi)a = fV_\Theta a$ it results that $\{\mathcal{E},\mathcal{F},\Theta(\lambda)\}$ is outer if and only if

$$\bigvee_{f\in A} fV_\Theta\mathcal{E} = H^2_{\mathcal{F}}(\alpha). \tag{2.8}$$

<u>Remark 2.</u> If $\{\mathcal{E},\mathcal{F},\Theta(\lambda)\}$ is an (α,β)-automorphic outer function then for any $\lambda \in D$, $\Theta(\lambda)\mathcal{E}$ is dense in $\mathcal{F}$. Indeed let $b \in \mathcal{F}$ such that $(\Theta(0)a,b)_{\mathcal{F}} = 0$ for any $a \in \mathcal{E}$. Let $h \in H^2_{\mathcal{F}}(\alpha)$ such that $h(0) = b$ (see Remark 1). Then for any $f \in A$ and $a \in \mathcal{E}$ we have

$$(fV_{\Theta}a,b)_{L^2_{\mathcal{F}}(m)} = (f(z_0)\Theta(0)a,b(0))_{\mathcal{F}} = f(z_0)(\Theta(0)a,b)_{\mathcal{F}} = 0.$$

Using (2.8) we obtain $h = b$, i.e. $b = 0$. It results $\Theta(0)\mathcal{E}$ is dense in $\mathcal{F}$. Since $\Theta(\lambda)\mathcal{E} = \alpha(A)\Theta(0)\beta(A)^*\mathcal{E}$ for some $A \in G$ it results $\Theta(\lambda)\mathcal{E}$ is dense in $\mathcal{F}$ for any $\lambda \in D$.

Proposition 1. An L^2-bounded (α,β)-automorphic function $\{\mathcal{E},\mathcal{F},\Theta(\lambda)\}$ is simultaneously inner and outer if and only if it is a unitary constant function. Its constant value U realizes a unitary equivalence between α and β.

Proof. In both cases Θ_+ is a unitary operator from $H^2_{\mathcal{E}}(\beta)$ onto $H^2_{\mathcal{F}}(\alpha)$ such that

$$\Theta_+ T_\beta = T_\alpha \Theta_+ .$$

As in the proof of Theorem 6 in [2] we can show that Θ_+ is the multiplication with a unitary constant function.

3. Factorization theorems.

For an (α,β)-automorphic function $\{\mathcal{E},\mathcal{F},\Theta(\lambda)\}$ which is L^2-bounded the corresponding operator V_Θ from $\mathcal{E}$ into $H^2_{\mathcal{F}}(\alpha)$ can be considered as operator from $\mathcal{E}$ into $L^2_{\mathcal{F}}(m)$. We shall define an $\mathcal{L}(\mathcal{E})$-valued semi-spectral measure F_Θ on $\partial\Omega$ (cf. [11]) by

$$F_\Theta(\sigma) = V_\Theta^* \chi_\sigma V_\Theta \tag{3.1}$$

where for any Borel subset σ of $\partial\Omega$, χ_σ is the operator of multiplication with the characteristic function χ_σ on $L^2_{\mathcal{F}}(m)$.

We shall call F_Θ the semi-spectal measure attached to $\{\mathcal{E},\mathcal{F},\Theta(\lambda)\}$.

Proposition 2. Let $\{\mathcal{E},\mathcal{F}_1,\Theta_1(\lambda)\}$ be (α_1,β)-automorphic and $\{\mathcal{E},\mathcal{F}_2,\Theta_2(\lambda)\}$ be (α_2,β)-automorphic and outer. Suppose that $F_{\Theta_1} \leq F_{\Theta_2}$. Then there exists a contractive $\{\mathcal{F}_2,\mathcal{F}_1,\Theta(\lambda)\}$ which is (α_1,α_2)-automorphic, such that

$$\Theta_1(\lambda) = \Theta(\lambda)\Theta_2(\lambda), \quad \lambda \in D . \tag{3.2}$$

If $F_{\Theta_1} = F_{\Theta_2}$, then $\{\mathcal{F}_2,\mathcal{F}_1,\Theta(\lambda)\}$ is inner. If moreover $\{\mathcal{E},\mathcal{F}_1,\Theta_1(\lambda)\}$ is outer then $\Theta(\lambda) \equiv U$ where U is a unitary operator from $\mathcal{F}_2$ onto $\mathcal{F}_1$ such that

$$\alpha_1(A)U = U\alpha_2(A), \quad A \in G . \tag{3.3}$$

Proof. Let $f_1,\dots,f_n \in A$ and $a_1,\dots,a_n \in \mathcal{E}$. We have

$$\|\sum_{k=1}^{n} f_k V_{\Theta_1} a_k\|^2_{L^2_{\mathcal{F}_1}(m)} = \sum_{k,j} \int_{\partial\Omega} f_j(z)\overline{f_k(z)}\,d(F_{\Theta_1}(z)a_j,a_k)_{\mathcal{E}}$$

$$\leq \sum_{k,j} f_j(z)\overline{f_k(z)}\,d(F_{\Theta_2}(z)a_j,a_k)_{\mathcal{E}} = \|\sum_{k=1}^{n} f_k V_{\Theta_2} a_k\|^2_{L^2_{\mathcal{F}_2}(m)}.$$

Hence

$$\|\sum_{k=1}^{n} f_k V_{\Theta_1} a_k\| \leq \|\sum_{k=1}^{n} f_k V_{\Theta_2} a_k\|. \tag{3.4}$$

Since $\{\mathcal{E},\mathcal{F}_2,\Theta_2(\lambda)\}$ is outer it results that we can define a contraction Q from $H^2_{\mathcal{F}}(\alpha_2)$ into $H^2_{\mathcal{F}_1}(\alpha_1)$ by

$$Q\sum_{k=1}^{n} f_k V_{\Theta_2} a_k = \sum_{k=1}^{n} f_k V_{\Theta_1} a_k. \tag{3.5}$$

Since for any $h \in D(\Theta_{2+})$ we clearly have $Q\Theta_{2+}h = \Theta_{1+}h$ and

$$[Q\Theta_{2+}h](\lambda) = [\Theta_{1+}h](\lambda) = \Theta_1(\lambda)h(\lambda) = [\Theta_{1+}\Phi^{\beta}\Psi^{\beta}(\lambda)h(\lambda)](\lambda)$$

$$= [Q\Theta_{2+}\Phi^{\beta}\Psi^{\beta}(\lambda)h(\lambda)](\lambda) = [Q\Phi^{\alpha_2}\Psi^{\alpha_2}(\lambda)\Theta_{2+}h](\lambda),$$

it results that Q intertwines the point evaluations on $H^2_{\mathcal{F}_2}(D;\alpha_2)$ and $H^2_{\mathcal{F}_1}(D;\alpha_1)$. Hence $Q = \Theta_+$ where $\{\mathcal{F}_2,\mathcal{F}_1,\Theta(\lambda)\}$ is a contractive (α_1,α_2)-automorphic function on D. We have

$$\Theta_1(\lambda)a = [\Theta_{1+}\Phi^{\beta}\Psi^{\beta}(\lambda)a](\lambda) = [Q\Theta_{2+}\Phi^{\beta}\Psi^{\beta}(\lambda)a](\lambda)$$

$$= \Theta(\lambda)[\Theta_+\Phi^{\beta}\Psi^{\beta}(\lambda)a](\lambda) = \Theta(\lambda)\Theta_2(\lambda)a.$$

If $F_{\Theta_1} = F_{\Theta_2}$, then clearly in (3.4) we have equality. It results that Q is an isometry and consequently Θ_+ is isometry. Hence $\{\mathcal{F}_2,\mathcal{F}_1,\Theta(\lambda)\}$ is inner. If moreover $\{\mathcal{E},\mathcal{F}_1,\Theta_1(\lambda)\}$ is outer then clearly $Q = \Theta_+$ has dense range and consequently $\{\mathcal{F}_2,\mathcal{F}_1,\Theta(\lambda)\}$ is outer. From Proposition 1 we conclude that $\Theta(\lambda)$ is unitary constant and its constant value U verifies (3.3).

The proof of the Proposition 2 is complete.

Proposition 3. Let $\{\mathcal{E},\mathcal{F},\Theta(\lambda)\}$ be an β-automorphic analytic function which is L^2-bounded and let F_Θ be its semi-spectral measure on $\partial\Omega$. There exists an α-automorphic analytic function $\{\mathcal{E},\mathcal{F}_1,\Theta_1(\lambda)\}$ which is L^2-bounded and outer such that $F_{\Theta_1} = F_\Theta$. The outer function $\{\mathcal{E},\mathcal{F}_1,\Theta_1(\lambda)\}$ is uniquely determined up to a unitary constant factor from the left.

Proof. Let $\mathcal{K}$ and $\mathcal{K}_+$ be the subspaces of $L^2_{\mathcal{F}}(m)$ given by

$$\mathcal{K} = \bigvee_{g\in C(\partial\Omega)} gV_\Theta\mathcal{E}\ , \quad \mathcal{K}_+ = \bigvee_{f\in A} fV_\Theta\mathcal{E}\ .$$

The multiplication by the coordinate function z defines a normal operator N on $\mathcal{K}$. Clearly $N\mathcal{K}_+ \subset \mathcal{K}_+$. Applying Theorem 0 we obtain a Hilbert space $\mathcal{F}_1$ and a unitary representation α of G on $\mathcal{F}_1$ such that

$$\mathcal{K} = L^2_{\mathcal{F}_1}(m) \oplus \mathcal{K}_1, \quad \mathcal{K}_+ = H^2_{\mathcal{F}_1}(\alpha) \oplus \mathcal{K}_1$$

and $\mathcal{K}_1$ reduces N to a normal operator. Since $\mathcal{K}_+ \subset H^2_{\mathcal{F}}(\beta)$ and $N|\mathcal{K}_+ = T_\beta|\mathcal{K}_+$ it results that $\mathcal{K}_1 = 0$ because T_β is pure subnormal. Hence we have

$$\mathcal{K} = L^2_{\mathcal{F}_1}(m), \quad \mathcal{K}_+ = H^2_{\mathcal{F}_1}(\alpha)$$

Since $V_\Theta\mathcal{E} \subset \mathcal{K}_+$ it results that there exists an α-automorphic analytic function $\{\mathcal{E}, \mathcal{F}_1, \Theta_1(\lambda)\}$ such that $V_{\Theta_1}a = V_\Theta a$ as elements in $\mathcal{K}_+$. Clearly $\{\mathcal{E}, \mathcal{F}, \Theta_1(\lambda)\}$ is L^2-bounded and $F_{\Theta_1} = F_\Theta$. Since

$$H^2_{\mathcal{F}_1}(\alpha) = \mathcal{K}_+ = \bigvee_{f\in A} fV_\Theta\mathcal{E} = \bigvee_{f\in A} fV_{\Theta_1}\mathcal{E}$$

it results that $\{\mathcal{E}, \mathcal{F}_1, \Theta_1(\lambda)\}$ is outer.

The unicity part of the Proposition 3 results from Proposition 2.

We can state now the Beurling type factorization theorem.

<u>Theorem.</u> Let $\{\mathcal{E}, \mathcal{F}, \Theta(\lambda)\}$ be an β-automorphic L^2-bounded analytic function of D. There exists an α-automorphic function $\{\mathcal{E}, \mathcal{F}_1, \Theta_1(\lambda)\}$ which is outer and an (β,α)-automorphic function $\{\mathcal{F}_1, \mathcal{F}, \Theta_2(\lambda)\}$ which is inner such that

$$\Theta(\lambda) = \Theta_2(\lambda)\Theta_1(\lambda), \quad \lambda \in D. \tag{3.6}$$

This factorization is unique in the following sense: If $\{\mathcal{E}, \mathcal{F}_1', \Theta_1'\}$ is an α'-automorphic outer function and $\{\mathcal{F}_1', \mathcal{F}, \Theta_2'(\lambda)\}$ is an (β,α')-automorphic inner function such that

$$\Theta(\lambda) = \Theta_2'(\lambda)\Theta_1'(\lambda)$$

then there exists a unitary operator U from $\mathcal{F}_1$ onto $\mathcal{F}_1'$ such that

$$U\alpha(A) = \alpha'(A)U, \quad A \in G \tag{3.7}$$

$$U\Theta_1(\lambda) = \Theta_1'(\lambda), \ \Theta_2(\lambda) = \Theta_2'(\lambda)U, \quad \lambda \in D. \tag{3.8}$$

<u>Proof.</u> Applying Proposition 3 we obtain an α-automorphic function $\{\mathcal{E}, \mathcal{F}_1, \Theta_1(\lambda)\}$ such that $F_{\Theta_1} = F_\Theta$. Using Proposition 2 we obtain an

inner (β,α)-automorphic function such that

$$\Theta(\lambda) = \Theta_2(\lambda)\Theta_1(\lambda), \quad \lambda \in D. \tag{3.9}$$

Suppose now that

$$\Theta(\lambda) = \Theta_2'(\lambda)\Theta_1'(\lambda), \quad \lambda \in D \tag{3.10}$$

where $\{\mathcal{E}, \mathcal{F}_1', \Theta_1'(\lambda)\}$ is an α'-automorphic outer function and $\{\mathcal{F}_1', \mathcal{F}, \Theta_2'(\lambda)\}$ is an (β,α')-automorphic inner function.

Then it is easy to see that $F_{\Theta_1'} = F_{\Theta_1}$. Using Proposition 2 we obtain a unitary operator U from $\mathcal{F}_1$ onto $\mathcal{F}_1'$ such that (3.7) is satisfied and

$$\Theta'(\lambda) = U\Theta(\lambda), \quad \lambda \in D. \tag{3.11}$$

From (3.9), (3.10) and (3.11) it results

$$\Theta_2(\lambda)\Theta_1(\lambda) = \Theta_2'(\lambda)U\Theta_1(\lambda)$$

and using Remark 2 we obtain

$$\Theta_2(\lambda) = \Theta_2'(\lambda)U, \quad \lambda \in D.$$

The proof of the Theorem is complete.

References

[1] Abrahamse, M. B., Douglas, R. G.: Operators on multiply connected domains. Proc. roy. Irish Acad., Sect. A 74 (1974), 135 - 141.

[2] Abrahamse, M. B., Douglas, R. G.: A class of subnormal operators related to multiply connected domains. Advances Math. 19 (1976), 106 - 148.

[3] Ahern, P., Sarason, D.: The H^p spaces of a class of function algebras. Acta math. 117 (1967), 123 - 163.

[4] Ball, J. A.: Operator extremal problems, expectation operators and applications to operators on multiply connected domains. To appear in J. Operator Theory.

[5] Beurling, A.: On two problems concerning linear transformation in Hilbert space. Acta math. 81 (1949), 239 - 255.

[6] Bungart, L.: On analytic fiber bundles I. Topology 7 (1968), 55 - 68

[7] Douglas, R. G.: On factoring positive operator functions. J. Math. Mech. 16 (1966), 119 - 126.

[8] Grauert, H.: Analytische Faserungen über holomorphvollständigen Räumen. Math. Ann. 135 (1958), 263 - 273.

[9] Lowdenslager, D. B.: On factoring matrix valued functions. Ann. of

Math, II. Ser. 78 (1963), 450 - 454.

[10] Sarason, D. E.: The H^p-space of an annulus. Mem. Amer. math. Soc. 56 (1956).

[11] Suciu, I.: Function algebras. Editura Academiei Republicii Socialiste România, Bucureşti and Noordhoff International Publishing, Leyden (1975).

[12] Suciu, I.: Non-bounded intertwining operators of bundle shifts. To appear in Proc. American-Roumanian Seminar on Operator Theory.

[13] Suciu, I.: Factorization theorems for spectral valued functions on multiply connected domains. To appear.

[14] Suciu, I., Valuşescu, I.: Factorization of semi-spectral measures. Revue Roumaine Math. pur. appl. 20 (1976), 773 - 793.

[15] Sz.-Nagy, B., Foiaş, C.: Sur les contractions de l'espace de Hilbert. IX. Acta Sci. math. 25 (1964), 283 - 316.

[16] Sz.-Nagy, B., Foiaş, C.: Harmonic analysis of operators on Hilbert spaces. North-Holland Publishing Co., Amsterdam - London, American Elsevier Publishing Co., New York and Akadémiai Kiadó, Budapest (1970).

[17] Voichik, M., Zalcman, L.: Inner and outer functions of Riemann surfaces. Proc. Amer. math. Soc. 16 (1965), 1200 - 1204.

[18] Wold, H.: A study in the analysis of stationary time series. Almqvist & Wiksell, Stockholm (1938).

INCREST
Bd. Păcii 220
77538 Bucureşti
Romania

DEFECT RELATIONS OF HOLOMORPHIC CURVES AND THEIR ASSOCIATED CURVES IN CP^m

Chen-Han Sung

§1. Introduction.

A holomorphic curve $f := (f^0,\ldots,f^m)$ in an m-dimensional complex projective space CP^m is nothing but a system of entire functions $\{f^j\}_{j=0}^m$ in the complex line C, which have no common zeros. It is non-degenerate if and only if there is no linear relations among the f^j's.

The defect relation for non-degenerate holomorphic curves in CP^m was obtained by Ahlfors [1] and Cartan [2]. This defect relation is in terms of a (generalized) Nevanlinna characteristic function for f. The characteristic function defined by Cartan [2] for a system of entire functions is essentially the same as the order function defined by Weyl [7] for the corresponding holomorphic curve, which was then used in [1]. A modern treatment of the work by Ahlfors and Weyl can be found in Wu [8].

However, a holomorphic curve in CP^m might lie in a projective linear subspace of CP^m. My first result is

Theorem I. Let $\{D_\alpha\}_{\alpha=1}^N$ be a family of hyperplanes in general position in CP^m. Let $f : C \to CP^m$ be a holomorphic curve which lies on a linear subspace CP^{k+1} but does not lie on any lower dimensional linear subspace of CP^m, nor on any D_α with $1 \leqq \alpha \leqq N$. Then

$$\sum_{\alpha=1}^{N} \delta(f,D_\alpha) \leqq (2m - k) \tag{1.1}$$

where $0 \leqq k \leqq m - 1$. Furthermore, this bound is sharp.

The above theorem with $k \geqq 0$ readily yields the following Picard-type result:

A non-constant holomorphic mapping $f : C^n \to CP^m$ can miss at most 2m hyperplanes in general position in CP^m.

On the other hand, the classical defect relation of Ahlfors [1] and Cartan [2] is nothing but the special case $k = m - 1$ of Theorem I.

Next, we consider, at each ξ of C, the sequence of derived vectors $f,f^{(1)},\ldots,f^{(m)}$. For a non-degenerate holomorphic curve, f and the m first derivatives are linearly independent almost everywhere in C. The exterior products

$$f_j := f \wedge f^{(1)} \wedge \ldots \wedge f^{(j)}, \; j = 1,2,\ldots,m \; ,$$

then define a family of curves

$$f_j : C \to Gr(m,j)$$

in $Gr(m,j)$, the Grassman manifold of projective j-planes in CP^m. We call these the associated curves of rank j, $j = 1,\ldots,m$. In particular, f_1 represents the tangent line and f_j the osculating j-plane which has a contact of order j with the curve. For each j, $Gr(m,j)$ can be embedded into some $CP^{n(m,j)}$ where

$$n(m,j) := \binom{m+1}{j+1} - 1 \; . \tag{1.2}$$

Hence f_j may be viewed as a holomorphic curve in $CP^{n(m,j)}$, although f_j may be degenerate in $CP^{n(m,j)}$ even if f is non-degenerate in CP^m. Nevertheless, we have

<u>Theorem II.</u> Let $f : C \to CP^m$ be a non-degenerate holomorphic curve and $f_j : C \to Gr(m,j) \subset CP^{n(m,j)}$ be its associated curve of rank j, $1 \leq j \leq m-1$. Let $\{D_\alpha^j\}_{\alpha=1}^N$ be a family of hyperplanes in general position in $CP^{n(m,j)}$ such that none of these D_α^j's contains $f_j(C)$. Then, we have

$$\sum_{\alpha=1}^N \delta(f_j,D_\alpha^j) \leqq 2 \cdot n(m,j) - (m-j), \quad \text{for} \quad 1 \leqq j \leqq \frac{m}{2} \; , \tag{1.3}$$

$$\sum_{\alpha=1}^N \delta(f_j,D_\alpha^j) \leqq 2 \cdot n(m,j) - j, \quad \text{for} \quad \frac{m}{2} \leqq j \leqq m-2 \; , \tag{1.4}$$

and

$$\sum_{\alpha=1}^N \delta(f_{m-1},D_\alpha^{m-1}) \leqq m+1, \quad \text{for} \quad j = m-1 \; . \tag{1.5}$$

The bound of (1.5) is sharp, but the sharpness of (1.3) and (1.4) is open.

<u>§2. Remarks.</u>

(1) It is clear that the "non-degeneracy" condition on curve in the Ahlfors-Cartan defect relation is completely relaxed in our Theorem I. Meanwhile, Theorem I does give the sharp defect relation for a non-degenerate holomorphic curve with any family of hyperplanes in CP^{k+1}. Hence, the condition "in general position" on the family of hyperplanes in the Ahlfors-Cartan defect relation is also relaxed completely. A more

detailed discussion concerning the variants and applications of Theorem I can be found in [5]. The bound of (1.1) is sharp by the example in [4].

(2) In case that f_j is non-degenerate in $CP^{n(m,j)}$, we have, by Theorem I,

$$\sum_{\alpha=1}^{N} \delta(f_j, D_\alpha^j) \leq n(m,j) + 1 \tag{2.1}$$

for any family of hyperplanes $\{D_\alpha^j\}_{\alpha=1}^N$ in general position in $CP^{n(m,j)}$.

In general, f_j is degenerate in $CP^{n(m,j)}$ even though f is non-degenerate in CP^m. But, if A^j is a decomposable unit $(j + 1)$-vector in C^{m+1}, then

f_j lies in the hyperplane in $CP^{n(m,j)}$ defined by A^j $\Leftrightarrow$
f is degenerate in CP^m.

Indeed, Ahlfors [1] proved that if f is non-degenerate in CP^m, then

$$\sum_{\alpha=1}^{N} \delta(f_j, D_\alpha^j) \leq n(m,j) + 1$$

for a family of hyperplanes $\{D_\alpha^j\}_{\alpha=1}^N$ in general position in $CP^{n(m,j)}$, defined by decomposable unit $(j + 1)$-vectors in C^{m+1}.

Using results of Toda [6] and the author [5], Niino [3] recently proved that if f_j is non-constant together with all hypotheses in our Theorem II hold, then

$$\sum_{\alpha=1}^{N} \delta(f_j, D_\alpha^j) \leq 2 \cdot n(m,j) - 1, \quad \text{for} \quad 1 \leq j \leq m - 2 \tag{2.2}$$

and

$$\sum_{\alpha=1}^{N} \delta(f_{m-1}, D_\alpha^{m-1}) \leq m + 1 . \tag{2.3}$$

It is clear that our bound of (1.3) and (1.4) is better than the bound of (2.2). Moreover, we will see that f_j is non-constant whenever f is non-degenerate. Thus, Niino's assumption that f_j is non-constant is redundant. In fact, we have

Theorem III. If f is non-degenerate in CP^m and $1 \leq j \leq m - 1$, then f_j can not lie on any projective linear subspace of dimension $(m - j)$ or j in $CP^{n(m,j)}$.

Notice that $(m-j) \geq j$ for $1 \leq j \leq m/2$, and $(m - j) \leq j$ for $m/2 \leq j \leq m - 1$. Finally, we give a necessary and sufficient condition which implies non-degeneracy.

Theorem IV. f is non-degenerate in $CP^m \Leftrightarrow f_{m-1}$ in non-degenerate in $CP^m \Leftrightarrow f_{m-1}$ is non-constant in CP^m.

Theorem III points out that Niino's bound for (2.2) cannot be achieved by the associated curves of any non-degenerated curve f. Indeed, the example [3, p. 85] used by Niino is actually a degenerate curve in CP^3. The proof of Theorem III and Theorem IV will appear later.

(5) Using Theorem I, we have (1.3), (1.4) and (1.5) from Theorem III, and Theorem IV respectively. The bound of (1.5) is sharp as the example

$$f(\xi) := \left[e^{\xi}, e^{2\xi}, \ldots, e^{(m+1)\xi}\right] \tag{2.4}$$

shows.

In view of Theorem I, it is natural to ask:

(*) what is the lowest dimension of possible linear (projective) subspaces containing $f_j(C)$ in $CP^{n(m,j)}$, while f is non-degenerate in CP^m?

Theorem III indicates some partial answers to (*). Theorem IV gives a firm answer for $j = m - 1$. However, the analogues of Theorem IV for f_j with $1 \leqq j \leqq m - 2$ is false by example (2.4).

§3. Proof of Theorem I.

A proof of Theorem I would inordinately lengthen this note. I shall give a brief outline of the proof, with details to appear in [4].

Our general philosophy is that the important geometrical properties about the curve $f : C \to CP^m$ are carried by the pull-back of the structure equations of CP^m to C through f.

In particular, let $Z_0, \ldots, Z_m$ be the Frenet frame along f (cf. [4], [8]). The Maurer-Cartan forms θ_{jk} are defined by

$$dZ_j = \sum_{k=0}^{m} \theta_{jk} Z_k \tag{3.1}$$

with

$$\theta_{jk} + \bar{\theta}_{kj} = 0 \tag{3.2}$$

for $0 \leqq j, k \leqq m$. We then introduce a family of volume forms

$$v_j := (i/2\pi)\theta^*_{j,j+1} \wedge \theta^*_{j,j+1} \tag{3.3}$$

for $0 \leqq j \leqq m - 1$ on C, where $\theta^*_{j,j+1}$ are the pull-back of $\theta_{j,j+1}$ to C through f. We can write these as

$$v_j := (i/2\pi)h_j^2 d\xi \wedge d\bar{\xi} \tag{3.4}$$

for $0 \leqq j \leqq m-1$, where h_j is real-valued and ξ is the local coordinate on C.

On the other hand, we study the contact of our curve $f(C)$ with D_α by considering

$$\phi_j(D_\alpha) := |(Z_0, A_\alpha)|^2 + \cdots + |(Z_j, A_\alpha)|^2 \tag{3.5}$$

for $0 \leqq j \leqq m$, while

$$D_\alpha := \{Z \in CP^m | (Z, A_\alpha) = \sum_{k=0}^{m} z^k \bar{a}_\alpha^k = 0\}$$

with

$$(A_\alpha, A_\alpha) = 1.$$

It is clear that we can cover C by an open covering $\{U_\lambda\}_{\lambda\in\Lambda}$ such that each $f(U_\lambda)$ meets at most m of the D_α's. We then construct a metric on C as

$$ds^2 := H^2 d\xi d\bar{\xi} \tag{3.6}$$

where

$$H := S \cdot h \tag{3.7}$$

consists of a special factor S which is C^∞, positive and varying from one local neighborhood U_λ to another one, and a basic factor h defined by

$$h^\beta := \left\{ \prod_{\alpha=1}^{N} \frac{\phi_{k+1}(D_\alpha)}{\phi_o(D_\alpha) \cdot \left\{ \prod_{j=0}^{k} \log \frac{\mu}{\phi_j(D_\alpha)} \right\}^2} \right\} \cdot h_o^{2m} \cdot \prod_{j=1}^{k} h_j^{2(k+1-j)} \tag{3.8}$$

with

$$\beta := 2m + k \cdot (k+1).$$

As (1.1) will be obvious if there are less than $(2m - k + 1)$ hyperplanes D_α, we may assume that

$$N > (2m - k). \tag{3.9}$$

Taking (3.6) to be our measure over C as $f(c)$ travels in CP^m, we claim that

<u>Lemma 1</u> Given $\varepsilon > 0$, there exists a constant $c > 0$ such that

$$\beta \cdot \Delta \log H \geqq c \cdot (H^2 \cdot \tilde{v}_o) + 2(N - 2m + k)v_o - \varepsilon(\sum_{j=0}^{k} v_j) \tag{3.10}$$

where

$$\Delta := (i/\pi)\partial\bar{\partial} \quad \text{and} \quad \tilde{v}_o := (i/2\pi)d\xi d\bar{\xi}.$$

We note that integration of (3.10) gives

$$\beta\cdot\int_1^u\Big(\int_{\Delta_r}\Delta\log H\Big)d\log r \geq c\cdot\int_1^u\Big(\int_{\Delta_r}H^2\cdot\tilde{v}_o\Big)d\log r + 2(N-2m+k)T_o(u)$$
$$-\varepsilon\cdot\Big(\sum_{j=0}^{k}T_j(u)\Big)$$

where

$$\Delta_r := \{\xi\in\mathbb{C} \mid |\xi| \leq r\}$$

and

$$T_j(u) := \int_1^u\Big(\int_{\Delta_r}v_j\Big)d\log r \quad \text{for } j = 0,\dots,k. \tag{3.11}$$

Next, by Gauss-Bonnet formula for metric with singularities and using property of logarithm, we find that for some constant c',

$$\sum_{\alpha=1}^{N}N_o(u,D_\alpha) \geq (N-2m+k)T_o(u) - \varepsilon\cdot\Big(\sum_{j=0}^{k}T_j(u)\Big) + c' \tag{3.12}$$

as $u\to\infty$, except for a set $I\subset[1,\infty)$ for which $\int_I d\log u < +\infty$, where

$$N_o(u,D_\alpha) := \int_1^u n(r,D_\alpha)d\log r \tag{3.13}$$

is the counting function, while $T_o(u)$ is called the order function.

Now, using the growth relations between $T_o(u)$ and $T_j(u)$, we are led to

Lemma 2. Given $\varepsilon > 0$, there exists constants c and c' such that

$$\sum_{\alpha=1}^{N}N_o(u,D_\alpha) \geq (N-2m+k)T_o(u) - \varepsilon\cdot T_o(u) - c\cdot\log T_o(u) - c' \tag{3.14}$$

as $u\to\infty$, except for a set $I\subset[1,\infty)$ for which $\int_I d\log u < +\infty$.

As $\sum(\limsup) \geq \limsup(\sum)$, it is clear that

$$\sum_{\alpha=1}^{N}\delta(D_\alpha) \leq N - \limsup_{u\to\infty}\sum_{\alpha=1}^{N}N_o(u,D_\alpha)/T_o(u). \tag{3.15}$$

Then (1.1) follows from this together with (3.14). The bound is sharp by the example given in [4].

References

[1] Ahlfors, L. V.: The theory of meromorphic curves. Acta Soc. Sci. Fenn. Nova Ser. A 3 (4) (1941), 3 - 31.

[2] Cartan, H.: Sur les zeros des combinaisons linéaires de p fonctions holomorphes données. Mathematica, Cluj 7 (1933), 5 - 31.

[3] Niino, K.: Deficiencies of the associated curves of a holomorphic curve in the projective space. Proc. Amer. math. Soc. 59 (1976), 81 - 88.

[4] Sung, C. H.: On H. Cartan's conjecture on a system of holomorphic functions. To appear.

[5] Sung, C. H.: On the associated curves of a holomorphic curve in the complex projective space. To appear.

[6] Toda, N.: Sur les combinaisons exceptionnelles de fonctions holomorphes. Tôhoku math. J., II. Ser. 22 (1970), 290 - 319.

[7] Weyl, H., Weyl, J.: Meromorphic curves. Ann. of Math., II. Ser. 39 (1938), 516 - 538.

[8] Wu, H.: The equidistribution theory of holomorphic curves. Princeton Univ. Press, Princeton, N.J. (1970).

Purdue University
Mathematics Department
West Lafayette, IN 47907
U.S.A.

BOUNDARY BEHAVIORS OF QUASIREGULAR MAPPINGS

Hiroshi Tanaka

In this paper we shall give a non function-theoretical proof to a well-known theorem in a cluster set theory of analytic functions on Riemann surfaces. In our discussion we use only modulus method of path families so that all of our results are also valid in higher dimensional cases.

1. Quasiconformal metric

Let R be an open Riemann surface. Let Γ be a path family on R. We refer to [1] and [13] for the definition and properties of modulus $M(\Gamma)$ of Γ. The following well-known properties of moduli are useful in our discussion:

(i) If $\Gamma_1 \subset \Gamma_2$, then $M(\Gamma_1) \leqq M(\Gamma_2)$.

(ii) $M(\Gamma_1 \cup \Gamma_2) \leqq M(\Gamma_1) + M(\Gamma_2)$.

(iii) If $\Gamma_1 > \Gamma_2$, then $M(\Gamma_1) \leqq M(\Gamma_2)$.

For a subset E of R we denote by $\Gamma(E) = \Gamma(E;R)$ the family of all paths on R starting at a point of E and clustering nowhere in R. Now we shall define a hyperbolicity and parabolicity of an open Riemann surface by using a modulus method.

Definition 1. An open Riemann surface R is said to be hyperbolic or parabolic accordingly as $M(\Gamma(K)) > 0$ or $= 0$ for some non degenerate continuum K in R.

Definition 2. (T. Kuusalo [7], H. Tanaka [12]) For $a,b \in R$ we set $C(a,b;R)$ the family of all continua in R containing both a and b. We define

$$\rho_R(a,b) = \rho(a,b;R) = \inf \{M(\Gamma(K));\ K \in C(a,b;R)\}.$$

Remarks. (1) The function ρ_R is called a quasiconformal metric by T. Kuusalo [7] and is also discussed in [3], [4] and [5].

(2) Let R be a hyperbolic Riemann surface and let $C(K)$ be the Green capacity of a compact subset K of R (cf. [2]). Then we can see that $C(K) = M(\Gamma(K))$. Hence $\rho_R(a,b) = \inf \{C(K);\ K \in C(a,b;R)\}$ (H. Tanaka [12]).

Theorem 1. (T. Kuusalo [7], H. Tanaka [12])

(i) If R is parabolic, then $\rho_R = 0$.

(ii) If R is hyperbolic, then ρ_R is a distance on R compatible with its topology.

Lemma 1. If D is a domain on a hyperbolic Riemann surface R, then $\rho_R(D) = \sup_{a,b\in D} \rho_R(a,b) \leqq M(\Gamma(D))$.

By the aid of Theorem 10.12 in [13] we can prove the following lemma.

Lemma 2. If D is a relatively compact domain on an open Riemann surface whose relative boundary ∂D consists of a finite number of non degenerate continua, then ρ_D is a complete distance.

2. Quasiregular mappings

Definition 3. (cf. [8], [10]) Let R, R' be two open Riemann surfaces and f be a continuous mapping from R into R'. We say that f is (K-)q quasiregular mapping $(1 \leqq K < \infty)$ if either f is a constant mapping or it is not a constant mapping such that f is an open, discrete mapping and $K \cdot M(\Gamma) \geqq M(f\Gamma)$ for all path families Γ in R, where $f\Gamma = \{f \circ \gamma;\ \gamma \in \Gamma\}$.

Theorem 2. Let R, R' be two open Riemann surfaces and f be a K-quasiregular mapping from R into R'. Then

$$K \cdot \rho(a,b;R) \geqq \rho(f(a),f(b);R')$$

for all $a,b \in R$.

Proof. We may assume that f is not a constant mapping. Let F be any continuum in $C(a,b;R)$. Since f is continuous, we have $f(F) \in C(f(a),f(b);R')$ and hence $M(\Gamma(f(F))) \geqq \rho(f(a),f(b);R')$. By the quasiregularity of f, we obtain that $K \cdot M(\Gamma(F)) \geqq M(f\Gamma(F))$. On the other hand it follows from S. Rickman's result [11] that $\Gamma(f(F)) > f\Gamma(F)$. Thus we have $M(f\Gamma(F)) \geqq M(\Gamma(f(F)))$ and $K \cdot M(\Gamma(F)) \geqq \rho(f(a),f(b);R')$. Since F is arbitrary, we complete the proof.

Corollary. ([12]) If $f : R \to R'$ is an onto K-quasiconformal mapping, then $(1/K) \cdot \rho(a,b;R) \leqq \rho(f(a),f(b);R') \leqq K \cdot \rho(a,b;R)$ $(a,b \in R)$.

3. Boundary behaviors of quasiregular mappings

Let R be an open Riemann surface. We refer to [2] for the definition of the Kerékjártó-Stoïlow's compactification R^*_{KS} of R. A domain G on R is said to be an end if it is not relatively compact and the relative boundary of G in R is not empty and consists of a finite number of closed analytic Jordan curves. We set $\beta(G) = \overline{G}^{KS} - (G \cup \partial G)$ where the closure $\overline{G}^{KS}$ of G is taken in R^*_{KS}. Then $G \cup \beta(G)$ is an open neighborhood of $\beta(G)$. For each $e \in \Delta_{KS} = R^*_{KS} - R$ there exists a sequence $\{G_n\}_{n=1}^{\infty}$ of ends on R such that $G_n \supset G_{n+1} \cup \partial G_{n+1}$ $(n = 1,2,\ldots)$ and $\bigcap_{n=1}^{\infty} (G_n \cup \beta(G_n)) = \{e\}$. We say that such a sequence $\{G_n\}_{n=1}^{\infty}$ is a <u>determining sequence</u> of e.

<u>Definition 3.</u> An end G on R is said to be parabolic if $M(\Gamma(\partial G;G)) = 0$ where $\Gamma(\partial G;G)$ is the family of all paths on $G \cup \partial G$ starting at a point of ∂G and clustering nowhere in R.

<u>Lemma 3.</u> If G is a parabolic end on R, then ρ_G can be extended to a complete distance on $G \cup \beta(G)$ compatible with its topology.

<u>Proof.</u> By the definition of parabolic end we see that for each $e \in \beta(G)$ there exists a connected neighborhood $U(e)$ of e in $G \cup \beta(G)$ such that $\rho_G(U(e) \cap G)$ is arbitrary small (cf. Lemma 1). Since $\beta(G)$ is compact, we can find a neighborhood D of $\beta(G)$ in $G \cup \beta(G)$ such that $D \cap G$ is totally bounded with respect to ρ_G. This completes the proof.

Let G be an end on R and f be a quasiregular mapping of G into a complex plane $\mathbb{C} = \{|w| < \infty\}$. For each $e \in \beta(G)$ we denote by $C_G(f,e)$ the full cluster set of f at e. If $\{G_n\}_{n=1}^{\infty}$ is a determining sequence of e, then $C_G(f,e) = \bigcap_{n=1}^{\infty} \overline{f(G_n)}$ where the closure $\overline{f(G_n)}$ of $f(G_n)$ is taken in the extended complex plane $\hat{\mathbb{C}} = \{|w| \leq \infty\}$.

<u>Theorem 3.</u> If G is a parabolic end on an open Riemann surface and f is a quasiregular mapping of G into $\mathbb{C}$, then either $C_G(f,e)$ is a single point or total, i.e., $C_G(f,e) = \hat{\mathbb{C}}$ for each $e \in \beta(G)$.

<u>Proof.</u> Suppose $C_G(f,e)$ is not total. Then by the aid of a suitable conformal mapping of $\hat{\mathbb{C}}$ onto itself, we may assume that $C_G(f,e) \subset \{|w| < 1/2\}$. Let $\{G_n\}_{n=1}^{\infty}$ be a determing sequence of e such that $G_1 \subset G$. Then $f(G_n) \subset U = \{|w| < 1\}$ for sufficiently large n. Let $\{z_k\}_{k=1}^{\infty}$ be any sequence in G tending to e. Then it follows from Theorem 2 that $K \cdot \rho(z_k, z_j; G_n) \geq \rho(f(z_k), f(z_j); U)$ for sufficiently large k and j. Since each G_n is a parabolic end, it follows from Lemma 3 that $\{z_k\}_{k=1}^{\infty}$ is a ρ_{G_n}-Cauchy sequence and hence is a ρ_U-Cauchy sequence. Since ρ_U is a complete distance by Lemma 2, $\{f(z_k)\}_{k=1}^{\infty}$ is a convergent sequence in $\mathbb{C}$. Thus $C_G(f,e)$ is a single point and we complete the proof.

As a corollary of Theorem 3 we obtain the following theorem.

A Theorem of M. Heins [6]. If G is a parabolic end on an open Riemann surface and f is a bounded analytic function on G, then $f(z)$ has a limit as $z \to e$ for each $e \in \beta(G)$.

Remark. In the case of n-dimensional Euclidean domain the corresponding result to Theorem 3 has been proved by O. Martio, S. Rickman and J. Väisälä [9].

References

[1] Ahlfors, L. V., Sario, L.: Riemann surfaces. Princeton Univ. Press, Princeton, N. J. (1960).

[2] Constantinescu, C., Cornea, A.: Ideale Ränder Riemannscher Flächen. Springer-Verlag, Berlin - Göttingen - Heidelberg, (1963).

[3] Gál, I. S.: Conformally invariant metrics on Riemann surface. Proc. nat. Acad. Sci. USA 45 (1959), 1629 - 1633.

[4] Gál, I. S.: Conformally invariant metrics and uniform structures I, II. Nederl. Akad. Wet., Proc., Ser. A 63 (1960), 218-231,232-244.

[5] Gol'dstein, V. M., Vodop'yanov, S. K.: Metric completion of a domain by using a conformal capacity invariant under quasi-conformal mappings (Russian). Doklady Akad. Nauk. SSSR 238 (1978), 1040 - 1042.

[6] Heins, M.: Riemann surfaces of infinite genus. Ann. of Math., II. Ser. 55 (1952), 296 - 317.

[7] Kuusalo, T.: Generalized conformal capacity and quasiconformal metrics, in "Proceedings of the Romanian-Finnish Seminar on Teichmüller spaces and quasiconformal mappings, Braşov 1969". Publishing House of the Akademy of the Socialist Republic of Romania, Bucharest (1971), 193 - 202.

[8] Martio, O., Rickman, S., Väisälä, J.: Definitions for quasiregular mappings. Ann. Acad. Sci. Fenn., Ser. A I 448 (1969).

[9] Martio, O., Rickman, S., Väisälä, J.: Distortion and singularity of quasiregular mappings. Ann. Acad. Sci. Fenn., Ser. A I 465 (1970).

[10] Poleckii, E. A.: The method of moduli for non homeomorphic quasiconformal mappings (Russian). Mat. Sbornik, n. Ser. 83 (125) (1970), 261 - 272.

[11] Rickman, S.: Path lifting for open discrete mappings. Duke math. J. 40 (1973), 187 - 191.

[12] Tanaka, H.: Metrics induced by capacities and boundary behaviors of quasiconformal mappings on open Riemann surfaces. Hokkaido math. J. 5 (1976), 145 - 154.

[13] Väisälä, J.: Lectures on n-dimensional quasiconformal mappings. Lecture Notes in Mathematics 229, Springer-Verlag, Berlin - Heidelberg - New York (1971).

Department of Mathematics
Hokkaido University
Sapporo, 060, Japan

ON THE MAXIMAL OUTER FUNCTION OF A SEMI-SPECTRAL MEASURE

Ilie Valuşescu

Let $\mathbf{T}$ be the unit circle and $\mathbf{D}$ be the unit disc in the complex plane C. If $\mathcal{H}$ is a Hilbert space and $\mathcal{L}(\mathcal{H})$ the C^*-algebra of all linear bounded operators on $\mathcal{H}$, then an $\mathcal{L}(\mathcal{H})$-valued semi-spectral measure on $\mathbf{T}$ is a map F from $\mathcal{B}(\mathbf{T})$-the family of Borel subsets σ of $\mathbf{T}$ into $\mathcal{L}(\mathcal{H})$ such that for any $h \in \mathcal{H}$ the map $\sigma \to (F(\sigma)h,h)$ is a positive Borel measure on $\mathbf{T}$. A semi-spectral measure E is spectral if $E(\mathbf{T}) = I_{\mathcal{H}}$ and for any $\sigma_1, \sigma_2 \in \mathcal{B}(\mathbf{T})$, $E(\sigma_1 \cap \sigma_2) = E(\sigma_1)E(\sigma_2)$.

If $\mathcal{K}$ is a Hilbert space, $V : \mathcal{H} \to \mathcal{K}$ a bounded operator and E an $\mathcal{L}(\mathcal{K})$-valued spectral measure on $\mathbf{T}$, then by

$$F(\sigma) = V^*E(\sigma)V \ , \quad \sigma \in \mathcal{B}(\mathbf{T}) \ , \tag{1}$$

an $\mathcal{L}(\mathcal{H})$-valued semi-spectral measure F on $\mathbf{T}$ is obtained. Conversely, the Naimark dilation theorem assures that for any semi-spectral measure F there exists such a triplet $[K,V,E]$, which verifies (1), the so called spectral dilation of the semi-spectral measure F. The spectral dilatation $[K,V,E]$ is minimal if

$$\mathcal{K} = \bigvee_{\sigma \in \mathcal{B}(\mathbf{T})} E(\sigma)V\mathcal{H}. \tag{2}$$

For a Hilbert space $\mathcal{F}$, the corresponding semi-spectral measure of the multiplication by e^{it} in $L^2(\mathcal{F})$ is denoted by $E_{\mathcal{F}}^{\times}$.

An $\mathcal{L}(\mathcal{H})$-valued semi-spectral measure F is called of analytic type if it has a spectral dilation of the form $[L^2(\mathcal{F}),V,E_{\mathcal{F}}^{\times}]$ and $V\mathcal{H} \subset L_+^2(\mathcal{F})$. The name arises from the fact that there exists an operator valued analytic function $\{\mathcal{H},\mathcal{F},\Theta(\lambda)\}$ (see [4], [7]) such that for any $h \in \mathcal{H}$ we have

$$\Theta(\lambda)h = (Vh)(\lambda) \ , \quad \lambda \in \mathbf{D} \ . \tag{3}$$

This attached analytic function is an L^2-bounded one, i.e, there exists a positive constant M such that

$$\sup_{0 \leq r<1} \frac{1}{2\pi} \int_0^{2\pi} \|\Theta(re^{it})h\|^2 \, dt \leq M^2 \|h\|^2, \quad h \in \mathcal{H}, \tag{4}$$

or equivalently

$$\sum_{n=0}^{\infty} \|\Theta_n h\|^2 \leq M^2 \|h\|^2, \quad h \in \mathcal{H}, \tag{5}$$

where Θ_n are the coefficients of $\Theta(\lambda)$.

Conversely, to any L^2-bounded analytic function $\{\mathcal{H},\mathcal{F},\Theta(\lambda)\}$ corresponds an $\mathcal{L}(\mathcal{H})$-valued semi-spectral measure F_Θ of analytic type, with a dilation as $[L^2(\mathcal{F}),V_\Theta,E^{\times}]$, such that (3) is verified.

An L^2-bounded analytic function $\{\mathcal{H},\mathcal{F},\Theta(\lambda)\}$ is outer if

$$\bigvee_{n=0}^{\infty} e^{int} V_\Theta \mathcal{H} = L_+^2(\mathcal{F}) \ . \tag{6}$$

Let F_1, F_2 be $\mathcal{L}(\mathcal{H})$-valued semi-spectral measures on $\mathbf{T}$. One say that $F_1 \leq F_2$ if $F = F_2 - F_1$ is an $\mathcal{L}(\mathcal{H})$-valued semi-spectral measure on $\mathbf{T}$. In [4] the following theorem was proved

Theorem 1. Let F be an $\mathcal{L}(\mathcal{H})$-valued semi-spectral measure on $\mathbf{T}$ and $[\mathcal{K},V,E]$ be its minimal spectral dilation. There exists an L^2-bounded outer function $\{\mathcal{H},\mathcal{F}_1,\Theta_1(\lambda)\}$ with the following properties:

(i) $F_{\Theta_1} \leq F$

(ii) For every other L^2-bounded analytic function $\{\mathcal{H},\mathcal{F},\Theta(\lambda)\}$ for which $F_\Theta \leq F$ we have also $F_\Theta \leq F_{\Theta_1}$. The properties (i) and (ii) determine the outer function $\Theta_1(\lambda)$ up to a constant unitary factor from the left. In order that equality holds in (i) it is necessary and sufficient that the condition

$$\bigcap_{n=0}^{\infty} U^n \mathcal{K}_+ = \{0\} \tag{7}$$

be satisfied, where $\mathcal{K}_+ = \bigvee_{n=0}^{\infty} U^n V\mathcal{H}$ and U is the unitary operator on $\mathcal{K}$ corresponding to the spectral measure E.

Therefore, for any $\mathcal{L}(\mathcal{H})$-valued semi-spectral measure F on $\mathbf{T}$ there exists a unique outer L^2-bounded analytic function $\{\mathcal{H},\mathcal{F},\Theta_1(\lambda)\}$ whose the corresponding semi-spectral measure F_{Θ_1} is maximal among the $\mathcal{L}(\mathcal{H})$-valued analytic type semi-spectral measures dominated by F. This unique outer L^2-bounded function is called the maximal outer function of the semi-spectral measure F. The maximal outer function obtained by the above factorization theorem, plays an important role, especially in prediction theory. In such a way a characterization of the prediction-error operator is obtained, an estimation of the predictible part of a process, and in the filtering theory, too. For more details see [4] and [5]. In what follows some properties will be analysed concerning the contraction case.

Let T be a contraction on the Hilbert space $\mathcal{H}$ and $\mathcal{K} \supset \mathcal{H}$ be the space of its unitary dilation U. If E is the $\mathcal{L}(\mathcal{K})$-valued spectral measure corresponding to U, then

$$F_T(\sigma) = P_{\mathcal{H}} E(\sigma)|\mathcal{H}, \quad \sigma \in \mathcal{B}(\mathbf{T}), \tag{8}$$

defines an $\mathcal{L}(\mathcal{H})$-valued semi-spectral measure. In general, the converse is not valid. A characterization of the semi-spectral measures which correspond to contractions is given in

Proposition 2. The $\mathcal{L}(\mathcal{H})$-valued semi-spectral measure F, with the minimal spectral dilation $[\mathcal{K},V,E]$, corresponds to a contraction T on $\mathcal{H}$ if and only if $V^*V = I_{\mathcal{H}}$ and

$$(V^*UV)^n = V^*U^nV, \tag{9}$$

where U is the unitary operator corresponding to the spectral measure E.

Now, let us find the maximal function of F_T. If we put

$$\mathcal{K}_+ = \bigvee_{n=0}^{\infty} U^n \mathcal{H}, \tag{10}$$

then the restriction U_+ of U on $\mathcal{K}_+$ is an isometry. Taking the Wold decomposition of U_+ on $\mathcal{K}_+$ it follows that

$$\mathcal{K}_+ = M_+(\mathcal{L}_*) \oplus \mathcal{R}, \tag{11}$$

where $\mathcal{L}_* = \mathcal{K}_+ \ominus U_+\mathcal{K}_+$, $M_+(\mathcal{L}_*) = \bigoplus_{n=0}^{\infty} U_+^n \mathcal{L}_*$ and $\mathcal{R} = \bigcap_{n=0}^{\infty} U_+^n \mathcal{K}_+$. Let $P^{\mathcal{L}_*}$ be the orthogonal projection of $\mathcal{K}_+$ onto $M_+(\mathcal{L}_*)$, $\Phi^{\mathcal{L}_*}$ be the Fourier representation of $M_+(\mathcal{L}_*)$ onto $L_+^2(\mathcal{L}_*)$ and V_1 be the bounded linear operator from $\mathcal{H}$ into $L_+^2(\mathcal{L}_*)$ defined by

$$V_1 = \Phi^{\mathcal{L}_*} P^{\mathcal{L}_*}|\mathcal{H} \tag{12}$$

Then the $\mathcal{L}(\mathcal{H})$-valued semi-spectral measure F_1 defined by

$$F_1(\sigma) = V_1^* \overset{\times}{E}_{\mathcal{L}_*}(\sigma) V_1, \quad \sigma \in \mathcal{B}(\mathbf{T}), \tag{13}$$

is of analytic type with the spectral dilation $[L^2(\mathcal{L}_*), V_1, \overset{\times}{E}_{\mathcal{L}_*}]$. Therefore there exists an L^2-bounded analytic function $\{\mathcal{H},\mathcal{L}_*,\Theta_1(\lambda)\}$ such that $V_{\Theta_1} = V_1$ and $F_{\Theta_1} = F_1$. This function is an outer one. Indeed, $\bigvee_{n=0}^{\infty} e^{int} V_{\Theta_1}\mathcal{H} = \bigvee_{n=0}^{\infty} e^{int} \Phi^{\mathcal{L}_*} P^{\mathcal{L}_*}\mathcal{H} = \Phi^{\mathcal{L}_*} P^{\mathcal{L}_*} \mathcal{K}_+ = L_+^2(\mathcal{L}_*)$.

Let us now verify that $\{\mathcal{H},\mathcal{L}_*,\Theta_1(\lambda)\}$ is the maximal function, i.e. $F_{\Theta_1} \leq F_T$ and for any $\{\mathcal{H},\mathcal{F},\Theta(\lambda)\}$ with $F_\Theta \leq F_T$ we have $F_\Theta \leq F_{\Theta_1}$. For analytic polynomial p one has

$$\int_0^{2\pi} |p|^2 d(F_{\Theta_1}h,h)_{\mathcal{H}} = \|pV_{\Theta_1}h\|^2_{L^2_+(\mathcal{L}_*)} = \|p\Phi^{\mathcal{L}_*}P^{\mathcal{L}_*}h\|^2_{L^2(\mathcal{L}_*)}$$

$$= \|p(U)P^{\mathcal{L}_*}h\|^2_{\mathcal{K}} = \|P^{\mathcal{L}_*}p(U)h\|^2_{\mathcal{K}} \leq \|p(U)h\|^2 = \int_0^{2\pi} |p|^2 d(F_T h,h) .$$

Thus $F_{\Theta_1} \leq F_T$. The equality holds iff $\Phi^{\mathcal{L}_*}P^{\mathcal{L}_*}|\mathcal{H} = P^{\mathcal{L}_*}|\mathcal{H}$ i.e. iff $\mathcal{R} = \{0\}$, or equivalently if T is of the class $C_{\cdot 0}$.

If $\{\mathcal{H},\mathcal{F},\Theta(\lambda)\}$ is another L^2-bounded analytic function such that $F_\Theta \leq F$, then $X:\mathcal{K}_+ \to L^2_+(\mathcal{F})$ defined by

$$X(\sum_{k=0}^{n} U^n h_k) = \sum_{k=0}^{\infty} e^{ikt} V_\Theta h_k \tag{14}$$

is a contraction. Indeed, we have

$$\|X(\sum_{k=0}^{n} U^k h_k)\|^2 = \|\sum_{k=0}^{\infty} e^{ikt} V_\Theta h_k\|^2 = \sum_{k,j=0}^{n} \int_0^{2\pi} e^{i(k-j)t} d(F_\Theta(t)h_k,h_j)$$

$$\leq \sum_{k,j=0}^{n} \int_0^{2\pi} e^{i(k-j)t} d(F(t)h_k,h_j) = \|\sum_{k=0}^{n} U^k h_k\|^2 .$$

By the fact that $XU = e^{it}X$ it follows that

$$X\mathcal{R} = X \bigcap_{n=0}^{\infty} U^n \mathcal{K}_+ = \bigcap_{n=0}^{\infty} e^{int} X\mathcal{K}_+ \subset \bigcap_{n=0}^{\infty} e^{int} L^2_+(\mathcal{F}) = \{0\}.$$

Therefore $XP^{\mathcal{L}_*} = X$, and for any analytic polynomial p we have

$$\int_0^{2\pi} |p|^2 d(F_\Theta h,h) = \|pV_\Theta h\|^2 = \|Xp(U)h\|^2 = \|XP^{\mathcal{L}_*}p(U)h\|^2$$

$$\leq \|P^{\mathcal{L}_*}p(U)h\|^2 = \|\Phi^{\mathcal{L}_*}P^{\mathcal{L}_*}p(U)h\|^2 = \|p(e^{it})\Phi^{\mathcal{L}_*}P^{\mathcal{L}_*}h\|^2$$

$$= \|pV_{\Theta_1} h\|^2 = \int_0^{2\pi} |p|^2 d(F_{\Theta_1} h,h)$$

which implies that $F_\Theta \leq F_{\Theta_1}$.

Hence $\{\mathcal{H},\mathcal{L}_*,\Theta_1(\lambda)\}$ is the maximal function of F_T or, the so called maximal function of the contraction T. In [8] an explicit form of the maximal function of T was found, namely $\{\mathcal{H},\mathcal{L}_*,\Theta_1(\lambda)\}$ coincides with $\{\mathcal{H},\mathcal{D}_{T*},\Theta(\lambda)\}$, where

$$\Theta(\lambda) = D_{T*}(I - \lambda T^*)^{-1}, \ \lambda \in \mathbf{D} . \tag{15}$$

From the fact that the characteristic function (see [7]) $\Theta_T(\lambda) = [-T + \lambda D_{T*}(I - \lambda T^*)^{-1}D_T]|\mathcal{D}_T$ verifies $\Theta_T(\lambda)D_T = D_{T*}(I - \lambda T^*)^{-1}(\lambda I - T)$, it results the following relation between the maximal function and the characteristic function of the contraction T:

$$\Theta_T(\lambda)D_T = \Theta(\lambda)(\lambda I - T), \quad \lambda \in \mathbf{D} . \tag{16}$$

As was seen, the semi-spectral measure of a contaction T is of analytic type if and only if $T \in C_{\cdot 0}$. In this case the contaction T is uniquely determined (up to unitary equivalence) by its maximal function. Moreover, in the $C_{\cdot 0}$ case, the maximal function of T gives an explicit form (see [8]) of the imbedding of $\mathcal{H}$ into the space $\boldsymbol{H} = H^2(\mathcal{D}_{T^*})\ominus\Theta_T H^2(\mathcal{D}_T)$ of Sz.-Nagy-Foiaş functional model of T. Namely the image of an element $h \in \mathcal{H}$ in the space of the functional model $\boldsymbol{H}$ is the function $u \in H^2(\mathcal{D}_{T^*})$ defined by

$$u(\lambda) = \Theta(\lambda)h, \quad \lambda \in \mathbf{D} . \tag{17}$$

In general, the maximal function of a contraction is not bounded. If $\{\mathcal{H},\mathcal{L}_*,\Theta_1(\lambda)\}$ is bounded, then there exists a.e. $\Theta_1(e^{it})$ as non-tangential strong limit of $\Theta_1(\lambda)$ and

$$dF_{\Theta_1} = \frac{1}{2\pi}\,\Theta_1(e^{it})^*\Theta_1(e^{it})dt \qquad \text{a.e.} \tag{18}$$

Concerning the boundedness of $\Theta_1(\lambda)$ one has the following

Proposition 3. The L^2-bounded analytic function $\{\mathcal{H},\mathcal{F},\Theta(\lambda)\}$ is bounded if and only if the corresponding semi-spectral measure F_Θ is boundedly dominated by the Lebesgue measure dt on $\mathbf{T}$.

A complete proof can be found in [8].

Hence the maximal function of T is bounded iff F_{Θ_1} is boundedly dominated by the Lebesgue measure on $\mathbf{T}$. A class of contractions with bounded maximal function is that of spectral radius less than one. This follows from the fact (see [2]) that the contraction T with the spectral radius $\rho(T) < 1$ is characterized by bounded derivative F_T.

In the $C_{\cdot 0}$ case we have the following

Corollary 4. A contraction $T \in C_{\cdot 0}$ has the spectrum in the open unit disc if and only if its maximal function $\{\mathcal{H},\mathcal{L}_*,\Theta_1(\lambda)\}$ is bounded.

Let T_1, T_2 be contractions on the Hilbert space $\mathcal{H}$. It is said [6] that T_1 is Harnack dominated by T_2 if there exists a constant $c > 0$ such that for any polynomial p with $\operatorname{Re} p \geqq 0$ one has

$$\operatorname{Re} p(T_1) \leqq c \operatorname{Re} p(T_2) . \tag{19}$$

From the density of the real part of the polynomials in the set of continuous real functions on $\mathbf{T}$, it results that T_1 is Harnack dominated by T_2 if and only if F_{T_1} is boundedly dominated by F_{T_2}. In particular, a contraction T is Harnack dominated by the null contraction 0 if and only if F_T is boundedly dominated by the Lebesgue measure dt on $\mathbf{T}$. Therefore, by the Proposition 3 it results the following

Corollary 5. If T is a contraction Harnack dominated by the null contraction 0, then its maximal function $\{\mathcal{H},\mathcal{L}_*,\Theta_1(\lambda)\}$ is bounded.

If, moreover, $T \in C_{\cdot 0}$, then the converse of the above Corollary is also true.

In the particular case of the strict contraction $(\|T\| < 1)$ one obtains more about its maximal function. Firstly, by the invertibility of the defect operators, the maximal function appears as a factor of the characteristic function $\Theta_T(\lambda)$, i.e.

$$\Theta_T(\lambda) = \Theta(\lambda)(\lambda I - T)D_T^{-1} \ . \tag{20}$$

Moreover we have

Proposition 6. A contraction T on a Hilbert space $\mathcal{H}$ has the semi-spectral measure F_T of analytic type and the maximal function $\{\mathcal{H},\mathcal{L}_*,\Theta_1(\lambda)\}$ bounded, with bounded inverse, if and only if T is a strict contraction.

Proof. Two contractions T_1 and T_2 are Harnack equivalent [3] if they are mutually Harnack dominated. If T is a strict contraction then (see [1]) T is Harnack equivalent with the null operator 0. Hence there exists a positive constant c such that

$$c\, dt \leqq dF_T \leqq c^{-1} dt \ . \tag{21}$$

A strict contraction T is of the class C_{00} so that F_T is of analytic type. From (21) it results that the maximal function $\{\mathcal{H},\mathcal{L}_*,\Theta_1(\lambda)\}$ is bounded and the associated operator Θ_1 from $L^2_+(\mathcal{H})$ into $L^2_+(\mathcal{L}_*)$ defined by

$$(\Theta_1 u)(e^{it}) = \Theta_1(e^{it})u(e^{it}), \quad u \in L^2_+(\mathcal{H}) \ , \tag{22}$$

is bounded with bounded inverse. By the fact that Θ_1 intertwines the shift operators in $L^2_+(\mathcal{H})$ and $L^2_+(\mathcal{L}_*)$, using Lemma 3.2. from [7], it follows that $\Theta_1(\lambda)$ has a bounded inverse.

Conversely, if T has the semi-spectral measure F_T of analytic type and the maximal function bounded with bounded inverse, then for any trigonometric polynomial p and $h \in \mathcal{H}$ we have

$$\int_0^{2\pi} |p(e^{it})|^2 d(F_T(t)h,h) = \int_0^{2\pi} |p(e^{it})|^2 d(F_{\Theta_1}(t)h,h)$$

$$= \int_0^{2\pi} |p(e^{it})|^2 (\Theta_1(e^{it})^*\Theta_1(e^{it})h,h)dt = \int_0^{2\pi} \|\Theta_1(e^{it})p(e^{it})h\|^2 dt$$

$$\leq \|\Theta_1\|^2 \int_0^{2\pi} |p(e^{it})|^2 \|h\|^2 \, dt .$$

Also we have

$$\int_0^{2\pi} \|\Theta_1(e^{it})p(e^{it})h\|^2 dt \geq \|\Theta_1^{-1}\|^{-2} \int_0^{2\pi} \|\Theta_1(e^{it})^{-1}\Theta_1(e^{it})p(e^{it})h\|^2 dt$$

$$= \|\Theta_1^{-1}\|^{-2} \int_0^{2\pi} |p(e^{it})|^2 \|h\|^2 dt.$$

It follows that for any positive continuous function g on $\mathbf{T}$ one has

$$\|\Theta_1^{-1}\|^{-2} \int_0^{2\pi} g\,dt \leq \int_0^{2\pi} g\,dF_T(t) \leq \|\Theta_1\|^2 \int_0^{2\pi} g\,dt .$$

Thus, there exists a positive constant c such that

$$c\,dt \leq dF_T \leq c^{-1}dt$$

and it follows that T is a strict contraction.

References

[1] Foiaş, C.: On Harnack parts of contractions. Revue Roumaine Math. pur. appl. 19 (1974), 315 - 318.

[2] Schreiber, M.: Absolutely continuous operators. Duke math. J. 29 (1962), 175 - 192.

[3] Suciu, I.: Analytic relations between functional models for contractions. Acta Sci math. 34 (1973), 359 - 365.

[4] Suciu, I., Valuşescu, I.: Factorization of semi-spectral measures. Revue Roumaine Math. pur. appl. 21 (1976), 773 - 793.

[5] Suciu, I., Valuşescu, I.: Factorization theorems and prediction theory. Revue Roumaine Math. pur. appl. 23 (1978), 1393 - 1423.

[6] Suciu, N.: On Harnack ordering of contractions. To appear in Bull. Polon. Math.

[7] Sz.-Nagy, B., Foiaş, C.: Harmonic analysis of operators on Hilbert spaces. North-Holland Publishing Co., Amsterdam - London, American Elsevier Publishing Co., New York and Akadémiai Kiadó, Budapest (1970).

[8] Valuşescu, I.: The maximal function of a contraction. Preprint INCREST 38 (1978).

INCREST
Bd. Păcii 220
77538 Bucureşti
Romania

COMMUTING SYSTEMS OF OPERATORS AND INTEGRAL HOMOMORPHISMS

F.-H. Vasilescu

1. Introduction

Let X be a complex Banach space, $\mathcal{L}(X)$ the algebra of all linear bounded operators on X and $b = (b_1,\ldots,b_n) \subset \mathcal{L}(X)$ a commuting system. If $Sp(b,X)$ denotes the (joint) spectrum of the system b when acting on X (in the sense of J. L. Taylor [3]) then $Sp(b,X)$ is a compact set in $\mathbb{C}^n$ and for any open set $U \supset Sp(b,X)$ there exists an algebra homomorphism from $A(U)$ into $\mathcal{L}(X)$, denoted by $f \to f(b)$ $(f \in A(U))$ such that for any complex polynomial p the value of this homomorphism in p coincides with the direct computation of $p(b)$; $A(U)$ is here the algebra of all complex-valued functions which are analytic in U. In fact, the value $f(b)$ depends only on the germ associated with f on $Sp(b,X)$ (see [3], [4], for details).

Recently, we have shown that the multiplicativity of the map $f \to f(b)$ $(f \in A(U))$ is a consequence of a more general property of $A(U)$-module homomorphism of a certain map, defined on a space of exterior forms, with values in X [7]. The structure of $A(U)$-module of X is that induced by $f \to f(b)$ $(f \in A(U))$. As we have proved this result only for Hilbert spaces, we shall try to extend the assertion for a general Banach space. Unlike in [7], we restrict ourselves to the smooth forms, which is the most meaningful case. An extension of the results to the case of the forms having derivatives in the theory of distributions sense which are integrable is also possible. Let us give a brief description of the notations and the definitions which will be used in the sequel (see [4], [7], for some other details).

For any system of indeterminates $\sigma = (\sigma_1,\ldots,\sigma_n)$ and every complex linear space L we denote by $\Lambda[\sigma,L]$ $(\Lambda^p[\sigma,L],\ p = 0,1,\ldots,n)$ the space of all exterior forms (homogeneous exterior forms of degree p in $\sigma_1,\ldots,\sigma_n$) whose coefficients are elements of L. If A is an algebra of endomorphisms of L then the algebra $\Lambda[\sigma,A]$ acts naturally on the space $\Lambda[\sigma,L]$. In particular, if $\alpha = (\alpha_1,\ldots,\alpha_n)$ is a commuting system of endomorphisms of L then

$$\delta_\alpha \xi = (\alpha_1\sigma_1 + \cdots + \alpha_n\sigma_n) \wedge \xi \qquad (\xi \in \Lambda[\sigma,L]) \tag{1.1}$$

in an endomorphism of $\Lambda[\sigma,L]$ with the property $(\delta_\alpha)^2 = 0$.

When X is a Banach space, we give $\Lambda[\sigma,X]$ a structure of Banach space, since $\Lambda[\sigma,X]$ may be identified with a direct sum of 2^n copies of X.

For any open set $U \subset \mathbb{C}^m$ and every Banach space X we denote by X_U the set of all indefinitely differentiable X-valued functions in U. However, for $X = \mathbb{C}$ we prefer the traditional notation $C^\infty(U)$.

Consider another system of indeterminates $\bar{\zeta} = (\bar{\zeta}_1,\dots,\bar{\zeta}_m)$. For any $\xi \in \Lambda[\bar{\zeta},X_U]$ we define

$$\bar{\partial}\xi(z) = \left(\frac{\partial}{\partial \bar{z}_1}\bar{\zeta}_1 + \dots + \frac{\partial}{\partial \bar{z}_m}\bar{\zeta}_m\right) \wedge \xi(z), \tag{1.2}$$

where $z = (z_1,\dots,z_m) \in U$. Of course, the operator (1.2) is the usual $\bar{\partial}$. We use the system $\bar{\zeta} = (\bar{\zeta}_1,\dots,\bar{\zeta}_m)$ instead of the common $d\bar{z} = (d\bar{z}_1,\dots,d\bar{z}_m)$ in order to stress the independence of the former on the points in U.

Any system $\alpha = (\alpha_1,\dots,\alpha_n) \subset A(U,\mathcal{L}(X))$ with the property $\alpha_j(z)\alpha_k(z) = \alpha_k(z)\alpha_j(z)$ for any $z \in U$ and $j,k = 1,\dots,n$ will be called a commuting system [4]. The set of all points $z \in U$ such that $\alpha(z) = (\alpha_1(z),\dots,\alpha_n(z))$ is singular as a system of linear operators on X [3] will be denoted by $\mathcal{S}_U(\alpha,X)$. The set $\mathcal{S}_U(\alpha,X)$ is relatively closed in U [3].

If $\alpha = (\alpha_1,\dots,\alpha_n)$ and $\beta = (\beta_1,\dots,\beta_m)$ are two systems of endomorphisms then (α,β) stands for the system $(\alpha_1,\dots,\alpha_n, \beta_1,\dots,\beta_m)$.

When $F : U \to \mathcal{L}(X)$ and $f : U \to X$ then we use occasionally the notation Ff for the function $z \to F(z)f(z)$ $(z \in U)$, therefore F acts as a linear operator on the spaces of X-valued functions on U.

2. Two calculation formulas

J. L. Taylor has pointed out in [4] two important calculation formulas for the integrals of Cauchy-Weil type. One of them is a Fubini type theorem and the other is a change of variables theorem. In this section we shall state and prove some slightly improvements of these results (see also [7]).

Consider a commuting system $\alpha = (\alpha_1,\dots,\alpha_n)$ in $A(U,\mathcal{L}(X))$, where X is a fixed Banach space and $U \subset \mathbb{C}^m$ an open set. The system α will be associated with the system of indeterminates $\sigma = (\sigma_1,\dots,\sigma_n)$. We consider also another system of indeterminates $\bar{\zeta} = (\bar{\zeta}_1,\dots,\bar{\zeta}_m)$ and the operator $\bar{\partial}$ given by (1.2). Then the operator $\delta_\alpha + \bar{\partial}$ acts in the space $\Lambda[(\sigma,\bar{\zeta}),X_U]$, which can be identified either with $\Lambda[\sigma,\Lambda[\zeta,X_U]]$ or with $\Lambda[\bar{\zeta},\Lambda[\sigma,X_U]]$, the operator δ_α being given by (1.1) for $L = \Lambda[\bar{\zeta},X_U]$.

Let us denote by $Z^p_\alpha[(\sigma,\bar{\zeta}),X_U]$ the set of those $\xi \in \Lambda^p[(\sigma,\bar{\zeta}),X_U]$

such that $(\delta_\alpha + \bar{\partial})\xi = 0$ in U. An important step in the construction of the Cauchy-Weil integral is the following result, essentially proved in [4].

2.1. Theorem. Consider an open set $U \subset \mathbb{C}^m$ and a commuting system $\alpha = (\alpha_1,\ldots,\alpha_n) \subset \Lambda(U,\mathcal{L}(X))$. Then for any $\eta \in Z^p_\alpha[(\sigma,\bar{\xi}),X_V]$ there exists a solution $\xi \in \Lambda^{p-1}[(\sigma,\bar{\zeta}),X_V]$ of the equation $(\delta_\alpha + \bar{\partial})\xi = \eta$ in $V \subset U \setminus \mathcal{S}_U(\alpha,X)$, V open, for any $p = 0,1,\ldots,n$, where $\Lambda^{-1}[(\sigma,\bar{\zeta}),X_V] = \{0\}$. Moreover, the support of ξ may be chosen in any neighbourhood of the support of η in V.

Outline of the proof. By a theorem of J. L. Taylor [4], the sequence

$$0 \to \Lambda^o[\sigma,X_V] \overset{\delta_\alpha}{\to} \Lambda^1[\sigma,X_V] \overset{\delta_\alpha}{\to} \cdots \overset{\delta_\alpha}{\to} \Lambda^n[\sigma,X_V] \to 0 \qquad (2.1)$$

is exact, with the remark that the space $\mathcal{B}(U,X)$ considered in [4] is actually X_U, as shown in [5].

Let us write $\eta = \eta_o + \eta_1 + \cdots + \eta_p$, where η_j is of degree j in $\bar{\zeta}_1,\ldots,\bar{\zeta}_m$. If $\xi = \xi_o + \xi_1 + \cdots + \xi_{p-1}$, with ξ_j of degree j in $\bar{\zeta}_1,\ldots,\bar{\zeta}_m$ then we have the system of equations

$$\begin{cases} \delta_\alpha \xi_o = \eta_o \\ \delta_\alpha \xi_j + \bar{\partial}\xi_{j-1} = \eta_j \quad (j = 1,\ldots,p-1) \\ \bar{\partial}\xi_{p-1} = \eta_p, \end{cases} \qquad (2.2)$$

with the conditions

$$\begin{cases} \delta_\alpha \eta_o = 0 \\ \delta_\alpha \eta_j + \bar{\partial}\eta_{j-1} = 0 \quad (j = 1,\ldots,p) \\ \bar{\partial}\eta_p = 0. \end{cases} \qquad (2.3)$$

The system (2.2) with the conditions (2.3) can be solved step by step, on account of the exactness of the sequence (2.1). From this iterative construction it follows easily that the support of the solution ξ can be chosen in any neighbourhood of the support of η.

From now on we shall denote by $B_\alpha\eta \in \Lambda^{p-1}[(\sigma,\bar{\zeta}),X_V]$ a fixed solution of the equation $(\delta_\alpha + \bar{\partial})\xi = \eta$ in V, where $\eta \in Z^p_\alpha[(\sigma,\bar{\zeta}),X_V]$ and V is an open subset of $U \setminus \mathcal{S}_U(\alpha,X)$. Let us denote also by P_σ the map on $\Lambda[(\sigma,\bar{\zeta}),X_U]$ annihilating any monomial containing at least one $\sigma_1,\ldots,\sigma_n$ and letting invariant the other terms.

We define now a concept of Cauchy-Weil integral with an additional term (see [4] for the original concept). Let us fix a function $\varphi \in C^\infty(U)$, $\varphi = 0$ in a neighbourhood of $\mathcal{S}_U(\alpha,X)$, which is supposed to be compact, such that the support of $1 - \varphi$ is compact. Then for any $\eta \in Z^m_\alpha[(\sigma,\bar{\zeta}),X_U]$ we define

$$\mu_\alpha(\eta) = \int_U ((\bar{\partial}\varphi P_\sigma B_\alpha\eta)(z) - P_\sigma\eta(z)) \wedge dz, \tag{2.4}$$

with $dz = dz_1 \wedge \cdots \wedge dz_m$; here we integrate the part of degree m in $\bar{\zeta}_1,\ldots,\bar{\zeta}_m$ of the integrand, by substituting $\bar{\zeta}_1,\ldots,\bar{\zeta}_m$ with $d\bar{z}_1,\ldots,d\bar{z}_m$, respectively.

2.2. Lemma. The integral (2.4) is an element of X which does not depend on the particular choice of $B_\alpha\eta$ and φ.

Proof. Since the form $(\delta_\alpha + \bar{\partial})\varphi B_\alpha\eta - \eta$ has compact support, the integrand of (2.4) has also compact support, hence the integral exists.

Take now two solutions ξ_1 and ξ_2 of the equation $(\delta_\alpha + \bar{\partial})\xi = \eta$ in $U \setminus \mathcal{S}_U(\alpha,X)$ and φ_1, φ_2 in $C^\infty(U)$ with the properties from the definition (2.4). Then the forms $\theta_j = (\delta_\alpha + \bar{\partial})\varphi_j\xi_j - \eta$ $(j = 1,2)$, extended with zero, have compact support and the difference $\theta_1 - \theta_2$ is null in a neighbourhood of $\mathcal{S}_U(\alpha,X)$. By Theorem 2.1 we can write $\theta_1 - \theta_2 = (\delta_\alpha + \bar{\partial})\lambda$, where the support of λ is compact. Since $P_\sigma\bar{\partial} = \bar{\partial}P_\sigma$, we have $P_\sigma\theta_1 - P_\sigma\theta_2 = \partial P_\sigma\lambda$, therefore, by the Stokes formula

$$\int (P_\sigma\theta_1 - P_\sigma\theta_2)(z) \wedge dz = 0,$$

i.e., the desired independence.

For our purpose we need a variant of (2.4) with parameters. We consider from now on an open set $U \subset \mathbb{C}^{n+m}$. The points in U will be written as pairs (z,w) with $z = (z_1,\ldots,z_n) \in \mathbb{C}^n$ and $w = (w_1,\ldots,w_m) \in \mathbb{C}^m$. The operator $\bar{\partial}$ will be defined in U with the system of indeterminates $(\bar{\zeta},\bar{\omega}) = (\bar{\zeta}_1,\ldots,\bar{\zeta}_n, \bar{\omega}_1,\ldots,\bar{\omega}_m)$, while $\bar{\partial}_z$ and $\bar{\partial}_w$ will be the operators $\bar{\partial}$ in $\mathbb{C}^n$ and $\mathbb{C}^m$, corresponding to $\bar{\zeta}$ and $\bar{\omega}$, respectively. We shall use another system $\beta = (\beta_1,\ldots,\beta_m)$ in $A(U,\mathcal{L}(X))$, which will be associated with the system of indeterminates $\tau = (\tau_1,\ldots,\tau_m)$ when defining the operator δ_β.

Assume now that $\alpha = (\alpha_1,\ldots,\alpha_n) \subset A(U,\mathcal{L}(X))$ is a commuting system and that $\mathcal{S}_U(\alpha,X)$ is $\mathbb{C}^n$-compact in U, i.e., for every compact $K \subset W$ the set $(\mathbb{C}^n \times K) \cap \mathcal{S}_U(\alpha,X)$ is compact, where W is the projection of U on the last m coordinates. Taking $\varphi \in C^\infty(U)$ which is zero in a neighbourhood of $\mathcal{S}_U(\alpha,X)$ such that the support of $1 - \varphi$ is $\mathbb{C}^n$-compact, we define

$$\mu_\alpha(\eta)(w) = \int_{\mathbb{C}^n} ((\bar{\partial}\varphi P_\sigma B_\alpha\eta)(z,w) - P_\sigma\eta(z,w)) \wedge dz, \tag{2.5}$$

for any $\eta \in Z^n_\alpha[(\sigma,\bar{\zeta},\bar{\omega}),X_U]$.

2.3. Lemma. The integral (2.5) is an element of X for every $w \in W$, which does not depend on the particular choice of $B_\alpha\eta$ and φ. Moreover, $\mu_\alpha(\eta)$ is analytic on W.

Proof. The existence of the integral (2.5) follows from the fact that the integrand has a $\mathbb{C}^n$-compact support. Note also that $\eta_\alpha(\eta)(w) \in X$ since the degree of the integrand is n, hence we cannot have significant terms in $\bar{\omega}_1,\ldots,\bar{\omega}_m$.

The fact that (2.5) does not depend on $B_\alpha\eta$ and φ follows as in the proof of Lemma 2.2. Indeed, with similar notations, we can find λ whose support is $\mathbb{C}^n$-compact and whose degree in $\bar{\zeta}_1,\ldots,\bar{\zeta}_n$ is at most $n-1$, according to Theorem 2.1. Then $\int(\bar{\partial}P_\sigma\lambda)(z,w)\wedge dz$ $= \int(\bar{\partial}_z P_\sigma\lambda)(z,w)\wedge dz = 0$, by the Stokes formula. Note also that we can write

$$\begin{aligned}\bar{\partial}_w\mu_\alpha(\eta)(w) &= \int_{\mathbb{C}^n} \bar{\partial}_w((\bar{\partial}\varphi P_\sigma B_\alpha\eta)(z,w) - P_\sigma\eta(z,w))\wedge dz\\ &= \int_{\mathbb{C}^n} \bar{\partial}((\bar{\partial}\varphi P_\sigma B_\alpha\eta)(z,w) - P_\sigma\eta(z,w))\wedge dz\\ &= -\int_{\mathbb{C}^n} P_\sigma\bar{\partial}\eta(z,w)\wedge dz = \int_{\mathbb{C}^n} P_\sigma\delta_\alpha\eta(z,w)\wedge dz = 0,\end{aligned}$$

since

$$\int_{\mathbb{C}^n} \bar{\partial}_z((\bar{\partial}\varphi P_\sigma B_\alpha\eta)(z,w) - P_\sigma\eta(z,w))\wedge dz = 0,$$

by the Stokes formula.

For a commuting system $\alpha = (\alpha_1,\ldots,\alpha_n) \subset A(U,\mathcal{L}(X))$ we denote by $(\alpha)^c$ the <u>commutant</u> of the set $\{\alpha_1(z,w),\ldots,\alpha_n(z,w);\ (z,w)\in U\}$ in $\mathcal{L}(X)$. If $\eta \in Z_\alpha^n[(\sigma,\bar{\zeta},\bar{\omega}),((\alpha)^c)_U]$ then we shall use sometimes the notation $\mu_\alpha(\eta)x$ for $\mu_\alpha(\eta x)$, where $x \in X$.

The announced Fubini type result is given by the following

2.4. Theorem. Let U be an open set in $\mathbb{C}^{n+m}$ and W its projection on the last m coordinates. Consider $\alpha = (\alpha_1,\ldots,\alpha_n) \subset A(U,\mathcal{L}(X))$ and $\beta = (\beta_1,\ldots,\beta_m) \subset A(W,\mathcal{L}(X))$ such that (α,β) is a commuting system. Assume that $\mathcal{S}_U(\alpha,X)$ is $\mathbb{C}^n$-compact in U and $\mathcal{S}_W(\beta,X)$ is compact. Then for any $\xi \in Z_\alpha^n[(\sigma,\bar{\zeta},\bar{\omega}),X_U]$ and $\eta \in Z_\beta^m[(\tau,\bar{\omega}),((\alpha,\beta)^c)_W]$ we have

$$\mu_{(\alpha,\beta)}(\eta\wedge\xi) = \mu_\beta(\eta\mu_\alpha(\xi)).$$

Proof. We shall use a method from [6] and [7]. Consider $\varphi \in C^\infty(U)$ with the properties from the definition (2.5) for $\mathcal{S}_U(\alpha,X)$. Analogously, take $\psi \in C^\infty(W)$ and $\chi \in C^\infty(U)$ as in the definition (2.4) for $\mathcal{S}_W(\beta,X)$

and $\mathscr{S}_U((\alpha,\beta),X)$, respectively. Note the inclusion

$$\mathscr{S}_U((\alpha,\beta),X) \subset \mathscr{S}_U(\alpha,X) \cap (\mathbb{C}^n \times \mathscr{S}_W(\beta,X)),$$

obtained from the projection property of the joint spectrum [3].

Let us define $\xi_o = (\delta_\alpha + \bar{\partial})\varphi B_\alpha \xi - \xi$, whose support is $\mathbb{C}^n$-compact, and $\eta_1 = \eta \wedge \xi_o$. Notice that

$$(\delta_\alpha + \delta_\beta + \bar{\partial})\eta_1 = ((\delta_\beta + \bar{\partial}_W)\eta) \wedge \xi_o + (-1)^m \eta \wedge ((\delta_\alpha + \bar{\partial})\xi_o) = 0,$$

therefore we may consider the forms

$$\eta_2 = -(\delta_\alpha + \delta_\beta + \bar{\partial})\psi B_{(\alpha,\beta)}\eta_1 + \eta_1,$$

$$\tilde{\eta}_2 = (\delta_\alpha + \delta_\beta + \bar{\partial})\chi B_{(\alpha,\beta)}(\eta \wedge \xi) - \eta \wedge \xi.$$

Since η_1 has compact support, η_2 has compact support too. Plainly, $\tilde{\eta}_2$ has also compact support. Moreover, both η_2 and $\tilde{\eta}_2$ are equal to $-\eta \wedge \xi$ in a neighbourhood of $\mathscr{S}_U((\alpha,\beta),X)$, therefore the support of $\eta_2 - \tilde{\eta}_2$ is compact and disjoint of $\mathscr{S}_U((\alpha,\beta),X)$. By Theorem 2.1 we have then in U

$$\eta_2 - \tilde{\eta}_2 = (\delta_\alpha + \delta_\beta + \bar{\partial})\theta_1, \tag{2.6}$$

where the support of θ_1 is compact.

Consider now

$$\eta_3 = -P_\sigma \eta_2 = (\delta_\beta + \bar{\partial})\psi P_\sigma B_{(\alpha,\beta)}\eta_1 - P_\sigma \eta_1$$

$$= (\delta_\beta + \bar{\partial})\psi P_\sigma B_{(\alpha,\beta)}\eta_1 - \eta \wedge (\bar{\partial}\varphi B_\alpha \xi - P_\sigma \xi).$$

Since we have $(\delta_\beta + \bar{\partial})P_\sigma \eta_1 = P_\sigma(\delta_\alpha + \delta_\beta + \bar{\partial})\eta_1 = 0$, we can consider the form

$$\tilde{\eta}_3 = (\delta_\beta + \bar{\partial})\psi B_\beta P_\sigma \eta_1 - P_\sigma \eta_1.$$

Both η_3 and $\tilde{\eta}_3$ have compact support and are equal to $-P_\sigma \eta_1$ in a neighbourhood of $\mathscr{S}_U(\beta,X) = \mathbb{C}^n \times \mathscr{S}_W(\beta,X)$. By applying Theorem 2.1 once more we obtain

$$\eta_3 - \tilde{\eta}_3 = (\delta_\beta + \bar{\partial})\theta_2, \tag{2.7}$$

where the support of θ_2 is compact.

From (2.6) and (2.7) we infer

$$P_{(\sigma,\tau)}\tilde{\eta}_2 = \bar{\partial}\chi P_{(\sigma,\tau)}B_{(\alpha,\beta)}(\eta \wedge \xi) - P_\tau \eta \wedge P_\sigma \xi = P_\tau \tilde{\eta}_3 + \bar{\partial}\theta_3$$

$$= \bar\partial\psi P_\tau B_\beta(\eta\wedge(\bar\partial\varphi P_\sigma B_\alpha\xi - P_\sigma\xi) - P_\tau\eta\wedge(\bar\partial\varphi P_\sigma B_\alpha\xi - P_\sigma\xi)) + \bar\partial\theta_3.$$

where the support of θ_3 is compact.

Let us remark that the difference

$$(\delta_\beta + \bar\partial_w)\psi\int B_\beta\eta\wedge(\bar\partial\varphi P_\sigma B_\alpha\xi - P_\sigma\xi)\wedge dz$$
$$- (\delta_\beta + \bar\partial_w)\psi B_\beta\eta\int(\bar\partial\varphi P_\sigma B_\alpha\xi - P_\sigma\xi)\wedge dz$$

is zero in a neighbourhood of $\mathscr{S}_W(\beta,X)$ and its support is compact, since

$$\int\bar\partial_z B_\beta\eta\wedge(\bar\partial\varphi P_\sigma B_\alpha\xi - P_\sigma\xi)\wedge dz = 0,$$

by the Stokes formula, the support of the integrand being $\mathbb{C}^n$-compact. From (2.8) and from Lemma 2.2 we obtain then

$$\mu_{(\alpha,\beta)}(\eta\wedge\xi) = \int(\bar\partial\chi P_{(\sigma,\tau)}B_{(\alpha,\beta)}(\eta\wedge\xi) - P_\tau\eta\wedge P_\sigma\xi)\wedge dz\wedge dw$$
$$= \int(\int\bar\partial\psi P_\tau B_\beta(\eta\wedge(\bar\partial\varphi P_\sigma B_\alpha\xi - P_\sigma\xi) - P_\tau\eta\wedge(\bar\partial\varphi P_\sigma B_\alpha\xi - P_\sigma\xi))\wedge dz)\wedge dw$$
$$= \int((\bar\partial_w\psi P_\tau B_\beta\eta\int(\bar\partial\varphi P_\sigma B_\alpha\xi - P_\sigma\xi)\wedge dz) - P_\tau\eta\int(\bar\partial\varphi P_\sigma B_\alpha\xi - P_\sigma\xi)\wedge dz)\wedge dw$$
$$= \mu_\beta(\eta\mu_\alpha(\xi)),$$

which finishes the proof of the theorem.

The next calculation rule for the integral (2.5) will be a change of variables theorem. Take a commuting system $\alpha = (\alpha_1,\dots,\alpha_n)$ in $A(U,\mathscr{L}(X))$ $(U\subset\mathbb{C}^{n+m})$ and consider also a matrix $u = (u_{jk})_{j,k=1}^n$ of commuting elements in $A(U,\mathscr{L}(X))$, which commute also with $\alpha_1,\dots,\alpha_n$. Then we can define the commuting system $\beta = (\beta_1,\dots,\beta_n)$, where $\beta_j = \sum_{k=1}^n u_{jk}\alpha_k$. We are looking for a relation between the integrals (2.5) corresponding to α and β. For, we define a map $\tilde u$ from U into $\mathscr{L}(\Lambda[\sigma,X])$ given by

$$\tilde u(z,w)(x\sigma_{j_1}\wedge\dots\wedge\sigma_{j_p}) = \sum_{k_1<\dots<k_p}\det(u_{k_hj_q})_{h,q}(z,w)x\sigma_{k_1}\wedge\dots\wedge\sigma_{k_p}$$

and $\tilde u(z,w)x = x$, for any $x\in X$ and $(z,w)\in U$ (here "det" stands for the determinant). Each map $\tilde u(z,w)$ preserves the exterior structure of $\Lambda[\sigma,X]$, therefore we can define naturally a map $\tilde u$ on $\Lambda[\sigma,\Lambda[(\bar\xi,\bar\omega),X_U]]$ preserving the exterior structure. Following [4], the map $\tilde u$ will be called the <u>special transformation</u> induced by the matrix u. Note that we have

$$(\delta_\beta + \bar\partial)\tilde u\xi = \tilde u(\delta_\alpha + \bar\partial)\xi, \tag{2.9}$$

for any $\xi \in \Lambda[(\sigma,\bar\zeta,\bar\omega),X_U]$.

2.5. Proposition. Let U be an open set in $\mathbb{C}^{n+m}$ and $\alpha = (\alpha_1,\dots,\alpha_n) \subset A(U,\mathcal{L}(X))$ a commuting system. Assume that $u = (u_{jk})_{j,k=1}^n$ is a matrix of commuting elements in $A(U,\mathcal{L}(X))$ which commute also with $\alpha_1,\dots,\alpha_n$. Denote by $\beta_j = \sum_{k=1}^n u_{jk}\alpha_k$, $\beta = (\beta_1,\dots,\beta_n)$ and let $\tilde u$ be the special transformation induced by the matrix u on $\Lambda[(\sigma,\bar\zeta,\bar\omega),X_U]$. If $\mathcal{S}_U(\alpha,X)$ and $\mathcal{S}_U(\beta,X)$ are both $\mathbb{C}^n$-compact in U then $\mu_\beta(\tilde u\xi) = \mu_\alpha(\xi)$, for any $\xi \in Z_\alpha^n[(\sigma,\bar\zeta,\bar\omega),X_U]$.

Proof. Take $\varphi \in C^\infty(U)$, $\varphi = 0$ in a neighbourhood of $\mathcal{S}_U(\alpha,X) \cup \mathcal{S}_U(\beta,X)$ φ as in the definition (2.5). If $\xi \in Z^n[(\sigma,\bar\zeta,\bar\omega),X_U]$ is arbitrary, by the relation (2.9) the forms $B_\beta\tilde u\xi$ and $\tilde u B_\alpha\xi$ are both solutions of the equation $(\delta_\beta + \bar\partial)\theta = \xi$, therefore, by Lemma 2.3, we can write

$$\mu_\beta(\tilde u\xi) = \int (\bar\partial\varphi P_\sigma B_\beta\tilde u\xi - P_\sigma\tilde u\xi)\wedge dz = \int (\bar\partial\varphi P_\sigma\tilde u B_\alpha\xi - P_\sigma\tilde u\xi)\wedge dz$$

$$= \int \tilde u(\bar\partial\varphi P_\sigma B_\alpha\xi - P_\sigma\xi)\wedge dz = \mu_\alpha(\xi),$$

since $P_\sigma\tilde u = \tilde u P_\sigma$ and $\tilde u$ is the identity on the forms not containing $\sigma_1,\dots,\sigma_n$.

3. Commuting systems of operators

The results of the previous section will be applied to commuting systems of operators $b = (b_1,\dots,b_n) \subset \mathcal{L}(X)$. Let us denote by $\alpha(z)$ the system $(z_1 - b_1,\dots,z_n - b_n)$, for every $z = (z_1,\dots,z_n) \in \mathbb{C}^n$. Then $\mathcal{S}_{\mathbb{C}^n}(\alpha,X) = Sp(b,X) = \mathcal{S}_U(\alpha,X)$, for each $U \supset Sp(b,X)$, where $U \subset \mathbb{C}^n$ is an open set. Let us consider a system of indeterminates $\sigma = (\sigma_1,\dots,\sigma_n)$ associated with b and a system of indeterminates $\bar\zeta = (\bar\zeta_1,\dots,\bar\zeta_n)$ associated with $\bar\partial$. Let us denote by $Z^n_{\alpha,c}[(\sigma,\bar\zeta),X_U]$ the subspace of $Z^n_\alpha[(\sigma,\bar\zeta),X_U]$ generated by the elements of the form ηx with $\eta \in Z^n_\alpha[(\sigma,\bar\zeta),((\alpha)^c)_U]$ and $x \in X$. Clearly $Z^n_{\alpha,c}[(\sigma,\bar\zeta),X_U]$ is an $A(U)$-module. We shall show that the restriction of the map μ_α given by (2.4) on $Z^n_{\alpha,c}[(\sigma,\bar\zeta),X_U]$ is an $A(U)$-module homomorphism, where the $A(U)$-module structure of X is induced by the map

$$\nu_\alpha(f)x = \mu_\alpha(fx\sigma_1\wedge\cdots\wedge\sigma_n), \tag{3.1}$$

where $f \in A(U)$ and $x \in X$ are arbitrary, The map (3.1) is a (continuous) homomorphism from the unital algebra $A(U)$ into $\mathcal{L}(X)$, as shown in [4], and we have $\nu_\alpha(f) = f(b)$, with the sense from the Introduction. In fact, the multiplicativity of the map (3.1) is a consequence of a more general result.

3.1. Theorem. Assume that $b = (b_1, \ldots, b_n) \subset \mathcal{L}(X)$ is a commuting system and define $\alpha(z) = (z_1 - b_1, \ldots, z_n - b_n)$, for every $z = (z_1, \ldots, z_n) \in \mathbb{C}^n$. If $U \supset Sp(b,X)$ is an arbitrary open set in $\mathbb{C}^n$ then for any $f \in A(U)$ and $\xi \in Z^n_{\alpha,c}[(\sigma,\bar{\zeta}),X_U]$ we have $\mu_\alpha(f\xi) = \nu_\alpha(f)\mu_\alpha(\xi)$.

Proof. Consider the commuting system $\beta(z,w) = (\alpha(z),\alpha(w))$, for $(z,w) \in U \times U$. The system $\alpha(z)$ will be associated $\sigma_1, \ldots, \sigma_n$ while the system $\alpha(w)$ with $\tau_1, \ldots, \tau_n$. It is sufficient to assume that $\xi = \eta x$, where $\eta \in Z^n_{\underline{\alpha}}[(\tau,\bar{\omega}),((\alpha)^c)_U]$ and $x \in X$. Since the values of $\nu_\alpha(f)$ $(f \in A(U))$ are in the double commutant of the system $(b_1, \ldots, b_n)$ in $\mathcal{L}(X)$ ([4], or an easy consequence of Lemma 2.2), then by Lemma 2.2 and Theorem 2.4 we can write

$$\nu_\alpha(f)\mu_\alpha(\eta x) = \mu_\alpha(\eta\nu_\alpha(f)x) = \mu_\alpha(\eta\mu_\alpha(fx\sigma_1 \wedge \cdots \wedge \sigma_n))$$

$$= \mu_\beta(\eta \wedge fx\sigma_1 \wedge \cdots \wedge \sigma_n).$$

Let us consider the matrix $u = (u_{jk})^{2n}_{j,k=1}$, where $u_{jj} = 1$ $(j = 1, \ldots, 2n)$, $u_{j,n+j} = -1$ $(j = 1, \ldots, n)$ and $u_{jk} = 0$ otherwise. Then the special transformation induced by the matrix u acts in the following way: The indeterminates σ_j $(j = 1, \ldots, n)$ are left invariant while the indeterminates τ_j are replaced with $\tau_j - \sigma_j$ $(j = 1, \ldots, n)$, in every exterior form. In particular, the form $\eta \wedge fx\sigma_1 \wedge \cdots \wedge \sigma_n$ is left invariant. Note also that the system $\beta(z,w)$ is transformed by the matrix u into the system $(\gamma(z,w),\alpha(w))$, where $\gamma(z,w) = z - w$. By Theorem 2.4 and Proposition 2.5 we have then

$$\mu_\beta(\eta \wedge fx\sigma_1 \wedge \cdots \wedge \sigma_n) = \mu_{(\gamma,\alpha)}(\eta \wedge fx\sigma_1 \wedge \cdots \wedge \sigma_n) = \mu_\alpha(\eta\nu_\gamma(f)x)$$

$$= \mu_\alpha(f\eta x) = \mu_\alpha(f\xi),$$

since $\mu_\gamma(f)(w) = f(w)$ [4], which completes the proof.

It may happen that Theorem 3.1 be true for every $\xi \in Z^n[(\sigma,\bar{\xi}),X_U]$. The present arguments seem not to be strong enough for such a result. In the one-dimensional case such an assertion, and actually a bit more, is true.

3.2. Proposition. Let U be an open set in $\mathbb{C}$ and $b \in \mathcal{L}(X)$ such that $Sp(b,X) \subset U$. Assume that for a fixed function $g \in X_U$ there is a locally integrable X-valued function h on U such that $(\partial g/\partial \bar{z})(z) = (z - b)h(z)$ almost everywhere in U. If we denote by ξ the form $g\sigma + h\bar{\zeta}$ and define

$$\tilde{\mu}_\alpha(\xi) = \frac{1}{2\pi i}\int_\Gamma (z - b)^{-1}f(z)dz + \frac{1}{2\pi i}\int_\Delta h(z)\, dz \wedge d\bar{z}, \tag{3.2}$$

where $\alpha(z) = z - b$ $(z \in U)$ and $\Delta \subset Sp(b,X)$ is an open set in U whose boundary Γ is a finite system of smooth curves, then the integral (3.2) does not depend on Δ and we have $\tilde{\mu}_\alpha(f\xi) = f(b)\tilde{\mu}_\alpha(\xi)$, for any $f \in A(U)$, where $f(b)$ is given by the Riesz-Dunford functional calculus.

Proof. $\xi = g\sigma + h\bar{\zeta}$ has the property $(\delta_\alpha + \bar{\partial})\xi = 0$, therefore the independence of the integral (3.2) on Δ can be obtained as in the proof of Lemma 2.2, by using also the Stokes formula.

Take now $\Delta_1 \supset \bar{\Delta} \supset \Delta$ such that the boundary Γ_1 of Δ_1 has similar properties with those of Γ. Then we can write

$$\frac{1}{2\pi i}\int_\Gamma (z-b)^{-1}f(z)g(z)dz = \left(\frac{1}{2\pi i}\right)^2\int_\Gamma (z-b)^{-1}\left(\int_{\Gamma_1}(w-z)^{-1}f(w)dw\right)g(z)dz$$

$$= \left(\frac{1}{2\pi i}\right)^2\int_{\Gamma_1} f(w)\left(\int_\Gamma (w-b)^{-1}((z-b)^{-1} + (w-z)^{-1})g(z)dz\right)dw)$$

$$= f(b)\left(\tilde{\mu}_\alpha(\xi) - \frac{1}{2\pi i}\int_\Delta h(z)\ dz \wedge d\bar{z}\right)$$

$$+ \left(\frac{1}{2\pi i}\right)^2\int_{\Gamma_1} f(w)(w-b)^{-1}\left(\int_\Gamma (w-z)^{-1}g(z)dz\right)dw.$$

On the other hand, as $w \notin \Delta$, we have by the Stokes formula,

$$\left(\frac{1}{2\pi i}\right)^2\int_{\Gamma_1} f(w)(w-b)^{-1}\left(\int_\Gamma (w-z)^{-1}g(z)dz\right)dw$$

$$= \frac{1}{2\pi i}\int_{\Gamma_1} f(w)(w-b)^{-1}\left(-\frac{1}{2\pi i}\int_\Delta (w-z)^{-1}(z-b)h(z)\ dz \wedge d\bar{z}\right)dw$$

$$= f(b)\frac{1}{2\pi i}\int_\Delta h(z)\ dz \wedge d\bar{z} - \left(\frac{1}{2\pi i}\right)^2\int_{\Gamma_1} f(w)\left(\int_\Delta (w-z)^{-1}h(z)\ dz \wedge d\bar{z}\right)dw$$

$$= f(b)\frac{1}{2\pi i}\int_\Delta h(z)\ dz \wedge d\bar{z} - \frac{1}{2\pi i}\int_\Delta f(z)h(z)\ dz \wedge d\bar{z},$$

and from this calculation we infer the equality (3.2).

Proposition 2.3 shows that in the one-dimensional case the formula (2.4) is strongly connected with a vector variant of the Cauchy-Pompeiu formula [2], [1].

References

[1] Hörmander, L.: An introduction to complex analysis in several variables. Van Nostrand, Princeton, N. J. (1966).

[2] Pompeiu, D.: Sur une classe de fonctions d'une variable complexe, Rend. Circ. mat. Palermo, II. Ser. 33 (1912), 108 - 113, 35 (1913), 277 - 281.

[3] Taylor, J. L.: A joint spectrum for several commuting operators.

J. functional Analysis 6 (1970), 172 - 191.

[4] Taylor, J. L.: The analytic functional calculus for several commuting operators. Acta math. 125 (1970), 1 - 38.

[5] Vasilescu, F.-H.: On a class of vector-valued functions (Romanian). Studii Cerc. mat. 28 (1976), 121 - 127.

[6] Vasilescu, F.-H.: A Martinelli type formula for the analytic functional calculus. Revue Roumaine Math. pur. appl. 23 (1978), 1587 - 1605.

[7] Vasilescu, F.-H.: Analytic perturbations of the $\bar{\partial}$-operator and integral representation formulas in Hilbert spaces. J. Operator Theory 1 (1979), 187 - 205.

INCREST
Bd. Păcii 220
77538 Bucureşti
Romania

LOWER BOUNDS FOR THE n-MODULI OF PATH FAMILIES WITH APPLICATIONS TO BOUNDARY BEHAVIOR OF QUASICONFORMAL AND QUASIREGULAR MAPPINGS

Matti Vuorinen

1. Introduction

If E_1, E_2, and $G \subset R^n$ then $\Delta(E_1,E_2;G)$ denotes the set of all closed paths joining E_1 and E_2 in G (cf. [9, p. 21]). Let E_1, $E_2 \subset R^n$ with $0 \in \bar{E}_1$, $\bar{E}_2$, let $B^n(r) = \{x \in R^n : |x| < r\}$ for $r > 0$, and let $\Gamma_r = \Delta(E_1,E_2;R^n \setminus \bar{B}^n(r))$ for $r > 0$. In this note we shall study the problem of finding a lower bound for the n-modulus $M(\Gamma_r)$ of Γ_r in terms of r and the densities of E_1 and E_2 at 0. One can derive estimates of this kind by means of the cap-inequality [9, 10.9] under the assumption that both E_1 and E_2 intersect $\partial B^n(r)$ for all $r \in (0,1]$. Hence it is of interest to study the situation under weaker density conditions on E_1 and E_2. For this purpose we employ the <u>lower capacity density</u> cap <u>dens</u> (E,x) of a set E at $x \in R^n$ (cf. Section 2 and Martio-Sarvas [6]).

The main result of this note, which is given in Theorem 3.1, states that if cap <u>dens</u> $(E_j,0) = \delta_j > 0$, $j = 1,2$, then

$$M(\Gamma_r) \geq c \log \frac{1}{r}$$

for small $r > 0$, where $c > 0$ is a constant depending only on n, δ_1, and δ_2. In the special case mentioned above this estimate follows from the cap-inequality (with a better constant c). Observe that in the general case E_1 and E_2 may be compact sets of zero Hausdorff dimension (cf. Remark 3.3). The proof of this lower bound is based on the so-called <u>comparison principle for the modulus</u> (cf. Näkki [8] and Martio, Rickman, Väisälä [5, 3.11]). We shall apply our results to the study of angular limits of quasiconformal mappings and to the study of distortion of quasiregular mappings. Complete proofs will appear elsewhere [11].

2. Preliminary results

For $x \in R^n$ $(n \geq 2)$ and $r > 0$ let $B^n(x,r) = \{y \in R^n : |y - x| < r\}$, $S^{n-1}(x,r) = \partial B^n(x,r)$, $B^n(r) = B^n(0,1)$, $S^{n-1}(r) = S^{n-1}(0,1)$, $B^n = B^n(1)$, and $S^{n-1} = S^{n-1}(1)$. If $r > s > 0$ then we write $R(r,s) = B^n(r) \setminus \bar{B}^n(s)$. If Γ is a path family so that each $\gamma \in \Gamma$ intersects both boundary components of $R(r,s)$, then the following upper bound holds for the n-modulus of Γ [9, 7.5]

$$M(\Gamma) \leq \omega_{n-1}(\log \frac{r}{s})^{1-n} \tag{2.1}$$

where ω_{n-1} is the surface area of S^{n-1}. Given $E \subset R^n$, $r > 0$, and $x \in R^n$ we introduce the abbreviation

$$M(E,r,x) = M(\Delta(S^{n-1}(x,2r), \overline{B}^n(x,r) \cap E; R^n)) .$$

The lower and upper capacity densities of E at x are defined by

$$\text{cap } \underline{\text{dens}} \ (E,x) = \liminf_{r\to 0} M(E,r,x) ,$$

$$\text{cap } \overline{\text{dens}} \ (E,x) = \limsup_{r\to 0} M(E,r,x) .$$

2.2. Remark. If E is compact, then one can give an alternative equivalent definition for capacity densities by making use of the n-capacity of a condenser. We refer the reader to Martio-Sarvas [6] and to the paper of Ziemer quoted in [6].

In [8] Näkki called the following lemma the comparison principle for the modulus. Martio, Rickman, and Väisälä have used the idea behind Lemma 2.3 in the proof of Lemma 3.11 in [5]. In the sequel we denote by $c_n > 0$ the constant of the cap-inequality [9, 10.9] depending only on n.

2.3. Lemma. (Näkki [8]) Let F_1, F_2, and F_3 be three sets in R^n and let $\Gamma_{ij} = \Delta(F_i,F_j;R^n)$, $1 \leq i, j \leq 3$. If there exist $x \in R^n$ and $0 < a < b$ such that $F_1, F_2 \subset \overline{B}^n(x,a)$ and $F_3 \subset R^n \setminus B^n(x,b)$ then

$$M(\Gamma_{12}) \geq 3^{-n} \min \{M(\Gamma_{13}), M(\Gamma_{23}), c_n \log \frac{b}{a}\} .$$

Let E_1 and E_2 be two sets in R^n with $M(E_j,s,0) = \delta_j > 0$, $j = 1,2$, for some $s > 0$. If we apply Lemma 2.3 with $F_1 = E_1 \cap \overline{B}^n(s)$, $F_2 = E_2 \cap \overline{B}^n(s)$, and $F_3 = S^{n-1}(2s)$, then we get a lower bound in terms of δ_1, δ_2, and n for $M(\Gamma_{12}) = M(\Delta(F_1,F_2;R^n))$ and hence also for $M(\Delta(E_1,E_2;R^n))$. If we apply twice the upper bound (2.1), then we can show that a lower bound of the same type holds for $M(\Delta(E_1,E_2;R(\lambda s,s/\lambda)))$ as well, when $\lambda \geq \lambda_o > 1$ and λ_o is a sufficiently large number depending only on δ_1, δ_2, and n. This is the idea of the proof of the following lemma.

2.4. Lemma. ([11, 3.1]) Let $E_1, E_2 \subset R^n$ with $M(E_j,s,0) = \delta_j > 0$, $j = 1,2$, for some $s > 0$. Then there are constants $\lambda > 1$, $t > 0$ depending only on δ_1, δ_2, and n such that $M(\Delta(E_1,E_2;R(\lambda s,s/\lambda))) \geq t$.

3. Lower bounds for the moduli of path families

3.1. Theorem. Let $E_j \subset R^n$, $j = 1,2$, with $M(E_j,r,0) \geq \delta_j > 0$ for $r \in (0,1]$, $j = 1,2$, and let $\lambda > 1$ be the constant given by Lemma 2.4. For $r \in (0,1]$ write $\Gamma_r = \Delta(E_1,E_2;R^n \setminus \bar{B}^n(r))$. Then there exists a constant $c > 0$ depending only on δ_1, δ_2, and n such that $M(\Gamma_r) \geq c \log \frac{1}{r}$ for $r \in (0,1/\lambda]$.

Proof. By Lemma 2.4 there are $\lambda > 1$ and $t > 0$ depending only on δ_1, δ_2, and n such that for every $r \in (0,1]$ $M(\tilde{\Gamma}_r) \geq t$, where $\tilde{\Gamma}_r = \Delta(E_1,E_2;R(\lambda r,r/\lambda))$. Let $r_k = \lambda^{2-2k}$, $k = 1,2,\cdots$. Fix $r \in (0,1/\lambda]$ as in the theorem. Let

$$m = \max\{k \in N : r_k/\lambda \geq r\} \geq 1 .$$

Then $\lambda^{-3m} \leq \lambda^{-1-2m} \leq r \leq \lambda^{1-2m}$ and hence $m \geq (\log \frac{1}{r})/(3 \log \lambda)$. Obviously $\tilde{\Gamma}_{r_1},\cdots,\tilde{\Gamma}_{r_m}$ are separate subfamilies of Γ_r and we get by [9, 6.7]

$$M(\Gamma_r) \geq \sum_{j=1}^{m} M(\tilde{\Gamma}_{r_j}) \geq \frac{t}{3 \log \lambda} \log \frac{1}{r} .$$

The following theorem can be proved in the same way.

3.2. Theorem. ([11 , 3.5]) Let $\delta > 0$ and let $E \subset R^n$ be a set with $M(E,r,0) \geq \delta > 0$ for $r \in (0,1]$. Then there is a constant $d^* > 0$ depending only on δ and n such that if $r \in (0,1)$ and F_r is a continuum joining $S^{n-1}(r)$ and S^{n-1} then $M(\Gamma_r^*) \geq d^* \log \frac{1}{r}$, where $\Gamma_r^* = \Delta(E,F_r;R^n)$.

3.3. Remark. The lower bound of Theorem 3.1 follows from the cap-inequality [9, 10.9] in some particular cases, e.g. when $E_j \cap S^{n-1}(r) \neq \emptyset$, $j = 1,2$, for all $r \in (0,1]$. Observe that the condition $M(E_j,r,0) \geq \delta_j > 0$, $j = 1,2$, for all $r \in (0,1]$ may be satisfied if $E_j \cap S^{n-1}(r) = \emptyset$ except for countably many r (cf. [10, 2.5(1)]) or even if the Hausdorff dimensions of E_1 and E_2 are zero (this follows from a result of H. Wallin, (cf. [10, 2.5(3)])).

Theorem 3.1 shows that $M(\Gamma_r)$ tends to infinity with a certain rapidity as $r \to 0$ if cap $\underline{\text{dens}}$ $(E_j,0) > 0$, $j = 1,2$. In the next theorem we study the behavior of $M(\Gamma_r)$ under the more general assumptions cap $\underline{\text{dens}}$ $(E_1,0) > 0$ and cap $\overline{\text{dens}}$ $(E_2,0) > 0$. When $r \to 0$ $M(\Gamma_r) \to \infty$ also in this case but the convergence may take place as slowly as we please (cf. [11, 3.9]). The next theorem follows from 2.4 and [9, 6.7].

3.4. Theorem. ([11, 3.8]) Let E_1 and E_2 be two sets with cap $\underline{\text{dens}}$ $(E_1,0) > 0$ and cap $\overline{\text{dens}}$ $(E_2,0) > 0$ and let

$\Gamma_r = \Delta(E_1,E_2;R^n \setminus \overline{B}^n(r))$. Then $M(\Gamma_r) \to \infty$ as $r \to 0$.

4. Applications

4.1. Quasiconformal and quasiregular mappings. For the definitions and basic properties of n-dimensional quasiregular mappings we refer the reader to Martio, Rickman, and Väisälä [4] - [5]. A mapping is quasiconformal if and only if it is quasiregular and injective. A quasiconformal mapping $f : G \to G'$, where G and G' are domains in R^n, satisfies for some $K \in [1,\infty)$ the following inequality (cf. [9, 13.1])

$$M(\Gamma)/K \leqq M(f\Gamma) \leqq K\, M(\Gamma) \tag{4.2}$$

for every path family Γ in G. Here $f\Gamma = \{f\circ\gamma : \gamma \in \Gamma\}$.

The next result was proved in [10, 6.5].

4.3. Lemma. Let $f : B^n \to G'$ be a quasiconformal mapping, let $b \in \partial B^n$, and let (b_k) be a sequence in B^n so that $b_k \to b$ and $f(b_k) \to \alpha$ as $k \to \infty$. If $t \in (0,1)$, then $f(x) \to \alpha$ as $x \to b$ through the set $\cup B^n(b_k,(1 - |b_k|)t)$.

Using this lemma and the results of Sections 2 and 3 we shall prove the following result.

4.4. Theorem. Let $f : B^n \to G'$ be a quasiconformal mapping, let $b \in \partial B^n$, and let $E \subset B^n$ be a set such that $b \in \overline{E}$ and $\lim_{x\to b, x\in E} f(x) = \alpha$ exists. If cap dens $(E,b) > 0$ then f has angular limit α at b.

Proof. Suppose that this is not the case. Then there exist $s \in (0,1)$ and a sequence (b_k) in B^n with $b_k \to b$ such that $1 - |b_k| > s|b_k - b|$ and $f(b_k) \to \beta \neq \alpha$.

From Lemma 4.3 it follows that $f(x) \to \beta$ as $x \to b$ and $x \in \cup B^n(b_k,(1 - |b_k|)/2) = F$. We may assume that $\alpha, \beta \neq \infty$ and that

$$fF \subset B^n(\beta,|\alpha - \beta|/3) \;;\; fE \subset B^n(\alpha,|\alpha - \beta|/3) \;. \tag{4.5}$$

If J_k is the segment joining b and b_k, then $m(J_k \cap F) \geqq |b_k - b|s/2$ for each k. This together with [9, 10.9] implies that cap dens $(F,b) > 0$. Let $\Gamma = \Delta(E,F;B^n)$. A symmetry principle for the modulus [10, Section 4] and Theorem 3.4 imply that $M(\Gamma) = \infty$. This contradicts (4.2) since $M(f\Gamma) < \infty$ by (2.1) and (4.5).

4.6. Remark. (1) Tord Hall has proved in [3, Thm II] a result related to Theorem 4.4 for bounded analytic functions (cf. also [10, Section 4]). In [2, p. 21] F. W. Gehring proved that a quasiconformal mapping of B^3 having an asymptotic value α at $b \in \partial B^3$ has angular limit α at b. Theorem 4.4 seems to be new even if $n = 2$ and f is conformal. It was shown in [10, Section 6] that the condition cap $\underline{\text{dens}}$ $(E,b) > 0$ of Theorem 4.4 cannot be replaced by cap $\overline{\text{dens}}$ $(E,b) > 0$.

(2) It is possible to weaken the hypotheses of Theorem 4.4 in other directions. E.g. a quasiconformal version of J. L. Doob's result [1, Thm 4] was proved in [10, Section 5].

R. Miniowitz proved in [7, Thm 3] a distortion result for quasiregular mappings $f : B^n \to R^n$ with $S^{n-1}(r) \cap (R^n \setminus fB^n) \neq \emptyset$ for all $r > 0$. Remark 3.3 shows that the condition on the omitted set is substantially weaker in the following theorem. On the other hand, the constant corresponding to α was better in [7, Thm 3].

4.7. Theorem. Let $f : B^n \to R^n$ be a quasiregular mapping such that $M(R^n \setminus fB^n, r, 0) \geq \delta > 0$ for all $r > 0$ and let $x \in B^n$. If $|f(x)| \geq |f(0)|$ then

$$|f(x)| \leq |f(0)| \; 2^{8\alpha} \left(\frac{1 + |x|}{1 - |x|}\right)^{\alpha}$$

where $\alpha = 2^{n-1} c_n K_I(f)/d^* > 0$ and the numbers $c_n > 0$ and $d^* > 0$ are as in Lemma 2.3 and Theorem 3.2.

Proof. The proof follows from the K_I-inequality for quasiregular mappings, from the lower bound of Theorem 3.2, and from the upper bound

$$M(\Delta([0,x], \partial B^n; B^n)) \leq 2^{n-1} c_n \log \left(\frac{1 + |x|}{1 - |x|} \cdot 2^8\right)$$

due to G. D. Anderson (cf. [7, Thm 3 and Lemma 1]). Here $[0,x]$ is the segment joining 0 and $x \in B^n$.

References.

[1] Doob, J. L.: The boundary values of analytic functions. Trans. Amer math. Soc. 34 (1932), 153 - 170.

[2] Gehring, F. W.: The Carathéodory convergence theorem for quasiconformal mappings in space. Ann. Acad. Sci. Fenn., Ser. A I 336/11 (1963).

[3] Hall, T.: Sur la mesure harmonique de certains ensembles. Ark. Mat. Astr. Fys. 25A, No. 28 (1937), 1 - 8.

[4] Martio, O., Rickman, S., Väisälä, J.: Definitions for quasiregular mappings. Ann. Acad. Sci. Fenn., Ser. A I 448 (1969).

[5] Martio, O., Rickman, S., Väisälä, J.: Distortion and singularities for quasiregular mappings. Ibid. 465 (1970).

[6] Martio, O., Sarvas, J.: Density conditions in the n-capacity. Indiana Univ. Math. J. 26 (1977), 761 - 776.

[7] Miniowitz, R.: Distortion theorems for quasiregular mappings. To appear in Ann. Acad. Sci. Fenn., Ser. A I.

[8] Näkki, R.: Extension of Loewner's capacity theorem. Trans. Amer. math. Soc. 180 (1973), 229 - 236.

[9] Väisälä, J.: Lectures on n-dimensional quasiconformal mappings. Lecture Notes in Mathematics 229, Springer-Verlag, Berlin - Heidelberg - New York (1971).

[10] Vuorinen, M.: On the existence of angular limits of n-dimensional quasiconformal mappings. To appear.

[11] Vuorinen, M.: Lower bounds for the moduli of path families with applications to non-tangential limits of quasiconformal mappings. To appear in Ann. Acad. Sci. Fenn., Ser. A I.

University of Helsinki
SF-00100 Helsinki 10
Finland

POTENTIAL THEORY AND APPROXIMATION OF ANALYTIC FUNCTIONS BY RATIONAL INTERPOLATION

Hans Wallin

Introduction.

The problem to approximate analytic functions by interpolation by means of polynomials or rational functions with poles at prescribed points, and to investigate how good this approximation procedure is, has been studied in great detail (see [22] for a survey). To some extent the interpolation by rational functions with prescribed poles leads to a theory which is very close to the one on interpolation by polynomials. If we want to interpolate by means of rational functions whose poles are allowed to be free and are determined by the interpolation conditions we get a quite different theory and are in a natural way led to the Padé approximants.

If f is analytic at zero, the (n,ν) Padé approximant of f is the unique rational function $R_{n\nu} = P_{n\nu}/Q_{n\nu}$ where the polynomials $P_{n\nu}$ of degree $\leq n$ and $Q_{n\nu}$ of degree $\leq \nu$, $Q_{n\nu} \not\equiv 0$, are determined so that $fQ_{n\nu} - P_{n\nu}$ has a zero of multiplicity at least $n + \nu + 1$ at the origin. Today quite a few results are known on the convergence problem for Padé approximation, i.e. the problem to study when and how fast $R_{n\nu} \to f$ as $n + \nu \to \infty$ (see [2] or [1] for a survey).

This paper treats the convergence problem for the more general rational interpolants $R_{n\nu} = P_{n\nu}/Q_{n\nu}$ obtained by requiring $fQ_{n\nu} - P_{n\nu}$ to be zero at a prescribed set of $n + \nu + 1$ points $\beta_{jn\nu}$, $1 \leq j \leq n + \nu + 1$, the interpolation points (see (1.3) in Section 1 for the detailed definition). Results on this more general problem have been given by Saff [18], Karlsson [10], [11], Warner [23], [24], Gončar-Lopez [8], Lopez [13], and Gelfgren [4]. The last three papers investigate the case when f is a function of so called Stieltjes type, a case which is not treated here. In particular, this paper is a study of the connection between the choice of the interpolation points $\beta_{jn\nu}$, the analyticity properties of f, and the convergence of $R_{n\nu}$ to f as $n + \nu \to \infty$. The potential theoretical notions relevant in this investigation are found in Sections 2.1 and 2.3. The paper gives a survey with unified proofs on some general convergence results including some generalizations and complementary results. In Section 2 convergence is studied in the case when $\nu = o(n)$ (the non-diagonal case). Section 3 contains the case when $\nu = n$ or $1/\lambda \leq \nu/n \leq \lambda$ for some constant $\lambda \geq 1$ (the diagonal case). In Section 4 the connection to the best rational approximation is studied.

1. The rational interpolants.

We shall approximate f by means of rational functions $P_{n\nu}/Q_{n\nu}$ of type (n,ν), i.e. $P_{n\nu}$ and $Q_{n\nu}$ shall be polynomials of degrees $\leq n$ and $\leq \nu$, respectively. For each pair of non-negative integers n, ν, let

$$\beta_{jn\nu},\ 1 \leq j \leq n + \nu + 1\ , \tag{1.1}$$

be $n + \nu + 1$ complex numbers, the interpolation points, belonging to a set E and put

$$\omega_{n\nu}(z) = \prod_{j=1}^{n+\nu+1} (z - \beta_{jn\nu})\ . \tag{1.2}$$

Let f be a function analytic at least at the points $\beta_{jn\nu}$ for all j, n and ν. Hermite's interpolation problem is to find a rational function $r_{n\nu}$ of type (n,ν) such that

$$(f(z) - r_{n\nu}(z))/\omega_{n\nu}(z) \text{ is analytic at } \beta_{jn\nu},\ 1 \leq j \leq n + \nu + 1\ .$$

This problem is not always solvable. However, we shall see that we can always find polynomials $P_{n\nu}$ of degree $\leq n$ and $Q_{n\nu}$ of degree $\leq \nu$, $Q_{n\nu} \not\equiv 0$, such that

$$(f(z)Q_{n\nu}(z) - P_{n\nu}(z))/\omega_{n\nu}(z) \text{ is analytic at } \beta_{jn\nu},\quad 1 \leq j \leq n + \nu + 1. \tag{1.3}$$

In fact, (1.3) is equivalent to a system of $n + \nu + 1$ linear equations stating that $fQ_{n\nu} - P_{n\nu}$ is zero at $\beta_{jn\nu}$, $1 \leq j \leq n + \nu + 1$; if k of the numbers $\beta_{jn\nu}$, $1 \leq j \leq n + \nu + 1$, coincide, the corresponding zero shall have multiplicity k. Since the number of unknown coefficients in $P_{n\nu}$ and $Q_{n\nu}$ is $n + \nu + 2$ we have more unknowns than equations. Hence, we can always determine $P_{n\nu}$ and $Q_{n\nu}$ satisfying (1.3). It is also easy to see that the corresponding rational function $R_{n\nu} = P_{n\nu}/Q_{n\nu}$ is unique (see for instance [10, p. 7] or [23, p. 24]) and that it solves Hermite's interpolation problem when this problem is solvable. We call $R_{n\nu} = P_{n\nu}/Q_{n\nu}$ satisfying (1.3) the rational interpolant (or multipoint Padé approximant) of type (n,ν) to f and the interpolation points $\beta_{jn\nu}$, $1 \leq j \leq n + \nu + 1$. If $\beta_{jn\nu} = 0$ for all j, n and ν we get the (n,ν) Padé approximant of f which, when $\nu = 0$, gives the Taylor polynomial around zero of degree n of f. If β_j, $j = 1,2,\ldots$, is a given sequence of complex numbers and $\beta_{jn\nu} = \beta_j$ for all j, n and ν, the rational interpolants have been called the Newton-Padé approximants [3] since they are connected to Newton series.

2. The non diagonal case.

2.1. Connection to potential theory. We denote the logarithmic potential of a measure μ by $u(z;\mu)$, i.e.

$$u(z;\mu) = \int \log \frac{1}{|z-\xi|}\, d\mu(\xi) \; .$$

The following lemma is of importance for the convergence theory (see Gontscharoff [9] or Krylov [12, Theorem 1, p. 246] for a special case and Warner [24, p. 112] for the general case and for the idea of proof; Warner gives a somewhat stronger version but his proof of that is incomplete).

Lemma 1. Let μ_n, $n = 1,2,\ldots$, be non-negative Borel measures with support in a compact subset E of the complex plane $\mathbb{C}$ so that $\mu_n(E) = 1$. Assume that $\mu_n \to \mu$ in the sense that $\int f\, d\mu_n \to \int f\, d\mu$ for all continuous functions f on E. Then

a) If $K \subset \mathbb{C}$ is a compact set, α a real number, and $u(z;\mu) > \alpha$ on K, then $u(z;\mu_n) > \alpha$ on K if $n > n(K,\alpha)$, and

b) $\lim_{n\to\infty} u(z;\mu_n) = u(z;\mu)$ uniformly on compact subsets of $\mathbb{C} \setminus E$.

2.2. Generalization of Montessus de Ballore's theorem. For a fixed non-negative integer ν we consider, for $n = 1,2,\ldots$, the interpolation points $\beta_{jn\nu} \in E$, $1 \leq j \leq n+\nu+1$, E compact, and introduce the associated measure μ_n to $\beta_{jn\nu}$, $1 \leq j \leq n+\nu+1$, which is a non-negative measure μ_n with total mass 1, i.e. $\mu_n(\mathbb{C}) = 1$, which assigns the point mass $1/(n+\nu+1)$ to each of the points $\beta_{jn\nu}$, $1 \leq j \leq n+\nu+1$. We assume that $\mu_n \to \mu$ and introduce the open sets

$$E_\rho = \{z \in \mathbb{C} : e^{-u(z;\mu)} < \rho\},\ \rho > 0 \; .$$

Theorem 1. Let $\beta_{jn\nu} \in E$, $1 \leq j \leq n+\nu+1$, $n = 1,2,\ldots$, $\nu \geq 0$ fixed, where $E \subset \mathbb{C}$ is compact and assume that the associated measures μ_n converge to a measure μ. Assume that, for some $\rho > 0$, f is meromorphic in an open set containing $\bar{E}_\rho \cup E$ with exactly ν poles $z_1, \ldots, z_\nu$ and that $z_j \in E_\rho$, $1 \leq j \leq \nu$. Let $P_{n\nu}/Q_{n\nu}$ be the rational interpolant of type (n,ν) to f and $\beta_{jn\nu}$, $1 \leq j \leq n+\nu+1$. Then

$$f - P_{n\nu}/Q_{n\nu} \to 0, \text{ as } n \to \infty \; ,$$

uniformly on compact subsets of $E_\rho \setminus \{z_j\}_1^\nu$, with geometric degree of convergence (see (2.4)). Moreover, for n sufficiently large $P_{n\nu}/Q_{n\nu}$ has ν poles and these converge to the poles z_j, $1 \leq j \leq \nu$, of f, as $n \to \infty$,

In the Padé case, i.e. when $\beta_{jn\nu} = 0$ for all j and n, this is a classical result by Montessus de Ballore [15] which has been generalized by Saff [18] to the case of a special μ and by Warner [24]. The theorem as stated here is slightly more general than Warner's version which requires that $z_j \in \complement E$, $1 \leq j \leq \nu$, i.e. that none of the poles z_j is a point of accumulation of the interpolation points $\beta_{jn\nu}$. The proof below is a combination of the ideas of Warner (part 2 of the proof) and an idea of H. S. Shapiro (part 3 of the proof) to use a certain normalization of $Q_{n\nu}$ to give a simple proof of Montessus de Ballore's theorem.

Proof. 1) If we put $h_\nu(z) = \prod_1^\nu (z - z_k)$, then $h_\nu(f\,Q_{n\nu} - P_{n\nu})/\omega_{n\nu}$ is analytic in an open set D containing $\bar{E}_\rho \cup E$. Then, by Cauchy's integral formula, for $z \in \bar{E}_\rho \cup E$,

$$\frac{h_\nu(z)(f\,Q_{n\nu} - P_{n\nu})(z)}{\omega_{n\nu}(z)} = \frac{1}{2\pi i} \int_\Gamma \frac{h_\nu(\xi)(f\,Q_{n\nu} - P_{n\nu})(\xi)}{\omega_{n\nu}(\xi)(\xi - z)}\, d\xi \,, \tag{2.1}$$

if Γ is a cycle in D with index (winding number) $\operatorname{ind}_\Gamma(a) = 0$ for $a \notin D$ and $= 1$ for $a \in \bar{E}_\rho \cup E$. By deforming the cycle Γ into infinity we find that the term in the right member of (2.1) which contains $P_{n\nu}$ is zero since the degree of the polynomial $h_\nu(\xi)P_{n\nu}(\xi)$ is $\leq \nu + n$ and the degree of $\omega_{n\nu}(\xi)(\xi - z)$ is $n + \nu + 2$. Hence, from (2.1) we conclude, if $K \subset E_\rho$, K compact,

$$\max_{z\in K} |h_\nu(z)(f\,Q_{n\nu} - P_{n\nu})(z)| \leq c(K,\Gamma,f) \max_{\xi\in\Gamma} |Q_{n\nu}(\xi)| \max_{\substack{z\in K\\ \xi\in\Gamma}} \left|\frac{\omega_{n\nu}(z)}{\omega_{n\nu}(\xi)}\right| \,, \tag{2.2}$$

where $c(K,\Gamma,f) < \infty$ is independent of n.

2) From the definition of μ_n as point masses it follows that

$$\left|\frac{\omega_{n\nu}(z)}{\omega_{n\nu}(\xi)}\right| = \exp\{-(n + \nu + 1)(u(z;\mu_n) - u(\xi;\mu_n))\} \,.$$

The potential $u(z;\mu)$ is lower semicontinuous [20, p. 45] and hence $\exp\{-u(z;\mu)\}$ is upper semicontinuous and assumes a maximum α on the compact set K. Also, $\alpha < \rho$ since $K \subset E_\rho$, and $\exp\{-u(\xi;\mu)\} \geq \rho$ for $\xi \in \Gamma \subset \complement E_\rho$. We now use the assumption that $\mu_n \to \mu$ and Lemma 1. Hence, given $\varepsilon > 0$, we can find $n(\varepsilon)$ so that, if $n \geq n(\varepsilon)$, $u(z;\mu_n) > \log(1/\alpha) - \varepsilon$, $z \in K$, and $u(\xi;\mu_n) < \log(1/\rho) + \varepsilon$, $\xi \in \Gamma$. Altogether, from the formula above we get, for $n \geq n(\varepsilon)$,

$$\max_{\substack{z\in K\\ \xi\in\Gamma}} \left|\frac{\omega_{n\nu}(z)}{\omega_{n\nu}(\xi)}\right| \leq \left(\frac{\alpha e^{2\varepsilon}}{\rho}\right)^{n+\nu+1} = a^{n+\nu+1} , \qquad (2.3)$$

where $a = \alpha e^{2\varepsilon}/\rho < 1$ if ε is chosen small enough.

3) Now we normalize $|Q_{n\nu}|$ by multiplying $P_{n\nu}$ and $Q_{n\nu}$ by the same constant so that $\max_{\Gamma}|Q_{n\nu}(\xi)| = \max_{\Gamma}|h_\nu(\xi)|$. By combining this with (2.2) and (2.3) we conclude that

$$\max_{z\in K}|h_\nu(z)(f\, Q_{n\nu} - P_{n\nu})(z)| \to 0, \; n \to \infty .$$

Since $Q_{n\nu}$ are polynomials of degree $\leq \nu$ we can choose a subsequence $Q_{n_j\nu}$, $j = 1,2,\ldots,$ converging pointwise in $\mathbb{C}$ to a polynomial Q of degree $\leq \nu$. Hence, by the last formula, $P_{n_j\nu} \to P$, $j \to \infty$, where P satisfies $h_\nu f\, Q = h_\nu P$ on E_ρ. This formula shows, since $h_\nu(z_j) = 0$ and $(h_\nu f)(z_j) \neq 0$, that $Q(z_j) = 0$, $1 \leq j \leq \nu$. Hence, $Q(z) = ch_\nu(z)$, c constant. By a suitable normalization in the argument of $Q_{n\nu}$ at a certain fixed point we can obtain that $Q(z) = h_\nu(z)$. Analogously, we realize that every subsequence of $Q_{n\nu}$, $n = 1,2,\ldots,$ contains a subsequence converging to h_ν in $\mathbb{C}$. By a theorem of Vitali this means that $Q_{n\nu} \to h_\nu$, $n \to \infty$, uniformly on compact sets, i.e. the zeros of $Q_{n\nu}$ approach the poles z_j of f. Since $P_{n\nu} \to P$ uniformly on K and $P(z_j) \neq 0$, $1 \leq j \leq \nu$, $P_{n\nu}/Q_{n\nu}$ has ν poles if n is large and these poles approach the poles of f.

4) Let K_1 be a compact subset of $E_\rho \setminus \{z_j\}_1^\nu$. By going back to (2.2) and using (2.3) and the fact that $Q_{n\nu} \to h_\nu$ uniformly on K_1, and that $h_\nu \neq 0$ on K_1, we finally get what remains to prove:

$$\limsup_{n\to\infty} \{\max_{z\in K_1} |(f - P_{n\nu}/Q_{n\nu})(z)|\}^{1/n} = a < 1 . \qquad (2.4)$$

<u>Remark.</u> If none of the poles $z_1,\ldots,z_\nu$ is a point of accumulation of the interpolation points $\beta_{jn\nu}$, $1 \leq j \leq n + \nu + 1$, $n = 1,2,\ldots,$ it is easy to see from the proof that, for n large, $P_{n\nu}/Q_{n\nu}$ interpolates to f at $\beta_{jn\nu}$, $1 \leq j \leq n + \nu + 1$, i.e. solves Hermite's interpolation problem.

<u>2.3. Convergence in capacity.</u> If K is a compact set in $\mathbb{C}$, the (logarithmic) capacity of K is defined as

$$\operatorname{cap} K = \liminf_{n\to\infty} \inf_{P_n} \max_{z\in K} |P_n(z)|^{1/n} ,$$

where the infimum is taken over all polynomials P_n of degree n and leading coefficient one. The capacity of an arbitrary set A in the extended complex plane is defined as $\operatorname{cap} A = \sup\{\operatorname{cap} K: K \subset A,\ K$ bounded and closed$\}$. The capacity of a continuum is positive [20, p. 56], the capacity is zero of a denumerable union of Borel sets of capacity zero [20, p. 57], and the capacity of a disk with radius r is r [20, p. 84]. A sequence of functions f_n is said to converge in capacity on a set A to f if, for every $\varepsilon > 0$, $\operatorname{cap}\{z \in A: |f_n(z) - f(z)| > \varepsilon\} \to 0$, $n \to \infty$. If we consider the rational interpolants of type (n,ν) and an f with p poles where $0 \leq p < \nu$ we can only prove a weaker result than Theorem 1.

Theorem 2. For $k = 1,2,\ldots,$ and E compact, let $\beta_{jn_k\nu_k} \in E$, $1 \leq j \leq n_k + \nu_k + 1$, have associated measures μ_k which converge to a measure μ. Assume that, for some $\rho > 0$, f is meromorphic with p poles $z_1,\ldots,z_p$ in an open set containing $\bar{E}_\rho \cup E$. Suppose that the non-negative integers n_k and ν_k satisfy $\nu_k \geq p \geq 0$, for $k \geq k_0$, and

$$(\nu_k + 1)/n_k \to 0,\ k \to \infty .$$

Let P_k/Q_k be the rational interpolant of type (n_k,ν_k) to f and $\beta_{jn_k\nu_k}$, $1 \leq j \leq n_k + \nu_k + 1$. Then, if $\delta > 0$ and $K \subset E_\rho$, K compact, there exists an $\eta < 1$ and a k_1 so that, for $k > k_1$,

$$|f - P_k/Q_k| < \eta^{n_k} \quad \text{on} \quad K \setminus A_k \quad \text{where} \quad \operatorname{cap} A_k \leq \delta .$$

In particular, P_k/Q_k converges in capacity to f on compact parts of E_ρ. Special cases of this result have been given in the Padé case by Stahl [19, p. 15] (ν_k constant), Gončar [7, p. 503] ($\nu_k \to \infty$), and Baker [1, Th. 14.2].

Proof. We start as in the proof of Theorem 1 and note that formula (2.2) may be proved in the same way in the present situation. That is, if $h_p(z) = (z - z_1) \cdots (z - z_p)$, we get for $z \in K$, $K \subset E_\rho \subset \operatorname{int} \Gamma$,

$$|(f - P_k/Q_k)(z)| \leq c(K,\Gamma,f) \left|\frac{1}{h_p(z)}\right| \max_{\xi\in\Gamma}\left|\frac{Q_k(\xi)}{Q_k(z)}\right| \max_{\xi\in\Gamma}\left|\frac{\omega_{n_k\nu_k}(z)}{\omega_{n_k\nu_k}(\xi)}\right| . \quad (2.5)$$

The last factor of (2.5) is estimated by means of (2.3) and $|Q_k(\xi)/Q_k(z)|$ in the following way. Choose $R \geq 1$ so large that $\Gamma \subset \{|z| \leq R\}$. We normalize P_k and Q_k so that Q_k has leading coefficient 1 and we use the following lemma (which is proved later) with $Q = Q_k$.

Lemma 2. Let $Q(z) = (z - w_1) \cdots (z - w_m)$ and $R \geq 1$ be given. Assume that $|w_j| \leq 2R$ for $1 \leq j \leq m^*$, $m^* \leq m$ and put $Q^*(z) = (z - w_1) \cdots (z - w_{m^*})$. Then

$$\max_{|\xi|\leq R} |Q(\xi)|/|Q(z)| \leq (3R)^m/|Q^*(z)| \quad \text{for} \quad |z| \leq R .$$

This produces, in the estimation of the right member of (2.5), a factor majorized by $(3R)^{\nu_k}/|Q_k^*(z)|$ since Q_k has degree $\leq \nu_k$. However, $1/Q_k^*$ may be estimated together with the factor $1/h_p$ in (2.5) in the following way. The polynomial $h_p Q_k^*$ has leading coefficient 1 and degree $p + \nu_k^*$, where $\nu_k^* \leq \nu_k$. Hence, from the definition of capacity it follows that $|h_p Q_k^*| > \delta^{p+\nu_k^*} \geq \delta^{p+\nu_k}$ except on a set A_k with cap $A_k \leq \delta$, if $\delta \leq 1$. Taken together, we get for large k from (2.5) with the number a in (2.3), $a < 1$, that, for $z \in K \setminus A_k$,

$$|(f - P_k/Q_k)(z)| \leq c(K,\Gamma,f)((3R)^{\nu_k}/\delta^{p+\nu_k}) \cdot a^{n_k+\nu_k+1} . \qquad (2.6)$$

Since $\nu_k + 1 = o(n_k)$ and $a < 1$, the last number is less than η^{n_k} for some $\eta < 1$ and $k > k_1$ which is what we wanted to prove.

Proof of Lemma 2. Let $|\xi| \leq R$ and $|z| \leq R$. If $|w_j| \leq 2R$, then $|\xi - w_j|/|z - w_j| \leq 3R/|z - w_j|$. If $|w_j| > 2R$, then $|z - w_j| \geq R$ and $|\xi - w_j|/|z - w_j| \leq 3R/R = 3 \leq 3R$. Taken together, these estimates for the different factors in $|Q(\xi)/Q(z)|$ prove the lemma.

Remark. Instead of using the fact that if $Q(z)$ is a polynomial of degree m with leading coefficient 1, then $|Q(z)| \geq \delta^m$ except on a set of capacity $\leq \delta$, we could use a lemma by Cartan [21, p. 241] stating that, if $\delta, \alpha > 0$, then $|Q(z)| > \delta^m$ except on a set which may be covered by at most m disks with radii r_i such that $\sum r_i^\alpha \leq e(2\delta)^\alpha$. This would give another way to measure the exceptional set A_k in Theorem 2.

2.4. Pointwise convergence. We use the same notation as in Theorem 2.

Theorem 3. Except for the assumptions in Theorem 2 we furthermore assume that

$$\sum_{k=1}^{\infty} (\nu_k + 1)/n_k < \infty .$$

Then P_k/Q_k converges to f on E_ρ except on a subset of capacity zero.

Proof. We use (2.6) with $\delta = \delta_k < 1$, i.e.

$$|(f - P_k/Q_k)(z)| \leq c(K,\Gamma,f)[(3R)^{\nu_k}/\delta_k^{p+\nu_k}] \cdot a^{n_k+\nu_k+1} , \quad a < 1 , \qquad (2.7)$$

except on a subset A_k of K satisfying cap $A_k \leq \delta_k < 1$. If this right member tends to zero when $k \to \infty$, the divergence set A is a subset of $\bigcup_{k_0}^{\infty} A_k$ for all k_0. Hence, if $\cup A_k \subset K$ has diameter $<d$ we have for all k_0 [20, p. 63]

$$\frac{1}{\log\frac{d}{\text{cap } A}} \leq \sum_{k_0}^{\infty} \frac{1}{\log\frac{d}{\text{cap } A_k}} \leq \sum_{k_0}^{\infty} \frac{1}{\log\frac{d}{\delta_k}} . \tag{2.8}$$

Consequently, cap A = 0 if the last sum is finite. Now, since $\sum(\nu_k + 1)/n_k < \infty$ there exist a_k, $a_k \to \infty$, so that $\sum a_k(\nu_k + 1)/n_k < \infty$. Hence, if we define δ_k by putting $1/\log(d/\delta_k) = a_k(\nu_k + 1)/n_k$, then cap A = 0. It is easy to check that the right member of (2.7) tends to zero. This proves the desired convergence on all compact subsets K of E_ρ and hence the theorem.

As an example, we may in Theorem 3 take ν_k = constant and $n_k = k^{1+\varepsilon}$ with the constant $\varepsilon > 0$ but not with $\varepsilon = 0$. If we would had started from the version of Theorem 2 obtained by means of Cartan's lemma (see the remark at the end of Section 2.3) we would have got a version of Theorem 3 giving convergence except on a set having Lebesgue measure zero or what is called α-dimensional Hausdorff measure zero. It is also possible to generalize Cartan's lemma to introduce other Hausdorff measures than the α-dimensional one. Results of the above type have been proved, for the Padé case, by many authors including Wallin [21], Zinn-Justin [25], Gončar [7], and Lubinsky [14]. Lubinsky proves [14, p. 16] convergence in the case ν_k = constant, $n_k = k$ except on a set having what is called logarithmic dimension at most 1 (and, in particular, Hausdorffdimension 0).

3. The diagonal case.

3.1. Generalization of a theorem by Pommerenke.

Theorem 4. Let E and F be closed sets in the extended complex plane $\overline{\mathbb{C}}$ such that $E \cap F = \phi$ and cap F = 0. Let f be analytic in $\overline{\mathbb{C}} \setminus F$. Let $P_{n\nu}/Q_{n\nu}$ be the rational interpolant of type (n,ν) to f and given interpolation points $\beta_{jn\nu} \in E \cap \mathbb{C}$, $1 \leq j \leq n + \nu + 1$. Then, for $\varepsilon > 0$, $\delta > 0$, $R > 0$, and $\lambda \geq 1$, there exists n_0 such that

$$|(f - P_{n\nu}/Q_{n\nu})(z)| < \varepsilon^n \quad \text{if} \quad n \geq n_0,\ 1/\lambda \leq n/\nu \leq \lambda , \tag{3.1}$$

for $|z| \leq R$ except on a set $A_{n\nu}$ where cap $A_{n\nu} \leq \delta$.

In particular, if $n = n_k$ and $\nu = \nu_k$ where $1/\lambda \leq n_k/\nu_k \leq \lambda$, we get convergence in capacity to f on bounded sets of the corresponding sequence P_k/Q_k, $k = 1,2,\ldots$, of rational interpolants. In the Padé case Theorem 4 was proved by Pommerenke [17] in a different formulation as a generalization of a theorem by Nuttall [16]. A special case of Theorem 4 was given by Karlsson [10, p. 29] in the case with different interpolation points. By making stronger assumptions it is like in Section 2.4 possible to prove results on pointwise convergence in the diagonal case (see Wallin [21], Zinn-Justin [25], and Karlsson [10 , p. 23]). However, the sequence of (n,n) Padé approximants may diverge in $\mathbb{C} \setminus \{0\}$ even for entire functions (Wallin [21]).

In the proof of Theorem 4 we shall need the following lemma.

Lemma 3. Let $K \subset \mathbb{C}$ be a compact set and put $1/K = \{w = 1/z : z \in K\}$. Then

a) $\operatorname{cap} \frac{1}{K} \leqq (\max\{|w|^2 : w \in \frac{1}{K}\}) \cdot \operatorname{cap} K$, if $0 \notin K$;

b) $\operatorname{cap} \frac{1}{K} = 0$ if and only if $\operatorname{cap} K = 0$.

Proof of Lemma 3. a) follows easily if we use the fact that the capacity equals the transfinite diameter [20, p. 73], i.e. $\operatorname{cap} K = \lim_{n\to\infty} \delta_n(K)$ where

$$[\delta_n(K)]^{n(n-1)/2} = \max_{z_j \in K} \prod_{1 \leqq j<k \leqq n} |z_j - z_k| .$$

In fact, by choosing points w_{jn}, $1 \leqq j \leqq n$, which realize the maximum in the definition of $\delta_n(1/K)$ and putting $z_{jn} = 1/w_{jn}$ we find that $\delta_n(1/K) \leqq [\max\{|w|^2 : w \in 1/K\}] \cdot \delta_n(K)$ which gives a). b) follows immediately from a) in the case $0 \notin K$ and the case $0 \in K$ follows if we use the case $0 \notin K$ on the sets $K_n = K \cap \{|z| \geqq 1/n\}$ and $1/K_n = (1/K) \cap \{|z| \leqq n\}$.

3.2. Proof of Theorem 4 when E and F are bounded and closed. Since $\operatorname{cap} F = 0$ there exists, by definition, for any given number $\eta > 0$, an integer $k > 0$ and a polynomial h of degree k with leading coefficient 1 so that $F \subset D$ where $D = D_\eta = \{z : |h(z)| < \eta^k\}$. We assume that $n > k$ and introduce a positive integer ℓ such that $n - k < k\ell \leqq n$. Since $h^\ell(fQ_{n\nu} - P_{n\nu})/\omega_{n\nu}$ is analytic in CF and zero at infinity we can use Cauchy's theorem to conclude that, for $z \in CD$,

$$\frac{h(z)^\ell (f\, Q_{n\nu} - P_{n\nu})(z)}{\omega_{n\nu}(z)} = \frac{1}{2\pi i} \int_\Gamma \frac{h(\xi)^\ell (f\, Q_{n\nu} - P_{n\nu})(\xi)}{\omega_{n\nu}(\xi)(\xi - z)}\, d\xi \, , \qquad (3.2)$$

where Γ is a cycle in D with index $\operatorname{ind}_\Gamma(a) = -1$ for $a \in F$. Furthermore, we define 2ρ as the distance between E and F, i.e. $2\rho = \operatorname{dist}(E,F) > 0$, and assume that Γ is chosen as a subset of $\{z : \operatorname{dist}(F,z) \leqq \rho\}$. Since the interpolation points $\beta_{jn\nu}$ belong to E

and $\omega_{n\nu}$ is defined by (1.2) this means that

$$|\omega_{n\nu}(\xi)| \geq \rho^{n+\nu+1} \ , \ \xi \in \Gamma \ . \tag{3.3}$$

It also means that $\mathrm{ind}_\Gamma(a) = 0$ if a is an interpolation point and, hence, that the term in the right member of (3.2) containing $P_{n\nu}$ is zero for $z \in \complement D$. From (3.2) and the fact that $|h(\xi)| < \eta^k$, $\xi \in \Gamma$, we therefore get the following estimate for $z \in \complement D$ (compare (2.5)):

$$|(f - P_{n\nu}/Q_{n\nu})(z)| \leq c(\eta,\Gamma,f)\frac{\eta^{k\ell}}{|h(z)|^\ell}\max_{\xi\in\Gamma}\left|\frac{Q_{n\nu}(\xi)}{Q_{n\nu}(z)}\right|\max_{\xi\in\Gamma}\left|\frac{\omega_{n\nu}(z)}{\omega_{n\nu}(\xi)}\right| \ , \tag{3.4}$$

where $c(\eta,\Gamma,f)$ is independent of n. Now we assume that $R \geq 1$ is chosen so large that $\Gamma \subset \{|z| \leq R\}$ and that $P_{n\nu}$ and $Q_{n\nu}$ are normalized so that $Q_{n\nu}$ has leading coefficient 1. We estimate the right member of (3.4) for $z \in (\complement D) \cap \{|w| \leq R\}$ in the following way: $\eta^{k\ell} \leq \eta^{n-k}$ if $\eta \leq 1$; $\omega_{n\nu}(\xi)$ is estimated by (3.3); $|\omega_{n\nu}(z)| \leq (|z| + \max\{|t|;\ t \in E\})^{n+\nu+1} \leq M^{n+\nu+1}$ where M depends on R and the compact set E; $Q_{n\nu}(\xi)/Q_{n\nu}(z)$ is estimated by means of Lemma 2 which in the denominator produces a polynomial $Q^*_{n\nu}(z)$ with leading coefficient 1 and degree $\nu^* \leq \nu$; $(h^\ell Q^*_{n\nu})(z)$ has leading coefficient 1 and degree $k\ell + \nu^* \leq n + \nu$ and hence $|h^\ell Q^*_{n\nu}| > \delta_1^{k\ell+\nu^*} > \delta_1^{n+\nu}$ except on a set $B_{n\nu}$ with $\mathrm{cap}\, B_{n\nu} \leq \delta_1$, if $0 < \delta_1 \leq 1$. Altogether we get from (3.4)

$$|(f - P_{n\nu}/Q_{n\nu})(z)| \leq c(\eta,\Gamma,f)\eta^{n-k}(3R)^\nu(\tfrac{1}{\delta_1})^{n+\nu}(\tfrac{M}{\rho})^{n+\nu+1} \ , \tag{3.5}$$

if $z \in \complement D$, $|z| \leq R$, $z \notin B_{n\nu}$, where $\mathrm{cap}\, B_{n\nu} \leq \delta_1$. Now we put $A_{n\nu} = (B_{n\nu} \cup D) \cap \{|z| \leq R\}$ and use the fact that $\mathrm{cap}\, D \leq \mathrm{cap}\{z : |h(z)| \leq \eta^k\} \leq \eta$ and the following inequality which is a consequence of the first inequality in (2.8),

$$\mathrm{cap}(A_j) \leq a,\ j = 1,\ 2 \Rightarrow \mathrm{cap}(A_1 \cup A_2) \leq \{a\ \mathrm{diam}(A_1 \cup A_2)\}^{1/2} \ . \tag{3.6}$$

This gives, if $\eta \leq \delta_1$, $\mathrm{cap}\, A_{n\nu} \leq (\delta_1 \cdot 2R)^{1/2}$ which is equal to δ if we choose $\delta_1 = \delta^2/2R$. Having fixed numbers ρ, R, M, δ_1, and k we get, by choosing η small enough, that the right member of (3.5) is less that ε^n for $n \geq n_0$, $1/\lambda \leq n/\nu \leq \lambda$. This proves Theorem 4 when E and F are compact subsets of $\mathbb{C}$.

Remark. In the next step of the proof we shall need a slightly different version of what we have just proved. Suppose that instead of working with the rational interpolant $P_{n\nu}/Q_{n\nu}$ we determine a rational function $p_{n\nu}/q_{n\nu}$ of type (n,ν) such that

$$(f(z)\, z^n\, q_{n\nu}(z) - z^\nu\, p_{n\nu}(z))/\omega_{n\nu}(z)$$

is analytic at $z = \beta_{jn\nu}$, $1 \leqq j \leqq n + \nu + 1$. This is possible by the same reason that it was possible to define $P_{n\nu}/Q_{n\nu}$. Then it is very easy to check that Theorem 4 in the case we have just proved remains true if we replace $P_{n\nu}(z)/Q_{n\nu}(z)$ by $z^{\nu} p_{n\nu}(z)/z^{n} q_{n\nu}(z)$. In fact, the factor z^{ν} in $z^{\nu} p_{n\nu}(z)$ does not matter since the corresponding term vanishes in the proof and instead of the polynomial $Q_{n\nu}(z)$ we consider $z^{n} q_{n\nu}(z)$ which does not give any essential difference in the estimation.

3.3. Proof of Theorem 4 in the general case. We shall reduce the general case to the special case treated in Section 3.2 by means of an inversion transformation $L(z)$.

1) We choose $\alpha \in \mathbb{C} \setminus (E \cup F)$, put $w = L(z) = 1/(z - \alpha)$, and introduce the function $g(w) = f(L^{-1}(w))$, the sets $F' = L(F)$ and $E' = L(E)$, and the points $\gamma_{jn\nu} = L(\beta_{jn\nu})$, $1 \leqq j \leqq n + \nu + 1$. Then the sets F' and E' are bounded and closed, $E' \cap F' = \phi$, and g is analytic in $\overline{\mathbb{C}} \setminus F'$. Furthermore, we claim that $\operatorname{cap} F' = 0$. In fact, $z = L^{-1}(w) = \alpha + 1/w$ gives $F = \alpha + 1/F'$ and $\operatorname{cap} 1/F' = \operatorname{cap} F = 0$ from which we conclude, by Lemma 3, that $\operatorname{cap} F' = 0$. Hence, we may use the result from Section 3.2, as formulated in the remark, on F', E', g, and the interpolation points $\gamma_{jn\nu}$.

2) We determine a rational function $p_{n\nu}/q_{n\nu}$ of type (n,ν) such that

$$(g(w)w^{n} q_{n\nu}(w) - w^{\nu} p_{n\nu}(w))/\prod_{j=1}^{n+\nu+1} (w - \gamma_{jn\nu})$$

is analytic at $\gamma_{jn\nu}$, $1 \leqq j \leqq n + \nu + 1$. By putting $w = L(z)$ and $\gamma_{jn\nu} = L(\beta_{jn\nu})$ into this expression we see that

$$[f(z)(z - \alpha)^{\nu} q_{n\nu}(L(z)) - (z - \alpha)^{n} p_{n\nu}(L(z))]/\omega_{n\nu}(z)$$

is analytic at $z = \beta_{jn\nu}$, $1 \leqq j \leqq n + \nu + 1$. However, this means that

$$P_{n\nu}(z)/Q_{n\nu}(z) = (z - \alpha)^{n} p_{n\nu}(L(z))/(z - \alpha)^{\nu} q_{n\nu}(L(z)) , \qquad (3.7)$$

since the right member is a rational function of type (n,ν) and the rational interpolant $P_{n\nu}/Q_{n\nu}$ is unique.

3) Let $\varepsilon > 0$, $\delta > 0$, $R > 0$, and $\lambda \geqq 1$ be given and assume that $1/\lambda \leqq n/\nu \leqq \lambda$. According to the case treated in the remark in Section 3.2 we get, for $\delta' > 0$, $\varepsilon' > 0$, that there exists n_0 such that

$$|g(w) - w^{\nu} p_{n\nu}(w)/w^{n} q_{n\nu}(w)| < \varepsilon \quad \text{if} \quad n \geqq n_0 ,$$

for $|w| \leqq 1/\varepsilon'$, $w \notin A'_{n\nu}$ where $\operatorname{cap} A'_{n\nu} \leqq \delta'$. We consider, for a certain $r' \geq \varepsilon'$, those w only which satisfy $|w| \geq 1/r'$ and we may assume that $A'_{n\nu}$ is closed and a subset of $\{w : 1/r' \leq |w| \leq 1/\varepsilon'\}$. By going from the variable w to $z = L^{-1}(w)$ we conclude by means of (3.7) that (3.1) is true for $\varepsilon' \leq |z - \alpha| = 1/|w| \leq r'$, $z \notin L^{-1}(A'_{n\nu})$.

4) Now we finally introduce the exceptional set $A_{n\nu}$ as $A_{n\nu} = \{L^{-1}(A'_{n\nu})\} \cup \{z : |z - \alpha| \leq \varepsilon'\}$ and assume that $r'^2\delta' < \varepsilon'$. We want to estimate $\operatorname{cap} A_{n\nu}$. Since $L^{-1}(A'_{n\nu}) = \alpha + 1/A'_{n\nu}$ we get, by Lemma 3,

$$\operatorname{cap}(L^{-1}(A'_{n\nu})) = \operatorname{cap}(1/A'_{n\nu}) \leq r'^2 \operatorname{cap} A'_{n\nu} \leq r'^2\delta' < \varepsilon' . \tag{3.8}$$

Also, $L^{-1}(A'_{n\nu}) \subset \{|z| \leq \alpha + r'\}$ and the set $\{|z - \alpha| \leq \varepsilon'\}$ has capacity ε' and is a subset of a disk with centre at 0 and radius $\leq |\alpha| + \varepsilon' \leq |\alpha| + r'$. Hence, we conclude from (3.8) and (3.6) that

$$\operatorname{cap} A_{n\nu} \leq \{\varepsilon' \cdot 2(|\alpha| + r')\}^{1/2} . \tag{3.9}$$

Now we assume that $r' \geq \varepsilon'$ was chosen so large that $\{|z| \leq R\} \subset \{|z - \alpha| \leq r'\}$ and that ε' is so small that the right member of (3.9) is $\leq \delta$. Having fixed r' and ε' we choose δ' so that $r'^2\delta' < \varepsilon'$. This proves that (3.1) is valid on $|z| \leq R$ except on $A_{n\nu}$, $\operatorname{cap} A_{n\nu} \leq \delta$ and Theorem 4 is proved in the general case.

4. Connection to best rational approximation.

4.1. Statement of the theorem. There is a very close connection between the approximation properties of the rational interpolants $P_{n\nu}/Q_{n\nu}$ of type (n,ν) and the best rational approximation of degree (n,ν),

$$\rho_{n\nu}(f,F) = \inf \sup_{z \in F} |(f - p_{n\nu}/q_{n\nu})(z)| ,$$

where the infimum is taken over all rational functions $p_{n\nu}/q_{n\nu}$ of type (n,ν). In the diagonal case and the approximation given by Theorem 4 the connection is given by the following two theorems (see Section 4.4 for the other cases) whose proofs are given in Sections 4.3 and 4.2, respectively.

Theorem 5. Let f be an analytic function in an open set D and $P_{n\nu}/Q_{n\nu}$ rational functions of type (n,ν) for $1/\lambda \leq n/\nu \leq \lambda$ where $\lambda \geq 1$ is given. Assume that, for $\varepsilon > 0$, $\delta > 0$, and $K \subset D$, K compact, there exists n_0 so that

$$|(f - P_{n\nu}/Q_{n\nu})(z)| < \varepsilon^n \quad \text{if} \quad n \geq n_0,\ 1/\lambda \leq n/\nu \leq \lambda , \tag{4.1}$$

for $z \in K$ except on a set $A_{n\nu}$ where $\text{cap}\, A_{n\nu} \leq \delta$. Then, for every $K_1 \subset D$, K_1 compact,

$$\max_{n/\lambda \leq \nu \leq \lambda n} \{\rho_{n\nu}(f,K_1\}^{1/n} \to 0 \quad \text{as} \quad n \to \infty . \tag{4.2}$$

Conversely, we have for the rational interpolants to f.

Theorem 6. Let f be analytic in an open set D and assume that (4.2) holds for every $K_1 \subset D$, K_1 compact, and some fixed $\lambda \geq 1$. Let $E \subset D$, E compact, be given and let $P_{n\nu}/Q_{n\nu}$ be the rational interpolant of type (n,ν) to f and a set of interpolation points $\beta_{jn\nu} \in E$, $1 \leq j \leq n + \nu + 1$. Then, for $\varepsilon > 0$, $\delta > 0$, and $K \subset D$, K compact, there exists n_0 such that (4.1) holds for $z \in K \setminus A_{n\nu}$ where cap $A_{n\nu} \leq \delta$.

By combining the Theorems 4 and 5 we see that if f is analytic outside a closed set F of capacity zero, then (4.2) holds for every compact $K_1 \subset CF$. Gončar has observed [5, p. 161] that this last result may be obtained also from a result by Walsh [22, § 8.7]. By combining this fact with Theorem 6 we obtain an alternative proof of Theorem 4 in the case when E is compact. In fact, the proof of Theorem 6 is very similar to the proof of the first part of Theorem 4. The Theorems 5 and 6 with $n = \nu$ are due to Gončar [6, Th. 3] in the Padé case and to Karlsson [11, Th. 2 and Th. 1] in the general case. Gončar noticed the connection between Padé approximation and best rational approximation when investigating the class R_0 of functions f satisfying, for some disk Δ, $\{\rho_{nn}(f,\Delta)\}^{1/n} \to 0$. Clearly, if (4.2) holds (with $n = \nu$) for every compact $K_1 \subset D$, then $f \in R_0$. The fact that the converse statement holds was proved Gončar [5, Th. 1 and Remark 4].

4.2. Proof of Theorem 6. Let $\varepsilon > 0$, $\delta > 0$, and $K \subset D$, K compact, be given and let Γ be a cycle in D with index $\text{ind}_\Gamma(a) = 0$ for $a \notin D$ and $= 1$ for $a \in E \cup K$. Since $\text{dist}(\Gamma,E) = \rho > 0$, (3.3) holds. We take a rational function $p_{n\nu}/q_{n\nu}$ of type (n,ν) such that

$$\max_{\xi\in\Gamma} |(f - p_{n\nu}/q_{n\nu})(\xi)| \leq 2\rho_{n\nu}(f,\Gamma) ,$$

which is possible if $\rho_{n\nu}(f,\Gamma) > 0$; if $\rho_{n\nu}(f,\Gamma) = 0$ an obvious modification is needed. Since $(f\, Q_{n\nu} - P_{n\nu})/\omega_{n\nu}$ is analytic in D we can, as usual, use Cauchy's integral formula. We get, for $z \in K$,

$$\frac{q_{n\nu}(z)(f\, Q_{n\nu} - P_{n\nu})(z)}{\omega_{n\nu}(z)} = \frac{1}{2\pi i} \int_\Gamma \frac{[q_{n\nu}\, Q_{n\nu}(f - p_{n\nu}/q_{n\nu})](\xi)}{\omega_{n\nu}(\xi)(\xi - z)}\, d\xi$$

$$+\frac{1}{2\pi i}\int_{\Gamma}\frac{(Q_{n\nu}\,P_{n\nu}-q_{n\nu}\,P_{n\nu})(\xi)}{\omega_{n\nu}(\xi)(\xi-z)}\,d\xi\ .$$

By deforming Γ into infinity we find that the last term is zero since the degree of the denominator is larger than or equal to 2 plus the degree of the numerator. Hence, by the choice of $p_{n\nu}/q_{n\nu}$, we get the following estimate (compare (3.4)) for $z\in K$,

$$|(f-P_{n\nu}/Q_{n\nu})(z)|\leqq c(\Gamma)\rho_{n\nu}(f,\Gamma)\max_{\xi\in\Gamma}\left|\frac{q_{n\nu}(\xi)\;Q_{n\nu}(\xi)}{q_{n\nu}(z)\;Q_{n\nu}(z)}\cdot\frac{\omega_{n\nu}(z)}{\omega_{n\nu}(\xi)}\right|\ . \qquad (4.3)$$

Now, $|\omega_{n\nu}(z)/\omega_{n\nu}(\xi)|$ is estimated like in Section 3.2. We normalize so that $q_{n\nu}\,Q_{n\nu}$ has leading coefficient 1 and estimate $q_{n\nu}(\xi)\,Q_{n\nu}(\xi)/q_{n\nu}(z)\,Q_{n\nu}(z)$ in the same way as $Q_{n\nu}(\xi)/Q_{n\nu}(z)$ was estimated in Section 3.2. This gives, if $\Gamma\subset\{|z|\leqq R\}$ where $R\geqq 1$, for some M depending on E and K,

$$|(f-P_{n\nu}/Q_{n\nu})(z)|\leq c(\Gamma)\rho_{n\nu}(f,\Gamma)(3R)^{2\nu}(\tfrac{1}{\delta})^{2\nu}(\tfrac{M}{\rho})^{n+\nu+1}\ , \qquad (4.4)$$

if $z\in K\setminus A_{n\nu}$ where $\operatorname{cap} A_{n\nu}\leqq\delta$. The theorem follows from the last estimate.

<u>4.3. Proof of Theorem 5.</u> The following proof is, except for details, the one given in [11, Section 2]. For a given compact $K_1\subset D$, choose a cycle Γ with index $\operatorname{ind}_\Gamma(a)=0$ for $a\notin D$ and $=1$ for $a\in K_1$ and a compact set $K\subset D$ containing Γ in its interior. We put $\operatorname{dist}(K_1,\Gamma)\geqq\rho>0$, normalize so that $Q_{n\nu}(z)$ has leading coefficient 1, and put $Q_{n\nu}=Q'_{n\nu}\cdot Q''_{n\nu}$ with $Q'_{n\nu}(z)=\pi(z-z_j)$ where the product is taken over those zeros $z_j=z_{jn\nu}$ of $Q_{n\nu}$ having distance $\leqq\rho/2$ to K_1. Now, let $P_{n\nu}$ be the Taylor polynomial of degree n of $Q''_{n\nu}f$ around a point $z_0\in K_1$; for convenience in notation we assume that $z_0=0$. By Cauchy's integral formula we get, for $z\in K_1$,

$$\frac{(Q''_{n\nu}f-P_{n\nu})(z)}{z^{n+1}}=\frac{1}{2\pi i}\int_{\Gamma}\frac{(Q_{n\nu}f-P_{n\nu})(\xi)}{Q'_{n\nu}(\xi)\xi^{n+1}(\xi-z)}d\xi+\frac{1}{2\pi i}\int_{\Gamma}\frac{(P_{n\nu}-Q'_{n\nu}P_{n\nu})(\xi)}{Q'_{n\nu}(\xi)\xi^{n+1}(\xi-z)}d\xi.\ (4.5)$$

Again we find that the last integral is zero by deforming Γ into infinity. The numerator of the integrand of the first integral of (4.5) is estimated by means of (4.1) which gives

$$|Q_{n\nu}f-P_{n\nu}|\leq\varepsilon^{n}\,|Q_{n\nu}|\quad\text{if}\quad n\geqq n_0,\ 1/\lambda\leq n/\nu\leq\lambda\ , \qquad (4.6)$$

on $K\setminus A_{n\nu}$ where $\operatorname{cap} A_{n\nu}\leq\delta$. By the maximum principle $A_{n\nu}=\phi$ or each component of $A_{n\nu}$ meets the boundary of K. Since the diameter of a component of $A_{n\nu}$ is $\leq 4 \operatorname{cap} A_{n\nu} \leq 4\delta$ (see [20, p. 85]) we may choose δ so small that $A_{n\nu} \cap \Gamma = \phi$. By using (4.6) in (4.5) we then obtain, for $z \in K_1$, $n \geq n_0$,

$$|(f - p_{n\nu}/Q''_{n\nu})(z)| \leq c(K_1,\Gamma)\varepsilon^n \max_{\xi\in\Gamma} \frac{|Q_{n\nu}(\xi) z^n|}{|Q''_{n\nu}(z)\ Q'_{n\nu}(\xi)\xi^n|} .$$

Here we use the estimations $|z| \leq c$, $z \in K_1$; $|\xi| \geq \rho$, $\xi \in \Gamma$; $|Q'_n(\xi)| \geq (\rho/2)^\nu$, $\xi \in \Gamma$; $|Q''_{n\nu}(z)| \geq (\rho/2)^\nu \pi(|z_j| - R)$, $z \in K_1$, if $\rho/2 \leq 1$ and the product is taken over those zeros $z_j = z_{jn\nu}$ of $Q_{n\nu}$ for which $|z_j| > 2R$, $R = \max(1, \operatorname{diam} K)$; and $|Q_{n\nu}(\xi)| \leq \pi(R + |z_j|)$, $\xi \in \Gamma$. This gives

$$\rho_{n\nu}(f,K_1) \leq \max_{z\in K_1} |f(z) - \frac{p_{n\nu}(z)}{Q''_{n\nu}(z)}| \leq c(K_1,\Gamma)\ \frac{(c\varepsilon)^n \cdot (3R)^\nu}{(\rho/2)^{2\nu} \cdot \rho^n}$$

which proves the theorem since ε may be chosen arbitrarily small.

4.4. Some further comments. By starting from (4.3) and (4.5), respectively, we may give analogues of the Theorems 6 and 5 in other situations. For instance, we may consider the non-diagonal case and use (2.3) or the convergence degree given by Theorem 2. It is clearly also possible to get a connection between the size of $\rho_{n\nu}(f,\Gamma)$ and the degree of convergence in capacity of $f - P_{n\nu}/Q_{n\nu}$. This problem was considered by Karlsson in [11]. In this case it is of interest to note that it is possible, by an easy argument, to replace the factor $(3R/\delta)^{2\nu}$ in (4.4) by $(3R/\delta)^\nu$.

It is also possible to give results on pointwise convergence in the same way as in Section 2.3. For instance, if, except for the assumptions in Theorem 6, $\rho_{nn}(f,K_1) \leq c(K_1)\rho_n^n$ for all compact sets $K_1 \subset D$ and

$$\sum_{n=1}^{\infty} (\log \frac{1}{\rho_n})^{-1} < \infty ,$$

then $f - P_{nn}/Q_{nn} \to 0$ in D, except on a subset of capacity zero.

References

[1] Baker, G. A., Jr.: Essentials of Padé approximants. Academic Press, New York - San Francisco - London (1975).

[2] Chui, C. K.: Recent results on Padé approximants and related problems, in "Approximation Theory II". Academic Press, New York - San Francisco - London (1976), 79 - 115.

[3] Gallucci, M. A., Jones, W. B.: Rational approximations correspondin to Newton series (Newton-Padé approximants). J. Approximation

Theory 17 (1976), 366 - 392.

[4] Gelfgren, J.: Rational interpolation to functions of Stieltjes type. Preprint, Department of Math., Univ. of Umeå 5 (1978).

[5] Gončar, A. A.: A local condition of single-valuedness of analytic functions. Math. USSR, Sbornik 18 (1972), 151 - 167. (Russian original, Mat. Sbornik 89 (131) (1972)).

[6] Gončar, A. A.: On the convergence of Padé approximants. Math. USSR Sbornik 21 (1973), 155 - 166 (Russian original, Mat. Sbornik 92 (134) (1973)).

[7] Gončar, A. A.: On the convergence of generalized Padé approximants of meromorphic functions. Math. USSR Sbornik 27 (1975), 503 - 514 (Russian original, Mat. Sbornik 98 (140) (1975)).

[8] Gončar, A. A., Lopez, G.: On Markov's theorem for multipoint Padé approximation (Russian). Mat. Sbornik 105 (147) (1978), 512 - 524.

[9] Gontscharoff, V. L.: Theory of interpolation and approximation (Russian). Moscow (1954).

[10] Karlsson, J.: Rational interpolation with free poles in the complex plane. Preprint, Department of Math., Univ. of Umeå 6 (1972).

[11] Karlsson, J.: Rational interpolation and best rational approximation. J. math. Analysis Appl. 53 (1976), 38 - 52.

[12] Krylov, V. I.: Approximate calculation of integrals. Macmillan, New York - London (1962).

[13] Lopez, G.: On convergence for multipoint Padé approximants to functions of Stieltjes type (Russian). Doklady Akad. Nauk SSSR 239 (1978), 793 - 796.

[14] Lubinsky, D. S.: On convergence of non-diagonal Padé approximants. Manuscript, Univ. of the Witwatersrand, Johannesburg (1978).

[15] de Montessus de Ballore, R.: Sur les fractions continues algébriques. Bull. Soc. math. France 30 (1902), 28 - 36.

[16] Nuttall, J.: The convergence of Padé approximants of meromorphic functions. J. math. Analysis Appl. 31 (1970), 147 - 153.

[17] Pommerenke, C.: Padé approximants and convergence in capacity. J. math. Analysis Appl. 41 (1973), 775 - 780.

[18] Saff, E. B.: An extension of Montessus de Ballore's theorem on the convergence of interpolating rational functions. J. Approximation Theory 6 (1972), 63 - 67.

[19] Stahl, H.: Beiträge zum Problem der Konvergenz von Padéapproximierenden. Thesis, Technische Universität Berlin (1976).

[20] Tsuji, M.: Potential theory in modern function theory. Maruzen, Tokyo (1959).

[21] Wallin, H.: The convergence of Padé approximants and the size of the power series coefficients. Appl. Analysis 4 (1974), 235 - 251.

[22] Walsh, J. L.: Interpolation and approximation by rational functions in the complex domain. Amer. Math. Soc., Providence, R. I. (1969).

[23] Warner, D. D.: Hermite interpolation with rational functions. Thesis, Univ. of California at San Diego (1974).

[24] Warner, D. D.: An extension of Saff's theorem on the convergence of interpolating rational functions. J. Approximation Theory 18 (1976), 108 - 118.

[25] Zinn-Justin, J.: Convergence of Padé approximants in the general case. Rocky Mountain J. Math. 4 (1974), 325 - 329.

Matematiska institutionen
Umeå Universitet
S-90187 Umeå
Sweden

Vol. 580: C. Castaing and M. Valadier, Convex Analysis and Measurable Multifunctions. VIII, 278 pages. 1977.

Vol. 581: Séminaire de Probabilités XI, Université de Strasbourg. Proceedings 1975/1976. Edité par C. Dellacherie, P. A. Meyer et M. Weil. VI, 574 pages. 1977.

Vol. 582: J. M. G. Fell, Induced Representations and Banach *-Algebraic Bundles. IV, 349 pages. 1977.

Vol. 583: W. Hirsch, C. C. Pugh and M. Shub, Invariant Manifolds. V, 149 pages. 1977.

Vol. 584: C. Brezinski, Accélération de la Convergence en Analyse Numérique. IV, 313 pages. 1977.

Vol. 585: T. A. Springer, Invariant Theory. VI, 112 pages. 1977.

Vol. 586: Séminaire d'Algèbre Paul Dubreil, Paris 1975–1976 (29ème Année). Edited by M. P. Malliavin. VI, 188 pages. 1977.

Vol. 587: Non-Commutative Harmonic Analysis. Proceedings 1976. Edited by J. Carmona and M. Vergne. IV, 240 pages. 1977.

Vol. 588: P. Molino, Théorie des G-Structures: Le Problème d'Equivalence. VI, 163 pages. 1977.

Vol. 589: Cohomologie l-adique et Fonctions L. Séminaire de Géométrie Algébrique du Bois-Marie 1965–66, SGA 5. Edité par L. Illusie. XII, 484 pages. 1977.

Vol. 590: H. Matsumoto, Analyse Harmonique dans les Systèmes de Tits Bornologiques de Type Affine. IV, 219 pages. 1977.

Vol. 591: G. A. Anderson, Surgery with Coefficients. VIII, 157 pages. 1977.

Vol. 592: D. Voigt, Induzierte Darstellungen in der Theorie der endlichen, algebraischen Gruppen. V, 413 Seiten. 1977.

Vol. 593: K. Barbey and H. König, Abstract Analytic Function Theory and Hardy Algebras. VIII, 260 pages. 1977.

Vol. 594: Singular Perturbations and Boundary Layer Theory, Lyon 1976. Edited by C. M. Brauner, B. Gay, and J. Mathieu. VIII, 539 pages. 1977.

Vol. 595: W. Hazod, Stetige Faltungshalbgruppen von Wahrscheinlichkeitsmaßen und erzeugende Distributionen. XIII, 157 Seiten. 1977.

Vol. 596: K. Deimling, Ordinary Differential Equations in Banach Spaces. VI, 137 pages. 1977.

Vol. 597: Geometry and Topology, Rio de Janeiro, July 1976. Proceedings. Edited by J. Palis and M. do Carmo. VI, 866 pages. 1977.

Vol. 598: J. Hoffmann-Jørgensen, T. M. Liggett et J. Neveu, Ecole d'Eté de Probabilités de Saint-Flour VI – 1976. Edité par P.-L. Hennequin. XII, 447 pages. 1977.

Vol. 599: Complex Analysis, Kentucky 1976. Proceedings. Edited by J. D. Buckholtz and T. J. Suffridge. X, 159 pages. 1977.

Vol. 600: W. Stoll, Value Distribution on Parabolic Spaces. VIII, 216 pages. 1977.

Vol. 601: Modular Functions of one Variable V, Bonn 1976. Proceedings. Edited by J.-P. Serre and D. B. Zagier. VI, 294 pages. 1977.

Vol. 602: J. P. Brezin, Harmonic Analysis on Compact Solvmanifolds. VIII, 179 pages. 1977.

Vol. 603: B. Moishezon, Complex Surfaces and Connected Sums of Complex Projective Planes. IV, 234 pages. 1977.

Vol. 604: Banach Spaces of Analytic Functions, Kent, Ohio 1976. Proceedings. Edited by J. Baker, C. Cleaver and Joseph Diestel. VI, 141 pages. 1977.

Vol. 605: Sario et al., Classification Theory of Riemannian Manifolds. XX, 498 pages. 1977.

Vol. 606: Mathematical Aspects of Finite Element Methods. Proceedings 1975. Edited by I. Galligani and E. Magenes. VI, 362 pages. 1977.

Vol. 607: M. Métivier, Reelle und Vektorwertige Quasimartingale und die Theorie der Stochastischen Integration. X, 310 Seiten. 1977.

Vol. 608: Bigard et al., Groupes et Anneaux Réticulés. XIV, 334 pages. 1977.

Vol. 609: General Topology and Its Relations to Modern Analysis and Algebra IV. Proceedings 1976. Edited by J. Novák. XVIII, 225 pages. 1977.

Vol. 610: G. Jensen, Higher Order Contact of Submanifolds of Homogeneous Spaces. XII, 154 pages. 1977.

Vol. 611: M. Makkai and G. E. Reyes, First Order Categorical Logic. VIII, 301 pages. 1977.

Vol. 612: E. M. Kleinberg, Infinitary Combinatorics and the Axiom of Determinateness. VIII, 150 pages. 1977.

Vol. 613: E. Behrends et al., L^p-Structure in Real Banach Spaces. X, 108 pages. 1977.

Vol. 614: H. Yanagihara, Theory of Hopf Algebras Attached to Group Schemes. VIII, 308 pages. 1977.

Vol. 615: Turbulence Seminar, Proceedings 1976/77. Edited by P. Bernard and T. Ratiu. VI, 155 pages. 1977.

Vol. 616: Abelian Group Theory, 2nd New Mexico State University Conference, 1976. Proceedings. Edited by D. Arnold, R. Hunter and E. Walker. X, 423 pages. 1977.

Vol. 617: K. J. Devlin, The Axiom of Constructibility: A Guide for the Mathematician. VIII, 96 pages. 1977.

Vol. 618: I. I. Hirschman, Jr. and D. E. Hughes, Extreme Eigen Values of Toeplitz Operators. VI, 145 pages. 1977.

Vol. 619: Set Theory and Hierarchy Theory V, Bierutowice 1976. Edited by A. Lachlan, M. Srebrny, and A. Zarach. VIII, 358 pages. 1977.

Vol. 620: H. Popp, Moduli Theory and Classification Theory of Algebraic Varieties. VIII, 189 pages. 1977.

Vol. 621: Kauffman et al., The Deficiency Index Problem. VI, 112 pages. 1977.

Vol. 622: Combinatorial Mathematics V, Melbourne 1976. Proceedings. Edited by C. Little. VIII, 213 pages. 1977.

Vol. 623: I. Erdelyi and R. Lange, Spectral Decompositions on Banach Spaces. VIII, 122 pages. 1977.

Vol. 624: Y. Guivarc'h et al., Marches Aléatoires sur les Groupes de Lie. VIII, 292 pages. 1977.

Vol. 625: J. P. Alexander et al., Odd Order Group Actions and Witt Classification of Innerproducts. IV, 202 pages. 1977.

Vol. 626: Number Theory Day, New York 1976. Proceedings. Edited by M. B. Nathanson. VI, 241 pages. 1977.

Vol. 627: Modular Functions of One Variable VI, Bonn 1976. Proceedings. Edited by J.-P. Serre and D. B. Zagier. VI, 339 pages. 1977.

Vol. 628: H. J. Baues, Obstruction Theory on the Homotopy Classification of Maps. XII, 387 pages. 1977.

Vol. 629: W. A. Coppel, Dichotomies in Stability Theory. VI, 98 pages. 1978.

Vol. 630: Numerical Analysis, Proceedings, Biennial Conference, Dundee 1977. Edited by G. A. Watson. XII, 199 pages. 1978.

Vol. 631: Numerical Treatment of Differential Equations. Proceedings 1976. Edited by R. Bulirsch, R. D. Grigorieff, and J. Schröder. X, 219 pages. 1978.

Vol. 632: J.-F. Boutot, Schéma de Picard Local. X, 165 pages. 1978.

Vol. 633: N. R. Coleff and M. E. Herrera, Les Courants Résiduels Associés à une Forme Méromorphe. X, 211 pages. 1978.

Vol. 634: H. Kurke et al., Die Approximationseigenschaft lokaler Ringe. IV, 204 Seiten. 1978.

Vol. 635: T. Y. Lam, Serre's Conjecture. XVI, 227 pages. 1978.

Vol. 636: Journées de Statistique des Processus Stochastiques, Grenoble 1977, Proceedings. Edité par Didier Dacunha-Castelle et Bernard Van Cutsem. VII, 202 pages. 1978.

Vol. 637: W. B. Jurkat, Meromorphe Differentialgleichungen. VII, 194 Seiten. 1978.

Vol. 638: P. Shanahan, The Atiyah-Singer Index Theorem, An Introduction. V, 224 pages. 1978.

Vol. 639: N. Adasch et al., Topological Vector Spaces. V, 125 pages. 1978.

Vol. 640: J. L. Dupont, Curvature and Characteristic Classes. X, 175 pages. 1978.

Vol. 641: Séminaire d'Algèbre Paul Dubreil, Proceedings Paris 1976–1977. Edité par M. P. Malliavin. IV, 367 pages. 1978.

Vol. 642: Theory and Applications of Graphs, Proceedings, Michigan 1976. Edited by Y. Alavi and D. R. Lick. XIV, 635 pages. 1978.

Vol. 643: M. Davis, Multiaxial Actions on Manifolds. VI, 141 pages. 1978.

Vol. 644: Vector Space Measures and Applications I, Proceedings 1977. Edited by R. M. Aron and S. Dineen. VIII, 451 pages. 1978.

Vol. 645: Vector Space Measures and Applications II, Proceedings 1977. Edited by R. M. Aron and S. Dineen. VIII, 218 pages. 1978.

Vol. 646: O. Tammi, Extremum Problems for Bounded Univalent Functions. VIII, 313 pages. 1978.

Vol. 647: L. J. Ratliff, Jr., Chain Conjectures in Ring Theory. VIII, 133 pages. 1978.

Vol. 648: Nonlinear Partial Differential Equations and Applications, Proceedings, Indiana 1976–1977. Edited by J. M. Chadam. VI, 206 pages. 1978.

Vol. 649: Séminaire de Probabilités XII, Proceedings, Strasbourg, 1976–1977. Edité par C. Dellacherie, P. A. Meyer et M. Weil. VIII, 805 pages. 1978.

Vol. 650: C*-Algebras and Applications to Physics. Proceedings 1977. Edited by H. Araki and R. V. Kadison. V, 192 pages. 1978.

Vol. 651: P. W. Michor, Functors and Categories of Banach Spaces. VI, 99 pages. 1978.

Vol. 652: Differential Topology, Foliations and Gelfand-Fuks-Cohomology, Proceedings 1976. Edited by P. A. Schweitzer. XIV, 252 pages. 1978.

Vol. 653: Locally Interacting Systems and Their Application in Biology. Proceedings, 1976. Edited by R. L. Dobrushin, V. I. Kryukov and A. L. Toom. XI, 202 pages. 1978.

Vol. 654: J. P. Buhler, Icosahedral Golois Representations. III, 143 pages. 1978.

Vol. 655: R. Baeza, Quadratic Forms Over Semilocal Rings. VI, 199 pages. 1978.

Vol. 656: Probability Theory on Vector Spaces. Proceedings, 1977. Edited by A. Weron. VIII, 274 pages. 1978.

Vol. 657: Geometric Applications of Homotopy Theory I, Proceedings 1977. Edited by M. G. Barratt and M. E. Mahowald. VIII, 459 pages. 1978.

Vol. 658: Geometric Applications of Homotopy Theory II, Proceedings 1977. Edited by M. G. Barratt and M. E. Mahowald. VIII, 487 pages. 1978.

Vol. 659: Bruckner, Differentiation of Real Functions. X, 247 pages. 1978.

Vol. 660: Equations aux Dérivée Partielles. Proceedings, 1977. Edité par Pham The Lai. VI, 216 pages. 1978.

Vol. 661: P. T. Johnstone, R. Paré, R. D. Rosebrugh, D. Schumacher, R. J. Wood, and G. C. Wraith, Indexed Categories and Their Applications. VII, 260 pages. 1978.

Vol. 662: Akin, The Metric Theory of Banach Manifolds. XIX, 306 pages. 1978.

Vol. 663: J. F. Berglund, H. D. Junghenn, P. Milnes, Compact Right Topological Semigroups and Generalizations of Almost Periodicity. X, 243 pages. 1978.

Vol. 664: Algebraic and Geometric Topology, Proceedings, 1977. Edited by K. C. Millett. XI, 240 pages. 1978.

Vol. 665: Journées d'Analyse Non Linéaire. Proceedings, 1977. Edité par P. Bénilan et J. Robert. VIII, 256 pages. 1978.

Vol. 666: B. Beauzamy, Espaces d'Interpolation Réels: Topologie et Géometrie. X, 104 pages. 1978.

Vol. 667: J. Gilewicz, Approximants de Padé. XIV, 511 pages. 1978.

Vol. 668: The Structure of Attractors in Dynamical Systems. Proceedings, 1977. Edited by J. C. Martin, N. G. Markley and W. Perrizo. VI, 264 pages. 1978.

Vol. 669: Higher Set Theory. Proceedings, 1977. Edited by G. H. Müller and D. S. Scott. XII, 476 pages. 1978.

Vol. 670: Fonctions de Plusieurs Variables Complexes III, Proceedings, 1977. Edité par F. Norguet. XII, 394 pages. 1978.

Vol. 671: R. T. Smythe and J. C. Wierman, First-Passage Perculation on the Square Lattice. VIII, 196 pages. 1978.

Vol. 672: R. L. Taylor, Stochastic Convergence of Weighted Sums of Random Elements in Linear Spaces. VII, 216 pages. 1978.

Vol. 673: Algebraic Topology, Proceedings 1977. Edited by P. Hoffman, R. Piccinini and D. Sjerve. VI, 278 pages. 1978.

Vol. 674: Z. Fiedorowicz and S. Priddy, Homology of Classical Groups Over Finite Fields and Their Associated Infinite Loop Spaces. VI, 434 pages. 1978.

Vol. 675: J. Galambos and S. Kotz, Characterizations of Probability Distributions. VIII, 169 pages. 1978.

Vol. 676: Differential Geometrical Methods in Mathematical Physics II, Proceedings, 1977. Edited by K. Bleuler, H. R. Petry and A. Reetz. VI, 626 pages. 1978.

Vol. 677: Séminaire Bourbaki, vol. 1976/77, Exposés 489–506. IV, 264 pages. 1978.

Vol. 678: D. Dacunha-Castelle, H. Heyer et B. Roynette. Ecole d'Eté de Probabilités de Saint-Flour. VII-1977. Edité par P. L. Hennequin. IX, 379 pages. 1978.

Vol. 679: Numerical Treatment of Differential Equations in Applications, Proceedings, 1977. Edited by R. Ansorge and W. Törnig. IX, 163 pages. 1978.

Vol. 680: Mathematical Control Theory, Proceedings, 1977. Edited by W. A. Coppel. IX, 257 pages. 1978.

Vol. 681: Séminaire de Théorie du Potentiel Paris, No. 3, Directeurs: M. Brelot, G. Choquet et J. Deny. Rédacteurs: F. Hirsch et G. Mokobodzki. VII, 294 pages. 1978.

Vol. 682: G. D. James, The Representation Theory of the Symmetric Groups. V, 156 pages. 1978.

Vol. 683: Variétés Analytiques Compactes, Proceedings, 1977. Edité par Y. Hervier et A. Hirschowitz. V, 248 pages. 1978.

Vol. 684: E. E. Rosinger, Distributions and Nonlinear Partial Differential Equations. XI, 146 pages. 1978.

Vol. 685: Knot Theory, Proceedings, 1977. Edited by J. C. Hausmann. VII, 311 pages. 1978.

Vol. 686: Combinatorial Mathematics, Proceedings, 1977. Edited by D. A. Holton and J. Seberry. IX, 353 pages. 1978.

Vol. 687: Algebraic Geometry, Proceedings, 1977. Edited by L. D. Olson. V, 244 pages. 1978.

Vol. 688: J. Dydak and J. Segal, Shape Theory. VI, 150 pages. 1978.

Vol. 689: Cabal Seminar 76–77, Proceedings, 1976–77. Edited by A.S. Kechris and Y. N. Moschovakis. V, 282 pages. 1978.

Vol. 690: W. J. J. Rey, Robust Statistical Methods. VI, 128 pages. 1978.

Vol. 691: G. Viennot, Algèbres de Lie Libres et Monoïdes Libres. III, 124 pages. 1978.

Vol. 692: T. Husain and S. M. Khaleelulla, Barrelledness in Topological and Ordered Vector Spaces. IX, 258 pages. 1978.

Vol. 693: Hilbert Space Operators, Proceedings, 1977. Edited by J. M. Bachar Jr. and D. W. Hadwin. VIII, 184 pages. 1978.

Vol. 694: Séminaire Pierre Lelong – Henri Skoda (Analyse) Année 1976/77. VII, 334 pages. 1978.

Vol. 695: Measure Theory Applications to Stochastic Analysis, Proceedings, 1977. Edited by G. Kallianpur and D. Kölzow. XII, 261 pages. 1978.

Vol. 696: P. J. Feinsilver, Special Functions, Probability Semigroups, and Hamiltonian Flows. VI, 112 pages. 1978.

Vol. 697: Topics in Algebra, Proceedings, 1978. Edited by M. F. Newman. XI, 229 pages. 1978.

Vol. 698: E. Grosswald, Bessel Polynomials. XIV, 182 pages. 1978.

Vol. 699: R. E. Greene and H.-H. Wu, Function Theory on Manifolds Which Possess a Pole. III, 215 pages. 1979.